FITNESS OF THE COSMOS FOR LIFE

Biochemistry and Fine-Tuning

This highly interdisciplinary book highlights many of the ways in which chemistry plays a crucial role in making life an evolutionary possibility in the universe. Cosmologists and particle physicists have often explored how the observed laws and constants of nature lie within a narrow range that allows complexity and life to evolve and adapt. Here, these anthropic considerations are diversified in a host of new ways to identify the most sensitive features of biochemistry and astrobiology. Celebrating the classic 1913 work of Lawrence J. Henderson, *The Fitness of the Environment*, this book looks at the delicate balance between chemistry and the ambient conditions in the universe that permit complex chemical networks and structures to exist. It will appeal to scientists, academics, and others working in a range of disciplines.

JOHN D. BARROW is Professor of Mathematical Sciences in the Department of Applied Mathematics and Theoretical Physics at the University of Cambridge and Director of the Millennium Mathematics Project. He is the author of *The Artful Universe Expanded* (Oxford University Press, 2005) and *The Infinite Book: A Short Guide to the Boundless, Timeless and Endless* (Cape, 2005), as well as co-editor of *Science and Ultimate Reality: Quantum Theory, Cosmology and Complexity* (Cambridge University Press, 2004).

SIMON CONWAY MORRIS is Professor of Evolutionary Palaeobiology at the Earth Sciences Department, University of Cambridge. He is the author of *Life's Solution: Inevitable Humans in a Lonely Universe* (Cambridge University Press, 2003).

STEPHEN J. FREELAND is Associate Professor of Biological Sciences at the University of Maryland, Baltimore County. His research focuses on the evolution of the genetic code.

CHARLES L. HARPER, JR. is an astrophysicist and planetary scientist and serves as Senior Vice President of the John Templeton Foundation. He is co-editor of *Science and Ultimate Reality: Quantum Theory, Cosmology and Complexity* (Cambridge University Press, 2004); *Visions of Discovery: New Light on Physics, Cosmology, and Consciousness* (forthcoming from Cambridge University Press).

Cambridge Astrobiology

Series Editors

Bruce Jakosky, Alan Boss, Frances Westall, Daniel Prieur and Charles Cockell

Books in the series

1. Planet Formation: Theory, Observations, and Experiments
   Edited by Hubert Klahr and Wolfgang Brandner
   ISBN 978-0-521-86015-4

2. Fitness of the Cosmos for Life: Biochemistry and Fine-Tuning
   Edited by John D. Barrow, Simon Conway Morris, Stephen J. Freeland and
   Charles L. Harper, Jr.
   ISBN 978-0-521-87102-0

3. Planetary Systems and the Origins of Life
   Edited by Ralph Pudritz, Paul Higgs and Jonathon Stone
   ISBN 978-0-521-87548-6

FITNESS OF THE COSMOS FOR LIFE

Biochemistry and Fine-Tuning

Edited by

JOHN D. BARROW
University of Cambridge

SIMON CONWAY MORRIS
University of Cambridge

STEPHEN J. FREELAND
University of Maryland, Baltimore County

CHARLES L. HARPER, JR.
John Templeton Foundation

CAMBRIDGE
UNIVERSITY PRESS

CAMBRIDGE UNIVERSITY PRESS
Cambridge, New York, Melbourne, Madrid, Cape Town, Singapore, São Paulo

Cambridge University Press
The Edinburgh Building, Cambridge CB2 8RU, UK

Published in the United States of America by Cambridge University Press, New York

www.cambridge.org
Information on this title: www.cambridge.org/9780521871020

© Cambridge University Press 2008

First published 2008

Printed in the United Kingdom at the University Press, Cambridge

A catalog record for this publication is available from the British Library

ISBN 978-0-521-87102-0 hardback

Contents

v

Contributors

Jayanth R. Banavar
Box 262, 104 Davey Laboratory, Pennsylvania State University, University Park, PA 16802-6300, USA

John D. Barrow
Department of Mathematical Sciences, University of Cambridge, Wilberforce Road, Cambridge CB3 0WA, UK

Julian Chela-Flores
The Abdus Salam International Centre for Theoretical Physics, Strada Costiera 11, 34104 Trieste, Italy
   Instituto de Estudios Avanzados, Apartado Postal 17606, Parque Central, Caracas 1015-A, Venezuela
   School of Theoretical Physics, Dublin Institute for Advanced Studies, 10 Burlington Road, Dublin 4, Ireland

Simon Conway Morris
Department of Earth Sciences, University of Cambridge, Downing Street, Cambridge CB2 3EQ, UK

Paul C. W. Davies
College of Liberal Arts and Sciences, Arizona State University, 300 E. University/PO Box 876505, Foundation Bldg, Suite 2470, Tempe, AZ 85287-6505, USA

Christian de Duve
de Duve Institute and Louvain Medical School, Catholic University of Louvain, Avenue Hippocrate 75-B. 7550, B-1200 Brussels, Belgium
   The Rockefeller University, 1230 York Avenue, Box 282, New York, NY 10021, USA

Michael J. Denton
Department of Zoology, University of Sindh, Jamshoro, Sindh, Pakistan

Albert Eschenmoser
Laboratorium für Organische Chemie, ETH Hönggerberg, HCI H309, CH-8093
Zürich, Switzerland

J. J. R. Fraústo da Silva
Fundação Oriente, Rua do Salitre, 66/68, 1269-065 Lisboa, Portugal
Centro de Química Estrutual, Instituto Superior Técnico, Av. Rovisco Pais,
1049-01, Lisboa, Portugal

Stephen J. Freeland
Department of Biological Sciences, University of Maryland, Baltimore County,
1000 Hilltop Circle, Room 115, Baltimore, MD 21250, USA

Owen Gingerich
Harvard-Smithsonian Center for Astrophysics, 60 Garden Street, Cambridge, MA
02138, USA

John F. Haught
Department of Theology, Georgetown University, Washington, D.C. 20057, USA

William Klemperer
Department of Chemistry and Chemical Biology, Harvard University, 12 Oxford
Street, Cambridge, MA 02138, USA

Mario Livio
Space Telescope Science Institute, 3700 San Martin Drive, Baltimore, MD 21218,
USA

Amos Maritan
Instituto Nazionale per la Fisica della Materia, Dipartimento da Fisica G. Galilei,
Universita di Padova, Via Marzolo 8, 35131 Padova, Italy

Ernan McMullin
Program in History and Philosophy of Science, University of Notre Dame,
Box 1066, Notre Dame, IN 46556, USA

Everett Mendelsohn
Harvard University, Science Center 371, Cambridge, MA 02138, USA

Harold J. Morowitz
Krasnow Institute for Advanced Study, East Building 207 (MS 1D6), George
Mason University, Fairfax, VA 22030, USA

Edward T. Oakes
University of St. Mary of the Lake/Mundelein Seminary, 1000 East Maple Avenue, Mundelein, IL 60060, USA

Guy Ourisson
Centre de Neurochimie, Université Louis Pasteur, 5 rue Blaise Pascal, F-67084 Strasbourg Cedex 9, France*

Jeffrey P. Schloss
Department of Biology, Westmont College, 955 La Paz Road, Santa Barbara, CA 93108, USA

D. Eric Smith
Santa Fe Institute, 1399 Hyde Park Road, Santa Fe, NM 87501, USA

George M. Whitesides
Department of Chemistry, Harvard University, 12 Oxford Street, Cambridge, MA 02138, USA

R. J. P. Williams
Inorganic Chemistry Laboratory, University of Oxford, South Parks Road, Oxford OX1 3QR, UK

Shuguang Zhang
Center for Biomedical Engineering and the Center for Bits and Atoms, Massachusetts Institute of Technology, 500 Technology Square, NE47-379, Cambridge, MA 02139-4307, USA

* Professor Ourisson passed away while this book was in production.

Foreword: The improbability of life

George M. Whitesides

How did life begin?

I (and most scientists) would answer, "By accident." But what an absolutely unlikely accident it must have been! The earth on which life first appeared – prebiotic earth – was most inhospitable: a violent place, wracked by storms and volcanoes, wrenched by the pull of a moon that was much closer than the one we know now, still battered by cosmic impacts. On its surface and in its oceans were myriads of organic compounds, some formed in processes occurring on earth, some imported by infalls from space. Out of this universe of tumult and molecules, somehow a small subset of chemical processes emerged and accidentally replicated, thus stumbling toward what became the first cells. How could such a chaotic mixture of molecules have generated cells? Order usually decays toward disorder: *Why do the tracks that led to life point in the opposite direction?*

The origin of life is one of the biggest of the big questions about the nature of existence. *Origin* tends to occur frequently in these big questions: the *origin* of the universe, the *origin* of matter, the *origin* of life, the *origin* of sentience. We, scientists and non-scientists alike, have troubles with such "origins" – we were not there watching when the first events happened, we can never replicate them, and, when those first events happened, there was, in fact, no "we." I believe that one day we will be able to describe life in physical terms – that is, we will rationalize life satisfactorily in molecular detail based on accepted scientific law and scientific theory using the scientific method. But we certainly do not know yet how to do it.

Understanding how organized living cells emerged from disorganized mixtures of molecules is an entrancingly, seductively difficult problem – so difficult, as we now understand it, that science does not even have well-formulated, testable hypotheses about how it might have happened, only guesses and intuitions. This problem deserves our most careful thought. Its solution will tell us about our origins and describe how disorder can spontaneously become order. It will also test the capability of current science to understand systems comprising many interacting parts.

Before trying to answer the question *How did life begin?*, we must first think about what the question really is that we are trying to answer: What is the "life" whose origins we are trying to understand? What are the characteristics of a cell, the simplest embodiment of life, that might allow us to trace back to its origins? How do we recognize an "origin"? When does a set of molecules, and of processes that convert these molecules into one another, cross a line separating "not-alive" from "alive"? And what is the tool – the "scientific method" – that science will use to try to address this problem?

Let us begin with the scientific method, a very useful and quite reliable strategy for doing science. Although it sometimes seems plodding, the scientific method can tease apart astonishingly difficult and complicated problems by careful attention to detail. It starts with rigorously reproducible empirical observations: "Things fall down, not up." "Two objects at different temperatures, when placed in contact, reach the same temperature." "Hydrogen atoms absorb only light that has specific frequencies." The scientific method codifies and quantifies these observations as "physical laws," builds theories (Newtonian mechanics, thermodynamics, quantum mechanics) based on those laws, and then tests new observations or hypotheses for their compatibility with these theories. Based on these theories, science rationalizes the physical world and predicts aspects of it not previously observed. The tools of the scientific method are the millstones and the oven that science uses to grind observations into theory and bake theory into prediction.

The scientific method works most rigorously when it identifies observations that are incompatible with current hypotheses. Faced with a new observation, scientists list all hypotheses that might explain it and then discard those that are incompatible with accepted physical law. Hypotheses that are not discarded as incompatible remain possibilities. If only one remains, it is promoted to theory. If disproving all hypotheses but one is not possible, we may retreat to demonstrating compatibility with theory, recognizing that compatibility is weaker than proof. In science, we use the phrases "I think . . ." and "I believe . . ." as synonyms, both implying ". . . based on known physical law." In other words, "This theory accommodates all the observations that we currently know."

So, what is life? We can describe what it looks like and what it does, but not how it works (most of us are in the same situation even with much simpler systems: computers, electric toothbrushes, refrigerator magnets). I suggest that life has five major physical attributes (other scientists may suggest other lists, but the general principles will usually be the same):

1. *Life is compartmentalized.* All life that we know is embodied in cells, and all cells have a continuous, closed membrane that separates "inside" from "outside."
2. *Life is dissipative, or out-of-equilibrium.* Life requires a flow of energy. If the chemical and physical processes in living cells reach equilibrium, and there is no flux of energy through the cell, it is, so far as we know, dead (or, at least, "not-alive").
3. *Life is self-replicating.* The most evident characteristic of the cell is that it was produced by the division of a parent cell, and, in many cases, it too will divide and produce daughter cells.
4. *Life is adaptive.* The cell can adapt its internal environment so that it functions even when the outside environment changes; in some circumstances, it can even modify the outside environment to make its inside more comfortable.
5. *Life occurs in water.* All life, so far as we know, involves molecules and salts dissolved or organized in a medium that is mostly water. We do not know whether water is essential to all life or just to life as we know it. But, at this time, we know no exceptions: life occurs in water.

So, according to this view, life is a spatially distinct, highly organized network of chemical reactions that occur in water and is characterized by a set of remarkable properties that enable it to replicate itself and to adapt to changes in its environment. We can, thus, describe what we are still ignorant about, but not much more.

How remarkable is life? The answer is: *very*. Those of us who deal in networks of chemical reactions know of nothing like it. We understand some – but only some – of the characteristics of the network that make it so remarkable. One key to its behavior is catalysis. The rates of essentially all cellular reactions – the processes that convert one molecule into another – are controlled by other molecules (usually by a class of protein catalysts called enzymes). The catalysts are (in some sense) like valves in a chemical plant (which, in some sense, is what a cell is): they control the rate at which one kind of molecule becomes another in a way loosely analogous to that in which a valve controls the rate at which fluid flows through a pipe. The complexity of the network becomes clear when one realizes that the catalysts – the valves – are themselves controlled by the molecules they produce: the products of one reaction can control the rate at which another reaction takes place.

The catalysts provide plausible connections among the elements of the network. The conversations among catalysts – conversations controlled by the very molecules the catalysts are controlling – allow the components of the network to form a single, coherent, interconnected, albeit very complicated, entity rather than an inchoate collection of independent processes. And how intricate these "conversations" are! The molecules whose production is required for the cell to live and to replicate itself modify the activities of the catalysts that make them. These already very complex interactions are further modulated by additional signals that come from outside the cell and by signals generated by an internal "clock." (This clock – the "cell cycle" – is itself a set of chemical reactions that oscillates spontaneously in time and defines the sequence of stages through which the cell progresses as it replicates.) Many molecules in the cell also have multiple roles: intermediates in one or many synthetic pathways, controllers of the activity of catalysts, signals for generating the catalysts and other molecules, sources of energy, and components of the physical structure of the cell.

Today, we understand many aspects of the behavior of the cell and many fragments of the network, but not how it all fits together. We particularly do not understand the stability of life and of the networks that compose it. Our experience with other very complicated networks (e.g. the global climate, air-traffic-control systems, the stock market) is that they are puzzlingly unstable and idiosyncratic. But unlike these and other such networks, life is stable – it is able to withstand, or adapt to, remarkably severe external jolts and shocks; and its stability is even more puzzling than the instability of the climate. We have a hard enough time understanding even simple sets of coupled chemical reactions. And we have, at this time, no idea how to understand (and certainly not how to construct) the network of reactions that make up the simplest cell.

So, at least for now, the cell is beyond our ability to understand it. The community of people working on the nature of life has, nonetheless, great (and probably warranted) confidence that understanding life in purely physical terms is a tractable, if difficult, problem. This confidence is enormously bolstered by two facts.

First, we are surrounded by uncountable varieties of life, especially by multitudes of different types of living cells; we thus have many examples of different forms of life. We ourselves are communities of cells with the added complexities of hierarchical organization of these cells into tissues, of tissues into organs, and of organs into the organisms that are "we."

Second, the tools of modern molecular biology have given us an astonishing capability to examine, modify, deconstruct, and reconstruct the molecular components of cells to see how they respond to our tinkering. The simplest cells (such as those of the primitive intracellular parasite *Mycoplasma genitalium*) appear to have fewer than a thousand proteins. That number of catalysts is still very complicated, and we have as yet no conceptual tools for understanding a network of reactions of such complexity. But this level of complexity does not, in principle, seem unreachably beyond our understanding. A cellular network of a thousand proteins (catalysts and molecules that sense, signal, and control passage across membranes; act as the structural skeleton; and perform many other functions) talking to one another in groups through the compounds they produce seems to be something that we will be able to disentangle. Certainly, those who call themselves "systems biologists" believe we will. Still, the path that scientists are now following in trying to understand the molecular basis of life will test their creativity and strain their endurance:

first, understanding the pieces of the networks as thoroughly as possible; then, perhaps, devising a computer model of a cell; and ultimately, in some distant future, validating the correctness of the principles suggested by this model by designing a set of reactions entirely different from those in the cells we now know.

It is one thing to analyze a Bach fugue; it is quite a different thing to play one, or to write one, or to create the kind of communication between humans that we call "music." We shall, I confidently believe, eventually analyze the fugue of life – the interplay of metabolic processes in the cell – as a network of compartmentalized, adaptive chemical reactions that can, astonishingly, replicate repeatedly into identical, distinct, separate networks. This is a very difficult job, but one that we humans can accomplish. But where did the cell come from? How did this wonderfully, astonishingly complex system come into existence? We do not know. If it is very difficult to understand the operation of cellular life as we observe it today, it is even more difficult to understand how it might have originated in the past.

Thoughtful, deeply creative people from a wide range of backgrounds have been captivated by the question of the origin of life. There is no shortage of ideas about pieces of this puzzle. We know how the surfaces of minerals might have provided elementary, non-biological catalysts to start the process and how heat or sunlight might have contributed other reactions. We can guess why certain types of molecules and reactions tend to occur in metabolism. We understand how any number of plausible natural events occurring in a conceivable prebiotic earth – events that formed complex mixtures of chemicals in geothermal vents, in lightning, on impacts, and under intense solar irradiation – might have contributed relevant bits of chemistry. But we do *not* understand how something as subtle and complicated as the network of reactions that we recognize the cell to be – a network both responsive and robust – might have emerged from these rudimentary processes. How could a chemical sludge spontaneously become a rose, even with billions of years to try?

We can take two approaches in our research directed toward the origin of life: reasoning *backward* and reasoning *forward*. "Backward" starts with life as we know and characterize it now – cells, DNA, RNA, enzymes, membranes, metabolites, membrane receptors, channels, and import/export proteins – and extrapolates back to simpler and simpler systems to try to infer an origin. This approach has been spectacularly successful in "reverse engineering" evolution, at least part of the way; but it has always been guided by examples provided by the types of cells that are now alive. Still, there seems little doubt that evolution could proceed once there was a primitive cell, with RNA or an RNA-like molecule, and reactions that used RNA as a catalyst and also translated RNA into protein or protein-like catalysts that were part of the network of reactions. Several hundreds of millions of tidal pools, together with enormous volumes of lakes and oceans, over several hundreds of millions of years provided many opportunities to produce cellular and organismic complexity. This part of the development of the complexity of life no longer seems to be a serious issue, at least conceptually. And the anatomical and physiological structures that now so enthrall us – the eye, the ear, the kidney, tentacles, muscles – these all seem to me transfixingly interesting products of evolution, but not ones whose origins are incomprehensibly improbable. If we and the squid have the same camera eye, why not? With enough tries, "best" solutions are bound to emerge many times. If some creatures walk on two legs, some on four, some on six or eight – again, why not? Many solutions may work well enough to survive the rigors of evolutionary selection.

Reasoning "forward" is much more problematic. Although we can imagine many possible mangers for the birth of life – deep smokers in the abyssal depths, tidal pools, hot springs, and many others – and although each could plausibly produce primitive precursors to many of the reactions that now constitute cellular metabolism, we have (in my opinion) no idea how these simple reactions might have blundered together to make the first protocell. Monkeys sitting at typewriters pecking out Shakespeare seems child's play by comparison. For example, we still do not know:

- *What were the first catalysts?* Were they protein-analogs or RNA-analogs or minerals or some other species of which there is now no trace?
- *How did the first networks form, and why did they persist?* One can imagine countless catalytic reactions that might have occurred, but how some of these reactions became self-sustaining networks is entirely obscure.
- *How could the process that stores the information that specifies the catalysts – the RNA or precursor of the primitive cells – have evolved?* The connection between RNA (or its younger, more evolved cousin, DNA) and the proteins that are catalysts, the enzymes, is not at all obvious; how the two co-evolved is even less clear.
- *How did the energetic cycles that power every cell emerge?* Why is there potassium ion on the inside of the cell and sodium ion on the outside? What was the origin of chemiosmosis? Given the extraordinary complexity of the ATPases – the complicated aggregates of proteins that generate ATP using the free energy that derives from differences in the concentration of ions across membranes – how could they have evolved? We simply do not know.

Nothing in the cell violates the fundamental laws of physical science. The second law of thermodynamics, the law that describes everything that occurs in the range of sizes relevant to life, can sleep untroubled. The flux of energy – now (although not necessarily originally) produced in nuclear reactions in our sun, transferred to the surface of earth as sunlight, absorbed by plants in photosynthesis, captured as glucose and other compounds, used in the cell to generate the intermediates that make metabolism possible, and ultimately dissipated to space by radiation as heat – can evidently support life. But how life originated is simply not apparent. It seems so improbable! The complexity of the simplest cell eludes our understanding – how could it be that any cell, even one simpler than the simplest that we know, emerged from the tangle of accidental reactions occurring in the molecular sludge that covered the prebiotic earth? We (or, at least, I) do not understand. It is not impossible, but it seems very, very improbable.

This improbability is the crux of the matter. The scientific method can be paralyzed by problems that require understanding the very improbable occurrences that result from very, very large numbers of throws of the dice. Sometimes we can understand the statistics of the problem; sometimes we cannot. How likely is it that a comet will hit the earth? We now have good geological records. How likely is it that a star will explode into a nova? There are many, many observable stars, and we now understand the statistics of nova formation quite well.

But how likely is it that a newly formed planet, with surface conditions that support liquid water, will give rise to life? We have, at this time, no clue, and no convincing way of estimating. From what we do now know, the answer falls somewhere between "impossibly unlikely" and "absolutely inevitable." We cannot calculate the odds of the spontaneous emergence of cellular life on a plausible prebiotic earth in any satisfying and convincing way.

What to do? For all its apparent improbability, life does seem to have happened here (or perhaps on some similar planet that transferred life to here). Rationalizing the origin of life is a problem that chemists are probably best able to solve. Life is a molecular phenomenon. The possibilities of alternative universes and different distributions of the elements are irrelevant from the vantage point of the particular universe and planet – our earth – that we share with so many other forms of life. We understand the chemical elements (we do not need to know about exotic forms of matter or energy in this enterprise), the molecules they form, and their reactivities. We know the players in the game, and we understand the game they play. We can guess (albeit only roughly) the distribution of the elements on the surface of the earth in the epoch in which we believe that life emerged, and we can infer the abundances of the molecules that were probably present. We understand how catalysts function. But we do not see how it all fits together.

Is this a problem in which science can make progress? Yes, and perhaps no. Those researchers who have taken the approach of reasoning "backward" to infer how life might have been born have made rapid progress. They have used the tools of molecular biology to trace the early stages of evolution back to the point where DNA gave way to RNA, which in turn probably gave way to some more primitive molecule whose composition we don't know, but which was probably related to RNA. The paths are fainter and fainter as the trail becomes older and colder and as we move from fact into speculation beyond RNA. We still do not understand the connections between RNA, or its forgotten ancestor, and enzymes, or their also forgotten ancestors, and the metabolic web that supports and constitutes life. Moving "forward" – spinning and weaving the threads that connect "molecules" to "life" – has been technically and conceptually more difficult.

Still, compelling connections are apparent between what might have existed on the prebiotic earth and the molecules of surprising complexity that are now vital to life. We understand, for example, how molecules of astonishing sophistication, such as the porphyrins – the precursors to the "green" of the pigments that serve plants in photosynthesis and the "red" of the hemoglobin that transports oxygen in our blood – could have arisen from aqueous solutions of hydrogen cyanide, one of the simplest of molecules and a possible component of the atmosphere of prebiotic earth. But these demonstrations, marvelous as they are, do not bridge the gap between "forward" pathways from prebiotic molecules to life and "backward" pathways from modern cells to possible progenitors, those emerging from the gray area between "alive" and "not-alive." As yet, no step goes from solutions of molecules to the networks of interconverting molecules that make up living cells. I believe that no one yet knows how to bridge that gap.

How to progress? The best lead to the hardest part of the problem – the "forward" problem – is the hypothesis that life evolved, somehow, from autocatalytic reactions (that is, reactions whose products are themselves catalysts for the reactions that produce them). We know something about autocatalytic reactions: flames are autocatalytic, and so are explosions (and one speaks, sometimes, of the "explosion" of life). We also know other reactions that are autocatalytic, although the subject of "autocatalysis" has not been a particular preoccupation of chemistry or biochemistry. Autocatalysis offers, I believe, a plausible trail into the wilderness.

Here, I suggest, is a *process* that science can use to examine this question. Let us build and understand autocatalytic reactions; extend that understanding to other networks of catalytic reactions; and develop simple, and then more complex, networks of autocatalytic and catalytic reactions. If, in time, we can trace a pathway from "chemical sludge" to "life," we shall have provided an argument based on plausibility, if not on proof, for the origin of life.

If, in time, we cannot trace such a path, what then? In science, until it has been proven that something cannot be done, it is always possible that it can be done. Proving that life did not originate by accident in tidal pools or black smokers will be more difficult than proving that it might have done so. Also, patience may be in order. What is impossible for science today may be trivial for science in the future.

There is still much that we do not understand about nature. As we learn more, I believe that we will ultimately see a path – based on principles of chemistry and physics and geology – that could plausibly have led from disorganized mixtures of inanimate chemicals to the astonishingly ordered, self-replicating networks of reactions that provide the basis for life. The fact that I cannot yet understand how an inconceivably large number of tries at an extraordinarily improbable event might lead to "life" is more a reflection of my limited ability to understand than evidence of a requirement for some new principle. But, having said all of that, I do not know, and in some sense do not care, whether physical science as I now know it ultimately explains the origin of life or whether the explanation will require principles entirely new to me. I do care that science makes every effort to develop the explanation.

Although I believe that science will ultimately be successful in rationalizing the origin of life in terms of physical principles, it should be cautious and claim credit only for the puzzles it has already solved, not those whose solutions still lie in the future. The central conundrum about the origin of life – that, as an accidental event, it seems so very improbable – is not one that science has yet resolved. Claiming credit prematurely – claiming, in effect, that current science holds all the answers – may stunt the growth of the new ideas that a resolution may require.

What, then, do I know? I know that I do not, yet, understand how life originated (and that I may not live long enough to do so). Order from disorder! How could it have happened?

I also know that my father never imagined cloning, and his father would not have believed television. Go far enough back, and the wheel was beyond comprehension. Difficult problems may take time – lots of time – to solve.

And so now, after I wake in the morning – at least on a good morning after I've had my coffee and am not distracted by the countless midges that constitute most of reality-as-we-know-it – my overwhelming response to existence, and to life, remains one of delight in its wonderfully wild improbability.

For now, call it what you will. *L'Chaim!*

Preface

This book is part of a two-part program focused on the broad theme of "biochemistry and fine-tuning." *Fitness of the Cosmos for Life* began with a symposium held at Harvard University in October 2003[1] in honor of the 90th anniversary of the publication of Lawrence J. Henderson's *The Fitness of the Environment*.[2] The symposium was an interdisciplinary, exploratory research meeting of scientists and other scholars that served as a stimulus for the creative thinking process used in developing the content of this book. The chapters in this volume were developed following the symposium and take advantage of the rich technical and interdisciplinary exchange of ideas that occurred during the in-person discussions.

The *Fitness of the Cosmos* program has provided a high-level forum in which innovative research leaders could present their ideas. In the spirit of multidisciplinarity, the fields represented by the meeting participants and book contributors are diverse. From the sciences, the fields of physics, astronomy, astrophysics, cosmology, organic and inorganic chemistry, biology, biochemistry, earth science, medicine, and biomedical engineering are represented; the humanistic disciplines represented include the history of science, philosophy, and theology.

This volume explores in greater depth issues around which the 2003 meeting was convened. It addresses the broad inquiry *Is the cosmos "biocentric" and "fitted" for life?* Keeping this question in mind, the authors presented their thoughts in the context of their own research and knowledge of others' writings on topics of "fitness" and "fine-tuning." This work pays tribute to the groundbreaking inquiry of L. J. Henderson.

---

[1] *Fitness of the Cosmos for Life: Biochemistry and Fine-Tuning – An Interdisciplinary, Exploratory Research Project Commemorating the 90th Anniversary of the Publication of Lawrence J. Henderson's* THE FITNESS OF THE ENVIRONMENT,[2] held at the Harvard–Smithsonian Center for Astrophysics, October 11–12, 2003. See http://www.templeton.org/archive/biochem-finetuning.

[2] Henderson, L. J. (1913). *The Fitness of the Environment: An Inquiry into the Biological Significance of the Properties of Matter*. New York: MacMillan. Repr. (1958) Boston, MA: Beacon Press; (1970) Gloucester, MA: Peter Smith.

The editors sought to develop in this collection of essays a variety of approaches to illuminating ways in which the sciences address questions of purpose with respect to the nature of the universe and our place within it. The chapters offer a range of insights reflecting themes and questions around which the meeting was organized and cover key areas of debate and uncertainty. In addition to George Whitesides' thought-provoking Foreword, twenty-four distinguished authors contributed twenty-one chapters, grouped according to four broad thematic areas:

Part I *The fitness of "fitness": Henderson in context*
Part II *The fitness of the cosmic environment*
Part III *The fitness of the terrestrial environment*
Part IV *The fitness of the chemical environment*

The various research agendas engaging questions of "fitness" and "fine-tuning" applied to the cosmos stress that important future opportunities exist for continued and expanded inquiry into areas where the sciences touch on wider, deeper issues of human interest. It is important to note that the preliminary discussion recorded here represents relatively early-stage exploration into what may in time become a much larger and more coherent area of research.

We hope that we have produced a book that will serve to stimulate thinking and new investigations among many scientists and scholars concerned with "really big questions," such as *Why can and does life exist in our universe?* If we have succeeded in any way, *Fitness of the Cosmos for Life* will serve as a stimulus to the creative thinking of people who can take the inquiry much farther.[3]

---

[3] A follow-up symposium, *Water of Life: Counterfactual Chemistry and Fine-Tuning in Biochemistry*, took place in Varenna, Italy, in April 2005; a research volume based on that symposium is currently in development. See http://www.templeton.org/archive/wateroflife.

Acknowledgments

The editors acknowledge the John Templeton Foundation,[1] and Sir John Templeton personally, for making this project possible.

We also thank:

Owen Gingerich, who contributed a chapter to this volume, and George Whitesides, who contributed the Foreword, for hosting the symposium at Harvard University in 2003 and for helping to develop the symposium and this book;

Hyung Choi, for assuming an important role in developing the academic program for the symposium in conjunction with Charles Harper; and

Pamela Bond Contractor, working in conjunction with the John Templeton Foundation and the volume editors, for organizing the 2003 symposium at Harvard and for serving as developmental editor of this book.

Finally, we thank Cambridge University Press for supporting this book project and, in particular, Jacqueline Garget and Vincent Higgs for their editorial management.

---

[1] See http://www.templeton.org/.

Part I

The fitness of "fitness": Henderson in context

1

Locating "fitness" and L. J. Henderson

Everett Mendelsohn

Crane Brinton, Harvard historian, friend of Lawrence J. Henderson, and fellow member of The Saturday Club, wrote the obituary for Henderson in the Club's third commemorative volume (Brinton, 1958, p. 207). Noting that Henderson was somewhat out of the ordinary – crossing the Charles River on several occasions to keep appointments at the Medical School (Boston) and the College (Cambridge) and then recrossing it to get to the Business School (Boston) – Brinton went on to note Henderson's other non-traditional characteristics: "Ticketed as a biological chemist, he later took the title *physiologist* and, although he would not have liked the name, at the end of his career he was a *sociologist* [emphasis added]."

Brinton went on: "A cross section of his publications may indeed be so drawn up as to seem an academic scandal." Brinton ran through the publications, from the well-known *The Fitness of the Environment* (1913) and *The Order of Nature* (1917); the more esoteric *On the Excretion of Acid from the Animal Organism* (1910, 1911); the simple volume *Blood: A Study in General Physiology* (1928); the unexpected transcript of an interview on the experiments in the Liberty Bread Shop (Brinton, 1958, p. 208); in his later life, *The Study of Man* (1941); to *Pareto's General Sociology: A Physiologist's Interpretation* (1935). Brinton jocularly added that a piece by Henderson – a biographical memoir on the life of the poet Edwin Arlington Robinson (a close friend from his student days) written as a memoir for the American Academy of Arts and Sciences – is to be found in the Woodberry Poetry Room of Harvard's Lamont Library.

To Brinton, "the conclusion is inescapable: Henderson, who was so much else, was also a philosopher." But Brinton also modified his praises: Henderson did not have the gifts of a popularizer. He was not a polymath, despite his interests in many areas. Nor was he a Renaissance figure; he had no interest in music or in the fine

Fitness of the Cosmos for Life: Biochemistry and Fine-Tuning, ed. J. D. Barrow *et al.*
Published by Cambridge University Press. © Cambridge University Press 2007.

arts. And – almost mockingly – Brinton noted Henderson's very high regard for "the art of eating and drinking."

So who was this man whose *The Fitness of the Environment*, published some ninety years before, was chosen as the emblem of the project, *Fitness of the Cosmos for Life*?[1]

Who was L. J. Henderson?

Lawrence Joseph Henderson was born in Lynn, Massachusetts, an industrial city just north of Boston, on June 3, 1878. The son of a businessman, he received his early education in Salem, Massachusetts, the more upscale town of his father's family, before going to Harvard as a sixteen-year-old – actually not that unusual in the late nineteenth century. His father's business connections in the St. Pierre and Miquelon Islands of the Gulf of St. Lawrence, where the young Henderson spent his vacations, stimulated his interest in learning French.

After graduating in 1898, he went on to Harvard Medical School, receiving his M.D. degree in 1902 (although he never intended to be a physician). He followed the path of those Americans interested in advanced scientific training by spending two years in the Strasbourg (then in Germany) laboratory of the biochemist Franz Hofmeister. After returning to Harvard, he spent a year in the chemistry laboratory of Theodore W. Richards (his former teacher and later brother-in-law). In 1905, he was appointed Lecturer in Biochemistry at the Harvard Medical School. He then moved to the college and, rising through the ranks, became a professor in 1919. In 1934, he was appointed the Abbott and James Lawrence Professor of Chemistry, a post he held until his death on February 10, 1942.

Henderson was a key figure in establishing the Department of Physical Chemistry in the Medical School (1920), and seven years later he helped establish the Fatigue Laboratory at the Graduate School of Business Administration. Together with Alfred North Whitehead (whom he helped bring to Harvard) and President Abbott Lawrence Lowell, he founded the Society of Fellows at Harvard. As early as 1911, Henderson started teaching a general course in the history of science (one of the earliest in any university) and played an instrumental role in bringing the Belgian George Sarton, the pre-eminent historian of science, to Harvard in 1916. He received the obvious forms of scientific recognition, including election to the National Academy of Sciences (becoming its Foreign Secretary) and the American Academy of Arts and Sciences, and was also decorated with the French *Légion d'honneur*.

But Henderson was not a good experimenter, did not like manipulating the complex apparatus of his field (he later confessed to this in his unpublished series

[1] See www.templeton.org/biochem-finetuning/participants.html.

of "Memories" [1936–39]), was judged by colleagues to be incapable of writing or speaking simply, was known for making "passionate and intolerant assertions and suffered fools not at all." He consciously took the role of gadfly, (often rudely) wanting to shake people out of their comfort zone and stimulate them to respond. Brinton noted that despite his warmth, which he hid from the world, he appeared to many as "a cold scientist, pompous, even pedantic" (Brinton, 1958, pp. 211–12).

Many of those who recounted episodes from Henderson's life or who had encounters with him noted special characteristics. His very fair-minded former student and colleague John T. Edsall, the Harvard biochemist, noted in his entry on Henderson in the *Dictionary of American Biography* that

his mind and temperament were complex. Especially in his later years, he spoke often with intense distrust of "intellectuals," liberals, and uplifters, who he felt failed to understand the deep non-rational sentiments that are an essential foundation for a satisfactory and stable society . . . he could infuriate some of his hearers . . .   *(Edsall, 1973, p. 352)*

George Homans, Harvard professor of sociology and young disciple of Henderson's later work on the social theorist Vilfredo Pareto, put it more bluntly in his own autobiographical volume: "Henderson was always an extreme and outspoken conservative . . . his manner in conversation was feebly imitated by a pile driver" (Homans, 1984, p. 90). Or, as he put it in another context: "Henderson never lost his tastelessness" (p. 117). This, from a deep admirer of his work, a close younger colleague, and the co-author with Charles P. Curtis of a volume on Pareto's sociology.

Where did *The Fitness of the Environment* come from and where did L. J. Henderson go with it? In spite of the several fields in which Henderson worked, a number of commentators, his contemporaries, and later analysts noted a markedly similar approach in many of his endeavors. Looking back at his work later in life, Henderson himself noted more unity than he had been aware of at the time. His focus was on organization and system: the organism, the universe, and society. John Parascandola, the author of a doctoral dissertation and several important articles on Henderson, put it succinctly: "The emphasis in his work was always on the need to examine whole systems and to avoid the error of assuming that the whole was merely the sum of its parts" (1971, p. 63).

But if that is the general outlook – and there is no real contest about this among the commentators on Henderson's work – what were the proximate causes and immediate contexts of Henderson's first full statements of the system of organism and environment? What were its visible and tacit sources? A connected sub-question examines how Henderson's ideas compared with those of other contemporary biologists who were similarly examining the ideas of life and matter: Walter Bradford

Cannon, a Harvard colleague and author of *The Wisdom of the Body* (1932) and of a very full biographical memoir published by the National Academy of Sciences (Cannon, 1945), and Jacques Loeb, a Rockefeller Institute protagonist whose classic essay "The Mechanistic Concept of Life" (1912) stood in sharp contrast to the organicism of the two Harvard scientists.

The obvious first sources for Henderson's fitness argument were the studies he began in 1905 on the equilibrium between acids and bases achieved in the organism. These studies represented some of his most sustained scientific work. The buffer systems he noted served to maintain neutrality in physiological fluids. What he saw in this was "a remarkable and unsuspected degree of efficiency [and] a high factor of safety" (Parascandola, 1968, p. 70). In his 1908 paper "The theory of neutrality regulation in the animal organism," Henderson noted that, in part, this efficiency depended on the properties of some of the substances involved in physiological reactions: that is, the dissociation constants of carbonic acid and monosodium phosphate and the gaseous nature of carbon dioxide, which allows easy excretion. This buffer action is a key to the stability of all living organisms – but, even more, it served to stabilize hydrogen ion concentrations in oceans and other waters. Henderson realized that water, with its extraordinary properties, together with carbon dioxide seemed uniquely fit to serve as the basis for all living systems (Edsall, 1973, p. 350).

Reflecting on this early work in "Memories," Henderson cited this as the point at which he became interested in the "fitness" of those substances for physiological processes (1936–9, p. 134). According to Cannon (1945), the discovery of the "extraordinary capacity" of carbonic acid to preserve neutrality had "far-reaching influences in Henderson's thinking." Henderson extended research into neutrality-maintenance capacity, which became a key element in his later work on physico-chemical systems (Cannon, 1945, p. 35).

In his report on Henderson's early work, younger colleague John Edsall noted that these "basic facts pointed clearly to a 'teleological order' in the universe." But Edsall immediately went on to indicate that Henderson "explicitly disavowed any attempt to associate this order with notions of design or purpose in nature, and considered his views fully compatible with a mechanistic outlook on the problems of biology" (Edsall, 1973, p. 350).

Henderson also credited John Theodore Merz' *History of European Thought in the Nineteenth Century* for its influence on the philosophical sections of the *Fitness* volume. Merz' four-volume study, with a whole volume devoted to the sciences, is fundamentally organismic in its outlook, and Merz was quite adept at identifying scientific and philosophical interactions (Henderson, 1936–9, p. 173).

Retrospectively, Henderson also identified a "eureka moment" that occurred on or about Washington's Birthday, 1912, while he was walking down the slopes of

Monadnock (a southern New Hampshire mountain) and thinking about the history of science course he was teaching. He recounted: ". . . it occurred to me suddenly, unexpectedly, and without any preliminary symptoms that I was aware of what I had been looking for in thinking about the fitness of the environment; [it remained] vivid and unforgettable" (1936–9, p. 175). It seemed to come together for him when he saw phosphate systems as very efficient buffers; he pondered the "usefulness of substances" and wondered whether "usefulness was an accident" (p. 177).

But to make sure that he would not be misunderstood, Henderson hurriedly assured his readers (and himself?) "that at this stage, I knew nothing of the literature of natural theology." Although he vaguely recollected William Paley and the watchmaker, he confessed that there was nothing in the history of thought "of which I was more ignorant and to which I was more indifferent." Having grown up in a period dominated by Darwin, he had known nothing of the *Bridgewater Treatises* (in which natural theology was explored at length by nineteenth-century scientists), and he had not been worried by the introduction of final causes into science. He was aware of, but not thoroughly knowledgeable about, the teleological literature and arguments (pp. 170–9).

By February 1912, however, having become fully convinced of the primacy of carbonic acid and water in the environment and the importance of the buffer concept, he set about writing *The Fitness of the Environment*. He claimed that he made no outline of the book (or of later ones, for that matter, including the treatise on *Blood*) and spent less than sixty days (and probably closer to fifty) writing the volume (p. 186).

In structuring his argument in *Fitness*, Henderson pointed to the Darwinian view of fitness as involving a mutual relationship between the organism and the environment and stressed the essential role of the environment as being of equal importance to the evolution of the organism. He opened his argument with the following paragraph:

Darwinian fitness is compounded of a mutual relationship between the organism and the environment. Of this, fitness of [the] environment is quite as essential a component as the fitness which arises in the process of organic evolution; and in fundamental characteristics the actual environment is the fittest possible abode of life. Such is the thesis which the present volume seeks to establish. This is not a novel hypothesis. In rudimentary form it has already a long history behind it, and it was a familiar doctrine in the early nineteenth century. It presents itself anew as a result of the recent growth of the science of physical chemistry.   *(p. v)*

His strong claim was that the actual environment is the fittest one possible for living organisms. Let me now locate Henderson's claims.

Locating Henderson's claims

Even as a sophomore at Harvard, Henderson confided in his "Memories" that he had "a vague feeling that there are not only many undiscovered simple uniformities behind the complexities of things, but also undiscovered unifying principles and explanations" (1936–9, p. 16). But there was more. Alongside this explanation, he recounted that he came upon William Prout's hypothesis (1815–16) concerning the periodic classification of chemical elements (all are multiples of the atomic mass of hydrogen) and felt the order involved must have an explanation. Was he retrospectively claiming that *he had himself become "fit"* to search for an understanding of the "fitness principle"? He was certainly willing to stray beyond the boundaries of the laboratory and the conceptual borders of the sciences.

By 1908, just as he was embarking on the construction of the fitness theory, Henderson began attending the philosophy and logic seminars of Josiah Royce in Harvard's Department of Philosophy. Through this channel, he came to know the works of Alfred North Whitehead, Bertrand Russell, and other contemporary philosophers. He continued to sit in on philosophy seminars in subsequent years. In the preface to *Fitness*, he generously acknowledged Royce: "His learning and generosity have in the past aided me to reach an understanding of the philosophical problems of science, and in the preparation of this book have repeatedly guided me aright" (p. xi). Royce himself had expressed belief in a form of universal teleology in his 1901 book *The World and the Individual*, and he enthusiastically called Henderson's work to the attention of other philosophers. In a long footnote at the conclusion of *Fitness*, Henderson cited Royce's teleological vision from the 1896 volume *The Spirit of Modern Philosophy* (Henderson, 1913, p. 311). The two joined with other Harvard faculty to discuss issues in the history and philosophy of science. These meetings went on for a full decade (1936–9, pp. 209–12; Parascandola, 1968, p. 71).

In his work, Henderson's ideas of fitness developed along with a growing interest in regulation of the physiological processes of the organism. Although he only later referred to this work, it was very much in accord with the concept of maintaining the *milieu intérieur* developed in the later decades of the nineteenth century by Claude Bernard and other contemporaries. (Henderson wrote a preface to an English translation of *Experimental Medicine* [Henderson, 1927] and made significant use of Bernard in setting out the problem he explored in *Blood: A Study of General Physiology* [1928]). But in his paper on the excretion of acids (1911), Henderson zeroed in on the seeming fitness of certain substances for physiological processes, pointing to the excretion of phosphoric acid as an indicator of renal action needed to maintain an acid–base balance: "There seems to be nothing in evolutionary theory to explain it and for the present it must be considered a happy chance . . ." (1911, p. 21; Parascandola, 1968, p. 73).

In "Memories," Henderson looked back and noted that he had questioned whether the role of carbon dioxide and phosphates was somehow linked in retrospect to special properties that made them more appropriate for physiological processes. As noted earlier, he located the moment at which the idea of the reciprocal nature of biological fitness came to him on Washington's Birthday, 1912:

I saw that fitness must be a reciprocal relation, that adaptations in the Darwinian sense must be adaptations to something, and that complexity, stability, and intensity and diversity of metabolism in organisms could not have resulted through adaptation unless there were some sort of pattern in the properties of the environment that, as I now partly knew, is both intricate and singular. *(1936–9, pp. 177–80)*

His research focus became water, carbon dioxide, and other carbon compounds (see the bibliography in Cannon, 1945, pp. 52–3. At the level of theory, he looked for a single order that linked biological and cosmic evolution. (He addressed this latter theme at length in his second fitness book, *The Order of Nature*, 1917.) Was the explanation he sought mechanical or teleological? But teleology, as he used the term, was limited. There were *no final causes, no entelechy* (emphasis added). The "teleological principle" in his understanding was inherent in matter and energy. These natural phenomena have original principles "essentially not by chance." But Henderson was consciously agnostic and refused to seek or find religious links for teleology. (His aversion to religious thought went back to his boyhood and was described vividly in "Memories" [1936–9, pp. 31–3].) For Henderson, teleology stood in parallel to mechanism, not as a replacement for it. As he put it in the preface to *The Order of Nature*: "Beneath all the organic structures and functions are the molecules and their activities . . . [they] . . . have been moulded by the process of evolution . . . and have also formed the environment" (1917, p. iv).

Henderson was struggling not to be misunderstood, and he concluded his preface with a plea:[2]

I beg the reader to bear this in mind and constantly to remember one simple question: What are the physical and chemical origins of diversity among inorganic and organic things, and how shall the adaptability of matter and energy be described? He may then see his way through all the difficulties which philosophical and biological thought have accumulated around a problem that in the final analysis belongs only to physical science, and at the end he will find a provisional answer to the question.

But misunderstood he was. At least he thought he was. His correspondence was filled with letters attempting to clarify and define teleology. I include a long excerpt from a letter to Paul Lawson (Henderson, 1918b) so that the reader can better understand what Henderson was attempting to achieve:

---

[2] He returned directly to this issue in his review of J. S. Haldane's *Mechanism, Life and Personality*, 1913, discussed later in this chapter.

It is a little difficult for me to reply to your remarks concerning my two books and the idea of teleology. My own opinion is that what I have said is considerably less philosophical than your interpretation of it. If you will look at a living organism, or at a watch, you will find that it possesses, like many other things in the world, a pattern. There is a certain peculiarity, however, about the pattern of the watch which resembles the peculiarity of the pattern of the living organism, and differs from the peculiarity of the pattern of certain other things possessing other well-marked patterns, such as, for instance, the orbit of a planet, or a geometrical figure. This seems to me to be an objective characteristic of the watch which we know to have been an excellent proof of the fact that the watch was designed. It seems to me also to be an objective characteristic of the organism, and, in the case of the organism, the current interpretation of explanations of it is that it is natural selection.

What I maintain is that there is a pattern in the ultimate properties of the chemical elements and in the ultimate physico-chemical properties of all phenomena considered in relation to each other. I do not mean to say that this pattern is exactly of the same nature as the pattern of the watch or an organism. Still less do I mean to say or to imply anything about design or mind. The only minds that I know are the minds of the individual organisms that I encounter upon the earth. But I feel perfectly justified, in spite of a certain unavoidable vagueness and ambiguity, in using the word "teleology" for the pattern in which I am interested.

The important thing to my mind is, nevertheless, not any doubtful talking about the proper name to discuss such a thing, but the fact itself. That is to say, the objective fact that the properties of the elements bear a certain very curious relationship to the process of evolution.

In *The Order of Nature*, Henderson's philosophical explorations came farther forward as he recounted the ideas of natural organization and teleology in a wide array of earlier authors from Aristotle through Descartes, Leibniz, Kant, Goethe, Bernard, Dreisch, J. S. Haldane, and Bosanquet. But the problem of reconciling mechanism in nature with indications of purpose was the way Cannon had set out the problem in his biographical memoir: There was indeed "a teleological appearance of the world . . . It is something that is real . . ." The solar system, meteorological cycle, and organic cycle seem to imply "a harmony which corresponds to an order in nature." As for Henderson's question "What is the mechanistic origin of the present order of nature?" the answer, Cannon suggested, "may be approximately solved by discovering, step by step, how the general laws of physical science work together upon the properties of matter and energy so as to produce that order" (1945, p. 38).

Henderson had already indicated in the closing pages of *Fitness* what he thought he had achieved and what limits he had set on teleology:

At length we have reached the conclusion which I was concerned to establish. Science has finally put the old teleology to death. Its disembodied spirit, freed from vitalism and all material ties, immortal, alone lives on, and from such a ghost science has nothing to fear. The man of science is not even obliged to have an opinion concerning its reality, for it dwells in another world where he as a scientist can never enter.   *(1913, p. 311)*

But Henderson had struggled to reach this point in his argument. As he summed up his thinking, he again asked the question "What then becomes of fitness?" He had already banished all metaphysical teleology from science and was left to explore two possibilities: "An unknown mechanistic explanation" of both cosmic and organic evolution exists – or it does not. While Henderson found it hard to credit such an "unknown" explanation, he added, with the historian's eye, that before Darwin's enunciation of natural selection it was hard to imagine a mechanical explanation of biological fitness. Therefore, at the end of *Fitness* he warned: "We shall do well not to decide against such a possibility" (1913, pp. 305–6). But let me be clear. When Henderson was composing *Fitness*, he had rejected the then current theories of vitalism and that of a designer for nature; but he had insisted on maintaining the term "teleology," albeit adjusted as he saw "fit." Was there ambiguity in his text? Let us turn to Henderson's contemporaries for a response.

What did Henderson's contemporaries say about his work?

Henderson's two early books, *Fitness* (1913) and *The Order of Nature* (1917), were reviewed by contemporary scientists and philosophers. Their reception, not dramatic by any standard, gives a good indication of the role of his ideas. It is interesting to note that Henderson's "reflective" and philosophically structured presentations antedated his fuller theoretical-scientific volume on *Blood: A Study in General Physiology* (1928), which itself developed from a sequence of papers in the *Journal of Biological Chemistry*, entitled "Blood as a physico-chemical system," beginning in 1921 and concluding in 1931.

One of the earliest, but also the fullest, reviews of *Fitness* appeared in *Science* (the journal of the American Association for the Advancement of Science) in September 1913 by the physiologist Ralph S. Lillie, who was at the time teaching at Clark University and later taught at the University of Chicago. His opening lines set out his view: "This book is essentially a discussion of the nature and implications of organic adaptation, that is, of the relation between the living organism and the environment, but is written from an unusual point of view." Lillie took the time and space to follow Henderson through his argument chapter by chapter with the full identification of carbon, hydrogen, and oxygen and their unique characteristics "which make possible the production of living protoplasm." They demonstrate "the greatest possible fitness for life" Lillie (1913), p. 337.

But Lillie was not completely satisfied with the adaptive teleology that Henderson had developed. He noted the transfer of the conception of fitness from the organic to the inorganic environment, which thereby achieves the reciprocal nature of biological adaptation. However, Lillie countered that Henderson had not dealt in detail with the organism itself and the interrelation between organisms and the environment:

. . . in other words, what adaptation is, as a general condition or process . . . Of course, the universe is a fit environment for life because it continues to exist in it. Granted, systems having the properties of living beings could not have arisen had the properties of carbon, hydrogen, and oxygen, and of their combinations, been other than they are, but what does this prove?

Most biologists, Lillie asserted, would see the central thesis Henderson advanced "as either self-evident or inherently unprovable." He seemed to mock Henderson in a footnote by saying, sure, this world is the best possible environment for the organisms that came to live in it – almost a truism, he implies – but what of other organisms in a different cosmos? Biologists may well see the book as an essay on the elements and compounds that form protoplasm, thus calling attention to often overlooked "facts and principles" (p. 340).

But Lillie was not satisfied with this reading; instead, he wanted to probe the questions "of the final significance of biological adaptations and the novel and interesting manner in which they are raised." He was amazed at Henderson's surprise that the environment and the organism possess similar characteristics. The surviving organic forms are those that have been able to maintain equilibrium with their environment. If conditions change and organisms can't compensate, they will fail. That, after all, is what natural selection is all about. "The task of biological science is thus left where we found it: to account for the characteristics of organisms on the basis of the physico-chemical characteristics of their component elements and compounds . . ." and to demonstrate how these living characteristics are formed by the environment (p. 341). Does that mean that life was somehow potential or implicit in matter, in the universe? "To the scientific investigator," Lillie announced, "such a statement can have little meaning, since it is remote from the possibility of verification" (p. 341).

J. D. Bernal, the materialist, in his book *The Origin of Life* (1967) summed it up succinctly: all of Henderson's evidence shows that "life had to make do with what it had, for if it failed to do so it would not have been there at all" (p. 169). Is there a way out by postulating a universe biocentric from its inception? Lillie joined Henderson in a cautious welcome to this view, in that the complexity, peculiarities, and stability of organisms would be unintelligible except for something of this sort.

For the final question posed by his reading of *Fitness*, Lillie asked: "How then is it possible to reconcile teleology and the existence of will and purpose in nature with the existence of a physico-chemical determinism which appears the more rigid the further scientific analysis proceeds?" This question, which he did not answer in the review, Lillie admitted (and which is often pushed to the side by scientists), would require biological knowledge for a solution – if one is ever achieved. Lillie concluded that Henderson's book points biologists to the "importance and urgency of these questions (p. 342)." A polite, friendly, but hardly full endorsement.

Writing in *The Dial, A Fortnightly Journal of Literary Criticism, Discussion and Information*, Raymond Pearl, the population biologist, opened his 1913 review with reference to a metaphysical diversion "of my academic and intellectually irresponsible youth," in which orthodox Darwinism was turned on its head. "Is there not quite as much justification, so far as the objective facts of nature are concerned, for one to say that the environment is adapted to the organism as there is for him to make the converse propositions?" (Pearl 1913, p. 111). Could natural selection, "or any other mechanistic hypothesis," stand up to the task? It would utterly fail, Pearl argued. Before Henderson's *Fitness*, no systematic efforts had been made to examine the fitness of the elements of the environment for sustaining life.

Henderson's own examination, Pearl opined, was in many ways a remarkable one. He showed "conclusively" that the known environment of the earth is better adapted to the needs of organisms than any other that could be constructed. He praised the collection and critical digestion of a great mass of data, describing it as a "masterly contribution to scientific synthesis that establishes the now well-known conclusions." But having recited those findings, Pearl announced: "At this point the book as a contribution to natural *Science* [in original] comes to an end." Turning to the final chapter, "Life and the Cosmos," which Pearl called "a consideration of the philosophical consequences" of the earlier scientific material, he was much less kind. While this part of the book was well done, "[I]t seems to this reviewer, at least, to fall short in compelling logical force of the purely scientific part of the work" (p. 112). Henderson showed, Pearl noted, that "existing science" was unable to give any "satisfactory mechanical explanation" to the reciprocal fitness of organism and environment while not ruling out its possibility. Pearl was clearly not enthralled by Henderson's proposal of a "devitalized teleology in the form of a purposive 'tendency' working steadily through the whole process of evolution." The objection was direct: "This 'tendency' is not something which can be weighed or measured" but is rather an original property of matter "assumedly not by chance, which organized the universe in space and time." In other words, it falls beyond the bounds of science. But Pearl's overall commentary on *Fitness* was adulatory. Notwithstanding his assessment of the concluding philosophical chapter, he conferred on the book the highest of honors, calling it a "logical sequel to the *Origin of Species*" (p. 112).

An array of additional reviews appeared both in scientific journals, such as *Nature*, and in philosophical ones, such as *Mind*, with the *Hibbert Journal* generally praising the scientific data brought forward but scattering various interpretations of the philosophical conclusion throughout. The mechanism, vitalism, and teleology debates current in the opening decades of the twentieth century had already been rehearsed in the responses to Henderson's own attempts to reconcile the mechanical and the vital in a single system.

One interaction in the review literature, however, adds an additional element to Henderson's ideas among other philosophically oriented biologists: the exchange between Henderson and J. S. Haldane, the physiological vitalist. Haldane's own entry into the discussion came in his earliest book on the debate, *Mechanism, Life and Personality: An Examination of the Mechanistic Theory of Life and Mind*, published in 1913, the same year as Henderson's own contribution to the philosophical discourse. In *Science*, September 17, 1915, Henderson produced an extensive review, opening in what almost might be considered an "airy" fashion: "Dr. J. S. Haldane has long been known as a philosophic physiologist. Indeed it is now for more than three decades that he has occasionally relieved the labors of an orthodox and eminent scientific investigator with the pleasures of idealistic metaphysics" (Henderson 1915, p. 378). Henderson recounted at length Haldane's understanding of the claims of mechanism and the failings inherent in them, as well as the fundamental claim that Haldane finally reached: "The phenomena of life are of such a nature that no physical or chemical explanation of them is remotely conceivable" (p. 379). If the concept of "organism" had been the first major stumbling block for mechanism in Haldane's view, psychology, or mind, raised the bar for mechanism even higher.

Henderson would have none – or very little – of it: "It is no light task for a man of science to form a critical judgment of this book, for I believe that its weakness is on the philosophical side" (p. 381). Henderson had, of course, recently been put through some criticisms of his own philosophical endeavors. While he was quite willing to quickly accept the critique of childish or crude mechanistic explanations, he by no means gave way to Haldane's broad rejection: "When we turn to Haldane's philosophical objections to the mechanistic standpoint we encounter, I believe, grave inconsistencies in his argument" (p. 381). Henderson was unwilling to accept Haldane's claim of the prior impossibility of providing a mechanistic explanation. He referred to T. H. Morgan's work in developing a mechanistic theory of heredity, called "inconceivable" by Haldane. Henderson also referred to Darwin's feat of making a mechanistic explanation of evolution conceivable.

The structure of Henderson's arguments was cast very much in the mode of Claude Bernard's earlier use of levels of explanation and referred to Cannon's work on fear and rage, which adopted this Bernardian outlook. Henderson vigorously rejected Haldane's claim that "all attempts to trace the ultimate mechanism of life must be given up as meaningless." Instead, he countered with his own stand: "And for my own part I am obliged to say regarding [Haldane's] statement, 'The phenomena of life are of such a nature that no physical or chemical explanation of them is remotely conceivable,' that is true only in a sense quite different from its apparent meaning and is of no *scientific* interest." In having to confront the antimechanism of Haldane, Henderson further identified his own location as he attempted to reconcile the worlds of life and matter.

In 1917, Haldane undertook a review in *Nature* of Henderson's second book, *The Order of Nature* (1917), which he saw as a follow-up to *Fitness*. He noted that with the wide adoption of natural selection the nineteenth-century conception of teleology had largely dropped from scientific discourse. He further noted that Henderson accepted natural selection, yet wanted to maintain a version of teleology based on the physical properties of matter in the universe and the organisms existing in a functional relationship – the teleological arrangement: "[Henderson] avoids all theological inference, and leaves us with teleological arrangement as an ultimate and mysterious empirical fact" (Haldane 1917, p. 263).

But Haldane was not satisfied. Must we assume, he asked, that the universe is composed at the outset of matter – eternal, unchangeable, and independent? He was unhappy with the concept of system that Henderson proffered: "Biology deals, not merely with the 'efficient' causes of ordinary physics and chemistry, but also with what Aristotle called 'final' causes." It is in the biological facts that "teleology is revealed as immanent in nature – as of its essence and no mere accident" appearing in the physical environment – and not only in organisms. Biological concepts, Haldane believed, must be extended to the inorganic world. While knowledge of how this would work is not now present, it requires only a further extension of knowledge. Haldane's hope for the future was that physics and chemistry would be penetrated by conceptions akin to those of biology. If this occurs, "teleological reasoning will take a natural place in the physical sciences" (p. 263). As I understand it, this is not where Henderson was going; and in a later criticism of Henderson's book *Blood: A Study in General Physiology* (1928), Haldane stressed how his and Henderson's divergent views and also the extent to which Henderson's commitment to the understanding that living things (for example, protoplasm) are physico-chemical systems further separated them (Haldane, 1929).

In the years following publication of *Fitness of the Environment* and *The Order of Nature*, Henderson reported in "Memories" that he stepped back even farther from teleological guides. He also stated that after his work on the sociologist Pareto, he became significantly more skeptical of metaphysics – to the extent that he regretted some of his earlier writings, seeing the discussion of "teleology, vitalism, and so forth, more or less irrelevant and immature." He noted that he had been less skeptical than he should have been and claimed that much of what he wrote in attempting to explain fitness in metaphysical and teleological terms was meaningless (Parascandola, 1968, p. 107; Henderson, 1936–9: pp. 173ff.). But he did not reject fitness as a concept and continued to see it as a valuable, and perhaps even the most interesting, part of his scientific work.

As he moved to the close of *Blood* (1928), Henderson restated the claims he originally made in *The Order of Nature* (1917) for the critical role of carbon, hydrogen, and oxygen, which "make up a unique ensemble of properties . . . [which are] of the highest importance in the evolutionary process," making diversity possible.

These elements, he emphasized, provide the "fittest ensemble of characteristics for durable mechanism." In 1928, he still claimed: "For these facts I have no explanation to offer. All that I can say is that they exist, that they are antecedent to organic adaptations, that they resemble them, and that they can hardly be due to chance" (1928, pp. 355–6; 1917, pp. 184–5).

Did *Fitness* challenge and provoke his contemporaries to take up the concept and use it as a guide to further scientific work? Reviews do not suggest this. By comparison, his later work on *Blood* as a physiological system much more clearly evoked the laboratory labors of his contemporaries. Its detailed analysis of what he referred to as "an immensely complex system in equilibrium" served as a vigorous stimulant to further experiment and explanation. *Fitness* remains to this day a symbol of attempts to provide broader explanation of the complexity of the worlds of the living and the non-living. When George Wald, the Harvard biochemist, was asked to write the introduction to the 1970 reprint of *Fitness*, he tried to set Henderson's book in time: pre-World War I, a time when the atom was gaining its redefinition at the hands of Rutherford, Rydberg, Mosley, and Bohr. This was before important new forms of chemical bonding had been established, and biochemistry was still in its infancy. What Wald did not suggest was that Henderson's book stimulated new scientific endeavors. Instead, he alluded to the significant advances that had been made in the sciences, often obviating some of Henderson's questions. He pointed to one conjecture: "A possible abode of life not unlike the earth apparently must be a frequent occurrence in space" and that perhaps even "'thousands' of such planets" exist. He further noted the current expectation of there being "many thousand million millions" of such possible abodes for life." This conjecture should arise, in Wald's view, wherever it can (1970, p. xxii). It is in this sense that Henderson's "fitness" takes on an expansive meaning. It has fueled renewed interest in the origin of life and the obvious extension: the synthesis of life in the laboratory.

Concluding remarks

As other chapters in this volume indicate, "Fitness" and "Order" have taken on other meanings, perhaps meanings that are more expansive than Henderson himself intended. But it has always been clear that a book once published no longer belongs to the author, and its interpretation is no longer controlled by him. As indicated in the pages above, Henderson tried in his response to reviewers to limit what he saw as some of the metaphysical turns given to their readings. In some ways, these views were unavoidable given Henderson's own often imprecise ideas and his choice to use a term like "teleology" and attempt to give it his own meaning.

From early on in the years after *Fitness*, Henderson kept making clear his lack of sympathy with ideas of vitalism; and although he resisted announcing himself a committed mechanist, he clearly indicated his receptiveness to its explanatory outlook. In his paper "Mechanism, from the standpoint of physical science" (1918a) he once again revisited the debate begun as early as 1915 in his review of J. S. Haldane's *Mechanism, Life and Personality* (see above) and rejected the vitalism proposed by Hans Driesch and Haldane: ". . . for my part, I can only come back to the conviction that Driesch is talking too confidently about things that none of us understand, and that . . . the weight of the evidence is greatly against him" (1918a, p. 574). As for ". . . Haldane's conviction that it is impossible to conceive organization in physical and chemical terms, this seems by no means impossible to most physiologists. . . . I accept the mechanistic hypothesis as, upon the whole, most consistent with the evidence" (1918a, pp. 575–6).

Even when Henderson turned to the organismic views of Alfred North Whitehead he was cautious. Henderson liked Whitehead, was influential in bringing him to Harvard, and together with Harvard's President A. Lawrence Lowell involved him as one of the three founders of Harvard's Society of Fellows. Yet in his review of Whitehead's *Science and the Modern World,* the Lowell Lectures for 1925, while clearly appreciative of the development of the concept of organization, the ". . . doctrine Whitehead calls the theory of organic mechanism," Henderson is not fully enthusiastic. He notes a "lack of unity in the exposition," with the author "still engaged in working out his theories." Although he can "dimly . . . perceive" the possibility of overcoming the difficulties ". . . that have produced the conflicts between mechanism and vitalism, and between freedom and determinism . . . hope has been so long deferred . . . it is natural to be a skeptic" (1926, pp. 292–3).

By the end of his life, Henderson had turned his interest from biological systems to social systems. Complexity, interpretation, and organization were still very present, but the early challenges of "fitness" seemed largely absent.

Bibliography

Bernal, J. D. (1967). *The Origin of Life*. London: Weidenfeld & Nicolson.
Brinton, C. (1958). Lawrence Joseph Henderson, 1878–1942. In *The Saturday Club, A Century Completed, 1920–1956*, ed. E. W. Forbes and J. H. Finley, Jr. Boston, MA: Houghton Mifflin, pp. 207–14.
Cannon, W. B. (1932). *The Wisdom of the Body*. New York, NY: W. W. Norton.
Cannon, W. B. (1945). Biographical memoir of Lawrence Joseph Henderson, 1878–1942 presented to the National Academy of Sciences in the fall of 1943. In *Biographical Memoirs*. Washington, DC: National Academy of Sciences, vol. 23, 1945, pp. 31–58. [Contains the fullest bibliography of Henderson's publications, but is not complete.]

Cross, S. J. and Albury, W. R. (1987). Walter B. Cannon, L. J. Henderson and the organic analogy. *Osiris*, 2nd series, **3**, 165–92.

Edsall, J. T. (1973). Henderson, Lawrence Joseph. *Dictionary of American Biography*. Suppl. III. New York, NY: Charles Scribner's Sons, pp. 349–52.

Fry, I. (1996). On the biological significance of the properties of matter: L. J. Henderson's theory of the fitness of the environment. *Journal of the History of Biology*, **29**, 155–96.

Fry, I. (2000). *The Emergence of Life on Earth: A Historical and Scientific Overview*. New Brunswick, NJ: Rutgers University Press.

Haldane, J. S. (1913). *Mechanism, Life and Personality: An Examination of the Mechanistic Theory of Life and Mind*. London: J. Murray.

Haldane, J. S. (1917). *The Order of Nature* [book review]. *Nature*, **100** (December 6), pp. 262–3.

Haldane, J. S. (1929). Claude Bernard's conception of the internal environment. *Science*, **69**, 453–4.

Henderson, L. J. (1908). The theory of neutrality regulation in the animal organism. *American Journal of Physiology*, **21**, 427–48.

Henderson, L. J. (1910). On the excretion of acid from the animal organism. *VIII. Internationaler Physiologen-Kongress*, Wien.

Henderson, L. J. (1911). A critical study of the process of acid excretion. *Journal of Biological Chemistry*, **9**, 403–24.

Henderson, L. J. (1913). *The Fitness of the Environment: An Inquiry into the Biological Significance of the Properties of Matter*. New York, NY: Macmillan. Repr. (1958) Boston, MA: Beacon Press.

Henderson, L. J. (1915). *Mechanism, Life and Personality: An Examination of the Mechanistic Theory of Life and Mind*. *Science*, September 17, **42**, 378–82.

Henderson, L. J. (1917) *The Order of Nature: An Essay*. Cambridge, MA: Harvard University Press.

Henderson, L. J. (1918a). Mechanism, from the standpoint of physical science. *Philosophical Review*, **27** (6), 571–6.

Henderson, L. J. (1918b). Letter to Paul Lawson. Henderson Papers, Harvard University Archives.

Henderson, L. J. (1926). A philosophical interpretation of nature, review of A. N. Whitehead, *Science and the Modern World* (1925). *Quarterly Review of Biology*, **1** (2), 289–94.

Henderson, L. J. (1927). Introduction. In *Introduction to the Study of Experimental Medicine*, Claude Bernard, transl. H. C. Greene. New York, NY: Macmillan Company.

Henderson, L. J. (1928). *Blood: A Study in General Physiology*. New Haven, CT: Yale University Press.

Henderson, L. J. (1935). *Pareto's General Sociology, A Physiologist's Interpretation*. Cambridge, MA: Harvard University Press.

Henderson, L. J. (1936–9). Memories (typescript of unpublished manuscript). Copies are found in the Harvard Archives, HUG 4450.7, and the Baker Library, Harvard Business School, Cambridge, MA.

Henderson, L. J. (1941). *The Study of Man*. Philadelphia, PA: University of Pennsylvania Press.

Heyl, B. S. (1968). The Harvard Pareto circle. *Journal of the History of Behavioral Science*, **4**, 316–34.

Homans, G. C. (1984). *Coming to My Senses, The Autobiography of a Sociologist.* New Brunswick, NJ: Transaction Books.

Lillie, R. S. (1913). *Fitness of the Environment* [book review]. *Science*, **38**, 337–42.

Loeb, J. (1912). *The Mechanistic Concept of Life: Biological Essays.* Chicago, IL: University of Chicago Press.

Parascandola, J. L. (1968). *Lawrence J. Henderson and the Concept of Organized Systems.* Ph.D. dissertation, University of Wisconsin.

Parascandola, J. L. (1971). Organismic and holistic concepts in the thought of L. J. Henderson. *Journal of the History of Biology,* **4** (1), 63–113.

Parascandola, J. L. (1973). Henderson, Lawrence Joseph. *Dictionary of Scientific Biography*, vol. 6, pp. 260–2.

Parascandola, J. L. (1992). L. J. Henderson and the mutual dependence of variables, from physical chemistry to Pareto. In *Science at Harvard University, Historical Perspectives*, ed. C. A. Elliott and M. W. Rossiter. Bethlehem, PA: Lehigh University Press.

Pearl, R. (1913). Natural theology without theistic implications. *The Dial*, **55** (Aug. 16), 111–12.

Prout, W. (1815, 1816). On the relation between the specific gravities of bodies in their gaseous state and the weight of their atoms. *Annals of Philosophy*, **6**, 321–30; **7**, 111–13.

Wald, G. (1970). Introduction to L. J. Henderson, *The Fitness of the Environment: An Inquiry into the Biological Significance of the Properties of Matter*. Gloucester, MA: P. Smith.

2

Revisiting *The Fitness of the Environment*

Owen Gingerich

In 1913, long after Charles Darwin had argued for the fitness of organisms for their environment, the Harvard chemist Lawrence J. Henderson pointed out that the organisms would not exist at all except for the fitness of the environment itself. "Fitness there must be, in environment as well as in organism," he declared near the outset of his classic work, *The Fitness of the Environment* (1913, p. 6). While most of Henderson's contemporaries ignored the philosophical implications of this work, as John Barrow and Frank Tipler have noted, it "still comprises the foundation of the Anthropic Principle as applied to biochemical systems" (1986, p. 143).

Henderson pointed out the uniqueness of hydrogen, carbon, and oxygen in the chemistry of living organisms. Another two decades would pass before astronomers would establish that these were three of the four most abundant elements in the cosmos; but Henderson was at least aware that these atoms were commonly found in the stars and planets. In his treatise, he began with the properties of water, just as William Whewell had done eight decades earlier in his far more teleologically oriented *Bridgewater Treatise* (1833).

Henderson grouped the notable qualities of water under two headings: (1) thermal properties and (2) interaction with other substances. As far as he was concerned, these were empirical, observed properties with minimal theoretical explanation. (Remember that Rutherford's nuclear atom was still a future concept, while quantum mechanics and the nature of the hydrogen bond lay many more years ahead.)

Let me list water's notable properties in a somewhat different order and present a variety of specific examples. Water comes closer to being a universal solvent than any other known substance, a basic property familiar to anyone putting sugar into a cup of coffee. In the human body, the digestive process takes place after nourishment has been dissolved into a liquid – water – solution. Even rocks can be subject to water's dissolving powers: witness the ocean's salinity. The solubility of

Fitness of the Cosmos for Life: Biochemistry and Fine-Tuning, ed. J. D. Barrow *et al.*
Published by Cambridge University Press. © Cambridge University Press 2007.

carbon dioxide in water is particularly remarkable and so important in its consequences that Henderson devoted an entire chapter specifically to carbonic acid
(formed when CO_2 joins with H_2O to form H_2CO_3), a topic akin to his own
research interests.

The significance of this particular solubility is spectacularly demonstrated by
comparing the earth with its sister planet, Venus. The atmospheric pressure at the
surface of Venus is nearly a hundred times greater than at the surface of the earth,
and the Cytherean atmosphere itself is more than 96 percent carbon dioxide. The
earth's atmosphere would be similar if the oceans had not dissolved the carbon
dioxide and precipitated the excess in the form of limestone. One can scarcely
begin to imagine the tons of Indiana limestone resting on our shoulders if the earth,
like Venus, had no oceans.

Carbon dioxide is highly soluble not only in water, but also in air, capable of
dissolving essentially to the same extent in equal volumes of either substance. This
closely balanced solubility is vividly demonstrated to anyone opening a carbonated
beverage after vigorously shaking the can. In human metabolism, a complex series
of enzymes enables carbohydrates to be "burned" to produce the energy for life. In
this process, the oxygen combines with the carbon and hydrogen to form carbon
dioxide and water, waste products that must be eliminated. Fortunately, carbon
dioxide can be dissolved in the blood, and equally fortunately it can be released
into the air by the lungs, a process Henderson described clearly:

In the course of a day a man of average size produces, as a result of his active metabolism,
nearly two pounds of carbon dioxide. All this must be rapidly removed from the body. It
is difficult to imagine by what elaborate chemical and physical devices the body could rid
itself of such enormous quantities of material were it not for the fact that, in the blood, the
acid can circulate partly free . . . and in the lungs [carbon dioxide] can escape into the air
which is charged with but little of the gas. Were carbon dioxide not gaseous, its excretion
would be the greatest of physiological tasks; were it not freely soluble, a host of the most
universal existing physiological processes would be impossible.   *(1913, pp. 139–40)*

Michael Denton, who quoted this passage in *Nature's Destiny*, went on to say:

As every medical student learns, it can be shown, from estimates of the total amount of
carbon dioxide dissolved in the blood and from estimates of the difference in the amount
of dissolved carbon dioxide in arterial and venous blood, that most of the 200 milliliters of
carbon dioxide produced per minute in an average adult human cannot be transported in
simple physical solution to the lungs.   *(1998, p. 132)*

In fact, the carbonic acid formed when carbon dioxide dissolves in the water gradually ionizes to form an acidic H^+ ion and a bicarbonate base, HCO_3^-:

$$CO_2 + H_2O \rightarrow H_2CO_3 \rightarrow H^+ + HCO_3^-$$

In the lungs, the process is reversed, releasing the carbon dioxide. If an excess of H^+ ions accrues through metabolism, this excess also drives the process toward H_2O and CO_2, and the acidity is removed essentially by exhaling the carbon dioxide. Thus, this process not only eliminates the waste product of metabolism, but also preserves the neutrality of the blood. Henderson lauded the accuracy of this system, and Denton remarked that "It is a solution of breathtaking elegance and parsimony" (1998, p. 133). The same buffering that plays such a remarkable role for large air-breathing organisms also preserves the neutrality of the oceans.

Henderson pointed out another property of water: its high surface tension, which is substantially higher than that of any other common liquid except mercury. This curious property allows water striders to walk across the surface of ponds or, as a parlor trick, partyers to float a double-edged razor blade in a bowl of water. More important, this property helps water to flow upward, against the force of gravity, in the tiny veins of even tall plants.

Because of the very high specific heat of water, a comparatively large amount of heat energy is required to raise its temperature. This property accounts for the general constancy of ocean temperatures and keeps the earth's oceans in a liquid state. Coupled with this high specific heat is the remarkably high latent heat of vaporization – the amount of energy required to turn water into steam – "by far the highest known," as Henderson described it. More than five times more energy is required to vaporize a given quantity of water than to raise the temperature of the water from its freezing point to its boiling point.

At the other side of the temperature scale, water has a most peculiar property: it expands as it freezes, contrary to most known substances. Anyone who has suffered the misfortune of frozen water pipes in the winter will be all too familiar with this property. Were it not for this anomalous expansion, ice would sink when it freezes and form a frozen reservoir at the bottom of the oceans. Because of the low thermal conductivity of water, the oceans would not thaw out in the summer. "Year after year the ice would increase in winter and persist through the summer, until eventually all or much of the body of water, according to the locality, would be turned to ice" (Henderson, 1913, p. 109). Henderson further stated that "[t]his unique property of water [the anomalous expansion on freezing] is the most familiar instance of striking natural fitness of the environment, although its importance has perhaps been overestimated"; but he added that "on the basis of its thermal properties alone . . . water is the one fit substance for its place in the process of universal evolution, when we regard that process biocentrically" (1913, p. 107).

The crucial role of carbon in the formation of life, so obvious to any organic chemist, comes later in Henderson's treatment. He was obviously fascinated by the environment, writ large – the oceans and the atmosphere – more than with the circumstances of life itself. Perhaps this was because the latter topic was and is shrouded in so much mystery, all the more so in 1913, compared with the present. Nevertheless, he extolled the virtues of the complexity afforded by carbon chemistry. Fundamental to carbon's versatility is its central location in the first long row of the periodic table:

3	4	5	6	7	8	9	10
Li	Be	B	C	N	O	Fl	Ne
1	1	7	*c.* 2300	7	2	1	0

Below the elements, I have listed the number of hydrides formed by each of them. (Because this table is taken from my chemistry notes of a half-century ago, the numbers for carbon probably need updating. A Google search gives "thousands," "vast," and "near infinite" as the number of hydrocarbons.) Whatever the current number – which of course does not count the numerous compounds of carbon with oxygen or nitrogen – it is clear that carbon greatly exceeds any atom, other than hydrogen, in the number of different molecules it can make.

In the years following the publication of Henderson's book, insights into atomic structure made the role of carbon much clearer and the unusual properties of water more understandable, without in any way diminishing Henderson's arguments or the awe that accompanies appreciation of this fine-tuning of our environment.

From the astrophysical perspective concerning carbon, oxygen, and hydrogen, the massive nucleus is the chief consideration. For chemistry, the much lighter surrounding swarm of electrons is the key to a deeper understanding of these fitness properties. With the discovery of the nuclear structure of the atom and the subsequent development of quantum mechanics, a number of ways to envision atomic structure developed. Here I shall adopt one of the chemist's favorite models, using a tetrahedron to model the carbon atom. Two of carbon's six electrons fill the inner shell, and four are distributed at the corners of the tetrahedron. It is only these outer electrons in the second shell that are modeled by the tetrahedron, but this is the only part of the atom normally of interest to chemistry. (I disregard here, for example, the use of radioactive isotopes as tracers in determining chemical structures.)

Eventually, to make sense of the data accumulating in the 1920s, physicists proposed another degree of freedom (called "electron spin") in arranging the electrons, suggesting that two electrons (with opposite spin) could occupy the same position. Thus, in the carbon tetrahedral representation, each vertex could accommodate an additional electron, which could be an electron shared with a partner – for example, a hydrogen atom with its single electron – provided the electrons were paired with opposite spins. Such a sharing is known as a covalent bond. Four hydrogen atoms, one at each corner, would give the carbon tetrahedron its full complement of eight electrons, and, by sharing, each hydrogen atom would have

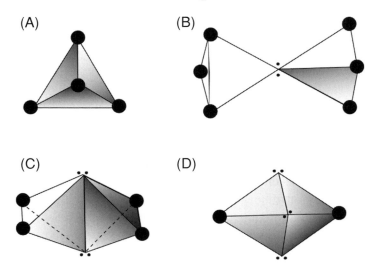

Figure 2.1. (A) The tetrahedral model for methane. The small black spheres represent the positions of the hydrogen atoms surrounding the tetrahedron of carbon. (B) Ethane, C_2H_6, with a pair of singly bonded carbon atoms. (C) Ethylene, C_2H_4, with a pair of doubly bonded carbon atoms. (D) Acetylene, C_2H_2, with a pair of triply bonded carbon atoms.

its full complement of two electrons in its inner shell. This stable configuration, shown in Figure 2.1A, is the molecule methane. Carbon, with its half-full quota of electrons, is as willing to lend as to receive; therefore, one carbon can bond with another, again provided that the shared electrons in each pair have opposite spins. The tetrahedral geometry allows a single, double, or triple covalent bond between two carbon atoms, as shown in Figure 2.1B–D. When the remaining vertices are filled with hydrogen atoms, the resulting gases are ethane, ethylene, and acetylene. This self-bonding property of carbon is the key to its prodigious fecundity.

Oxygen can also be approximated with a tetrahedral structure. This example, because of the light it sheds on the structure of water, is actually more informative than that of carbon. With six electrons for the outer shell, the oxygen atom will have two full vertices (with two electrons each) and two partly filled vertices (with a single electron each). Oxygen, like carbon, can form covalent bonds with itself, with either a single or a double bond. With a double bond (Figure 2.2C), no partly filled vertex remains, and the stable binary molecule that results is the normal form of oxygen gas found in the atmosphere. With a single bond, the two partly filled vertices can each join a hydrogen atom to form hydrogen peroxide (HO_2H) (Figure 2.2B) or join an additional oxygen atom to form a tight ring, ozone (O_3) (Figure 2.2D).

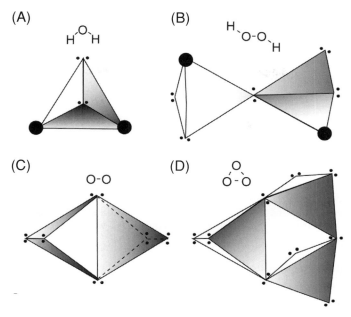

Figure 2.2. (A) Water, with the oxygen atom represented by the central tetrahedron. The small dots represent electron pairs that fill the open vertices. The model clearly shows the bent nature of the water molecule. (B) Hydrogen peroxide, with a pair of oxygen atoms forming a single covalent bond between them. (The atoms can rotate around the covalent bond.) (C) An oxygen molecule, O_2, with two double-bonded oxygen atoms. (D) An ozone molecule, O_3, with three double-bonded oxygen atoms.

The two partly open vertices of a single oxygen atom can be filled by the electrons shared with two hydrogen atoms to make a water molecule. The model shows that the hydrogen atoms will not lie on a straight line with the oxygen atom, a crucial fact if one is to understand the special properties of water. The tetrahedral angle is a reasonable approximation to the measured $104°30'$ angle in the water molecule.

In what was undoubtedly the most significant chemical treatise of the twentieth century, *The Nature of the Chemical Bond* (1939), Linus Pauling highlighted another extraordinarily important type of chemical bond, the so-called hydrogen bond. Although the double-bonding property of hydrogen was hinted at as early as 1912, Pauling used the principles of quantum mechanics and atomic modeling to show that hydrogen could participate in only a single covalent bond (that is, using shared electron pairs), so that a secondary bonding had to arise from something else, such as a weak electrostatic coupling. The "bent" model of the water molecule (Figure 2.2A) gives a qualitative idea of how this might work. The electron shared from each hydrogen atom is pulled toward the oxygen, leaving the flank of the positive hydrogen nucleus somewhat exposed. Meanwhile, the positive nucleus of

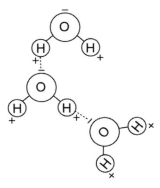

Figure 2.3. The negative shoulder of the oxygen atom in the water molecule couples briefly with the positive flank of a hydrogen atom in an adjoining molecule to produce a transitory hydrogen bond, giving liquid water its remarkably stable physical properties.

the oxygen atom is more than balanced by the extra shared electrons, leaving the broad shoulder of the water molecule with a residual negative charge. Therefore, an outlying positive hydrogen wing of one water molecule can be weakly and momentarily coupled with the negative oxygen shoulder of another water molecule, as shown in Figure 2.3. The duration of coupling is a second that is split very finely indeed, typically around 10^{-11} second.

Nevertheless, this continual coupling and recoupling give water its remarkable thermal properties. Although liquid at average terrestrial temperatures and pressures, the water molecules are subtly linked by the hydrogen bonds, and so an unusual amount of energy is required to raise the temperature of water, or to vaporize it. Similarly, the electrostatic (or ionic) properties of the water molecule act to make it a powerful solvent. Furthermore, as Pauling noticed, liquid water's statistical structure is akin to that of ice, and not just a single form of close packing. The temperature change in the equilibrium of two types of packing causes the anomalous expansion of water as it falls below 4 °C; the lower density of ice itself results from the more open crystal structure of the solid water (Pauling, 1939, p. 284).

Thus, we have been given a far deeper understanding of many of the extraordinary properties of water and its constituent atoms since *The Fitness of the Environment* was written. Also, totally unknown to Henderson in 1913 were the complex chemical shapes and coding in the structures of life: the DNA and protein molecules. These architectures depend critically not only on the presence of the key atoms, but also on the action of the hydrogen bond. The strands of the DNA double helix are joined by hydrogen bonds, strong enough to hold the long chromosomes together yet weak enough to permit the strands to unzip when replication begins. Similarly, the crucial folding of the proteins, where shape plays a vital role in their efficacy, is guided by the sites of hydrogen bonding.

Of course, these unique properties would have been of little avail had it not been for the substantial abundance of oxygen and carbon. But since hydrogen and oxygen rank numbers one and three, respectively, in cosmic abundance, water is guaranteed ubiquitous throughout the universe, while carbon is number four in the cosmic population. However, neither oxygen nor carbon emerged in the first three minutes of the Big Bang. At first glance, this might be labeled "God's Goof." That's how the physicist George Gamow felt when he discovered the presumed flaw in the nature of the light elements that prevented the heavier elements from forming. In the first minute of the Big Bang, energetic photons transformed into protons, and through collisions these protons fused into deuterium (nuclear particles of mass 2), tritium (nuclear particles of mass 3), and alpha particles (which would serve as mass 4 nuclei of helium atoms). But because there was no stable mass 5, at mass 4 the fusion process stopped – well short of the 12 mass units needed for carbon or the 16 for oxygen.

Gamow, with his impish wit, wrote his own version of Genesis 1, in which God, in His excitement at creating the universe, failed to call for a stable mass 5. Disappointed by the error, God "wanted to contract the universe again and start everything from the beginning. But, that would be much too simple. Instead, being Almighty, God decided to make heavy elements in the most impossible way":

And so God said: "Let there be Hoyle." And there was [Fred] Hoyle. And God saw Hoyle and told him to make heavy elements in any way he pleased.

And so Hoyle decided to make heavy elements in stars, and to spread them around by means of supernova explosions. But in doing so, Hoyle had to follow the blueprint of abundances which God prepared earlier when He had planned to make the elements from Ylem [the primordial soup of high-energy photons].

Thus, with the help of God, Hoyle made all heavy elements in stars, but it was so complicated that neither Hoyle, nor God, nor anybody else can now figure out exactly how it was done.
(Gamow, 1970, p. 127)[1]

Far from being a design flaw in our universe, however, the missing mass 5 seems essential to our existence. Suppose that mass 5 were stable. Then, with the overwhelming abundance of protons in the opening minutes of the universe, atom building could have taken place in mass steps of one, right up the nuclear ladder toward iron. This would have left no special abundance of carbon and oxygen, two essential building blocks of life. Because no stable mass 5 exists, element-building in the stars actually takes place in a two-step process. First, the hydrogen is converted into helium. Then, in a second process, the abundant helium is built up into atoms whose nuclei consist of integer numbers of helium nuclei of mass 4. This includes oxygen

---

[1] Gamow speculated that this parody might account for his not having received an invitation to the 1958 Solvay Congress on cosmology.

with nuclear mass 16 and carbon with mass 12, which, as I have indicated, are the most abundant atoms in the cosmos after hydrogen and helium. Without the missing mass 5 (as well as several other puzzling details in the structures of these lighter elements), not only might we not have the life-giving abundance of carbon and oxygen, we would also lack the long, slow hydrogen "burning" of main-sequence stars. It is this tedious process of long-drawn-out hydrogen burning that provides the stable solar environment in which the evolutionary sequences can work out.

What at first glance, then, appeared to be God's mistake turns out, in fact, to be one of God's most ingenious triumphs. Certainly, the way our universe works – that it takes a very long time to generate the heavier elements – depends critically on the lack of a stable mass 5. In the absence of a nuclear ladder with easy unit steps in mass, the ladder goes up in steps of four, so that the production of the various intervening heavier elements, such as nitrogen, is a complicated matter.

But another mystery presents itself: if the ladder goes up in steps of four, where is beryllium at mass step 8? Beryllium is naturally abundant, but it is seven orders of magnitude rarer than oxygen or carbon; this reflects the instability of the process of fusing two alpha particles (each of mass 4) into Be^8. A simultaneous triple collision – so that three alpha particles could fuse to produce carbon – is a highly improbable event. But in 1952, Edwin Salpeter pointed out an evolutionary process in more massive stars that had exhausted their hydrogen fuel. As the interior temperature rose because of the gradual gravitational collapse of the cores of such stars, the kinetic energy of the atomic nuclei eventually became hellish enough to produce a low equilibrium amount of Be^8. Then, sufficient collisions between beryllium and alpha particles (an exothermic reaction) could produce carbon and thereby fuel the stars' radiative output. Salpeter proposed the following reactions (1952, pp. 349–52):

$$He^4 + He^4 + 95 \text{ keV (kinetic energy input)} \rightarrow Be^8$$
$$He^4 + Be^8 \rightarrow C^{12} + 7400 \text{ keV (kinetic energy output)}$$

As alluded to in Gamow's parody, the late Fred Hoyle was a leading player in figuring out the subsequent processes. He noted that the stars seemed too young to have produced the observed high abundance of carbon unless some physical circumstances speeded up an otherwise extremely slow process. He therefore proposed that the collision cross-section of the beryllium was especially large, which is another way of saying that a resonance level for an appropriately excited level of carbon must be present. Because life, dependent on the high carbon abundance, does exist, he was able to predict the existence of a special resonance level in the carbon nucleus. Hoyle's prediction was a dramatic use of the strong Anthropic Principle, which states that the universe must have those properties that allow life to develop within it at some stage in its history (Barrow and Tipler, 1986, p. 21). The experimental work showing that the resonance really did exist eventually brought the 1983 Nobel Prize in Physics to Willy Fowler.

The stability of C^{12} makes it vulnerable to the reaction

$$He^4 + C^{12} \rightarrow O^{16} + 7100 \text{ keV (kinetic energy output)}$$

but in this case the closest resonance level is a half percent too low for the reaction to proceed with comparable efficiency. Had the resonance level in the carbon been 4 percent lower, there would be essentially no carbon. Had that level in the oxygen been only a half percent higher, virtually all of the carbon would have been converted to oxygen. Without that carbon abundance, none of us would be here now.

Hoyle later wrote:

Some supercalculating intellect must have designed the properties of the carbon atom, otherwise the chance of my finding such an atom through the blind forces of nature would be utterly minuscule... The numbers one calculates from the facts seem to me so overwhelming as to put this conclusion almost beyond question.   *(1981, pp. 8–12, esp. p. 12)*

These curious details of nuclear structure are among the many aspects of our universe – including the anomalous expansion of water and of ice – that make it remarkably fertile for the existence of intelligent life, so much so as to cry out for some explanation. Certainly, one logical possibility, along the lines of Aristotle's final causes, is that a supercalculating intellect has designed the universe to make it so. Such a conclusion can be coherent and satisfying in the framework of a general philosophical and religious view of the cosmos. With the rise of modern science in the seventeenth century, however, final causes tended to go out of fashion, and it may be argued that science's great success hinged on finding efficient causes – the "How" rather than the "Why." Newton's famous line in the General Scholium (added to the second edition of his *Principia*) concerning the ultimate nature of gravity, "I feign no hypotheses," is consonant with Henderson's own rather austere conclusion to his book:

Returning now to fitness, we may be sure that, whatever successes science shall in the future celebrate within the domain of teleology, the philosopher will never cease to perceive the wonder of a universe which moves onward from chaos to very perfect harmonies, and, quite apart from any possible mechanistic explanation of origin and fulfillment, to feel it a worthy subject of reflection . . . I cannot hope to have provided more than a very imperfect illumination of certain aspects of teleology in this venture upon the foreign field of metaphysics, and I should wish to be understood as very doubtful of my success in stating what seem to me some of the philosophical conclusions to be drawn from the fitness of the environment.

I wish, however, to put forward one scientific conclusion as a positive and, I trust, fruitful outcome of the present investigation. The properties of matter and the course of cosmic evolution are now seen to be intimately related to the structure of the living being and to its activities; they become, therefore, far more important in biology than has been previously suspected. For the whole evolutionary process, cosmic and organic, is one, and the biologist may now rightly regard the universe in its very essence as biocentric.   *(1913, pp. 310–12)*

Henderson's biocentric universe is very much with us today as astronomers and biologists join in the pursuit of the nascent science of astrobiology, speculating on the conditions and prospects for life elsewhere in the universe. Surely habitable environments abound, and the apparent fitness of the physico-chemical universe would seem to make it congenial to the existence of life that is chemically similar to the life we find on earth. Whether the universe teems with life, and even intelligent life, is an unanswered question – but quite possibly not unanswerable. The scientific strides made since 1913 suggest that a positive answer to this audacious question may come surprisingly soon; but *Homo sapiens* will never survive long enough for a negative answer, because it will be essentially impossible to establish that we are really alone.

References

Barrow, J. D. and Tipler, F. J. (1986). *The Anthropic Cosmological Principle*. Oxford: Oxford University Press.

Denton, M. (1998). *Nature's Destiny*. New York, NY: The Free Press.

Gamow, G. (1970). *My World Line: An Informal Biography*. New York, NY: Viking Press.

Henderson, L. J. (1913). *The Fitness of the Environment: An Inquiry into the Biological Significance of the Properties of Matter*. New York: Macmillan. Repr. (1958) Boston, MP: Beacon Press; (1970) Gloucester, MA: Peter Smith.

Hoyle, F. (1981). The Universe: past and present reflections. *Engineering and Science*, **XLV**, no. 2, November.

Pauling, L. (1939). *The Nature of the Chemical Bond*. Ithaca, NY: Cornell University Press.

Salpeter, E. (1952). Nuclear reactions in stars without hydrogen. *Astrophysical Journal*, **115**, 326–8. Repr. (1979) in *A Source Book in Astronomy and Astrophysics, 1900–1975*, ed. K. R. Lang and O. Gingerich. Cambridge, MA: Harvard University Press.

Whewell, W. (1833). *Astronomy and General Physics Considered with Reference to Natural Theology*. (*Bridgewater Treatise*, III.) London: William Pickering.

3

Is fine-tuning remarkable?

John F. Haught

My contribution to this volume is that of a theologian interested in the relationship between science and religion. I will be asking whether what we are calling "fitness for life" and "biochemical fine-tuning" are consistent with, and perhaps even supportive of, the ageless religious convictions that the universe is here for a reason and that life is the intended consequence of divine love, wisdom, and creative power.

Today, it is particularly striking to many scientists that cosmic constants, physical laws, biochemical pathways, and terrestrial conditions are just right for the emergence and *flourishing* of life. It is not surprising, of course, that, as life exists, the cosmic and chemical conditions for it *had to have been* formatted for such an emergence. It would be remarkable, however, if the format could have been otherwise, and hence not right for life. During the universe's history, it now seems that only a very restricted set of physical conditions operative at several major junctures of emergence could have opened the gateways to life (Hogan, 2000). So, what principles lie behind the narrowing of the gateways that allowed only those conditions preparatory to life to flow through while excluding any cosmological principles, physical parameters, and chemical laws that would not have permitted such an outcome?

In the long, unfolding story of nature's development, any conceivable series of physical conditions or constants other than those that would lead to life have been tossed aside. In fact, only the set that permitted life was allowed through a tightly constrictive filtering process. Close calls characterize the story throughout. And in our Big Bang universe, the fine-tuning of the specific trajectory that would lead to life began during the first moment of cosmic process. At that opening instant, all other sets of mathematically and physically conceivable expansion rates, gravitational constants, and densities of matter were put aside, and only a single set, one that would eventually sponsor life, was allowed entry into actuality. Later on,

Fitness of the Cosmos for Life: Biochemistry and Fine-Tuning, ed. J. D. Barrow *et al.*
Published by Cambridge University Press. © Cambridge University Press 2007.

the chemistry that would lead to life had to leave behind conceivable combinations that would prohibit the emergence of cellular and physiological complexity. Only a chemistry that would allow for a limited variety of three-dimensional protein folding, for example, could make way for the specific routes toward complexification that introduced sentience and consciousness into the Big Bang universe (Denton *et al.*, 2002). But can any of this fine-tuning be adequately understood without appealing to the idea of divine cosmic purpose?

Religion and theology cannot be indifferent to the question of whether the physical universe is the embodiment of an overarching meaning and purpose. However, cosmic purpose is not the same as design or fine-tuning, nor is the accumulation of scientific information about fine-tuning absolutely essential to a plausible theological affirmation of purpose in the universe. Theologically speaking, what purpose means, at the very minimum, is "the actualizing of value" – that is, of what appears self-evidently good – not necessarily the instantiation of design, even though design, order, or a pattern may be an aspect of the actualizing of value. Thus, a universe that appears to be in the process of bringing about such value-laden actualities as life, consciousness, freedom, creativity, and beauty, along with beings endowed with a capacity for reasonableness, selfless love, and promise keeping, could be said to have an overarching purpose, provided, of course, that these achievements have been intended.[1]

However, design and fine-tuning are ideas that are too narrow to capture the way in which careful theological reflection would conceive of divine intention or purposiveness in nature. After Darwin, moreover, it seems especially unwise to endow what is taken to be the Ultimate Source of the world's being exclusively, or even primarily, with the appellation "Designer." Similarly, the scientific cataloging of items suggestive of biochemical fine-tuning does not, by itself, offer much support to a theological argument for cosmic purpose. In fact, I believe it could be fatal for theology to focus only on design and fine-tuning. Such a concentration, after all, would only make the ancient and persistent theodicy question all the more intractable. Why would an allegedly designing deity, one capable of ordering natural processes in the beautiful arrangement of crystals, protein folds, and cellular mechanisms, refrain from arranging the larger world of life in such a way as to prevent disease, pain, and death? The notion that God is essentially an "Intelligent Designer" or "Fine-Tuner" has even led to the demise of natural theologies built too snugly on observation of natural forms of order alone (Buckley, 1987). The God of religious experience, moreover, is as much a *disturber* as a *distributer* of design.

[1] A reasoned theological discussion of purpose would also have to pay attention to the religious intuition that the world's attainment of value is in some way everlastingly preserved; I developed this concept elsewhere (2003).

The question of divine action

Whether research into biochemical fine-tuning could be theologically consequential inevitably brings up the more fundamental question of how to situate scientific explanations in relationship to theological understanding. In modern times, these two kinds of knowledge have often been seen as conflicting with each other. However, the best of our theologians have never taught that it is necessary to choose between natural and theological explanations. Theological explanations do not pretend to be an alternative to scientific accounts, just as good science does not appeal to supernatural causes.

Unfortunately, though, religious believers today still sometimes fear that as scientific explanations become more and more persuasive – for example, in accounting for the origin, evolution, and fine-tuning of life – the more they threaten to undermine a robust religious sense of the creative or purposive power of God. At the same time, scientists are sometimes apprehensive that the more room theology makes for the notion of divine action in accounting for natural phenomena, the less room will be available for meaningful scientific research (see my discussion of Cziko's 1995 book *Without Miracles*, below). For instance, the remarkable series of physical occurrences that have taken place in a chemistry attuned to life can be laid out as a fully natural process explicable in physical and chemical terms. So why do we need to invoke the idea of supernatural causation or cosmic purpose as essential anywhere in the unfolding chain of occurrences? Anyway, if a Creator had intended the universe to produce life, why did it take so long for the right chemistry to come along and for life to make its spontaneous debut relatively late in natural history? Science can give good physical explanations of emergent phenomena, and these accounts will surely become more detailed in the future. Thus, we must ask whether the idea of a wise Divine Designer (or Fine-Tuner) would be of much help in our attempts to understand the natural world.

This question is especially appropriate now that the story of nature shows itself to have been strewn not only with what seems to be, in the human scale of understanding, a wasted amount of time, but also a convoluted kind of engineering, such as the complicated stellar creation of carbon (as described, for example, by Burbidge *et al.*, 1957). And this is not to mention the often ruthless Darwinian processes that have shaped the story of life on earth. If the purpose of the universe is to produce life and mind, what do all the delay and Darwinian debris tell us about divine design, fine-tuning, and cosmic purpose? I believe that any research into biochemical fine-tuning – especially if the goal of such research is to shed light on the question of cosmic purpose – must not take place in abstraction from the larger picture of cosmic process and biological evolution. After all, Darwinian accounts of the evolution of life have made the idea of divine design seem increasingly shaky, and one cannot simply ignore this fact when reflecting on the question of fine-tuning (a notion that suggests design), whether at the level of cosmology or biochemistry.

Today, especially because of the apparent success and intellectual appeal of evolutionary explanations, the ideas of divine action in general and divine design in particular seem to have lost their former appeal in attempts to make ultimate sense of the natural world (see, for example, Dawkins, 1986, 1995, 1996; Dennett, 1995). Before Darwin, it was in the world of life and its adaptive complexity that natural theology ascertained the most dazzling displays of divine fine-tuning. The Boyle Lectures and Bridgewater Treatises, for example, sought characteristically to connect the intelligent activity of God to the complexities of life. But now that the idea of natural selection has become for many scientists the ultimate explanation of adaptive design in living beings (Mayr, 1997, pp. 64–78), it is tempting to look elsewhere for the primary evidence of divine action. For example, the physics of the early universe or segments of biochemistry seem to be exposing degrees of design or fine-tuning so improbable as to be suggestive of divine causation (Davies, 1992).

Even as a theologian interested in promoting good relations between science and religion, I confess to an uneasiness about this "regionalizing" of research into fine-tuning, especially if the ultimate objective of such a project is to see whether science is now opening up new areas, after (and perhaps in spite of) Darwin, for invoking the idea of divine action and purposiveness in nature. Concentrating separately on one or two chapters of apparent fine-tuning in the otherwise more ragged unfolding of nature may allow the natural theologian to momentarily ignore the disorder, waste, struggle, suffering, and death that accompany evolution's extravagant creativity. But the tormenting religious questions posed by Darwinian biology will continue to resist all attempts by natural theology to bracket them out. Therefore, speaking theologically, research into biochemical fine-tuning would strike me as artificial, evasive, and inconsequential as long as it gives any appearance of pushing aside the questions about divine action and cosmic purpose raised specifically by Darwinian biology. Such reserve seems especially appropriate because, these days, those most interested in finding evidence of intelligent design and fine-tuning in nature are often vociferous opponents of evolutionary biology (see, for example, Dembski, 1998, 1999; Wells, 2000).

At another extreme, however, it is also a matter of concern to me that much scientific thought has now concluded that the Darwinian notion of natural selection can provide an ultimately satisfying explanation of nearly everything in the world of life (Haught, 2000, 2003). The apparent success of evolutionary explanations of adaptive design has fostered suspicion that scientific enlightenment now renders the ideas of cosmic purpose and divine action altogether superfluous as far as a fundamental understanding of life is concerned. As one among many available examples of this exclusivist naturalism, I would note Gary Cziko's book *Without Miracles* (1995). Cziko maintains that the idea of divine influence, or what he calls "providence," is in competition with, rather than complementary to, scientific explanations. In a succession of chapters, he labors to show that it is Darwinian mechanisms rather than divine providence that explain any specific feature of life, such as sight, hearing, or intelligence. Hence, as far as Cziko is concerned, one is forced to choose either science or providential action, but not both.

Cziko's explanatory monism (in which only one explanatory slot is available for all) inevitably leads to the conclusion that theological appeals to divine influence are a threat to scientific explanation. Conversely, as mentioned previously, some religious believers – explanatory monists of another stripe – consider a successful scientific search for natural explanations of biochemical fine-tuning to be an implicit threat to the consequentiality of theological explanation. Both varieties of monocausal thinking, of course, raise the fascinating question of what it means to say that "God acts" in nature, an issue that lies at the heart of almost every contemporary controversy involving science and theology. I cannot give sufficient coverage to this discussion here, but I hope at least to demonstrate, by focusing on the topics of fitness for life and biochemical fine-tuning, that theological explanations can, at least in principle, co-exist quite comfortably and non-competitively with scientific explanations. I shall do so, however, only by also keeping in mind the special concerns about divine action raised by Darwinism.

Layered explanation

Theology, of course, does not strive for exactly the same kind of explanation in its own regions of inquiry as science does with respect to natural causes. Theology must take the discoveries of science seriously, but it cannot emulate the objectifying focus of scientific explanation without sacrificing its own substance. God, in other words, cannot become an object of scientific clarification without ceasing to be God. Theology, moreover, can be seen to have explanatory relevance only if it is first able to show that, in principle, a plurality of non-conflicting layers of explanation can exist for any particular set of natural phenomena. Thus, theology would have the role of ultimate explanation in an extended hierarchy of explanations that includes, and does not in any way compete with, scientific accounts. As long as one allows for layered explanation, in other words, theological discourse can be deeply explanatory of the universe without in any way interfering with the more lucid levels of scientific inquiry. Indeed, when it is taken as a deep explanation, theology even supports and promotes the more regional inquiries that physicists, chemists, or biologists undertake at their own proper planes of exploration. And in its refusal to compete with or intrude into scientific levels of explanation, theology may itself arrive at a more satisfactory understanding of the meaning and location of divine action or cosmic purpose than was possible before the emergence of modern science.

By "layered explanation," I mean that most things allow for more than one level of explanation. For example, if someone asks me to explain why a fire is burning in my backyard, one response might be, "Because the carbon in the wood is chemically combining with oxygen to make carbon dioxide." But another might be, "Because I applied a lit match to wood." And yet another could be, "Because I want to roast hot dogs." Different layers of explanation, in other words, can co-exist without competing with one another, and it would be a mistake to squeeze one kind of explanation into an explanatory groove that is appropriate to another. Explanatory pluralism, additionally, is more likely than monocausal accounting to put the human mind in touch with the full reality of what is being explained.

It would follow, then, that if a beneficent God, for some mysterious reason, freely intended to create a life-bearing universe, we would not look for evidence of this divine intentionality at the level of biochemical fine-tuning or natural selection of reproductively adaptive organisms. This would be like trying to find "I want hot dogs" in the smoke from the burning wood in my backyard. And yet, it is just such direct physical "evidence" that the scientific naturalist and other explanatory monists typically look for in theological accounts of the universe. The scientific naturalist's complaints about theology usually amount to disappointment that the latter is not very good at giving "scientific explanations." One of the clearest examples of this curious expectation is in E. O. Wilson's book *Consilience*, in which the noted author expresses his annoyance that by remaining silent on evolution "the biblical authors had missed the most important revelation of all!" (1998, p. 6). The underlying assumption here is that only one explanatory slot is available to all, and since it now belongs to science, no place remains for theology.

My point, though, is that good theology espouses an explanatory pluralism, as distinct from explanatory monism, and this means that a multiplicity of layers of explanation are available to quench the human desire to understand the world. Accordingly, reference to divine action or divine creativity would make sense only if one located it at levels of understanding other than where physical causes are being investigated. This decoupling of explanations is a function of the fact that every explanation is abstracted from the totality of causal factors that bring about any particular outcome. This does not mean that no connection at all exists between and among accounts. But theological explanations would stand in relation to nature – to such occurrences as biochemical fine-tuning, for example – analogously to the way in which "I want hot dogs" stands in relation to a chemical analysis of the burning logs in my backyard. That is, invoking divine creativity and purpose would lie at a distinct level of understanding, incommensurate but not incompatible with those of the sciences. And it could not be mapped directly onto any scientific accounts. Theology claims to explain things at a deeper level than physical causation, and the deeper explanation would simply not show up, nor should one expect it to do so, while one is setting forth, say, biochemical pathways leading to life.

Analogously, if one were to ask why life appeared on earth at all, even the most painstaking analysis of the physical and chemical fine-tuning required for this eventuality would uncover no purposiveness or divine intention at the many levels of scientific inquiry. Science abstracts from questions of purpose or ultimate meaning. Nevertheless, abundant logical space would still exist for a theological explanation at another level of understanding. Life, it could be said without any inconsistency, arose because of the properties of water (Henderson, 1913; Denton, 1998, pp. 22–46), the stellar formation of carbon (Burbidge *et al.*, 1957), precisely tuned cosmic constants (Rees, 2000), physical and chemical peculiarities (Northrop, 1979, pp. 168–205), *and also* because of the creative power of an Infinite Goodness. Just as it is not incoherent to say – all at the same time – that the fire is burning in my backyard because of chemical processes, *and* because it was ignited by a match, *and* because I want hot dogs, logically speaking plenty of non-competitive room exists, at least in principle, for a theological understanding of life to co-exist comfortably alongside scientific accounts. The only requirement is that the various layers not be logically inconsistent with one another.

Furthermore, contrary to E. O. Wilson's expectations, for theology to have deep explanatory value, its understanding of life *cannot* be spelled out in objectifying scientific language any more than "I want hot dogs" needs to be expressed in the idiom of chemical combustion. Indeed, the language of truly deep explanation *must* be that of symbol, metaphor, and analogy because it refers to a reality more encompassing than anything that we humans can bring into objectifying focus. Only a narrow commitment to explanatory monism – a contraction that can be justified by no available scientific evidence – would lead to a rejection of the richly layered model of understanding I am proposing here (and which Aristotle and other philosophers have proposed for centuries). Of course, the option to take the road of layered explanation is a fiduciary leap, but it seems to me to be a more reasonable kind of leap than the option for explanatory monism. As it casts a wider net, explanatory pluralism is less likely to leave out essential causal factors than explanatory monism, even if it still leaves the world looking a lot fuzzier at its foundations than a strict reductionist would prefer.

And yet, a theology comfortable with explanatory pluralism will still encourage all the sciences to push their own purely natural, and inevitably reductive, explanations as far as possible at their own appropriate levels of investigation. Good theology avoids any attempt to make room for divine action in the dark regions of yet uncharted scientific levels of inquiry. This is because its sense of layered explanation can, in principle, make ample room for theological explanation at levels more fundamental, metaphysically speaking, than those at which science functions. And so, without in any way contradicting elaborate physical, chemical, or evolutionary accounts of life, theology may be justified in claiming that chemistry is fine-tuned to life ultimately because of the infinite generosity and wisdom of God. But it does not expect to discover the finger of God anywhere in, or even at the temporal beginning of, the series of physical and biochemical causes that seem suggestive of a fine-tuning for life.

That room still exists for a deeper probing by theology after science has detailed the fitness of the cosmos and its complex chemistry for life is suggested by the fact that so many scientists themselves view fine-tuning as remarkable: that is, they have an intuition that something deeper is going on in nature than science itself can bring into focus. In recent years, as a matter of fact, it is scientists more than theologians who have noticed that the universe and earth appear to be remarkably fine-tuned for life. Modern thought, influenced by a mechanistic philosophy of nature, was not ready for this discovery. Up until the 1970s, science had increasingly pictured life as an anomaly in a pervasively lifeless and pointless universe. Jacques Monod's popular book *Chance and Necessity* (1971) epitomized the long-standing materialist claim that nature is essentially indifferent to life and intelligence. And much modern philosophy, art, and religious thought have taken shape in support of, or reaction to, that assumption.

But now, many scientists working at diverse scales from microphysics to astronomy have noted in a way they never did before that the universe, apparently from the beginning, was put together in a series of close calls that made it just right for the emergence of living and thinking beings (Davies, 1992). Hence, the universe no longer seems essentially lifeless and mindless. And, as other chapters in this volume point out, nature's biochemical details and the terrestrial environment as a whole provide a nest seemingly designed for life. So impressive is the still-accumulating information about the many emergent levels of the world's fine-tuning for life that some scientists can hardly suppress a suspicion that something momentous, perhaps even purposive, is afoot in the cosmos.

But one must be cautious. At what level should the explanation of the fine-tuning for life – in terms of the notion of purpose – be located? And what exactly does it mean to say that it is fine-tuned for life? Few scientists doubt that physics, chemistry, and terrestrial ecology are just right to sponsor the adventure of life. However, as I noted earlier, what is most interesting and most at issue, at least when all the dust clears, is whether nature has been *intentionally* prepared for life. And, is science as such ever really equipped to answer the deep question of whether intentionality or purpose is involved at any level?

I believe not, especially because the scientific method, as such, deliberately leaves out considerations of purpose. If, in their more popular essays on the implications of science, scientists hold forth on the question of purpose in nature, it is not as scientists, but as (usually amateur) philosophers that they do so. And that they attempt to draw philosophical conclusions directly out of the data of science is sometimes indicative more of a prior commitment to explanatory monism than of a genuine respect for the integrity of science.

And yet, an intriguing sense of the remarkableness of life persists among scientists themselves, including evolutionists. Remarkableness, as much as some scientists try to disown its power, still remains the muse of biology, as well as of all other sciences. In his extensive research on the phenomenon of evolutionary convergence, the paleobiologist Simon Conway Morris, an editor of this volume, has noted that "words like 'remarkable', 'striking', 'extraordinary', or even 'astonishing', and 'uncanny' are commonplace" in the literature (2003, p. 128). In other words, scientists have an incentive to embark on their research programs only if they somehow find nature astonishing. In fact, it was clearly his own experience of the remarkableness of convergence that led Conway Morris to challenge Stephen Jay Gould's conviction that chance was the determining factor in evolution. Gould had claimed that if the "tape" of terrestrial life's evolution were rewound and played again, the results would be completely different next time around (1989, p. 50). This is because, for Gould, the deepest causes of life's various forms were "contingent," or purely accidental events in natural history (such as climatic changes and mass extinctions). Like most evolutionists, Conway Morris gave a place to chance, but he disagreed with Gould's enshrinement of contingency as the ultimate cause of evolutionary outcomes. He tracked numerous instances of convergence, in which independent, ancestrally unrelated lines of evolutionary adaptation have led time and again to similar physiological outcomes. If the tape of evolution were replayed, the results might be different in some details the next time around, Conway Morris allowed, but generally speaking the consequences would be quite similar to what they turned out to be this time. In any case, life is not a simple lottery in which life forms follow no restrictive corridors of development.

It is "remarkable" to Conway Morris, for instance, that more than a dozen mole-like burrowing (fossorial) animals from around the world exhibit closely comparable bodily features, even though they belong to entirely different species (2003, pp. 139–44). The rapacious teeth of some predatory marsupials are nearly identical in structure to those of certain placental animals, although their respective evolutionary courses unfolded oceans apart. In species that are clearly unrelated, phylogenetically speaking, the eyes, snouts, auditory mechanisms, and other attributes have developed independently any number of times in nearly the same way, following similar biochemical and morphological pathways. Such similarities cannot be due either to sheer coincidence, since the likenesses are so pronounced, or to genetic inheritance, since they occur in unrelated phyla. Biology exhibits so many instances of convergence that something other than sheer accident must be involved; thus, Gould must be wrong. It is almost as though "evolutionary hyper-spaces" – the possible forms in which life can clothe itself – have been laid out in advance, and their number is finite. Clearly, constraints bear on the possible routes one might imagine life having taken. Organisms can try on any number of virtual outfits, and alterations can be made; but the tailoring is not infinitely malleable.

It is not my place to pass judgment on the accuracy or inaccuracy of Conway Morris' claims. However, it seems appropriate to ask for an explanation of the limits that constrain life to the point of permitting so much convergence. It does not seem sufficient to invoke the idea of adaptation by natural selection exclusively because, by itself, that idea is too wide-ranging to be helpful. No matter what outcomes occur in evolution, the ultra-adaptationist will insist that the explanation must be "natural selection." This may be true in a general sense, but appealing to the idea of selection lies at the same level of vagueness as a meteorologist's claiming that the laws of physics are the cause of this afternoon's thunderstorm.

Life, to those who have no stomach for layered explanation, is generalizable as merely "simplicity *masquerading* as complexity" (Atkins, 1992). But to those who find life remarkable, a search for a deeper explanation is in order, since explanatory monism seems too shallow and too vague to capture most of what is actually going on in the universe. However, one needs to avoid reaching for theological explanation too early in the search. Layered explanation allows us to avoid a precipitous rush to bring in theological or quasi-theological categories such as Intelligent Design while plenty of room is available for scientific clarification. Correspondingly, if theology is to offer any explanation as far as fine-tuning is concerned, it should not be introduced prematurely or on those levels of methodological self-restriction where scientific research is being carried out. Such an intrusion would be just one more appeal to "God of the gaps" (i.e. God's role is limited to accounting for the "gaps" in scientific explanations of nature).

However, if one takes the approach of layered explanation, theological under-standing might be relevant at vertically deeper layers of understanding than those at which the physical and biological sciences normally function. The sense of remark-ableness, I believe, arises from the fact that the scientist's consciousness, no less than that of other human beings, is somehow already in the grasp of a dimly intuited need for *ultimate* explanation. But theology cautions science not to be too hasty in identifying or clarifying nature's deepest dimension. It is better instead to allow the sense of uneasiness that "something more" is involved to remain alive and unresolved as a spur to ever-deeper exploration. I have argued elsewhere that some evolutionists prematurely give to the notion of natural selection a metaphysically foundational status, and that this assignment actually inhibits more penetrating inquiry into the depths of nature (Haught, 2003). I would urge the same caution here with respect to the temptation to give exclusively naturalist accounts of fine-tuning, fitness, and convergence. Allowing for theological or metaphysical explanation at deeper levels than those at which science works delivers science from the burden of having to come up with the ultimate explanation.

And yet, scientists are human beings, and it is natural for all of us to look for the ultimate explanation, especially when things seem surprising. The "frequency of adjectival surprise associated with descriptions of convergence," Conway Morris says, "suggests to me that there is almost a feeling of unease in these similarities" (2003, p. 128). He suspects that some biologists "sense the ghost of teleology looking over their shoulders," and he does not consider this to be an "unworthy sentiment" (2003, p. 128). Convergence has an "eeriness" that makes it plain that human inquiry still has a long way to go until it gets to the bottom of evolution (2003, p. 128). Something more than chance, selection, or the sheer passage of time must be involved in the outcomes of earth's life processes. What the "something else" might be, however, Conway Morris does not specify, and since other explanatory levels are available outside of science to account in a deeper way for remarkableness, it is not his task as a scientist to do so.

As a scientist, Conway Morris seems quite aware that he is not supposed to dabble in metaphysics. Yet, in his *Life's Solution* he provides more than a hint of the need for explorations of life by metaphysics and theology. Still, I believe Conway Morris considers these explorations to be supplemental to those of science as such, and he does not conflate religious allusions with his detailed scientific presentation of the evidence for convergence. Knowing that it is not his task to do theology, contrary to the suspicions of some of his critics (e.g. Prothero, 2003), he is *not* doing violence to science by pointing out, at least by implication, the need for alternative levels of explanation to account fully for life's remarkableness. In fact, by his explicitly allowing for a theology of evolution in addition to the science of evolution, one could make the case that he is safeguarding the integrity of science. By explicitly making room for theological investigation of life at a level distinct from science, he can more easily let science be just science than can some of his critics. Unlike evolutionists such as Gould, Dawkins, and Prothero, all of whom are content to force a materialist ideology into the single explanatory slot of scientific understanding, Conway Morris seems to be aware of the need to decouple strictly scientific work from statements of belief. By at least implicitly allowing for an explanatory pluralism, he can provide scientific inquiry with much more immunity to metaphysics than evolutionists that either unconsciously or deliberately conflate scientific ideas with materialist ideology.

Is fine-tuning remarkable?

How does this discussion of Conway Morris apply to the issue of fine-tuning: the fitness of the cosmos, earth, and chemistry to the emergence of life? Isn't this set of physical factors also "remarkable" enough to suggest that conventional scientific accounts are leaving something out by way of explanation? I believe that we also need to approach this question with an openness to layered explanation. Science alone, after all, has a habit of reducing remarkableness to mere routine at its own levels of inquiry. This reduction, when confined to the scientific method, strictly speaking, does not seem inappropriate. Science rightly tries to bring a largely hidden concatenation of natural causes out of darkness into the light of day, and once the unbroken sequence is exposed, the seemingly supernormal façade pales into bland normality. This is how science works.

However, the stronger claim that, in the end, nothing in nature can still be remarkable after science has done its work is not so much a characteristic of science itself as it is of the belief system known as "scientific naturalism." Scientific naturalists strongly believe that nature is all there is. If nothing else exists besides nature, it follows that nature must be self-originating. Moreover, if the natural world is not rooted in any creative agency beyond itself, there can be no end or goal that would give overarching purpose to the universe. According to scientific naturalists, the whole scheme of things is pointless, even considering local pockets of purely human meaning. Nor is there room for miracles or divine responsiveness to prayer. Finally, with no divine causation, all causes must be purely natural, and every natural event is the consequence of other natural events (Hardwick, 1996). Thus, all phenomena, however astonishing they may at first appear, are ultimately unremarkable – that is, natural. Otherwise, the naturalist belief system would be exposed as incoherent. Accordingly, cognitional satisfaction cannot occur until one has arrived at the cold consequence that nothing is really remarkable after all. "Remarkable" is at most a temporary sticker, to be removed once the purely natural causes of things have become manifest.

In our own time, scientific naturalism is alluring to scientists and philosophers alike. The self-avowed materialist Nicholas Humphrey clearly exemplifies the scientific naturalist's longing to reduce the remarkable to the routine. In his book *Leaps of Faith* (1996), he tries to show that what may at first seem supernormal, and hence an invitation to adopt a non-materialist understanding of the world, always turns out, on closer inspection, to be completely normal, and therefore just one more defeat for supernormalism. Like James Randi, the famous debunker of Uri Geller's allegedly paranormal feats of magic, Humphrey considers it not only the scientist's but also the philosopher's responsibility to demonstrate that any natural phenomena that seem to exhibit intelligent design and fine-tuning are really normal or routine. For Humphey, fine-tuning as described in the present volume would be just one of the many "appearances" that, once we see through them, are really not remarkable at all.

Humphrey has a lot of company in the intellectual world today. Owen Flanagan, a well-known philosopher at Duke University, even goes so far as to declare that the main purpose of academic philosophy is to "make the world safe for naturalism" (2002, pp. 167–8). Nevertheless, even the most naturalistic of scientists and philosophers at least begin their inquiries into specific areas of research only when they find certain things remarkable enough to engage their interest. Without the horizon of a yet unknown world stretching out tantalizingly ahead, scientific exploration could not even get started. Nor could it continue unless the summons to explore the unknown keeps reappearing on the horizon with each advance. It is still an open question, therefore, whether all that we find initially *remarkable* is destined eventually to become *routine*.

An appropriate question for theology, therefore, is why the world remains continually remarkable and often becomes even more so as science progresses in its understanding. And what, moreover, would make fine-tuning remarkable enough to be both an interesting topic for research and something deserving of ceaseless (theological) amazement? To many scientists, even after they have reduced it scientifically to purely natural processes, fine-tuning remains – again at a deeper level than science alone can reach – no less eerie and astonishing than evolutionary convergence is to the likes of Conway Morris. Otherwise, it would not be drawing so much attention. What is it, then, that arouses the scientist's – or, for that matter, anybody's – interest? What is it that invites theological comment?

As mentioned previously, one factor is a sense of contingency, a realization that many outcomes are possible, but that these possibilities are constrained at every step by pruning principles. A sense of contingency arises because the cosmic or biochemical fine-tuning prior to and underlying life might very well not have occurred at all. And the natural events leading up to life did not have to take place exactly the way they did. The scientist's capacity to entertain counterfactual worlds is possible only because of a prior sense that contingency is involved in the physical unfolding of the world. That is, alternative sets of physical occurrences are theoretically possible, even though most of them would not have led to life. One reason for sensing remarkableness, then, is the apparent lack of inevitability that scientists are now discerning in the physical prehistory of life.

Contingency means "absence of necessity," and so a contingent event is any occurrence that has actually taken place but did not *have* to take place. Contingency is inevitably remarkable as it gives rise to the question of why this or that happened if it did not have to happen at all. If one arrives at the point of realizing that something *had* to happen, and that it had to happen the way it did, then there would be no more room for questioning. Thus, understandably, one important wing of scientific naturalism – let us call it "right-wing naturalism" – tries to eliminate remarkableness by mentally transforming all impressions of contingency into a sense of inevitability. Acknowledging overriding necessity leaves no room for astonishment.

Right-wing naturalists strive to eliminate remarkableness from nature by demonstrating that all natural phenomena, beneath any initial appearance to the contrary, are the consequence of an underlying physical necessity. According to the philosopher Daniel Dennett, for example, the rich display of novelty and living diversity in evolution is really nothing more than the outcome of an algorithmic (mathematically inevitable) process (1995, pp. 48–60). Apparently, it is the vocation of the right-wing scientific naturalist to help us put aside all childish wonder and become resigned to the inherent unremarkableness of nature. In my opinion, however, such a project can be carried out successfully only where a sense of deeply layered explanation has already been suppressed. Just what motives lie behind the scientific naturalist's often passionate suppression of explanatory pluralism is a topic deserving of a separate study.

However, a left-wing variety of scientific naturalism is also operating in today's world. Looking for a world shorn of abiding wonder, some scientists and philosophers are inclined to enshrine absolute contingency rather than physical necessity in the role of final explanation. In contemporary debates about how to account for evolutionary outcomes, a major issue is whether the ultimate cause of evolutionary change is the rigorous necessity of adaptation by natural selection or perhaps the irrational contingencies in natural history. Gould, to cite the most prominent representative of contingency proponents, claims that it is especially accidental, undirected events such as meteorite impacts, earthquakes, and climatic changes that explain the uniqueness of living organisms and species. Yet he shares with his more deterministic adversaries the same compulsion to eliminate remarkableness at all costs. In Gould's case, the reduction is not so much to routine as to *irreducible* unintelligibility, but the result is the same: life is no longer remarkable because its absolute origin has now been revealed. In spite of what seem to be irresolvable conflicts between them, the respective devotees of Dawkins and Gould may appreciate that both evolutionists, though taking different routes, end up exorcising any open-ended mystery from the story of life. For both factions, an allegedly scientific category, whether chance or necessity, functions in fact as a metaphysical explanation, rendering theology obsolete.

Conclusion: nature as narrative

I believe that the notions of chance and necessity, however, are both lifeless mental abstractions that fail to do justice in any way to the deeper fact that nature is a story that blends contingency, law, and time into something truly remarkable and unrepeatable. It is this narrative aspect of nature that I would want to reflect on if I were in search of something irreducibly remarkable about the universe and evolution. For scientific naturalists, chance and necessity are each apotheosized, ironically not unlike the deities of an ancient mythic pantheon. In fact, "chance" and "necessity" are really abstract terms that have been reified in such a way as to miss altogether the concretely narrative quality of natural reality. On the one hand, "chance," "accident," and "randomness" are terms that point only obliquely to nature's concrete, contingent openness to indeterminate future outcomes. On the other hand, "necessity" is a misleading label for the underlying consistency of lawful constraints that limit possibility. In the real world, "contingent" openness never exists independently of the habitually constraining and lawful consistency (misnamed "necessity") that gives continuity to nature in its narrative passage through time.

What remains indelibly remarkable, therefore, is not so much life's or the environment's fine-tuning, or evolutionary convergence, for that matter, but the delicate blend of openness, constraint, and temporality that clothes the cosmos with drama. It is this combination that most appropriately invites theological comment. What is eerie, astonishing, and amazing – and utterly irreducible to routine – is that nature is narrative to the core, and that the story is not over. Nature is not a state, but a historical genesis, a process of becoming, an epic still being told. And so we will never get to the bottom of fine-tuning and evolution until we have understood why their matrix is narrative and where the story is leading. Contrary to the tenets of naturalism, I doubt that the natural sciences can answer this question at their own levels without also leaving ample room for theological conjecture at its own appropriate level.

Fine-tuning is remarkable, then, primarily because it is situated within the more foundational context of a cosmic story. The tuning, after all, is never really as fine as it initially seems. If it were, necessity and rigidity would have locked life into eternal stasis – a kind of death, in other words. Nature would no longer have a story to tell. Order without novelty is meager monotony. But, blessedly, an openness to possibility still exists (inaptly called "contingency") that pierces through the armored consistency in natural processes. In evolution, this openness consists in part of the very imperfections inherent in biological adaptations. Nature must be open to the future if it is to avoid metamorphosing into hard-rock necessity. Its imperfections assist in keeping it open to the future. To an earlier and now passé brand of Darwinism, it was a theological scandal that no adaptations were perfect, as imperfection spoiled the idea of an Intelligent, Divine Designer. But the idea of Intelligent Design is itself just another abstract idea originating in our human tendency to disassociate order from the openness that makes nature narrative. As it turns out, the imperfection of organic adaptation is essential if the story is to keep going and to remain interesting. If nature is narrative, we must remark at how fortunate it is that adaptation and design are not comfortably complete.

On the other hand, openness to transformation does not mean absolute inde-terminateness, as fine-tuning and biological convergence show. A finite range of possibilities and a channeling aspect to evolution exist and keep life from splashing out all over the place in completely unrestrained "hyperspace." The morphologies assumed by life, whether on earth or elsewhere, seem to be finite in number. Life is open to possibility, but possibility is not limitless. Otherwise, the story would have no continuity. Evolution arises in a narrative matrix, and narrative requires habitu-alness and redundancy, along with novelty, to keep the life-journey from collapsing at any capricious moment into complete confusion.

Contingency, if one wishes to use this abstract term, is remarkable because it adds historicity and dramatic suspense to recurrent natural processes. Right-wing naturalism looks for the strain of lawful necessity (and hence predictability) in all natural occurrences, and so it is naturally uneasy with contingency. Contin-gency means uniqueness, singularity, specificity, and unrepeatability, and these all defy sheer generality and reductive simplicity. To fulfill its objective of exposing remarkableness as mere routine, therefore, right-wing scientific naturalism must at least implicitly deny that contingency is anything more than necessity not yet understood.

When it appears in combination with nature's habitualness, contingency is remarkable. But when it is absolutized as an independent and ultimate explanation, contingency is equivalent to unintelligibility or absurdity. At the point of being thus maximized, contingency, no less than necessity, banishes remarkableness from the world. The typical way of exorcising remarkableness is the right-wing naturalist's reduction of contingency to necessity, but as this objective is unachievable, another way of muffling surprise and awe is found in left-wing naturalism – namely, to exaggerate contingency to the point of making pure chance the ultimate explana-tion of the most momentous natural occurrences. I have been proposing, however, that both kinds of naturalism, because of their explanatory monism, can thrive only in an illusory and imaginative world of ideas quite cut off from the actual narrative flow of nature itself. And it is this narrative, a story that wends we know not where, that remains forever remarkable and that gives theology a permanent place in the human quest for deep explanation.

References

Atkins, P. (1992). *Creation Revisited*. New York, NY: W. H. Freeman.

Buckley, M. J. (1987). *At the Origins of Modern Atheism*. New Haven, CT: Yale University Press.

Burbidge, E. M., Burbidge, G. R., Fowler, W. A. *et al.* (1957). Synthesis of the elements in stars. *Reviews of Modern Physics*, **29**, 547–650.

Conway Morris, S. (2003). *Life's Solution: Inevitable Humans in a Lonely Universe*. New York, NY: Cambridge University Press.

Cziko, G. (1995). *Without Miracles: Universal Selection Theory and the Second Darwinian Revolution*. Cambridge, MA: Massachusetts Institute of Technology Press.

Davies, P. (1992). *The Mind of God: The Scientific Basis for a Rational World*. New York, NY: Simon & Schuster.

Dawkins, R. (1986). *The Blind Watchmaker*. New York, NY: W. W. Norton.

Dawkins, R. (1995). *River Out of Eden*. New York, NY: Basic Books.

Dawkins, R. (1996). *Climbing Mount Improbable*. New York, NY: W. W. Norton & Company.

Dembski, W. A. (1999). *Intelligent Design: The Bridge between Science and Theology*. Downers Grove, IL: InterVarsity Press.

Dembski, W. A., ed. (1998). *Mere Creation: Science, Faith and Intelligent Design*. Downers Grove, IL: InterVarsity Press.

Dennett, D. C. (1995). *Darwin's Dangerous Idea: Evolution and the Meaning of Life*. New York, NY: Simon & Schuster.

Denton, M. (1998). *Nature's Destiny: How the Laws of Biology Reveal Purpose in the Universe*. New York, NY: Free Press.

Denton, M. J., Marshall, C. J. and Legge, M. (2002). The protein folds as platonic forms: new support for the pre-Darwinian conception of evolution by natural law. *Journal of Theoretical Biology*, **219**, 325–42.

Flanagan, O. (2002). *The Problem of the Soul: Two Visions of Mind and How to Reconcile Them*. New York, NY: Basic Books.

Gould, S. J. (1989). *Wonderful Life: The Burgess Shale and the Nature of History*. New York, NY: W. W. Norton.

Hardwick, C. (1996). *Events of Grace: Naturalism, Existentialism, and Theology*. Cambridge, UK: Cambridge University Press.

Haught, J. F. (2000). *God after Darwin: A Theology of Evolution*. Boulder, CO: Westview Press.

Haught, J. F. (2003). *Deeper Than Darwin: The Prospect for Religion in the Age of Evolution*. Boulder, CO: Westview Press.

Henderson, L. J. (1913). *The Fitness of the Environment: An Inquiry into the Biological Significance of the Properties of Matter*. New York: Macmillan. Repr. (1958) Boston, MA: Beacon Press; (1970) Gloucester, MA: Peter Smith.

Hogan, C. J. (2000). Why the universe is just so. *Reviews of Modern Physics*, **72** (4), 1149–61.

Humphrey, N. (1996). *Leaps of Faith: Science, Miracles and the Search for Supernatural Consolation*. New York, NY: Basic Books.

Mayr, E. (1997). *This Is Biology*, Cambridge, MA: Belknap Press.

Monod, J. (1971). *Chance and Necessity: An Essay on the Natural Philosophy of Modern Biology*, trans. A. Wainhouse. New York, NY: Knopf.

Northrop, F. S. C. (1979). *Science and First Principles*. Woodbridge, CT: Ox Bow Press.

Prothero, D. R. 2003. Inevitable humans? Or hidden agendas? A review of *Life's Solution* by Simon Conway Morris. *The Skeptic*, **10** (3), 54–7.

Rees, M. (2000). *Just Six Numbers: The Deep Forces That Shape the Universe*. New York, NY: Basic Books.

Wells, J. (2000). *Icons of Evolution: Science or Myth? Why Much of What We Teach about Evolution Is Wrong*. Washington, DC: Regnery.

Wilson, E. O. (1998). *Consilience: The Unity of Knowledge*. New York, NY: Knopf.

4

Complexity in context: the metaphysical implications of evolutionary theory

Edward T. Oakes

It would be a poor thing to be an atom in a universe without physicists. And physicists are made of atoms. A physicist is the atom's way of knowing about atoms.

> – George Wald
> Introduction to Lawrence Henderson's
> *The Fitness of the Environment*

A *conscious* fruit fly would have to confront exactly the same difficulties, the same kind of insoluble problems, as man. . . . To defy heredity is to defy billions of years, to defy the first cell.

> – E. M. Cioran
> *The Trouble with Being Born*

We keep forgetting to go right down to the foundations. We don't put our question marks down deep enough. . . . What a Copernicus or a Darwin really achieved was not the discovery of a true theory but of a fertile new point of view. . . . A curious analogy could be based on the fact that even the hugest telescope has to have an eye-piece no larger than the human eye.

> – Ludwig Wittgenstein
> *Culture and Value*

Science is the midwife of metaphysics. However much it might protest that it confines itself to physical realities only, science cannot help but provoke metaphysical questions in the human mind. To be sure, when it confines itself to its own specialized sphere, each science is strictly physical, physical both in the scope of its investigation and in the results and data those investigations produce. But science's ultimate import is always *meta*physical.[1] This trans-physical import of science

[1] I shall be defining the term "metaphysics" more exactly as these reflections proceed, but I want to start by saying that by metaphysics I am not speaking of something *extra*-physical (spirit, *élan vital*, Intelligent Designer, and

Fitness of the Cosmos for Life: Biochemistry and Fine-Tuning, ed. J. D. Barrow *et al.*
Published by Cambridge University Press. © Cambridge University Press 2007.

becomes most evident when scientists look up from their desks and seek to explain
the technical results of their work to the general public in easy-to-understand terms.
It is especially at this popularizing level that we find science, *nolens volens*, bumping
into metaphysics.[2]

But often enough the bump is not felt as such. For metaphysics is the spectral
science, silent and invisible, often slighted and frequently despised. But however
contemned, it is always hovering and lurking about the premises; and for that
reason it is not much welcome by scientists, most of whom greet metaphysics the
way Jane Eyre reacted to the screams she heard emanating from the attic. John
Barrow captures this uneasiness of physicists facing metaphysical questions very
neatly in a passage that also explains why the inherently philosophical questions
won't go away, no matter how uneasy scientists might feel about them:

[The] tendency of fundamental physics to move towards questions traditionally of interest
to philosophers and theologians has developed in parallel with an increased lack of interest
amongst physicists in the philosophical questions raised by these developments. To most
scientists [the phrase] "philosophical questions" has become a handy label to apply to any
collection of vague or apparently unanswerable questions which only become worthy of
serious consideration when they become scientific.   *(Barrow, 1990, p. 2)*

Nor is this uneasiness, which sometimes borders on outright contempt, utterly
implausible. Even a quick glance at the current state of philosophy will dis-
may the empirically trained scientist. When indeed has philosophical speculation

so forth), but of something *trans*-physical, that is, of certain "ideal" (or if that word itself sounds too spooky,
certain "logical") realities that ineluctably come into play when the implications of science are realized and
discussed. Something similar holds true for science's ethical and environmental impact as well, but the wider
implications of those issues do not fall within the focus of this chapter.

[2] In other words, the situation is now the reverse of the one that obtained at the dawn of modernity, when philosophy
served as the midwife to science. Even as late as René Descartes and Isaac Newton, the sciences they pursued
were still called "natural philosophy." In fact, the word "scientist" had not even been invented until the nineteenth
century, by William Whewell in 1840, in a deliberate act of coinage, although the word also appeared in passing
(only to be rejected for its ugliness) in an 1834 article in the *Quarterly Review*:

Science . . loses all traces of unity. A curious illustration of this result may be observed in the want of any name
by which we can designate students of the knowledge of the material world collectively. We are informed that
this difficulty was felt very oppressively by the members of the British Association for the Advancement of
Science at their meetings in the last three summers . . . *Philosophers* was felt to be too wide and too lofty a
term . . . *savants* was rather assuming . . . [s]ome ingenious gentleman proposed that, by analogy with *artist*,
they might form *scientist*, and added that there could be no scruple in making free with this termination when
we have such words as *sciolist, economist*, and *atheist* – but this was not generally palatable.
(Quarterly Review, vol. LI (1834), p. 59)

Also: "We need very much a name to describe a cultivator of science in general. I should incline to call him a
Scientist" (Whewell, 1840, vol. I, *Introduction*, p. 113). The result was that when the sciences finally felt liberated
from the apron strings of philosophy (and not just terminologically) they often took on an anti-metaphysical
polemic; and philosophy, to the extent that it suffered from what was facetiously called "physics envy," adopted
that same animus against metaphysics, especially in the school known as Logical Positivism. But I shall argue
here that, in one of those great ironies that constitute intellectual history, science is now giving birth to a
whole new range of metaphysical challenges, challenges that can in fact only be resolved through a specifically
metaphysical analysis.

ever inspired confidence in a scientist? As portrayed by the Polish writer Leszek Kolakowski the contrast between philosophy and science could hardly be starker:

For centuries philosophy has asserted its legitimacy by asking and answering questions inherited from the Socratics and Pre-Socratics: how to distinguish the real from the unreal, true from false, good from evil . . . There came a point, however, when philosophers had to confront a simple, painfully undeniable fact: that *of the questions which have sustained European philosophy for two and a half millennia, not a single one has been answered to general satisfaction.* All of them, if not declared invalid by the decree of philosophers, remain controversial. It is just as possible, culturally and intellectually, to be a nominalist or an anti-nominalist today as it was in the twelfth century; no odder now than in ancient Greece to believe or deny that phenomena can be distinguished from essences, no more unusual to hold that the distinction between good and evil is a contingent one, a matter of convention, than to claim that it is embedded in the necessary order of things. Belief and non-belief in God are equally respectable; no norms of our civilization prevent us from claiming that language creates reality or the other way around; we shall not be barred from good society because we embrace or reject the semantic conception of truth.

(Kolakowski, 2001, pp. 1–2 [emphasis added])

Obviously, scientists cannot endure such methodological and ideological chaos in their chosen specialties. That is also why, by the way, no matter how much postmodern epistemologists hurl challenges at the validity of science, scientists themselves rarely betray any lack of confidence in their work or in the results of their research and experimentation. Leave the death-wishes to philosophers, they seem to be saying, we need to get on with our work.[3] And who can blame them? Philosophy seems largely devoted to proclaiming its own death, rather like those theologians of the death-of-God school. But what science ever acts as philosophy does?

For well over a hundred years, a large part of academic philosophy has been devoted to the business of explaining that philosophy is either impossible or useless or both. Thus philosophy demonstrates that it can happily survive its own death [by keeping] itself busy trying to prove that it has indeed died . . . There is an immense variety of unconnected philosophical paths, all converging at one point – that of anti-philosophy. The farewell to philosophy, like the "bye-bye" in a famous Laurel and Hardy scene, never ends. The issues

---

[3] Needless to say, philosophers can sometimes reciprocate with a kind of "you can't fire me because I quit" attitude, as in this rather peremptory dismissal of the scientist's dismissal of philosophy from the journals of Ludwig Wittgenstein:

It is all one to me whether or not the typical scientist understands or appreciates my work, since he will not in any case understand the spirit in which I write. Our civilization is characterized by the word 'progress.' Progress is its form rather than making progress being one of its features. Typically it constructs. It is occupied with building an ever more complicated structure. And even clarity is sought only as a means to this end, not as an end in itself. For me on the contrary clarity, perspicuity are values in themselves. I am not interested in constructing a building so much as in having a perspicuous view of the foundations of possible buildings. So I am not aiming at the same target as the scientists and my way of thinking is different from theirs. *(Wittgenstein, 1980, p. 7e)*

On the other hand, Wittgenstein could be pretty harsh about philosophy too: "Reading the Socratic dialogues one has the feeling: what a frightful waste of time! What's the point of these arguments that prove nothing and clarify nothing?" (id., p. 14e).

that once formed the kernel of philosophical reflection – being and non-being, good and evil, myself and the universe – seem to have been shunted aside, relegated, except as subjects of historical enquiry, to a corner of academia, almost as dark as that occupied by God in schools of Divinity or sex in Victorian conversation. *(Kolakowski, 2001, pp. 7–8)*

There is also a further worry that needs to be mentioned: metaphysics is often religious in its claims.[4] In fact, Aristotle called "metaphysics" (the word itself was a coinage of post-Aristotelian librarians) not only "first philosophy," but more crucially "theology."[5] And the episodes of theology dictating to science (or to "natural philosophy," as it used to be called) are long, well antedating Galileo. One notices this in a remark of Moses Maimonides (AD 1135–1204) in his famous *Guide for the Perplexed*, where he scores this aggressiveness by theology to dictate results and which he detected in some of his Christian and Muslim predecessors:

It is not our object to criticize things which are peculiar to either creed, or books which were written exclusively in the interest of one community or the other. We merely maintain that the earlier theologians, both of the Greek Christians and of the Muslims, when they laid down their propositions, did not investigate the real properties of things; first of all, they considered what *must* be the properties of the things which should yield proof for or against a certain creed; and when this was found they asserted that the thing must be endowed with those properties; then they employed the same assertion as a proof for the identical arguments which had led to the assertion, and by which they either supported or refuted a certain opinion.[6]

Yet, despite both the simultaneous chaos of philosophy and the repeated acts of aggression perpetrated on science by metaphysical theology, I shall be claiming here that metaphysical claims and implications won't go away, no matter how anxious that makes scientists feel.[7] This uneasiness might seem to be most acute among physicists, for cosmology ineluctably raises question after question that, at

---

[4] Even among avowedly anti-metaphysical philosophers, one can find some astonishing admissions of religious intent. Wittgenstein again: "What is good is also divine. Queer as it sounds, that sums up my ethics. Only something supernatural can express the supernatural" (Wittgenstein, 1980, p. 3e).

[5] The phrase *"ta meta ta phusika"* merely means "what comes after [Aristotle's treatise] the *Physics*," which could imply only the placement of this book right after the *Physics* in the collection of Aristotle's corpus of collected works. Aristotle himself called this branch of philosophy "first philosophy" because it dealt with first causes and immovable substances, so that "the science which deals with them must be prior, and must accordingly be called 'first philosophy'" (Aristotle, 1984, *Metaphysics*, Book Epsilon 1, 1026a29–30). But for Aristotle that same first philosophy must also be called "theology" because "the science which it would be most proper for gods to possess is a divine science, and so too is any science which deals with the divine items. But this present science [of first causes] alone has these two features: the gods are held to be among the causes of all things and to be their first principles, and either the gods alone or the gods above all others can possess this science" (id., Book Alpha 2, 983a6–9).

[6] Maimonides, 1928, pp. 109–10. This observation represents an unusually prescient anticipation of the methodology of the Intelligent Design movement, as will become clear as these reflections proceed.

[7] Kolakowski again: "But such things, although we may shunt them aside, ban them from acceptable discourse and declare them shameful, do not simply go away, for they are an ineradicable part of culture.... Our sensibility to the traditional worries of philosophy has not withered away; it survives subcutaneously, as it were, ready to reveal its presence at the slightest accidental provocation" (Kolakowski, 2001, p. 8).

least at first blush, seem metaphysical: the place of chance in quantum physics, the nature of space and time as relative to each other, the role of observership in the constitution of reality, the paradoxes of the anthropic principle, and so forth. But even a quick glance at the debates raging inside biology shows how those debates are frequently determined by issues primarily philosophical, not biological – to the palpable unease of many biologists. In the wake of the discovery of the helical structure of DNA, biologists might well feel confident that they have finally expelled such allegedly metaphysical notions as *élan vital*, soul, and even "life" (as an extra-biological concept) from their conceptual armory; but uncomfortable issues remain lurking in the attics of the biological sciences.

Take, for example, the pesky problem of teleology. Before publication of Charles Darwin's *On the Origin of Species* in 1859, it was the common assumption of everyone from Aristotle to William Paley that the neck of the giraffe was "made for" reaching vegetation atop tall trees, that eyes are "made for" seeing, and so forth. And since such complex formations like the eye could hardly have come about by mere chance, they must have been designed by a kind of divine foresight, just as a watch requires a clever watchmaker. However, with the appearance of the *Origin*, that explanation fell into disarray and lost its hold on the scientific public, or so it is claimed. In fact the situation was far more complex than the received wisdom would have it. Even avid Darwinians were not so sure. For example, Thomas Huxley ("Darwin's bulldog") saw Darwinism as the perfect opportunity to set up a secular religion to rival Christianity, yet he still believed in "saltations" – big leaps in evolution to account for the transition from, say, fox to dog – and even claimed that "there is a wider Teleology, that is not touched by the doctrine of Evolution."[8] On the other side of the coin (and ocean), the Harvard botanist Asa Gray rightly recognized that saltations would mean the demise of the theory of natural selection and vigorously defended Darwin in the New World; but he always remained an orthodox Congregationalist. In contrast, his great rival at Harvard, the zoologist Louis Agassiz, attacked the *Origin* root and branch, but abandoned the Calvinist religion of his Swiss homeland and became a Unitarian while at Harvard.[9]

These catfights among the Darwinians continue right down to today. In order to drive home the point that evolution has no goal and that humans are a complete fluke, the late Harvard professor Stephen Jay Gould insisted that if we rewound the tape of evolution and started the whole process over again, the chances would be

---

[8] Much to Darwin's annoyance, Huxley asserted his belief in saltations as early as his instant review of the *Origin*, which appeared in the London *Times* on Boxing Day in 1859 – the first review of the *Origin* to see print, a mere thirty-two days after the *Origin*'s publication date of 24 November 1859. His review "The Darwinian Hypothesis" and his later essay of 1864, "Criticisms of *The Origin of Species*," may both be found in Huxley, 1893.

[9] Details in Ruse (2003), pp. 142–3.

vanishingly small of evolution producing, yet one more time, the same roaches and sharks, tulips and mushrooms, humans and crabs, that we know today. Of course if the outcome of evolution were *that* much of a fluke, then, as Daniel Dennett points out in *Darwin's Dangerous Idea*, the search for intelligent life on other planets would be as pointless as a search for extraterrestrial kangaroos (see more on Dennett's point about this below). The whole point of such a search rests on the assumption that evolution will inevitably tease out the potential for intelligence already lurking, however embryonically, in the prebiotic chemical soup of other life-potential planets. But if intelligence is well-nigh inevitable, then does not such inevitability say something about the inherent nature of the universe? If we were so inevitable from the start, why the paroxysms of self-abasement about our fluke emergence, these pseudo-lachrymose sneers at man as an insignificant worm in the cosmic slime?

What makes this debate, at root, a metaphysical one is its reliance on either implicit or openly avowed metaphysical presuppositions. Even to make the claim he did, Gould must have presupposed that it was meaningful to speak of "rewinding the tape of evolution" or that one may meaningfully speak, after the event, of the relative weight that should be given to chance over necessity in the discussion of biological causality. Even to speculate on "what might have been" had tectonic plates arranged themselves differently (would kangaroos ever have evolved if the Australian continent had stayed fused to the Antarctic land mass?) is to enter into a realm of thought determined by modal verbs ("might have," "could have," and so forth); and the role of modal verbs in determining the structure of reality is itself a heated area of metaphysical debate among analytic philosophers. Similarly, the positing of chance and necessity in causal explanation already represents the introduction of metaphysical categories.[10]

It is the purpose of this chapter to try to explicate these metaphysical implications, above all as they pertain to biological complexity. *Complexity leads to perplexity*, goes the old saw in Introduction to Epistemology classes. Or more exactly: complexity in *things* leads to perplexity in the *mind*. Even arrangements that are not, in and of themselves, terribly complex – such as a circle of ten stones forming a barrier for a campfire, for example – prompt the human mind almost spontaneously to imagine the presence of design. Let us take the case of Robinson Crusoe; but this time let us not have this shipwrecked survivor, washed ashore on what seems to be an uninhabited island, discovering a human footprint in the sand. Rather let us assume he comes across, in a forest clearing, a circle of stones: just as with the

---

[10] For example, is "chance" but an admission that a cause is unknown by people or unintended by them, as Aristotle (and Albert Einstein) held and as ordinary language assumes when it speaks of automobile "accidents" (auto accidents are hardly uncaused, just unintended)? Or are there really uncaused events in the universe, for example at the quantum level, and if so, how is that to be determined?

footprint, so too with the stones – Crusoe knows that there is (or was) at least one human besides himself on the island.

But far from resolving his initial perplexity after being washed ashore, the formation of the stones has still not really answered all his questions. For example, the footprint in the sand surely indicates the recent presence of Man Friday on the island (the sand would have obliterated an old footprint). But assuming there is no ash, no burning embers, in the center of the stone formation, Crusoe has no idea when the circle was made or what kind of human, hostile or friendly, made the campfire. All he knows, almost instinctively, as it were, is that such a circle could hardly have formed itself (which means, in modern parlance, that the circle is not just complex but "irreducibly" complex, a term discussed further below), and therefore that a designer must have introduced – or rather, imposed – a mentally conceived form upon the stones scattered about nearby. Anything beyond that, however, leads only to further perplexity.

I have decided to use this primitive example (primitive, obviously, in more ways than one) because it highlights why and how arguments based on design can lead us astray. No one denies that the universe is complex, staggeringly so. Thus it seemed eminently sensible to extrapolate Crusoe's logic to the universe: something as complex as the universe must have been so arranged as to allow not just the formation of stars, but also the chemical presuppositions of life – and adherents of the Intelligent Design (ID) movement go further and insist that the transition from the inorganic to the organic needed another "jump start" (which they call "abiotic infusion") to account for the even greater complexity of life. But in the wake of Charles Darwin's theory of natural selection, this argument from design has been shown to rest on an optical illusion, or rather, on a philosophical category mistake. "Irreducible complexity," in other words, is not necessarily the same thing as what I shall term here "staggering complexity."[11]

In my opinion, it is the great virtue of the theory of natural selection that it made clear, for the first time, the distinction between irreducible and staggering complexity. Unfortunately, in too many minds – both Darwinian and anti-Darwinian – that

---

[11] I have adopted the term "staggering complexity" from Dennett (1995), where he compares the Vast number of possible combinations of letters composing a book to the comparatively Vanishingly small number of letters that actually make up a book (he capitalizes Vast and Vanishing to highlight his idiosyncratic use of these terms, drawn from Jorge Luis Borges' tale about the Library of Babel, a universe that contains all the possible books in the world, a number so Vast that the currently existing books in the world are in comparison Vanishingly small). Similarly with the genetic code: the *possible* combinations are Vastly large, but the ones *currently obtaining* in the world are (in comparison) Vanishingly small. Why? Because just as not all books are worth reading (or writing), in fact Vanishingly few are when set against the field of all possible books, similarly, Vanishingly few genetically coded "readouts" will result in a viable organism. However, those that result in viable organisms will seem Staggeringly complex when set against such relative simplicity as a circle of stones, which easily leads to the logical fallacy that as the circle of stones was designed, then *a fortiori* the eye must have been too. But once a pathway can be mapped to show how mutations drawn from Vast possibilities have been selected for, the apparent connection between irreducible and staggering complexity can be seen to result from an optical illusion.

lesson has not been learned. For too long it has been assumed that Darwin had expelled *all* forms of teleological understanding in biology, but that is going too far.[12] Actually, in some ways the theory of natural selection gives new purchase to the teleological mode of understanding. Even Charles Darwin himself (at least in his better moments) knew as much. Try as he sometimes might (in his weaker moments) to expunge teleology from biological thought, he knew it couldn't be done; so he ended up with a rather muddled ambivalence:

> The term "natural selection" is in some respects a bad one, as it seems to imply conscious choice; but this will be disregarded after a little familiarity. No one objects to chemists speaking of "elective affinity"; and certainly an acid has no more choice in combining with a base, than the conditions of life have in determining whether or not a new form be selected or preserved.   *(Darwin, 1868, Vol. I, p. 6)*

This passage tells us more than even Darwin himself seems to have divined. For what he is highlighting here, perhaps without realizing it and certainly without the technical vocabulary necessary to make the point clear, is that one must distinguish between what ID advocates call "abiotic infusion" and what Aristotle called *en-telecheia* ("entelechy"). Because this term is hardly a household word these days, perhaps the reader will allow me to cite this definition from the standard reference work for Greek philosophical terminology:

> **Entelecheia**: *state of completion or perfection, actuality.* Although Aristotle normally uses *entelecheia*, which is probably his own coinage, as a synonym for *energeia*, there is a passage (*Metaphysics* 1050a) that at least suggests that the two terms, though closely connected, are not perfectly identical. . . . The state of functioning (*energeia*) "tends toward" the state of completion (*en-telecheia*).   *(Peters, 1967, p. 57)*

In other words, entelechy refers not just to an entity's completed state, but more crucially to its tendency to reach that state (whether that final state came to be as a result of chance, necessity, or a combination of the two is irrelevant to the positing of a prior inherent possibility of the initial state to reach a final state). According to standard Aristotelian metaphysics, the assertion of the presence of this so-called entelechy does not require the positing of some extra element (such as "abiotic infusion" or Henri Bergson's *élan vital*) in the structure. On the contrary, assertions of entelechy are but the recognition, in biological terms, of a prior *logical* tautology: *posse sequitur esse*. This phrase from medieval logic means: if something exists, then it was always possible for it to exist; and since a complex biological organism is now before our investigating eyes, it must be based on a prior structure that already had the capability (called "entelechy") of reaching its complex final state.

---

[12] "Did Darwin deal a 'death blow to Teleology', as Marx exclaimed, or did he show how the 'rational meaning' of the natural sciences was to be empirically explained . . . thereby making a safe home in science for functional or teleological discussion?" (Dennett, 1995, p. 126).

Chemists might well speak of "elective affinity," but what they mean by that rather anthropomorphic term is what Aristotle more soberly meant by "entelechy": that chemical structures, under the right conditions, have an inherent, innate tendency to actualize themselves into structures of greater complexity. For after all, if they did not have such an innate capacity in the first place, they never would have reached their final, complex stage in the last place. This is what I shall call, without further ado, the "metaphysical" structure of natural complexity. But I hasten to add that the term "*meta*physical" does not refer to some spooky "*extra*-physical" entity (like a designing mind, abiotic infusion, vitalistic vapor, *élan vital*, whatever). Rather, it refers simply to those logical prerequisites we must employ in order to understand change *and* identity through time, growth, and development. And note as well that these logical presuppositions are operative no matter what position one takes about the *likelihood* of the emergence of complex formations from simpler elements. Whether the complex organism is quasi-inevitable, as when acorns result in oak trees, or is highly unlikely, as when (by some accounts) carbon molecules first begin to replicate,[13] still the fact of later emergence always means there was a prior entelechy to make that later emergence possible.

These logical substructures might seem tritely obvious, and in fact they are. For the only thing that these logical insights do is to highlight the necessary conceptual presuppositions we must use when discussing change, growth, increasing complexity, identity through time in the midst of development, and so forth. But unless we advert to these conceptual substructures inherent in any process of change, we are in danger of letting slip into our arguments false assumptions and faulty logic, such as the elision made by William Paley in the late eighteenth and early nineteenth centuries and by today's ID advocates when they conflate "staggering" complexity with "irreducible" complexity. But an equivalent metaphysical confusion must also reign among contemporary Darwinians, otherwise the debate over the radical contingency of the human species (Gould) against the virtual inevitability of intelligent life emerging (Dennett) could never arise.

Let us look at the actual conduct of the debate between these two men, for it is above all in the specifics of their argument that their metaphysical presuppositions will become clear. As we saw, Gould held that if we were to "rewind the tape of evolution" and start the process all over again, the chances of something like *Homo*

---

[13] It is the entire burden of Robert Shapiro's *Origins: A Skeptic's Guide to the Creation of Life on Earth* (Shapiro, 1987) to defend the notion that the transition from primitive chemical bonds to self-replicating molecules is unlikely in the extreme. As an agnostic biochemist, Shapiro is not trying to introduce a design-argument by the back door; but he does insist that talk of Darwinian inevitability has no justification based on what we know of organic chemistry. Nor is he much impressed with most scenarios on offer and he even links them, in their formal logic and mode of argumentation, with the creationists: "Once the spirit of skepticism has been relaxed in the major paradigm of a scientific field, it is difficult to limit the process. Variants may then appear which proclaim even more fanciful and spectacular solutions. The content of mythology increases. In the case of the origin of life, we have seen that the Creationists mark the logical end point of this process" (Shapiro, 1967, p. 266).

sapiens emerging again are next to infinitesimal. But what exactly does this mean? Let me take the example of bipedalism to show how the argument works in practice. In his book *The Origin of Humankind*, Richard Leakey calls the advent of bipedalism not just a major biological, but also a major adaptive, transformation:

> The origin of bipedal locomotion is so significant an adaptation that we are justified in calling all species of bipedal ape "human". This is not to say that the first bipedal ape species possessed a degree of technology, increased intellect, or any of the cultural attributes of humanity. It didn't. My point is that the adoption of bipedalism was so loaded with evolutionary potential – allowing the upper limbs to be free to become manipulative implements one day – that its importance should be recognized in our nomenclature. These humans were not like us, but without the bipedal adaptation they couldn't have become like us.
>
> *(Leakey, 1994, p. 13)*

So far, so good. But what were the conditions that selected for bipedalism? Here is the way Leakey summarizes one scenario:

> About 12 million years ago, a continuation of tectonic forces further changed the environment, with the formation of a long, sinuous valley [in central Africa], running from north to south, known as the Great Rift Valley. The existence of the Great Rift Valley has had two biological effects: it poses a formidable east-west barrier to animal populations, and it further promotes the development of a rich mosaic of ecological conditions. The French anthropologist Yves Coppens believes the east-west barrier was crucial to the separate evolution of human and apes . . . [and] dubs this the "East Side Story."
>
> *(Leakey, 1994, pp. 15–16)*

Let us assume for the sake of argument here that M. Coppens' East Side Story is the correct one. If it is, then it immediately raises the question: what would have happened if these tectonic plates had *not* moved in the way they did? Surely the formation of continents, mountain ranges, coastlines, and so forth follows no hard-and-fast rule of inevitability, any more than the weather does. Thus, if we rewound only that one shift of plates, presumably the conditions that led to bipedalism – and thus later to humans – would not have obtained. That in fact is also Leakey's conclusion: "natural selection operates according to immediate circumstances and not toward a long-term goal. *Homo sapiens* did eventually evolve as a descendant of the first humans, but there was nothing inevitable about it" (Leakey, 1994, p. 20).

Dennett counters this speculation with the retort that if such a view were true, the SETI Project (the once government-funded but now privately funded effort to Search for Extra-Terrestrial Intelligence) would be pointless. Admittedly, that is no slam-dunk refutation, for perhaps such a project *is* pointless. But the assumption (hope? possibility?) that there is life in the universe besides our own could not even be entertained if one were simultaneously to subscribe, with Leakey, to the Gould rewind-the-tape hypothesis. But, more crucially to the point at issue, Gould elides two kinds of contingency: the contingency of each human being emerging into life, and the contingency of life as such *and of its formal properties*. Here is Dennett's argument, where the distinction I am attempting to draw is put most lucidly:

Just what *is* Gould's claim about contingency? He says that "the most common misunderstanding of evolution, at least in lay culture," is the idea that "our eventual appearance" is "somehow intrinsically inevitable and predictable within the confines of the theory." *Our* appearance? What does that mean? There is a sliding scale on which Gould neglects to locate his claim about rewinding the tape. If by "us" he meant something very particular – Steve Gould and Dan Dennett, let's say – then we wouldn't need the hypothesis of mass extinction to persuade us how lucky *we* are to be alive; if our two moms had never met our respective dads, that would suffice to consign us both to Neverland, and of course the same counterfactual holds true of every human being alive today. Had such a sad misfortune befallen us, this would not mean, however, that our respective offices at Harvard and Tufts would be unoccupied. It would be astonishing if the Harvard occupant's name in this counterfactual circumstance was "Gould," and I wouldn't bet that its occupant would be a habitué of bowling alleys and Fenway Park, but I *would* bet that its occupant would know a lot about paleontology, would give lectures and publish articles and spend thousands of hours studying fauna . . . If, at the other extreme, by "us" Gould meant something very general, such as "air-breathing, land-inhabiting vertebrates," he would probably be wrong . . . So we may well suppose he meant something intermediate, such as "intelligent, language-using, technology-inventing, culture-creating beings." This is an interesting hypothesis. If it is true, then contrary to what many thinkers routinely suppose, the search for extra-terrestrial intelligence is as quixotic as the search for extra-terrestrial kangaroos – it happened once, here, but would probably never happen again. But [Gould's book] *Wonderful Life* offers no evidence in its favor; even if the decimations of the Burgess Shale fauna were random, whatever lineages happened to survive would, according to standard neo-Darwinian theory, proceed to grope toward the Good Tricks in Design Space. *(Dennett, 1995, p. 307)*[14]

It cannot be the task of this chapter to adjudicate this Gould–Dennett debate. I am citing their debate only to point out that the issue could never even hope to be adjudicated without appeal to what I have called the "metaphysics of complexity." By that I mean that, unless we have already determined what really constitutes chance, change over time, complexification from simpler elements, inevitability vs. contingency in the emergence of life, and so on, we will never be able to understand, let alone adjudicate, debates among the Darwinians. In fact, most of those debates only arise because prior philosophical confusions have not been acknowledged and resolved.

Now, I would not dream of disputing the claim that the deliverances of the empirical sciences will largely determine the outcome of this debate, but I wish to

[14] By the term "Good Tricks in Design Space," Dennett is referring to the inevitable constraints, both physical and environmental, imposed by natural selection for an organism to thrive to reproductive age. Moreover, for him these Good Tricks (his capitalization) lead to a kind of "arms race" of increasing complexity, making the emergence of intelligence well-nigh inevitable. In other words, Design Space constrains the Vast genetic possibilities by the limits of viability in a particular environment; but more crucially, these constraints are *formal* constraints, which, combined with the "arms race," make intelligence, if not downright inevitable, certainly far likelier than Gould would countenance.

highlight how much the debate is also – and inevitably – governed by metaphysical presuppositions that often go unnoticed. Just as advocates of ID (in my opinion) make a mistake of philosophical categorization when conflating "staggering complexity" with "irreducible complexity," so too do the ultra-Darwinians when they assume that Darwin expelled all forms of teleology from biology when in fact he only refuted William Paley's version of the argument from design.

To elucidate and expand on what I am driving at with this observation, I would like to add one other point: why can't Dennett be aligned with Gould that intelligence is a "built-in" feature of evolutionary development yet also hold that intelligence is unlikely anywhere else in the universe? Certainly there is nothing inherently impossible, in the logical sense, with asserting both these positions. Such a thesis is, in fact, the gravamen of the recent book in evolutionary theory, *Life's Solution*, by Simon Conway Morris (2003).[15] Perhaps the greatest virtue of this extraordinarily virtuous book is the author's alertness to the massive *philosophical* muddle of the (his word) "ultra-Darwinists":

> I am driven to observe of the ultra-Darwinists the following features as symptomatic. First, to my eyes, is their almost unbelievable self-assurance, their breezy self-confidence. Second, and far more serious, are particular examples of a sophistry and sleight of hand in the misuse of metaphor, and more importantly *a distortion of metaphysics* in support of an evolutionary programme. Consider how ultra-Darwinists, having erected a naturalistic system that cannot by itself possess any ultimate purpose, still allow a sense of meaning mysteriously to slip back in . . . Third, as has often been noted, the pronouncement of the ultra-Darwinists can shake with a religious fervour. Richard Dawkins is arguably England's most pious atheist. Their texts ring with high-minded rhetoric and dire warnings – not least of the unmitigated evils of religion – all to reveal the path of simplicity and straight thinking. More than one commentator has noted that ultra-Darwinism has pretensions to a secular religion, but it may be noted that, however heartfelt the practitioners' feelings, it is also without religious or metaphysical foundations. Notwithstanding the quasi-religious enthusiasms of ultra-Darwinists, their own understanding of theology is a combination of ignorance and derision, philosophically limp, drawing on clichés, and happily fuelled by the idiocies of the so-called scientific creationists. It seldom seems to strike the ultra-Darwinists that theology might have its own richness and subtleties, and might – strange thought – actually tell us things about the world that are not only to our real advantage, but will never be revealed by science. In depicting the religious instinct as a mixture of irrational fundamentalism and wish-fulfilment they seem to be simply unaware that theology is not the domain of pop-eyed flat-earthers.   *(Conway Morris, 2003, pp. 314–16 [emphasis added])*

Philosophical howlers and whoppers as egregious as those perpetrated by the ultra-Darwinians were bound to catch up with their advocates. And the reason is because the ultra-Darwinists *are indulging in precisely the kind of metaphysical speculations that they claim they are trying to expel.* The case of Charles Darwin is once again

---

[15] Professor Conway Morris is a co-editor of this current volume.

relevant here, for the crucial point is that his gradual movement toward agnosticism happened not, as is too widely assumed, because of his theory of natural selection, but for more personal reasons: he could not reconcile the death of his daughter Anne with the Christian doctrine of providence. In other words, *theodicy* was the real issue that led him to adopt an agnostic stance. But for the longest time, he saw that natural selection could be reconciled with Christian theism by using the medieval doctrine of secondary causality, a metaphysical doctrine that began with the neo-Platonist figure Proclus (AD 410?–85) and was given classical expression by St. Thomas Aquinas (AD 1225–74). As is well known, Darwin concluded the *Origin* with a paean to secondary causes:

To my mind it accords better with what we know of the laws impressed on matter by the Creator, that the production and extinction of the past and present inhabitants of the world should have been due to secondary causes, like those determining the birth and death of the individual. When I view all beings not as special creations, but as the lineal descendants of some few beings which lived long before the first bed of the Cambrian system was deposited, they seem to me to become ennobled.   *(Darwin, 1928, p. 462)*

Less well recognized, however, is the philosophical pedigree of this notion of secondary causality and the crucial role it has played in natural theology from the Middle Ages to the present. As the medieval historian of philosophy, Armand Maurer, puts it:

What is the nature of Darwin's argument for evolution by secondary causes and what is its value? It does not belong to science but to natural theology, for it concerns God the creator and the laws he has implanted in matter. It should more properly be called metaphysical, for the argument turns on the distinction between primary and secondary causes, which are traditionally the concern of metaphysics.   *(Maurer, 2004, p. 497)*[16]

Although it would be going much too far to say that Darwin's theory of evolution by natural selection helps to establish theism (Darwin's own changing convictions in that regard prove otherwise), it must none the less also be said that in certain ways his theory provides added support for certain schools of natural theology that come out of the Thomist (and Maimonidean) tradition, which holds that God is better praised by attributing natural events to natural causes rather than to God's direct involvement. Previously, both St. Augustine (AD 354–430) and St. Bonaventure (AD 1217?–74) had held that secondary causality, at least when taken too far, would denigrate God's role in the world; so they posited something called "seminal powers" (*rationes seminales*), asserting that nature cannot produce new effects except by awakening the "seeds" God had originally planted in nature. They also subscribed to an epistemological equivalent of this doctrine, by asserting that there

---

[16] This article traces both Darwin's own personal exposure to the natural theology of secondary causes through his own reading and the genealogy of the concept from Proclus to the nineteenth century.

were also seminal powers in the human mind, so that the human mind cannot attain to the truth without a special illumination from God to awaken latent powers, like the sun turning seeds into plants. But Aquinas insisted that created beings do produce new substances by their own powers and do reach (some) truths by the light of reason alone. As the renowned historian of medieval thought, Etienne Gilson, put it so well: "In St. Thomas man receives from God everything he receives [from God] in St. Augustine, but not in the same way. In St. Augustine God delegates his gifts in such a way that the very insufficiency of nature constrains it to return toward him; in St. Thomas God delegates His gifts through the mediacy of a stable nature which contains *in itself* . . . the sufficient reason of *all* its operations" (quoted in Maurer, 2004, pp. 509–10 [emphasis added]).[17]

Thomas's criticism of the Augustinian version of natural theology is no doubt the reason so many Thomists appreciate Darwin's achievement and even see his theory as confirmation of Thomas's theory of secondary causality.[18] This is also why so many evolutionary biologists who are ardent Christians (or at least theists), such as Ronald Fisher and Theodosius Dobzhansky, explicitly rely on the notion of secondary causality.[19] Needless to say, the concept of secondary causality establishes only the possible compatibility of evolution by natural selection with theism, not its actual compatibility. But the crucial point is that a denial of secondary causality must be argued metaphysically, for nothing in evolution itself can imply the implausibility of secondary causes, and still less their impossibility. Darwin himself recognized this as well in his autobiography, where he explicitly linked his constantly shifting opinions on God and Christianity with his gradual despair over the human mind's ability to reach metaphysical truths and *not* with his conviction of the truth of evolution:

[I concede] the extreme difficulty or rather impossibility of conceiving this immense and wonderful universe, including man with his capacity of looking far backwards and far into futurity, as the result of blind chance or necessity. When thus reflecting, I feel compelled to look to a First Cause having an intelligent mind in some degree analogous to that of man; and I deserve to be called a Theist. This conclusion was strong in my mind about the time, as far as I can remember, when I wrote *The Origin of Species*, and it is since that time that it has, very gradually with many fluctuations, become weaker. But then arises the doubt – can the mind of man, which has, as I fully believe, been developed from a mind as low as that possessed by the lowest animals, be trusted when it draws such grand conclusions?

(Darwin, 1892, p. 66)

---

[17] Maurer is quoting Étienne Gilson's notes for a seminar in the Ecole Practique des Hautes Etudes, the University of Paris in 1920, printed as an Appendix in Shook, 1984, p. 297.

[18] Nogar (1962) and Moreno (1990).

[19] "Near the top of anyone's list of the 'ten greatest evolutionists since Darwin' will be the English statistician Ronald Fisher . . . and the Russian-born American Theodosius Dobzhansky. . . . Both were ardent Christians" (Ruse, 2001, pp. 8–9).

On other occasions, Darwin would speak of the human mind's capacity for metaphysics as being impaired by the fact of its origins in the primeval slime of the early earth, so that an evolution-produced mind trying to puzzle out metaphysical truth was like "puzzling at astronomy without mechanics" or like trying to "illuminate the midnight sky with a candle": just as likely are we to expect that we might "throw the light of reason on metaphysics" (Darwin, 1974, p. 71).

I mention these remarks not to subscribe to Darwin's views (to claim that the evolutionary origin of the human mind renders it incapable of attaining metaphysical truth requires argument, not mere assertion), but to point out that his variations of opinion in matters of religion were directly related to his fluctuating stance toward metaphysics, not evolution. Moreover, despite such diffidence, he never did succeed very well in expunging metaphysical interests from his mind, any more than skepticism in the history of philosophy has ever managed to expel metaphysical reflection from the human mind, no matter how often it tries. "Lamarck, Plato, Hume, and God jostled for attention," says Janet Browne of Darwin's reading after he got back to England after his five-year voyage on the *Beagle*. "He had no end of theories about humanity occupying his brain: theories about the origin of language, about morality, religion, and race" (Browne, 1995, pp. 364–439).

In a way, it is too bad that in his desultory reading of metaphysics he never sufficiently attended to the idealist tradition, which could have illumined how the environment testifies to the meaning of the emergence of humans as a species. If he had thought through some of the metaphysical implications of his theory, he might not have been so diffident about the human mind's ability to reach metaphysical conclusions. For evolution does not just say something about life, about humans, about struggle, and about the lowly origins of mind from mud. It also says something about the environment, *including the metaphysical environment!*

To the best of my knowledge, Lawrence Henderson (1878–1942), the American biochemist at Harvard, was the first to shift the terms of the debate among evolutionary biologists from specific biological forms to the testimony natural selection gives regarding the nature of the environment doing the selecting (Henderson, 1913). In what follows, I wish to stress that I am presenting here my own version of Henderson's shift of perspective. Clearly he was right that, in Darwinian terms (not to mention Aristotelian terms), life cannot emerge unless the environment is first fit for life.[20] But I shall take his argument one step further and claim that we also live in a *metaphysical* environment without which rational life would not be possible.

---

[20] Notice how his argument is perfectly willing to go in both directions, from investigating what in the environment makes life possible *and* from investigating the properties of life to determine what the environment must therefore be like (or have been like): "Meanwhile it should be noted that there are two different ways of illustrating the fitness of a physical property. Properly employed, both are free from fallacy, and it will be desirable for us to employ both" (Henderson, 1913, p. 70).

My way of putting this argument is as follows. Because Darwinism gave a new explanation for how and why each organism is so well adapted to its environment, the record of convergence (similar organs, such as eyes and wings, developing independently across many phyla and species) surely must say something as well about the environment. In other words, the fact that so many biological forms developed photosensitive cells and then eyes (or their equivalents) also testifies to the ubiquity of light, just as the fact that wings developed independently on insects, birds, bats, some dinosaurs, and so forth, testifies to the density and viscosity of the atmosphere of the earth. So too with complex brains: to be adaptive, *brains have to evolve in response to the environment*. Or as I have argued elsewhere (Oakes, 2004, pp. 28–30), if wings evolve against and in response to air, and eyes against and in response to light, then brains must evolve against and in response to something like "mental air." By that admittedly metaphorical term, I mean those *a priori* ideal structures already part of the universe that make a mathematics-capable brain possible in the first place. In other words, Darwinism not only is compatible with Platonism, but presupposes it.

This perhaps startling point of view is admirably explained and defended by Daniel Dennett in his lucid philosophical meditation on Darwinian theory, *Darwin's Dangerous Idea*, where he neatly explains the necessity for Platonic thinking by using this intriguing thought experiment:

Suppose SETI [the Search for Extra-Terrestrial Intelligence] struck it rich, and established communication with intelligent beings on another planet. We would not be surprised to find that they understood and used the same arithmetic that we do. Why not? Because arithmetic is *right* . . . The point is clearly not restricted to arithmetic, but to all "necessary truths" – what philosophers since Plato have called *a priori* knowledge . . . It has often been pointed out that Plato's curious theory of reincarnation and reminiscence, which he offers as an explanation of the source of our *a priori* knowledge, bears a striking resemblance to Darwin's theory, and this resemblance is particularly striking from our current vantage point. Darwin himself famously noted the resemblance in a remark in one of his notebooks. Commenting on the claim that Plato thought our "necessary ideas" arise from the pre-existence of the soul, Darwin wrote: "read monkeys for pre-existence."   *(Dennett, 1995, pp. 129–30)*

But this particular thought experiment does not pertain directly to evolutionary theory, even if it does neatly dispatch certain postmodern theories that claim that rules of arithmetic and the law of gravity vary according to the culture that acknowledges them. Another thought experiment, however, illuminates more directly the role certain ideal *a priori* structures play in the evolution of humans:

Any functioning structure carries implicit information about the environment in which its function "works." The wings of a seagull magnificently embody principles of aerodynamic design, and thereby also imply that the creature whose wings they are is excellently adapted for flight in a medium having the specific density and viscosity of the atmosphere within a thousand meters or so of the surface of the Earth . . . Suppose we carefully preserved the body of a seagull and set it off into space (without any accompanying explanation), to be discovered by Martians. If they made the fundamental assumption that the wings were functional, and that their function was flight (which might not be as obvious to them as we, who have seen them do it, think), they could use this assumption to "read off" the implicit information about an environment for which these wings would be well designed. Suppose they then asked themselves how all this aerodynamic theory came to be implicit in the structure, or, in other words: How did all this information get into these wings? The answer *must* be: By an interaction between the environment and the seagull's ancestors.

(Dennett, 1995, pp. 197–8)

All well and good, but what happens when we continue this logic and do some reverse engineering on the human brain? If wings testify to the density and viscosity of the earth's atmosphere, and if the eye (or its equivalent) testifies to the ubiquity of light, to what then does the brain testify? We have already seen that the SETI project presupposes the presence across intergalactic planetary systems of the *a priori* structures of intelligence, mathematics above all: we will be able to recognize the presence of intelligence on other planets only by seeing that other life forms recognize the same ideal truths that we do. And that is the clue we need: just as we will be able to spot intelligent life on other planets only by the common communication of intelligible truths across vast distances, so too we now realize, with Dennett's reverse-engineering thought experiment, that such mutual recognition of *a priori* intelligible truth will have been made possible only because the information of mathematics, logic, and so forth, has "selected for" intelligence in two different locations in the universe. In other words, brains evolve over, against, and in response to "mental air," just as eyes evolve over, against, and in response to light, and wings over, against, and in response to atmosphere, physical air. Or to put the matter perhaps less metaphorically, no intelligence without intelligibility, or more exactly, no evolved intelligence without a prior environmental intelligibility.[21]

I admit that the logic of this argument leads to the dread specter of Idealism (which posits that ideal realities such as number, logic, and forms precede and determine material realities); and Idealism in particular tends to spook out scientists even

---

[21] John Haught, who contributes a chapter to the current volume, makes this argument of the "fitness of the environment for intelligence" the key moment in his challenge to Darwinian atheists, whom he addresses in the second person in a kind of *j'accuse*:

Your mental activity . . . is not something that takes place outside of the natural world. But if your intellection is part of nature, then this tells you something important about the nature of the universe itself. . . . Think, then, about what kind of universe it must be that gives rise in its evolution to an intelligence that cannot help trusting that truth is worth seeking. Does your native trust in this world's intelligibility – and in your own mind's capacity to grasp this intelligibility – "fit" well into the ultimately absurd and mindless universe articulated by your evolutionary materialism? Or instead does it not fit much more satisfactorily a universe that, in its depths, bears an endless intelligibility, a universe that invites you to search deeper and deeper for truth, and that beckons you with a promise that such a pursuit is worthwhile? *(Haught, 2003, p. 100)*

more than metaphysics does in general.[22] In fact, I rather suspect that the reason scientists find metaphysics so disquieting is precisely because it keeps leading to idealist conclusions. But if I am right that science finds itself regularly bumping into metaphysics, and if evolution means that the information of the environment is always being "read into" the organism so that one can "read back" what the environment must have been like, *based solely on the information in the organism,* then I see no alternative but to hold that reality is first ideal before it is material.

As is well known by now, some cosmologists go further and insist, by using an argument known as the Strong Cosmological Anthropic Argument, that material reality such as atoms (or, more specifically, subatomic particles and therefore *a fortiori* atoms) cannot exist unless there are knowers to know them. (This paper, which concentrates on biological complexity, has not dealt with that issue and will not so do in its conclusion.[23]) However, it does seem appropriate to conclude these reflections with a final consideration of what the universe would be like without knowers. At first glance, it would seem that evolutionary theory presupposes the possibility – indeed actuality – of a world without knowers, since the evolution of consciousness came last of all. A more logical objection would ask whether it is even meaningful to speak of a knower-less universe, since the question of such a possibility can arise only in minds already "thrown into" the universe. Yet, leaving this logical conundrum aside, the enigma of consciousness is still a fundamental datum of the world, and the human mind cannot help but ask what that means, what it says about the nature of the universe.[24]

---

[22] Philosophers, too, are not immune to Idealism-phobia. For example, Thomas Nagel confesses his atheism this way: "The thought that the relation between mind and the world is something fundamental makes many people in this day and age nervous . . . I want atheism to be true and am made uneasy by the fact that some of the most intelligent and well-informed people I know are religious believers" (Nagel, 1997, p. 130). Still, one salutes Nagel's honesty in refusing to let his nerves get the better of his philosophical acumen, which acumen leads him to see the Idealist implications of Darwinian theory: "Whatever justification reason provides must come from the reasons it discovers, themselves. They cannot get their authority from natural selection. . . . This means that the evolutionary hypothesis is acceptable only if reason does not need its support" (id., pp. 136, 139).

[23] None the less, since this chapter began by showing how physics has lately also been serving as the midwife of metaphysics, a nod in the direction of quantum physics would not be out of place:

A leading quantum theorist, Eugene Wigner . . . suggests that it is the entry of the information about the quantum system into the mind of the observer that collapses the quantum wave and abruptly converts a schizophrenic, hybrid, ghost state into a sharp and definite state of concrete reality. Thus, when the experimenter himself looks at the apparatus pointer, he causes it to decide upon either one position or the other, and thereby, down the chain, also forces the electron to make up its mind. If Wigner's thesis is accepted it returns us to the old idea of dualism – that mind exists as a separate entity on the same level as matter and can act on matter causing it to move in apparent violation of the laws of physics. . . . Whatever the validity of Wigner's ideas, they do suggest that the solution of the mind–body problem may be closely connected with the solution of the quantum measurement problem, whatever that will eventually be.   *(Davies, 1983, p. 115)*

[24] Notice again the transition from physics to metaphysics when the question becomes one of consciousness:

The present author, having no competence whatsoever to assess the soundness of the theories (however metaphysical their content) that scientists develop on the basis of their specialized knowledge, cannot say whether or not some differential equations and numerical relationships are embedded in the universe rather than imposed on it, and supposes that *the question is, strictly speaking, outside the realm of physics as it defines itself.* Let

"There is the best of precedents for concluding a long and abstract discussion of difficult philosophical matters with a myth," says Arthur Lovejoy at the conclusion of his *Revolt against Dualism*, his history of the philosophical attempts to rid the world of Cartesian dualism by naturalizing the mind in the manner that Darwin attempted at the end of his life. After demonstrating in great detail and with marvelous verve that all attempts to regard the mind as entirely a component of natural processes have failed, he ends his book by trying to imagine how Plato might have recast his myth of creation in the *Timaeus* if he had been writing today, that is, "if he had been enough of an evolutionist to conceive of the production of living creatures as proceeding from lower to higher, and if he had been more definitely mindful of the problem of the character and genesis of natural knowledge" (Lovejoy, 1955, p. 399). In other words, what would the Demiurge have fashioned if he had been a Darwinist working via the process of natural selection?

In this delightful revision of Platonic myth, Lovejoy shows that the Demiurge, if he wanted to let consciousness emerge out of Darwinian processes, would still have to have created space and time, for they remain essential constituents of evolutionary theory no less than of Platonism. He then imagines what the Demiurge would have said after he had fashioned time as the mobile image of eternity and space as the home for all things in the world: with space and time alone, there would still be something missing in the universe. "The creatures," so Lovejoy's Demiurge would say, "confined within the narrow bounds of their separate being, are wholly strangers to one another, . . . and the things they suffer and the deeds they do endure but for an instant and are then lost in the non-being of the Past." A sad outcome for so noble an effort, no? But because time decrees the evanescence of all things temporal, which not even the Demiurge can countermand, and because space decrees that all extended things be external to each other, nothing can be present in its own time and place and yet located elsewhere at the same time. What is allotted to one can never belong to another. At which point the Demiurge hits upon a solution:

Let me, then, consider how I may remove this imperfection, as far as the nature of things allows . . . What is needful is that the nature which belongs to one region of being shall be not only reproduced in another, not merely, when so reproduced, be beheld by a creature

us be satisfied with saying that knowing, the very act of conceptually grasping a truth, and the very fact that our mind communes with a world that it is not, and can assimilate that world, or make it into a self-conscious event – that this most common fact is, when we think of it, the strangest thing one could imagine. If God is incomprehensible, the fact of our perceiving and knowing is no less so – at least on the common (Cartesian) assumption that I am an observer of the universe, and that this universe is radically, irreducibly alien. The possibility of "mine-ness" may be a miracle in itself, but the fact that I can make a foreign body mine, take something not previously present in me and convert it into an act of awareness, is surely a miracle of miracles. When we think of it we feel tempted to embrace the Platonic-Augustinian theory of anamnesis and say that what we know is what has always been in us. And this is indeed one way of approaching the Whole-in-each-part theory: it is only because the Whole is in us that we can know anything at all.

(Kolakowski, 2001, pp. 77–8 [emphasis added])

having its existence in that other region, but that it shall be reproduced and beheld *as if present in the region in which it first existed*, as belonging to a thing which had or has its being *there*; only so can any creature see another *as being another*, and the mutual blindness of the parts be in a certain measure overcome . . . So considering, the Demiurgus proceeded to add the gift of knowledge to the many other less excellent gifts which he had already distributed amongst the various grades of living things; and he created the animal man to receive and have the custody of this gift.   *(Lovejoy, 1955, pp. 400–1 [emphasis added])*

But what is our gift is also our dilemma, as Blaise Pascal knew so well and explained so unsparingly: "*Thinking reed*. It is not in space and time that I must look for my dignity, but in the organization of my thoughts. I shall have no advantage in owning estates. Through space the universe grasps and engulfs me like a pinpoint; but through thought I can grasp it" (Pascal, 1995, no. 145, p. 36). So E. M. Cioran is right, and one of the epigraphs to this chapter also brings us round to the same conclusion as his aphorism: "A *conscious* fruit fly would have to confront exactly the same difficulties, the same kind of insoluble problems, as man . . . To defy heredity is to defy billions of years, to defy the first cell" (Cioran, 1973, pp. 5, 31). And maybe that fruit fly would come to see what Friedrich Nietzsche meant when he said in *The Gay Science*: "Even we knowers today, we godless anti-metaphysicians, still take our fire, too, from the flame lit by a faith thousands of years old – that Christian faith which was also the faith of Plato, that God is the truth, and truth is divine" (Nietzsche, 1974, no. 344).

References

Aristotle (1984). *Metaphysics*. In *The Collected Works of Aristotle*, vol. 2, transl. W. D. Ross, ed. J. Barnes. Princeton, NJ: Princeton University Press.

Barrow, J. D. (1990). *The World within the World*. Oxford: Oxford University Press.

Browne, J. (1995). *Charles Darwin: A Biography,* vol. I *Voyaging*. New York, NY: Alfred A. Knopf.

Cioran, E. M. (1973). *The Trouble with Being Born*, transl. R. Howard. New York, NY: Arcade Publications.

Conway Morris, S. (2003.) *Life's Solution: Inevitable Humans in a Lonely Universe*. Cambridge, UK: Cambridge University Press.

Darwin, C. (1868). *The Variation of Animals and Plants under Domestication*. London: John Murray.

Darwin, C. (1892). *The Autobiography of Charles Darwin and Selected Letters*, ed. F. Darwin. New York, NY: D. Appleton and Company.

Darwin, C. (1928). *On the Origin of Species*, 6th edn. New York, NY: Dutton.

Darwin, C. (1974). *Metaphysics, Materialism, & the Evolution of Mind: Early Writings of Charles Darwin*, transcribed and annotated by P. H. Barrett, with a commentary by H. E. Gruber. Chicago, IL: University of Chicago Press.

Davies, P. (1983). *God and the New Physics*. Harmondsworth: Penguin Books.

Dennett, D. C. (1995). *Darwin's Dangerous Idea: Evolution and the Meaning of Life*. New York, NY: Simon and Schuster.

Haught, J. F. (2003). *Deeper than Darwin: The Prospect for Religion in the Age of Evolution*. Boulder, CO: Westview Press.

Henderson, L. J. (1913). *The Fitness of the Environment: An Inquiry into the Biological Significance of the Properties of Matter*. New York: Macmillan. Repr. (1958) Boston, MA: Beacon Press; (1970) Gloucester, MA: Peter Smith.

Huxley, T. (1893). *Darwiniana*. London: Macmillan.

Kolakowski, L. (2001). *Metaphysical Horror*. Chicago, IL: Chicago University Press.

Leakey, R. (1994). *The Origin of Humankind*. London: Weidenfeld and Nicolson.

Lovejoy, A. O. (1955). *The Revolt against Dualism: An Inquiry Concerning the Existence of Ideas*. La Salle, IL: Open Court Publishing Co.

Maimonides, M. (1928). *Guide for the Perplexed*, 2nd edn, transl. M. Friedländer. London: Routledge; New York, NY: Dutton.

Maurer, A. (2004). Darwin, Thomists, and Secondary Causality. *The Review of Metaphysics*, **57**, 491–514.

Moreno, A. (1990). Finality and intelligibility in biological evolution. *The Thomist*, **54**, 1–31.

Nagel, T. (1997). *The Last Word*. New York, NY, and Oxford: Oxford University Press.

Nietzsche, F. (1974). *The Gay Science*, transl. Walter Kaufmann. New York, NY: Vintage.

Nogar, R. J. (1962). *The Wisdom of Evolution*. New York, NY: Doubleday.

Oakes, E. T. (2004). The evolution of evolution, a review of Simon Conway Morris, *Life's Solution: Inevitable Humans in a Lonely Universe*. *Commonweal*, **131/2** (30 January), 28–30.

Pascal, B. (1995). *Pensées*, transl. H. Levi. Oxford: Oxford University Press.

Peters, F. E. (1967). *Greek Philosophical Terms: A Historical Lexicon*. New York, NY: New York University Press.

Ruse, M. (2001). *Can a Darwinian Be a Christian? The Relationship between Science and Religion*. Cambridge, UK: Cambridge University Press.

Ruse, M. (2003). *Darwin and Design: Does Evolution Have a Purpose?* Cambridge, MA: Harvard University Press.

Shapiro, R. (1987). *Origins: A Skeptic's Guide to the Creation of Life on Earth*. New York, NY: Bantam.

Shook, L. K. (1984). *Etienne Gilson*. Toronto: Pontifical Institute of Mediaeval Studies.

Whewell, W. (1840). *The Philosophy of the Inductive Sciences*, 2 vols. London: Parker.

Wittgenstein, L. (1980) *Culture and Value*, transl. P. Winch. Chicago, IL: University of Chicago Press.

5

Tuning fine-tuning

Ernan McMullin

The chapters in this volume, written from a wide variety of perspectives, explore the possibility of extending the theme of "fine-tuning" beyond the domain of cosmology, where it first entered into serious discussion in the mid-1970s, to other sciences such as biochemistry and biology. As a prelude to this investigation, it seems worthwhile to explore the theme of fine-tuning itself in some detail, given the ambiguities that still surround it and the vigor of the continuing disagreement as to what its implications are. How did fine-tuning make its way into the cosmological discussion? What precisely did – and does – it amount to? What were – and still are – the responses to it? How is one to evaluate those responses? Achieving a measure of clarity on these issues should make it easier to appreciate the search for fine-tuning or its analogs elsewhere in the sciences.

The infinities of space and time in Newtonian mechanics were not propitious to the formulation of a cosmology, a theory of the cosmic whole, although the notion of gravity gave a hint, at least, as to how material complexity could form. The unification of space and time by Einstein's general theory in a non-Euclidean geometrical framework offered new possibilities, and Hubble's subsequent confirmation of galactic expansion pointed Lemaître to a universe model that would, from a "primeval atom," expand into the universe we know. Now, for the first time in the modern era, the universe could conceivably be presented as having a definite age, a definite time-related size, and a tentative list of contents. In short, the construction of a theory of the whole – a cosmology – became possible. After a period of indecision in regard to the new and controversial expansion model of the universe, while the rival steady-state model still challenged the notion of an initial cosmic state, the discovery in 1964 of the isotropic microwave radiation that had earlier been thought to be a likely consequence of the expanding-universe theory helped to gain widespread acceptance for the new cosmological model.

Fitness of the Cosmos for Life: Biochemistry and Fine-Tuning, ed. J. D. Barrow *et al.*
Published by Cambridge University Press. © Cambridge University Press 2007.

Fine-tuning, anthropic explanation, and design

But to begin our story, it may be best to go back to 1937 when, in response to Arthur Eddington's controversial *a priori* derivation of four "cosmical constants" from an epistemological analysis of the act of experimental observation alone, Paul Dirac argued that even if this approach were to work for smaller numbers roughly of the order of unity, it could not possibly suffice for the very large dimensionless numbers that characterize cosmology (Dirac, 1937; see also Barrow, 2002, pp. 106–12). He had been struck by a coincidence between three such numbers, each an integral power of a very large number, N, of the order of 10^{40}: (1) the ratio of the gravitational to the electric force is $\sim N^{-1}$; (2) the ratio of the mass of the universe to the mass of the proton is $\sim N^2$; and (3) the Hubble age of the universe in atomic units is $\sim N$.

Dirac was convinced that the occurrence of N in these apparently unrelated cosmic features could not possibly be a coincidence. However, one of the three, the Hubble age, is a variable. Assuming the coincidence to be significant, the other two would therefore also have to be variables in order to maintain equality. The consequences would be that the numbers of protons and neutrons would have to increase over time, and the gravitational "constant" would have to decrease, both troubling implications for theory. Dirac focused on the latter of the two consequences, proposing that the gravitational factor should weaken over time. (Although he did not note this, such a revision would have the benefit of explaining the enormous value of N, the feature that had puzzled Dirac most. It would simply reflect the great age of the universe at this point.[1])

Robert Dicke (1961) later hit on an interesting rejoinder. As noted, Dirac had taken it to be more than mere coincidence that the Hubble age should happen to have the value $\sim N$, thus linking it to the other two cosmic features. Dicke showed that this was indeed not a coincidence, but his explanation was quite different from Dirac's. Observers can exist in the universe only after heavy elements have had time to form, and within the maximum lifetime of a massive star.[2] Calculations show that, given the components from which the other two constants are constructed, the Hubble age during the "epoch of man," as Dicke called it, would *necessarily* be of the order of N. The relationship between H and the other two does not therefore indicate an invariant relationship, as Dirac thought, but is a selection effect, a constraint set on the value of T by the conditions of human observership.[3] There is no reason, then, to make the gravitational constant or the mass of the universe vary.

[1] For technical detail on the matters discussed in this section and the next, by far the best source is Barrow and Tipler, 1986.

[2] More correctly, it would be before all massive stars would have burned out, a much longer time, but one still in the very rough "of the order of" that he is relying on.

[3] The familiar example of a selection effect is that if in a large catch of fish none is smaller than a certain size, this might seem surprising until it is noted that it merely reflects the dimension of the mesh of the net in which the fish were caught. For an exhaustive recent discussion of selection effects, including the effect claimed by many-universe proponents, see Bostrom, 2002.

This episode illustrates particularly well the relations between four notions that are of central concern to our story. Dicke's is an *anthropic* explanation because it calls on human presence as an explanatory factor. (Eddington's is even more directly anthropic; indeed, it may be the most audaciously anthropic explanation in the modern history of cosmology!) Dicke's explanation also points to a *selection effect* that involves tuning only in the rather strained sense that the value of *H* has to lie within certain limits for the "coincidence" to appear. But the tuning is in no sense "*fine*": the "of the order of" in Dicke's calculations could have *H* lying at any moment over billions of years! Nor does he suggest that *design* might be involved; what attracted Dirac's attention was simply a numerical coincidence, no more. So his speculation and Dicke's response attracted little attention outside of cosmology. But to some cosmologists, Dirac and Dicke suggested an ingenious way to treat apparent cosmic coincidence.

The impetus for this suggestion came from two different sources. After the discovery in 1967 that the cosmic microwave background was isotropic to at least the 0.1 percent level, many were asking: What sorts of initial conditions could have made this possible after billions of years had passed? Brandon Carter and others posed a different query: What if the fundamental dimensionless constants of nature, like the ratios between the strengths of the four fundamental forces or the value of the fine-structure constant, were to be slightly different? What difference would it make to the prospects of life? It was the direction ultimately taken in response to the former of these questions that drew the attention of the wider public to the phenomenon that would later be dubbed "fine-tuning."

However, the immediate preference for a response to this question was for a chaotic cosmic starting point. Here one might detect the influence of what may be called a "Principle of Indifference" that tends to prevail among cosmologists: the directive to avoid, if at all possible, setting a specific constraint on the initial cosmic boundary conditions. (The Principle of Indifference and fine-tuning are, of course, antithetical to each other.) Setting a constraint, whether fine or not, forces one to ask the awkward question: But why these boundary conditions rather than others?[4] In any event, a "chaotic cosmology," as its protagonist, Charles Misner, dubbed it, proved to be unable to yield anything close to the required present level of isotropy.

Barry Collins and Stephen Hawking (1973) came to the conclusion that a quite extraordinary degree of constraint would have to be imposed on the initial cosmic energy density for the present cosmic state to develop. For that state to be as close to "flatness" as the present state is, the initial state would have to be flat to an almost unimaginable precision. That is, it would have to have an energy density almost

---

[4] Or, as Barrow and Tipler put it: "The appeal of this type of explanation [chaotic cosmology] is obvious: it makes knowledge of the (unknowable!) initial conditions at the 'origin' of the Universe largely superfluous to our present understanding of its large-scale character" (1986, p. 422; see also McMullin, 1993).

exactly poised between values that would lead either to an indefinitely expanding ("open") universe or to a collapsing ("closed") one, the so-called critical density.[5] Hence the term "fine-tuning," which soon came into common use to describe this sort of restriction.

This choice of label has been vigorously contested by some: fine-tuning carries with it the suggestion of a Tuner, they charge (see, for example, Grunbaum, 1998). Since the label has become more or less accepted in this context, however, it seems best to retain it at this stage while underlining the neutral descriptive sense in which it is being used. The term "fine-tuning" will be taken to refer simply to any surprisingly tight restriction that has to be set on a cosmic feature, in this case the initial cosmic state, if the sort of universe we now have is later to develop. Restriction of a parameter alone is not enough to constitute tuning; the restriction must be required if some other feature of interest is to come about.

Collins and Hawking could not believe that so specific a constraint could simply be a coincidence. But how is this fine-tuning to be explained, to be made less surprising? Recalling the anthropic strategy then in the air, they proposed an infinite ensemble of actually existing universes, of which ours is one, an ensemble that exhibits the widest range of values of the critical parameter. In such an ensemble, it would not be surprising that one exists as close as necessary to the critical value of the energy density, the value that would allow that universe to remain flat at a much later time. And this is the kind of one in which we must necessarily find ourselves: the formation of inhomogeneities such as galaxies, over the course of time, appears to require cosmic flatness, which is thus a necessary condition for life (and thus observers like us) to develop. Their many-universe hypothesis thus renders the fact that *our* universe is fine-tuned in regard to initial energy density no longer surprising. It cannot *but* be fine-tuned, i.e. display an at first sight surprisingly severe restriction on its initial conditions![6]

An anthropic argument of this sort involves in effect four steps. First is to postulate a large (infinite?) ensemble of existent universes exhibiting a range of the desired characteristic. Second is to establish, somehow, a probability measure over the ensemble as a means of ensuring that at least one member will have the needed

---

[5] The ratio between the actual energy density at any given moment in the cosmic expansion and the critical density at that moment is designated by the Greek letter Ω. This is a pure number and must be close to unity for the universe to be "flat"; i.e. the actual density at any moment must be close to the critical density at that moment. The energy density itself is, of course, constantly decreasing as the universe expands.

[6] Collins and Hawking suggested that their hypothesis carried with it a further bonus. The title of their article asks: "Why is the universe isotropic?" With the help of a number of simplifying assumptions, they argue that cosmic flatness ensures present isotropy. Since their many-universe hypothesis renders it no longer surprising that the initial state of the universe in which we find ourselves should be fine-tuned, it would also explain why it should be isotropic. ("Why is the universe isotropic? Because we are here.") Thus, in effect, the known isotropy of the microwave radiation would serve as further evidence for the many-universe hypothesis. Unfortunately, the assumptions that underlie the claim that flatness is a sufficient condition for isotropy do not hold up. (See Barrow and Tipler, 1986, pp. 422–30.) This, of course, still leaves intact their main argument (that the many-universe hypothesis accounts for the fine-tuning), which is somewhat confusingly mingled with the isotropy issue.

value of that characteristic. (This is, perhaps, the weakest link in the argument.) Third is to show that this cosmic characteristic (e.g. flatness) is necessary if a universe is to be life-bearing. Fourth is to point out that our universe, the universe that occasioned the fine-tuning puzzle, will necessarily be the one (or one of the ones) in the ensemble possessing the required characteristic.

This has the effect of moving the Principle of Indifference up one level to an ensemble of universes, within which none is to be regarded as special in itself. That one of those *appears* to be special would then simply be a selection effect imposed by the conditions under which the question is asked rather than by the nature of what is studied. Advertence to the selection effect transforms the situation: the fine-tuning required to arrive at the present state of our universe, which initially seemed surprising, now turns out to be intelligible. And in that sense it is "explained," or (if one prefers) is explained away, the ontological price being, of course, very high.

For some time before this Brandon Carter had been discussing the potential role of selection effects in cosmology. In 1974, he published a paper (Carter, 1974) in which he proposed the idea of an "anthropic principle," and the term soon took on a life of its own. The principle could take either a "weak" or a "strong" form. The weak form, the one relevant to the Collins–Hawking strategy, is in effect a truism, but one that can serve as a useful reminder: what is observed is restricted by whatever conditions are necessary to allow the existence of observers.[7] As the term "anthropic principle" is now used by cosmologists, it normally refers to the device of appealing to a large ensemble of existent universes as a means of transforming a troubling instance of fine-tuning into a non-troubling selection effect. More useful for the purposes of this discussion is the notion of anthropic *explanation*, an explanation involving human agency or human presence. It is already applicable in such natural sciences as physical anthropology and meteorology (in explaining climate warming, for example). The appeal to a many-universe solution is evidently a form of anthropic explanation.

But Carter's proposal also had a different, more controversial field of application: what if the dimensionless constants that define our world were to be different? Here, it turned out, a surprise was once again in store. In many instances, even a small percentage change in the value of a single physical constant, holding the other constants unchanged, would yield a universe that is not hospitable to life. It became, indeed, a sort of parlor game among physicists to work out consequences of this sort. Some of their conclusions: If the electromagnetic force were to be even slightly stronger relative to the other fundamental forces, all stars would be red dwarfs, and planets would not form. Or if it were a little weaker, all stars would

---

[7] The "strong" form contains an ambiguous "must" and either reduces to the weak form or is too strong: "The universe *must* be such as to allow the eventual appearance within it of observers" (see McMullin, 1993, pp. 376–7).

be very hot, and thus short-lived. Other thought experiments bore on the chemical constitution of the imagined universe: If the strong nuclear force were to be just a little stronger, all of the hydrogen in the early universe would have been converted into helium. If it were to be slightly weaker in percentage terms, helium would not have formed, leaving an all-hydrogen universe. If the weak nuclear force were to have been just a little weaker, supernovas would not have developed, and thus heavier elements would not have been created. And so on.[8]

In each of these cases, the consequence of the change in the constant's value would be relevant to the possibility of a biotic (life-hospitable) universe. The changes hypothesized above, just a few among the many similar ones proposed, would lead to a universe in which life could not have developed. Of course, some large assumptions are built into this sort of counterfactual reasoning, notably that we know at least some of the necessary conditions for life and, even more fundamentally, that single constants can be meaningfully varied in this abstract way. Both assumptions have given rise to much debate, but we shall simply acknowledge the existence of this debate here and pass on.

Fine-tuning of this sort might be called "nomic," since it is the laws of nature that are at issue. Its status is clearly very different from that of fine-tuning of such boundary conditions as initial energy density. What caused surprise in the latter case was extreme fine-tuning of a parameter that would have been expected to be free-ranging. The fine-tuning is there prior to any consideration of a tie to biotic conditions, whereas in the nomic case the surprise arises only when the biotic relevance of the relative values of the physical constants is noted. The significance of the fine-tuning depends here on biotic considerations. It seems plausible on the face of it that the initial cosmic conditions might have been different; a wide continuum of possible values of the energy density, for example, would not be inconsistent with theory. In the nomic case, however, we have no assurance that the constants of nature in our observable universe can be meaningfully regarded as variables on some larger scale. Treating them in this way is, in effect, to extend the Principle of Indifference to the nomic level.

There is another way to deal with cosmic fine-tuning, one that is highly contentious in the scientific context. It too makes essential reference to the (weak) Anthropic Principle. From the perspective of the Western theological tradition, all three of the major Abrahamic faiths agree in viewing God as the Creator of all that is, although they might spell out the notion of creation in somewhat different ways. They would also agree in holding that the Creator has a special care for humankind,

---

[8] For a catalog of possible instances of nomic fine-tuning, see, for example, Leslie (1996, pp. 33–56). That we should find ourselves in a three-dimensional universe might already be described as "tuning" of the most general sort: "The alternatives are too simple, too unstable or too unpredictable for complex observers to evolve and persist within them. As a result we should not be surprised to find ourselves living in three spacious dimensions subject to the ravages of a single time. There is no alternative" (Barrow, 2002, pp. 225–6).

a care that has been demonstrated over and over according to the Western faith-traditions. With these two beliefs as premises, one can immediately infer that the Creator would fine-tune the original work of creation in whatever way would be necessary to ensure the coming-to-be of the human race.[9] This inference requires no additional premises beyond two that are already central to those different religious traditions. And it immediately explains (i.e. eliminates the element of surprise from) the fine-tuning that set off the original inquiry.

For those who share these beliefs, there just isn't any surprise in the fine-tuning story. They would say: Well, *of course*! This explanation is also an anthropic one, but in a quite different way from the many-universe explanation. It is anthropic because it attributes an anthropic motive to the Creator in the original act of creation. The Creator is assumed to have got it right from the beginning, without need for later miraculous intervention.

This last point distinguishes the anthropic way of dealing with cosmic fine-tuning from that of the recent proponents of "Intelligent Design" (ID) in the context of biological evolution. In the evolutionary context a "special" action on the part of the Creator is claimed to have been required at crucial moments in the development of terrestrial life: at its origin, at least, and perhaps also at various stages in the development of various "irreducible complexities" in the living body.[10] The ID approach competes with possible scientific explanations of these developments, often arguing not only that such explanations are currently not known, but also that they are in principle excluded. The theistic explanation of fine-tuning is of a quite different sort. It involves no challenge to science, no postulation of a gap in possible scientific explanation. The fine-tuning is simply a specification of the original creation and involves no "special" action on God's part. Nor is it *ad hoc*: it follows directly from the original theological premises, once fine-tuning presents itself as a problem in cosmology.

Further fine-tuning developments in cosmology and physics

Three further developments in cosmology and physics bore directly on the fine-tuning theme. In 1981, Alan Guth proposed a modification of the Big Bang theory,

---

[9] For an early assessment of this argument, see McMullin, 1981, pp. 47–52.

[10] Michael Behe argues for the presence in the living cell of various "irreducible complexities" that would require the agency of intelligence to bring them about, but leaves open "the simplest possible design scenario," which would posit "a single cell – formed billions of years ago – that already contained all information to produce descendent organisms" (1996, p. 231). He supposes that "irreducibly complex" systems that manifested themselves long afterward, such as that which produces blood-clotting, might have had their designs already present in potency in that first cell but not yet "turned on" (id., p. 228). The force of this suggestion depends on what "turning on" amounts to. The original ID argument would require it to involve a "special" action on the Designer's part, which would seriously dilute the notion of potency Behe is employing here. If such action is not required, the complexities hardly qualify as "irreducible."

one that would, he hoped, have the effect of eliminating some troublesome anomalies. He postulated an enormous inflation of the infant universe during the first fraction of a second of its existence, one that lasted no longer than perhaps 10^{-35} seconds but that, in a brief moment, multiplied the diameter of the universe by a mind-boggling factor of perhaps 10^{50} (Guth, 1981). Over the next few years, the idea underwent rapid further development, the anomalies yielding in impressive manner one by one.[11] One of these was what Guth called the "flatness problem," the tight constraint on the initial energy density that had given rise to the fine-tuning issue in the first place. It was shown that the inflationary expansion would be so great that it would force the density of the infant universe to the critical value, no matter what the initial density might have been. The need for fine-tuning of the density parameter would in this way be eliminated, or at least reduced, and the Principle of Indifference would once more reign for the preinflation state, which could have any of the possible values of the energy density.

Besides handling anomalies, inflation had two other major features in its favor. It could explain the isotropy of the cosmic microwave radiation, for which Collins and Hawking had attempted to develop an anthropic explanation. More significantly, as later developments would show, it could make use of quantum theory to provide an answer to a long-standing puzzle: where were the inhomogeneities needed for galactic formation to come from if the background radiation was isotropic? Inflation would have the effect of magnifying quantum fluctuations to such an extent that they could furnish the necessary galactic seeds. Of course, that would depend on finding that the radiation was not in fact perfectly isotropic, as it had originally seemed to be.

Sure enough, very tiny departures from isotropy have been found by means of extraordinarily precise satellite observations. Furthermore, the pattern of these departures across the width of the sky is claimed to offer strong support to the inflation hypothesis (Spergel *et al.*, 2003; Schwartz and Terrero-Escalante, 2004). Other support has come, for example, from the distribution of diffuse intergalactic gas in the very early universe. It has to be said, however, that as yet no adequate theory has been formulated that can explain *why* the inflation occurred. The hypothesis does not yet have as strong a mandate, therefore, as does the Big Bang theory to which it is an addendum. But as matters stand, it would be fair to say that the warrant for the original claim of fine-tuning for the initial cosmic conditions has been substantially weakened. The universe may have had a "chaotic" starting point after all – that is,

---

[11] The horizon problem had to do with the horizon distance over which causal signals can propagate. The magnetic monopole problem pointed to the apparent absence in our cosmic neighborhood of the very heavy particles that theory seemed to require. Inflation solved each of them in the same way: by almost instantly creating immense distances between causal regions and between monopoles. The region destined to become our observable universe might at that early time have been no more than 10 cm across (see Barrow and Tipler, 1986, pp. 432–4).

one that needed no particular specification of boundary conditions so that the kind of universe we have might later develop.

The recently discovered acceleration of the cosmic expansion has led to the reintroduction of the "cosmological constant" that Einstein long ago postulated in order to keep the gravitating universe from collapsing. It was abandoned as unnecessary after the recession of the galaxies led to the postulation of cosmic expansion and, ultimately, of the Big Bang. Now it is coming back into favor again but on quite different grounds. However, its quite extraordinarily tiny value in comparison with the other fundamental constants (about 120 orders of magnitude smaller) is a surprise. Why so off-the-scale a value? One possible answer: a tiny value of the constant could be required if the universe were not to undergo immediate runaway expansion and thus be unable to support the development of stars and galaxies, and hence of complex life. Once again, an anthropic explanation suggests itself: why not an immense ensemble of other existing universes with different values of the cosmological constant, including at least one with the requisite value? (See, for example, Linde, 1990a, pp. 158–60; Weinberg, 2000.)

A further development that bears on the fine-tuning theme is the proliferation of many-universe formulations in the wake of the initial successes of the infla- tion hypothesis. Andrei Linde formulated a modified version of the inflation idea, dubbing it "chaotic inflation" (1990a, p. 13). Encouraged by developments in uni- fied theories of particle physics, and especially by the introduction there of scalar (instead of the usual vector) fields, he proposed a "chaotic" distribution of initial scalar fields (inflatons) as the cosmic expansion began. Inflation leads each of these fields to create its own "domain"; the domains are separated from one another by domain "walls." Since the fields vary from one domain to the next, the fundamental forces in each may "break out" differently, and hence the laws of physics in each might also vary. Each domain would become its own "universe," as it were, without causal contact with the other "universes" other than at the moment of origination in the single inflation event.

Pushing this speculation a step further, Linde, Vilenkin, and others postulate a "multiverse" that is infinite both in space and in time, one in which inflation events occur frequently, the one that produced our universe being just another of these (see Linde, 1990b; Vilenkin, 1995).[12] By supposing the multiverse to be eternal, they hoped to bypass the problem of how it is to be generated in the first place. It is, in effect, a given, embodying the Principle of Indifference at the ultimate level. Since the individual "bubble universes" are at no time in causal contact with one another,

[12] Vilenkin's proposal is that "small closed universes spontaneously nucleate out of nothing, where 'nothing' refers to the absence of not only matter, but also of space and time" (1995, p. 846). He goes on to assert: "We are one of an infinite number of civilizations living in thermalized regions of the metauniverse" (1995, p. 847). This illustrates rather well what may happen when one allows an infinite number of entities or of events to be realized!

it is no longer a *"uni*-verse" with the accompanying suggestion of a causal unity of some sort.[13] Hence the term "multiverse," which is rapidly gaining acceptance.

Fine-tuning in chemistry

Until now our discussion has focused on cosmology and physics. Might claims of fine-tuning also originate in chemistry? Restricting the inquiry to the basic materials needed for life to develop, do they yield the sorts of necessary conditions that fine-tuning relies on? The presence of carbon compounds as building blocks and of water as a solvent are the two most often-cited candidates. Both have, in fact, been challenged as supposedly necessary conditions for any form of life (silicon instead of carbon? solvents other than water?). Nevertheless, both remain favored as being essential to the complex chemistry of life.

As far back as 1953, Hoyle pointed to a striking "what if?" involving carbon (Hoyle, 1954). For carbon to form within stars from the fusion of beryllium and helium and for it not to convert too rapidly into oxygen, he predicted that the relevant nuclear resonance level would have to lie within a very narrow range of values (Dunbar *et al.*, 1953a,b; Barrow and Tipler, 1986, pp. 252–4). Later experimental work confirmed this prediction, although there has been some debate recently about how "fine" that tuning actually is. Just as important is that the carbon not convert too rapidly into oxygen; here the resonance level in oxygen that would have led to rapid conversion is barely avoided. Were the values to be slightly different in either of the two cases, there would not be enough carbon to sustain the chemistry of life (Hoyle, 1965).

So, carbon fine-tuning shows itself in chemistry. But what makes it interesting here is the further presumed tie between the presence of carbon and the capacity for organic life. Fine-tuning links this capacity for organic life to tightly specified resonance levels in the nucleus. But these levels depend in turn on the basic physics of the nucleus, and ultimately on the fundamental physical constants, as in the other nomic examples given above. In other words, the fine-tuning inference begins in biochemistry, but ends in physics. It seems, in principle, possible for an apparent instance of fine-tuning in biochemistry to turn out to be *only* apparent once the underlying physics is worked out. A wide range of the relevant parameters at the level of physics might converge to give the appearance of fine-tuning at the level of chemistry.

---

[13] Terminology in this context is still somewhat fluid. The term "multiverse" is also often used where a single originating event brings about an ensemble of domains that do share a causal relation with one another at the moment of origin, although not later. I continue to use the more familiar "many-universe" label for this latter construction, while recognizing that this is a somewhat problematic usage. I restrict the term "multiverse" to the infinite domain with its infinite series of inflation events.

But this is likely to be the exception. At any rate, in the case of carbon, recent calculations seem to indicate that the fine-tuning first apparent in the chemistry of the various nuclei involved carries over to the level of basic physics: were the strong force to be greater by as little as 5 percent, or the electromagnetic force to be greater by perhaps 4 percent, the production of carbon and oxygen in the infant universe could have shut down.[14]

Chemistry plays a double role here, first in revealing the fine-tuning required in order that carbon should form, and second in implying the (plausibly) necessary role that carbon plays in the economy of life generally. These are two separate arguments. But the decisive element in establishing the fine-tuning itself still lies at the level of fundamental physics: must the relevant physical constants be constrained within a tight percentage range so that the chemistry of the universe should be open to the development of life? Do any other plausible instances of fine-tuning exist in the basic chemistry of the raw materials required for living processes? Other chapters in this volume will take up this question.

Fine-tuning in biology?

What are the prospects for discovering instances of fine-tuning in realms other than physics and chemistry – in biology, say? This question will be explored elsewhere in this volume, but it may still be helpful at this stage to draw some inferences first from what we have just learned about the structure of the fine-tuning argument in its native home, cosmology.

Biological considerations play an indispensable role in all discussions of cosmic fine-tuning. But in this role they are not part of the fine-tuning claim itself. Typically, as we have seen, proponents of fine-tuning allege that an unexpected degree of constraint on some physical parameter or basic constant is necessary if some other physical condition is to be realizable in the universe. Then, in a separate argument, this latter condition is held to be necessary for the universe's being open to the development of complex life.

This second element in every fine-tuning discussion draws on biology only in the most general and speculative way. This is not the sort of biology that one finds in a textbook, but the kind science-fiction writers without number have used in imagining living forms that escape almost any restriction that analogies with the terrestrial life we know would be likely to impose. So the search for conditions necessary for *any* form of life throughout the universe has to proceed very warily indeed (Feinberg and Shapiro, 1980). Still, the conditions that have so far played a part in the fine-tuning discussions (the need for elements more complex than

[14] This is quite tentative: the results of the difficult calculations entailed have to be regarded with caution. See the chapters in this volume by Barrow and Livio (Chapters 7 and 8).

hydrogen and helium to form the material basis of living process, the need for several generations of stars to generate the heavier elements, etc.) seem reasonably secure.

But the problem is that we do not have a cosmic biology in the sense in which we *do* have a cosmic physics and chemistry. The life forms of earth, and consequently the biology built on them, have been shaped by the innumerable particularities of the terrestrial environment over billions of years. Countless independent causal lines have crossed and recrossed to form the unique biota of the earth of today and yesterday. The limited teleonomic order imposed by natural selection on individual evolutionary lineages is nowhere near enough to yield anything that could plausibly count as a cosmic biology. The evidence of convergence in the evolutionary record, so persuasively presented in recent work, does carry some weight in suggesting universals of various kind (Conway Morris, 2003). But this still falls far short of providing an abstract framework within which *all* living processes would have to fit.

The prospect of terrestrial biology's suggesting a direct fine-tuning relationship on its own account might therefore be regarded as problematic. It would seem that some feature of the complex living processes described in our biology would have to demand fine-tuning on the part of some underlying physical parameter. The immediate problem here is that living processes, as these are understood in terrestrial biology, do not necessarily have cosmic scope. Other worlds could well be found in our universe where biological processes of a substantially different sort flourish, calling into question the significance of an earth-related "fine-tuning." The second problem, of course, would be to establish a link to a physical parameter requiring fine-tuning for the biological process to go forward.

Still, it is worth carrying the matter a stage further. Take the issue of inevitability in evolutionary biology, for instance. A good case can be made for holding that evolution tends to converge over the course of time on certain favored forms (see Conway Morris, 2003, for an extended defense of the inevitability thesis). Some would go even further, starting with the origin of life in a suitable environment and moving easily to the eventual appearance of a technologically competent civilization capable of communicating with us. Others, however, would emphatically reject any suggestion of overall directionality in the evolutionary process, pointing to the many sources of contingency along the way – notably the occurrence of a major catastrophe on a global scale.[15] In a different, although related, context, the origin of life is taken by some to be of common, even inevitable, occurrence across the cosmos, whereas others would see complex life, at least, as a rare event, dependent on a host of contingent factors.[16]

---

[15] Stephen J. Gould (1989) would be regarded as the main defender of this position.

[16] Here the contrast could be, for example, between the views of Christian de Duve (1995) and those of, say, Francis Crick (1981).

Does this issue bear on considerations of fine-tuning? Strictly speaking, this hardly seems to be the case. Fine-tuning connotes a certain precariousness, a condition barely met. Features such as inevitability or convergence in the story of evolution convey just the opposite. Evolutionary sequences are sometimes said to converge on forms possessing specific functions with no hint of the environmental constraints required for this to happen. Claims of inevitability or convergence could be either (broadly) empirical, based on scrutiny of the historical record of terrestrial evolution, or (broadly) theoretical, based on how Darwinian evolution is supposed to work in general terms. But in neither case does anything like fine-tuning appear to be involved – quite the reverse, it might seem. Nor is the broader claim any better: that the cosmos is pregnant with life and the origin of life faces no particular barrier in the widest variety of cosmic circumstances. One might argue that a universe of this biophilic sort is antecedently unlikely and that some specific sort of fine-tuning at the physical level is responsible for our universe's being biophilic rather than more sparing in offering life support. This would seem to reduce to the fine-tuning argument in physics that we have already seen, although this could depend on what connotation, exactly, one attaches to "biophilic."

So far the issue has been whether fine-tuning, in the relatively precise sense that that term has acquired in cosmology, is likely to find fresh application in biology. The answer has been, on the whole, negative. But suppose one were to address the issue more broadly. Fine-tuning is an instance of design, real or apparent; but design is itself a much broader category. The indications of design in the living world described in loving detail by seventeenth-century exponents of natural theology, such as Robert Boyle and John Ray, did not point to fine-tuning as the clue to the involvement of intelligence. Fine-tuning is a highly specific sort of clue, as we have seen, leading to a specific set of alternative responses. Searching for evidence of *design* in the living world today, on the other hand, might perhaps have a higher likelihood of success, judging by the record, than trying to find analogs of fine-tuning there.

Might one infer, then, to something like design from claims of evolutionary inevitability or convergence? One could, of course, if one were to postulate a directive agency of some sort at work within the evolutionary process, itself impelling the process onward to higher and higher levels of complexity. Bergson's *élan vital* and Teilhard de Chardin's radial energy would be of this sort. But it has to be said that suggestions of this kind have, on the whole, met with a frosty reception from evolutionary biologists. Such views challenge the overall adequacy of the neo-Darwinian explanatory framework, not a popular stand; critics respond that this framework is perfectly capable of handling whatever directions evolution can be shown to have taken.

It was the introduction of this framework that undermined the case for design in the first place – design, that is, that would be associated with specific complex features of the living world. Its modern proponents often present the neo-Darwinian framework as having potentially universal applicability: wherever in the universe a certain set of conditions, centered on descent with modification, is to be found, evolution is bound to occur (see, for example, Dawkins, 1983). The only contingencies that arise in this scenario have to do with the satisfying of the required conditions and the presence in advance of reproducing organisms. (How well a Darwinian type of schema applies to the origin of the first cells is still a matter of debate.)

Holding that the contingencies are easily satisfied so that life must be commonplace throughout the vast spaces and times of the universe (the "pregnancy" thesis) has suggested to some that such a profusion would indicate the agency of design in some form (see, for example, Davies, 1999). Yet one wonders why frequency of occurrence, as against rarity, should of itself be significant in this context. If anything, extreme rarity would appear to be a more likely lead, if it could be established. But this is to barely scratch the surface of the burgeoning life sciences.

The prospects for finding indications in those sciences for fine-tuning proper may not be bright. But perhaps the notion of design might still prove capacious enough to extend into current discussions of such themes as optimality and fitness. It would need to do so at the cosmic level. At the local level of terrestrial biology, the many-world alternative would be all too obvious so that even the appearance of design would hardly be countenanced. It may be worth recalling at this point, for the historical morals that might be drawn, an earlier appeal to cosmic fitness that elicited many of the same questions as those we have just been wrestling with.

Fitness of the environment

L. J. Henderson's celebrated book, *The Fitness of the Environment* (1913), is often cited as a harbinger of the fine-tuning theme in discussions of the place of life in the universe. Having worked through the complexities of that theme, we are now in a better position to judge how close his contributions came to the contemporary notion of fine-tuning, and in what ways and for what reasons they fell short of it.[17] Henderson is not talking about the fitness of life for the environment in which it finds itself. He makes it clear that this can be explained along Darwinian lines: natural selection has shaped life to survive and, when possible, thrive in the particular environment it inhabits and to compensate for changes in that environment.[18]

---

[17] I am indebted in this section to Iris Fry's perceptive analysis of Henderson's book (1996).

[18] One could also have recourse to a broadly anthropic consideration, although Henderson did not take this tack. If (as Henderson believed) a multitude of planetary worlds exist where life could possibly be found, one would expect to find it in an environment for which it is fitted.

But this was not what Henderson meant by "fitness of the environment." It was not where he believed a complement to the Darwinian account needed to be given. What had struck him was the multiple "fitnesses" of the physical environment for the biochemistry of life in general. And by "environment," he meant to include both the "external" and "internal," with emphasis on the latter, following the practice initiated by Claude Bernard, who had had much to say about the *milieu intérieur*, as he termed it. Henderson was intrigued by the crucial roles that water and carbon dioxide (and at a deeper level their three elemental constituents, hydrogen, oxygen, and most especially carbon) played in all living processes. Those processes required complexity, as well as the capacity for self-regulation and metabolism. The properties of the three elements and the two compounds made from them seemed to him so ideally suited to these requirements (a conclusion worked out by him in impressive empirical detail) that some sort of explanation of this was called for.

Henderson's concern was not with particular environments, such as that of earth, but with the cosmic environment (as he saw it) that is revealed by astronomy. He went to some lengths to argue the optimality of the advantages offered by the two key compounds and, thus, of their three component elements as well. He discussed the possibility that water and carbon might be of frequent occurrence on other planets, arguing for the superiority of water over any other known solvent in the biotic context, as well as for carbon over silicon. He did not go as far as to propose them as necessary conditions for life, a more difficult case to make. Nevertheless, some resonances with the account of fine-tuning sketched above were already apparent.

His main theme was the remarkable fitness of the chemical environment for life, and his problem was how to account for this. He ruled out chance: the coincidences between the properties of the favored elements and the needs of life are too remarkable for this to be a plausible response. Likewise, he dismissed design as an alternative explanation. The notion of design was outside the realm of science; besides, to an agnostic like himself, it held no attraction. What is particularly striking is that he also excluded the many-universe alternative, the response favored (as we have seen) by many recent cosmologists. He realized that talking about the superior merits in the biotic context of the three favored elements might suggest comparison with other possible chemistries that had a different repertoire of elements. But this would introduce "other possible worlds in which matter may have different properties and energy different forms" (Henderson, 1916, p. 326; quoted in Fry, 1996, p. 163). And he will have nothing to do with such purely speculative "hypothetical worlds," restricting himself quite explicitly to the chemistry of the cosmos that we know. This was sufficient, in his estimate, to make his point about the remarkable biotic optimality of the three elements and their compounds.

Nowhere in his discussion did Henderson suggest the specifically anthropic factor of a selection effect caused by our necessarily finding ourselves in the only one of a large ensemble of universes in which we *could* find ourselves. It clearly would not have occurred to him, any more than to anyone else of his time, to take seriously the possibility of an ensemble of other actually existing universes governed by different physical laws as a means of solving his problem. He had already made his negative attitude clear in regard to hypothetical alternative chemistries. But more to the point, as far as the comparison with fine-tuning is concerned, his argument was not specifically anthropic in the first place. He talked about the conditions for life in general, about what makes life flourish. *Human* life was not central to the point he was making, as it is in the fine-tuning argument.

But if this was the case, what were his remaining options? How was he to account for the type of "fitness" he found so intriguing? He was left, in his view, with only one possible alternative, although he candidly recognized its shortcomings. Like many others of his day, he was a strong believer in evolution as a cosmic process, of which Darwinian natural selection governing living things would be one major manifestation. So (he suggested) a natural property must have been inherent in the original unorganized matter and energy bringing about an evolutionary process that in turn shaped the differentiated properties of the elements, making three of these optimally ready for the origin and flourishing of life.

Henderson described his view as the "new teleology," since in some sense life was described as a goal of the broader cosmic process. But he also insisted that the process was purely mechanistic; he wanted no part of the popular vitalisms of the day. Could he have it both ways? Could he expand the notion of a mechanism sufficiently to allow it to accomplish purpose-like goals without involving conscious purpose? Aristotle might have had advice to offer! If a mechanistic process accomplished goals or valued ends (as a thermostat does), it would of course prompt a new series of queries as to how this itself came about. Although he was never quite satisfied with his explanation of the fitness of the chemical environment, he remained convinced that an explanation of *some* sort was needed.

Yet Henderson has been hailed by some as an early defender of the type of argument labeled today as fine-tuning. How accurate is this identification? There is clearly some basis for it; but the tuning metaphor does not reflect the thrust of his argument particularly well. First, the positives. He recognized how well on a *cosmic* scale certain properties of the chemical environment fitted the requirements of the life process in general. And he found this degree of fit striking enough to warrant an explanation. Further, the alternative explanations he considered (and rejected) are similar to those found in contemporary accounts of fine-tuning. His own preferred explanation, as we have seen, was interestingly different from these: a process of a supposedly mechanistic sort that shaped the chemical properties of an initially unorganized chaos (the Principle of Indifference?) so as to make it possible for life to come into being and flourish.

Might this process be described, even broadly, as fine-tuning? It does not seem so. Fine-tuning implies a precise standard toward which the tuned activity is to be adjusted. Henderson did not suggest a standard of this sort that would permit life to go forward. Rather, the focus of his argument was on the multiplicity of ways in which the properties of the favored elements and their compounds conspire together, as it were, to provide a suitable environment for life. That environment was sketched in only a general way, with no suggestion of a specific quantitative parameter, the close approach to which in the constitution of nature is the occasion of surprise. Thus, his argument does not have the logical structure we have seen the fine-tuning argument to have.

In the Henderson approach, what is regarded as significant is the manifold fitness for life of the cosmic chemical environment in general. One might perhaps describe this as "tuning," in an extended sense of that term. But it would be more illuminating to situate the "fitness" to which Henderson pointed as the sort of consideration that, more generally, had already prompted arguments of the classical design sort, although of course with a novel twist and a resolute refusal to follow through to the classical design conclusion.

Four alternative responses to claims of cosmic fine-tuning

In conclusion, it may be helpful to evaluate the four alternative responses to claims of cosmic fine-tuning outlined above.[19]

Chance: One might simply dismiss as chance the cosmic constraints required for the universe to be life bearing. The parameter in question had to have *some* value, one might say, so why not this one? It need have no further significance. The universe, it might be said, is a contingent affair, and fine-tuning should be seen in that light. Such a response is impossible to refute, strictly speaking. The challenge to it comes, as we have seen, from two alternatives that present fine-tuning as potentially significant. Scientists on the whole seem uneasy about setting aside the quest for explanation, as the recourse to chance does. One can detect a note of urgency, almost, in many of the discussions of the original fine-tuning claim in the late 1970s and early 1980s: leaving the initial tight cosmic setting of the energy density simply as a given was obviously a troubling prospect. Just as troubling, perhaps, to some was the leap to an infinity of universes advocated by Collins and Hawking. And even more troubling to others was undoubtedly the rapidly growing interest of the broader community in the dramatic reappearance of indications of design in natural science at the cosmic level. In their view, an alternative response *had* to be found.

---

[19] Edward Harrison suggests a fifth alternative: intelligent observers able to fine-tune cosmic constants on their own account (1995). On this, see Barrow (2002, p. 286).

Consequence of an advance in theory: This reaction defined a second way to deal with the fine-tuning challenge: work, as Alan Guth did, toward a theoretical advance that would eliminate, or at least significantly reduce, the need for fine-tuning of the initial energy density. In the interim, simply assume that such an advance will someday be made. At the level of cosmic beginnings, scientific theory is more than ordinarily fragile – and quite evidently incomplete in the absence of a unifying successor to relativity theory and quantum theory. There is reason to think that a re-evaluation of fine-tuning claims in general may well lie somewhere down the road. This is obviously less likely in the case of nomic fine-tuning than in the simpler case of the tuning of an apparently contingent parameter such as energy density.

The two anthropic responses are, of course, the ones that have caught people's attention. Both involve ontological assertions that far transcend the observable universe. Both make essential reference to the human factor. Both draw implicitly or explicitly on the weak Anthropic Principle: human questioners like us are bound, of necessity, to find themselves in a universe that satisfies whatever cosmic conditions are required for life to be able to develop in at least one region within it. As these questioners regard their universe, they will inevitably discover it to be able to satisfy these conditions. This response of itself does not, of course, relieve the puzzlement occasioned by the discovery of fine-tuning. The original question still has to be faced: is the fact that our universe is fine-tuned in regard to a particular parameter, and thus life hospitable, significant or not? Is it no more than chance, a piece of good luck for us, a quirk in our current physical theories? At this point the two anthropic responses diverge.

Many-universe postulate: The first of these (the "scientific" one) in its strongest form postulates a vast ensemble of existent universes featuring a wide distribution of values of the relevant parameter. This response has a quite complex structure, a point that can easily be missed. Some philosopher–critics have charged that although the many-universe hypothesis would give reason to believe that *some* universe would be life hospitable, it would of itself give no reason to suppose that that universe would be *ours*, in all its particularity. Hence, the hypothesis does not explain why *this* universe turns out to be fine-tuned, and the fact that *our* universe is fine-tuned thus fails to support the hypothesis (see, for example, White, 2000).[20]

One need not enter into contested issues about individuation to see what is wrong with this argument. Nor does one need to call on Bayesian reasoning to be convinced that the many-universe hypothesis does not, of itself, explain why

---

[20] At the end of his analysis, White concludes: "Postulate as many other universes as you wish, they do not make it any more likely that *ours* should be life-permitting, or that we should be here. So our good fortune to exist in a life-permitting universe gives us no reason to suppose that there are many universes" [emphasis added]. For a critique, see Manson and Thrush (2003, pp. 67–83).

this universe in its particularity rather than another is fine-tuned for life. Rather, one has to call on a second, additional strand of argument along anthropic lines to show how the many-universe response as a whole works. That we should live in a life-hospitable universe rather than elsewhere obviously requires no explanation: it is a necessary truth. In the many-universe line of argument, what *does* require explanation is that there should be a universe of this *general* kind available. And this the many-universe hypothesis secures. Our existence ensures the existence of some life-hospitable universe and thus supports that hypothesis, which has such a universe as its consequence. This attack on the logic of the many-universe response fails.

However, the multiverse version of the postulate is still not much more than a fascinating, but highly speculative, mathematical exercise at this point. It faces many testing challenges (Ellis *et al.*, 2003, 2006; Manson, 2003). Is the notion of an infinitely realized domain of existent universes even coherent?[21] This issue has been debated since Aristotle's time and it is still not resolved. Doesn't making the multiverse eternal simply serve to evade the normal scientific question of how it was itself generated? How is one to ensure that the "bubble" universes brought about by the unending series of inflation events exhibit lawlike behavior of *any* sort, let alone the probabilistic distribution of such behaviors that the fine-tuning argument requires? If the notion of a multiverse is prompted primarily by the need to explain cosmic fine-tuning, does it not itself pose numerous demands for explanation in its own right?

A many-universe theory involving only a single inflation event would seem more accessible to scientific test at least. But even there, many issues remain unresolved. Do scalar fields exist? How is one to set up and validate a distribution function over the ensemble of universes? A generating process would have to be specified, but how is this process itself to be explained? That is, why *this* process rather than some other one? And might there not be some special conditions required to initiate the sort of inflation under consideration, leaving open the possibility that an element of fine-tuning might find its way back into the process that was supposed to banish it once and for all?

Further, the invocation of a many-universe solution does not of itself necessarily lay fine-tuning to rest: every time fine-tuning makes its appearance, the argument seems to be carried a level higher in order to rescue the Principle of Indifference. But this introduction of a higher-level ensemble, if it can be made specific enough, seems in practice itself to involve a further round of fine-tuning, and so the dialectic continues, at each level becoming more ontologically extravagant.

---

[21] Barrow and Tipler point out that if the notion of an infinitely realized universe were to be allowed, the original grounds for postulating inflation would no longer hold good. Why not argue instead that in a chaotically random infinite universe "there *must* exist a large, virtually homogeneous, and isotropic region, expanding sufficiently close to flatness . . . so that after fifteen billion years it looks like our universe?" (1986, p. 437).

One final objection that is often voiced may be less serious than it looks. Can the invocation of a multiplicity of universes permanently inaccessible to our instruments qualify as physics? The answer is yes, in principle it can if certain conditions are met. Retroduction, effect to cause inference, is central to the natural sciences generally: the physicist who infers to the structure of the atom or molecule on the basis of an emission spectrum or the geologist who infers to the inner structure of the earth on the basis of seismographic data is employing standard retroductive reasoning (see McMullin, 1992). This is the reasoning that underlies explanatory theories that account for inductively arrived at empirical regularities and that give cognitive access to causal structures that are, more often than not, not directly accessible to us. Such theories are warranted by their ability to account for the data that originally called them forth, but even more by their performance over time, their ability to overcome anomaly, their successfully predicting novel results, their continued survival in the face of severe test, their ability to unify previously disparate domains. The list of the desirable epistemic virtues is long (McMullin, 2007).

A many-universe theory is retroductive in form; the existence of the ensemble of universes that constitute its postulated causal structure will be warranted only to the extent that the theory itself behaves well over time. An adequate theory of inflationary origination over an ensemble, with our own universe as the (admittedly sole) observational warrant, would be enough, provided that the theory could be made sufficiently specific (there are doubts about the present candidates in that regard) and that it comes to be supported by something more than accounting for the original puzzling fine-tuning, a slender reed when considered on its own.

A caring Creator: The fourth alternative, the other anthropic response, is of a much more familiar kind; it comes from a quarter remote from scientific cosmology. Its premise, a Creator who has a care for human concerns, is a standard feature of the Western religious tradition. This response is the only one of the four that can be said to "explain" the fine-tuning phenomenon in something like the familiar sense of "explain": ascribing the fine-tuning to the motive of an agent. But, of course, putting it that way tends to mask the unfamiliarity of the agency and the problematic character of ascribing the fine-tuning to a Creator's motives. In its favor is that, unlike the many-universe alternative, it requires no new postulation, and it can appeal to a long philosophical tradition stretching back to Augustine and beyond, with highly developed metaphysics in its support. Against it is an equally familiar array of objections to the whole idea of a Creator and to the demand for an explanation of the universe's existence.

Sorting through the four alternatives

A decision between the four alternatives depends, then, in significant part on prior philosophical and theological commitments. To call it a decision between alternatives ought not to be taken to suggest that each explains the fine-tuning in the same sense of "explains." The first alternative, that fine-tuning is simply chance, maintains that no explanation is needed. The second recommends holding back, in the expectation that, in the long run, the supposed fine-tuning may prove illusory. The third takes fine-tuning seriously and explains it as a selection effect. The fourth also takes it seriously but draws instead on a prior belief in a Creator for whom humanity holds a special place.

Someone for whom neither of the anthropic explanations is attractive might insist on fine-tuning's being no more than a chance phenomenon, or will counsel caution and delay in regarding fine-tuning itself as established. Someone to whom neither of these responses seems at all plausible but who also rejects any sort of appeal that transcends, in principle, the possible reach of natural science will be inclined to opt for the daring ontological enlargement that goes with the many-universe hypothesis. If there is some indication that this can actually be brought within the range of physical theory, so much the better. Someone who is already comfortable with belief in a Creator will find fine-tuning in no way a challenge; quite the reverse.

Interestingly, the theistic response undercuts the need for a many-universe one because, if accepted, it makes the other redundant. In the absence of an already reasonably credible many-universe theory, a theist is likely to be skeptical about the need to add an ensemble of unobservable but real universes just to reduce fine-tuning in our own universe to a selection effect. From the theistic perspective, this is simply not called for. On the other hand, adopting the multiverse alternative would leave the theistic position still in place: we would still need to ask why the multiverse itself exists, even if it were eternal, and would refuse to accept the possibility of a universe's coming to be from nothing pre-existent, all talk of a quantum vacuum notwithstanding. On the other side, many-universe proponents are likely to be skeptical of venturing outside the possible range of cosmological theory, or indeed to be suspicious of any kind of recourse to theological solutions in explaining a feature of the natural world.

What is fascinating about this epistemic situation is that it may set theistic belief and natural science in partial competition with each other, without either venturing outside its proper area of competence. Many writers have held this to be impossible, that the realms of science and religion are in principle "non-overlapping magisteria," to recall Gould's well-known formulation of the "NOMA" principle (Gould, 1999). It is not as though the two are offering competing explanations here for the same feature of nature in the same sense of "explanation," so that if one is right, the other is wrong. Rather, if one is right, the other is unnecessary, although not refuted. The proponent of a multiverse does not (or at least should not) claim to explain how a multiverse could of itself come to be, or why it should exist in the first place.

Believers in a Creator would, of course, see this as the work of creation, although they might also want to emphasize that it was not as a cosmic explanation that they came to belief in God in the first place. So the two sides of this dialectic are not in strict contradiction with each other, although under certain circumstances they are still in competition.

If the phenomenon of fine-tuning can be assumed to be real, does the ability of the theistic position to handle it so neatly count as confirmatory evidence for that position? Here religious believers should tread very warily. The assumption above has been that belief in a Creator is for many a given, in need of no further confirmation. For them, the many-universe extrapolation is simply redundant. But if the two anthropic alternatives were to be regarded as rivals, each in search of confirmation, the epistemic situation would change. The independent plausibility of the many-universe response, however that may be assessed, could limit the confirmatory force of fine-tuning for belief in a Creator. And, of course, one would have to keep in mind the other two possible responses to fine-tuning, both quite difficult to evaluate in practice. The apparent fine-tuning might not be significant to begin with if either of these two were to be correct. Yet it is quite striking that some notable physicists take it seriously enough to warrant their calling into existence an infinity of universes co-existent with ours.

Still, the epistemic situation is so difficult to assess that it is clearly premature to make fine-tuning the key to a new natural theology (Corey, 1993, 2001). It seems best then, for the moment, at least, to fall back on the weaker notion of consonance. Fine-tuning is quite evidently *consonant* with belief in a Creator. To some, this conclusion might appear too weak: if theistic belief explains fine-tuning, they might argue, this should count epistemically in its favor. Strictly speaking this is true (assuming of course that fine-tuning is in fact the case.) However, the epistemic issues surrounding the fine-tuning argument are so intricate and so difficult to assess that making fine-tuning an independent motive for theistic belief may invite more trouble for its proponents than it is worth (McMullin, 1988, p. 71).

The fine-tuning debate has directed the attention of physicists to issues of an unfamiliar sort. This of itself has been a major contribution. How much weight, for example, should cosmologists give to the Principle of Indifference? That is, how serious a problem would it be to leave a constraint on critical cosmic conditions unexplained? Suppose (*per impossibile*, I suspect) that some future theory were to explain why the fundamental constants of nature have the values they do, would this eliminate fine-tuning? On the assumption that the universe had a temporal beginning, would the transition from (literally) nothing to an inconceivably energetic beginning be subject in principle to explanation in terms of physical theory? On the assumption that the universe did not have a temporal beginning, is the demand for explanation of why it should exist in the first place still a legitimate one? If the choice were to be between the two anthropic alternatives, there being no evidence for the many-universe one other than fine-tuning, on what basis would one presume to make the choice?

Questions like these do not fall into any of the categories to which we are accustomed. Yet they are real questions, ones that insistently pose themselves as we explore the boundary-lands of contemporary cosmology. They are quite surely not going to go away, so we had best give serious thought as to how they should be approached.

Acknowledgments

I am indebted to Iris Fry for her helpful comments on a first version of this essay, to Grant Mathews and Bill Stoeger for their help in illuminating some of the dark corners in the physics of the story, and to John Barrow for effective editorial advice.

References

Barrow, J. D. (2002). *Constants of Nature: From Alpha to Omega*. London: Cape.
Barrow, J. D. and Tipler, F. J. (1986). *The Anthropic Cosmological Principle*. Oxford: Oxford University Press.
Behe, M. (1996). *Darwin's Black Box*. New York, NY: Free Press.
Bostrom, N. (2002). *Anthropic Bias: Observation Selection Effects in Science and Philosophy*. New York, NY: Routledge.
Carter, B. (1974). Large number coincidences and the anthropic principle in cosmology. In *The Confrontation of Cosmological Theory with Astronomical Data*, ed. M. S. Longair. Dordrecht: Reidel, pp. 291–8.
Collins, C. B. and Hawking, S. W. (1973). Why is the universe isotropic? *Astrophysical Journal*, **180**, 317–34.
Conway Morris, S. (2003). *Life's Solution: Inevitable Humans in a Lonely Universe*. Cambridge, UK: Cambridge University Press.
Corey, M. A. (1993). *God and the New Cosmology*. Lanham, MD: Rowman and Littlefield.
Corey, M. A. (2001). *The God Hypothesis: Discovering Design in Our "Just Right" Goldilocks Universe*. Lanham, MD: University Press of America.
Crick, F. (1981). *Life Itself*. New York, NY: Simon and Schuster.
Davies, P. (1999). *The Fifth Miracle*. New York, NY: Simon and Schuster.
Dawkins, R. (1983). Universal Darwinism. In *Evolution from Molecules to Man*, ed. D. S. Bendal. Cambridge, UK: Cambridge University Press, pp. 403–25. [I owe this reference to Iris Fry.]
de Duve, C. (1995). *Vital Dust: Life as a Cosmic Imperative*. New York, NY: Basic Books.
Dicke, R. (1961). Dirac's cosmology and Mach's principle, *Nature*, **192**, 440–1.
Dirac, P. A. M. (1937). The cosmological constants. *Nature*, **139**, 323.
Dunbar, D. W. F., Pixley, R. E., Wenzel, W. A. *et al.* (1953a). The 7.68 MeV state in C^{12}. *Physical Review*, **92**, 649–50.

Dunbar, D. N. F., Wenzel, W. A. and Whaling, W. (1953b). A state in C^{12} predicted from astrophysical evidence. *Physical Review*, **92**, 1095.

Ellis, G. F. R., Kirchner, U. and Stoeger, W. R. (2003). Multiverses and physical cosmology. *Monthly Notices of the Royal Astronomical Society*, **347**, 921–36.

Ellis, G. F. R., Kirchner, U. and Stoeger, W. R. (2006). Multiverses and cosmology: philosophical issues. http://arXiv/org/abs/astro~ph/0407329.

Feinberg, G. and Shapiro, R. (1980). *Life beyond Earth*. New York, NY: Morrow.

Fry, I. (1996). On the biological significance of the properties of matter: L. J. Henderson's theory of the fitness of the environment. *Journal of the History of Biology*, **29**, 155–96.

Gould, S. J. (1989). *Wonderful Life*. Cambridge, MA: Harvard University Press.

Gould, S. J. (1999). *Rocks of Ages*. New York, NY: Ballantine.

Grunbaum, A. (1998). Theological misinterpretations of current physical cosmology. *Philo*, **1**, 15–34.

Guth, A. (1981). Inflationary universe: a possible solution to the horizon and flatness problems. *Physical Reviews*, **D23**, 347–56.

Harrison, E. (1995). The natural selection of universes capable of containing intelligent life. *Quarterly Journal of the Royal Astronomical Society*, **36**, 193–203.

Henderson, L. J. (1913). *The Fitness of the Environment: An Inquiry into the Biological Significance of the Properties of Matter*. New York: Macmillan. Repr. (1958), Boston, MA: Beacon Press; (1970), Gloucester, MA: Peter Smith.

Henderson, L. J. (1916). Teleology in cosmic evolution: a reply to Professor Warren. *Journal of Philosophy, Psychology, and Scientific Method*, **13**, 326.

Hoyle, F. (1954). The nuclear reactions occurring in very hot stars. *Astrophysical Journal*, **1** (suppl.), 121–46.

Hoyle, F. (1965). *Galaxies, Nuclei, and Quasars*. London: Heinemann.

Leslie, J. (1996). *Universes*. New York, NY: Routledge.

Linde, A. D. (1990a). *Inflation and Quantum Cosmology*. Boston, MA: Academic Press.

Linde, A. D. (1990b). *Particle Physics and Inflationary Cosmology*. New York, NY: Harwood.

Manson, N., ed. (2003). *God and Design: The Teleological Argument and Modern Science*. New York, NY: Routledge.

Manson, N. and Thrush, M. J. (2003). Fine-tuning, multiple universes, and the "this universe" objection. *Pacific Philosophical Quarterly*, **84**, 67–83.

McMullin, E. (1981). How should cosmology relate to theology? In *The Sciences and Theology in the Twentieth Century*, ed. A. Peacocke. Notre Dame, IN: University of Notre Dame Press, pp. 17–57.

McMullin, E. (1988). Natural science and belief in a Creator. In *Physics, Philosophy, and Theology*, ed. R. J. Russell, W. R. Stoeger and G. V. Coyne. Rome: Vatican Observatory Press, pp. 49–79.

McMullin, E. (1992). *The Inference That Makes Science*, Milwaukee, WI: Marquette University Press.

McMullin, E. (1993). Indifference Principle and Anthropic Principle in the history of cosmology. *Studies in the History and Philosophy of Science*, **24**, 359–9.

McMullin, E. (2007). The virtues of a good theory. In *The Routledge Companion to the Philosophy of Science*, ed. S. Psillos and M. Curd. London: Routledge, in press.

Schwartz, D. J. and Terrero-Escalante, C. A. (2004). Primordial fluctuations and cosmological inflation after WMAP 1.0: http://arXiv.org/abs/hep-ph/0403129 and *Journal of Cosmology and Astroparticle Physics*, 0408, 003–22.

Spergel, D. N., Verde, L., Peiris, H. V., *et al.* (2003). First year WMAP observations: determination of cosmological parameters. *Astrophysical Journal Supplement*, **148**, 175–94.

Vilenkin, A. (1995). Predictions from quantum cosmology. *Physical Review Letters*, **74**, 846–9.

Weinberg, S. (2000). A priori probability distribution of the cosmological constant. *Physical Review*, **D61**, 103905, 1–4.

White, R. (2000). Fine tuning and multiple universes. *Nous*, **34**, 260–76.

Part II

The fitness of the cosmic environment

6

Fitness and the cosmic environment

Paul C. W. Davies

The problem of what exists: why *this* universe?

Einstein reportedly said: "What really interests me is whether God had any choice in the creation of the world".[1] What he meant by this informal remark was whether the physical universe must necessarily exist as it is or whether it could have been otherwise (or could have not existed at all). Today, almost all scientists believe that the universe could indeed have been otherwise; no logical reason exists why it *has* to be as it is. In fact, it is the job of the experimental scientist to determine *which* universe actually exists, from among the many universes that might possibly exist. And it is the job of the theoretician to construct alternative models of physical reality, perhaps to simplify or isolate a particular feature of interest. To be credible, these models must be mathematically and logically self-consistent. In other words, they represent possible worlds.

Let me give one example from my own research (Birrell and Davies, 1978). The equations of quantum field theory describing a system of interacting subatomic particles are often mathematically intractable. But several "toy models" exist, the equations for which may be solved exactly because of special mathematical features. One of these, known as the Thirring model, describes a two-spacetime-dimensional world inhabited by self-interacting fermions. This impoverished model of reality is designed to capture some features of interest to physicists in the real world. It is not, obviously, an attempt to describe the real world in its entirety. Nevertheless, it is a *possible* real world. The universe could have been the Thirring version. But it is not. Given, then, that alternative possible worlds exist, what is it that decides which universe among this (probably infinite) array of possibilities is to be dignified as

---

[1] See, for example, www.humboldt1.com/~gralsto/einstein/quotes.html. No definitive source of this quotation has been determined as of this writing.

Fitness of the Cosmos for Life: Biochemistry and Fine-Tuning, ed. J. D. Barrow *et al.*
Published by Cambridge University Press. © Cambridge University Press 2007.

the *actual* world – the one that "really exists"? Or, to express it more poetically by using Stephen Hawking's (1988) words, "What is it that breathes fire into the equations and makes a universe for them to govern?"

Only two "natural" states of affairs commend themselves in this regard. The first is that nothing exists; the second is that everything exists. The former we may rule out on observational grounds. So might it be the case that everything that can exist, does exist? That is indeed the hypothesis proposed by some cosmologists, most notably Max Tegmark (2003, 2004), although as we shall see some dispute remains over the definition of "everything." Obviously, if *everything* existed, human observations would merely sample an infinitesimal subset of the whole of physical reality.

At first sight, the hypothesis "everything exists" appears extravagant – even absurd. The problem, however, for those who would reject this thesis is that if less than everything exists, then there must be some rule that divides those things that actually exist from those that are merely possible but are in fact non-existent. One is bound to ask: what would this rule be? Where would it come from? And why *that* rule rather than some other?

Another objection to the hypothesis that everything exists is that it seems to be an extraordinarily complicated explanation for what is observed, and hence undesirable on the grounds of Occam's razor. But this can be misleading. In certain circumstances, everything can be simpler than something. Consider an infinite crystal lattice. That regular periodic structure is very simple and can be described by specifying a few bits of data such as the periodicity and orientation of the crystal planes. Now remove a random subset of atoms from this array. By the algorithmic definition of "randomness" (Chaitin, 1988), this subset requires a lot of information to describe it (because it is not "algorithmically compressible"). What remains is, by definition, also random. So each subset, by itself, needs a lot of information to specify it – but the two subsets, when combined, require very little information. In that sense, the whole is simpler than the sum of its parts. In the same way, the set of all possible universes may be (algorithmically) simpler than one or a finite collection of universes. (There is considerable scope for these informal statements to be placed on a sound mathematical footing.)

Historically, most scientists and philosophers have assumed that only one real universe exists, and this should probably remain the default position today in the absence of direct evidence for the existence of any other universes. Monotheistic theologians sought to explain the specific nature of the universe (that is, why *this* universe rather than that) by appealing to divine selection: God made this universe (rather than some other universe that God had the power to make) as a free choice, perhaps with certain outcomes in mind (such as the emergence of sentient beings) (Haught, 1986).

On the whole, scientists have ignored the matter, it being deemed that the proper job of a scientist is to take this particular universe as given and get on with the job of figuring out what is going on in it. Some scientists have taken a more proactive stance by positively denying that there is *any* significance in the specific nature of the universe, taking the attitude that there is *no reason* why it is as it is rather than otherwise. The problem with this position is that science is supposed to explain the world in logical and rational terms. That is, scientists offer reasons for why things are as they are. Normally, this involves chains of reasoning that ultimately lead back to the laws of physics – considered (at least by physicists) to be the bedrock of physical reality. Thus, in answer to the question of, say, *Why did the snow melt?*, a (partial) scientific explanation might go as follows: because it was warmed by the sun, which was heated by thermonuclear reactions, which were triggered by the high temperatures of the solar core, which was produced by the gravitational attraction of the solar material, which moved according to the laws of gravity. If we now ask, *Why that law of gravity?*, the scientist might well respond, "No reason; that's just the way it is!" So a chain of reasoning in which each step is carefully linked logically and rationally to a level below terminates abruptly with the claim that the chain as a whole *exists reasonlessly*. This bizarre backflip cuts the ground from under the entire scientific enterprise, because it roots the rationality of physical existence in the absurdity of reasonless laws (Davies, 1991, 2007).

Problems of a divine selector

If the scientist is in trouble on this score, the theologian is not without problems either. To offer a credible theistic explanation for the specific nature of the world, or for the actual form of the laws of physics, something must be said about the nature of God. (Merely declaring "God did it!" tells us nothing useful at all.) An obviously necessary property of such a divine Creator is freedom of choice. Now, Christian theologians traditionally assert that God is a *necessary* being. If God exists necessarily (that is, if it is logically impossible for God to not exist), then we are invited to believe that this necessary being did not necessarily create the universe as it is (otherwise there is no element of choice, and nature is reduced to a subset of the divine being rather than a creation of this being). But can a necessary being act in a manner that is not necessary? It is far from obvious to this writer that the answer is yes. On the other hand if, counter to classical Christian theology, God is regarded as not necessary but *contingent*, then on what, precisely, are God's existence and nature contingent? If we don't ask, we gain nothing by invoking such a contingent God, whose existence would then have to be accepted as a brute fact. One might as well simply accept a contingent universe and be done with it. If we do ask, then we accept that reality is larger than God, and that an account of the universe must involve explanatory elements beyond God's being. But if we accept the existence of such explanatory elements, why is there any need to invoke divine elements too?

Attempts to reconcile a necessary God with a contingent single universe have a long tradition.[2] I am not a good enough philosopher to know whether these attempts are coherent; but I am bound to ask, even if such reconciliation were possible, *why God freely chose to make this universe rather than some other.* If the choice is purely whimsical, then the universe is absurd and reasonless once more. On the other hand, if the choice proceeds from God's nature (for example, a good god might make a universe inhabited by sentient beings capable of joy), then one must surely ask, *Why was God's nature such as to lead to* this *choice of universe rather than some other?* This further worry would be addressed in turn by proving not only that God exists necessarily, but that God's entire nature is also necessary. Such a conclusion would entail proving that, for example, an evil creator capable of making a world full of suffering is not merely undesirable, but *logically impossible.*

The fitness of the multiverse

Let me then turn to the now fashionable notion of the "multiverse" theory as an explanation for why the universe is as it is. Does this fare any better? The multiverse is defined to be an ensemble of universes within which the members may differ from one another. These "universes" might be completely disconnected spacetimes, or spaces joined occasionally, perhaps only prior to some epoch or following one another sequentially (separated by some clear physical bounding event such as a big crunch), or merely widely separated contiguous spatial regions of a single spacetime. (I use the word "universe" to denote a single member of this ensemble.) Crucially, these universes differ not just by "rearranging the furniture," but in the underlying laws of physics. As Martin Rees expresses it, what were hitherto believed to be universal, absolute, god-given laws are treated instead as merely local bylaws, valid only in our restricted cosmic patch (Rees, 2001).

The multiverse idea seems to offer progress. For example, one feature of the universe we would like to explain is why the underlying laws are biofriendly. This was Henderson's great insight: that nature is fit for life even as life is adapted to nature. He concluded his book with the clear statement: "The biologist may now rightly regard the universe in its very essence as biocentric" (Henderson, 1913). Given this now-uncontentious observation, we may then ask, *why do the laws of physics have the requisite mathematical form and assume the relevant values of various parameters within those laws, such that the universe may bring forth life and consciousness?* The multiverse theory seeks to account for this remarkable

---

[2] See, for example, Swinburne (1993).

and important property as an "observational selection effect." The universe – *this* universe – is observed by us to be biofriendly because it could not be observed at all, by us or anybody else, if it were biohostile.

Something else in favor of the multiverse theory is that some sort of cosmic ensemble is predicted naturally by combining two very fashionable branches of science: inflationary cosmology and string/M theory. The former is the standard model for how the universe emerged from the Big Bang displaying certain key properties, such as large-scale uniformity and (relatively) small-scale irregularity.[3] The latter is a mathematical theory that attempts to unify the various forces and particles of nature into a final fundamental theory.[4] According to one interpretation of string/M theory, what were previously regarded as fixed parameters, such as the masses of various particles or the strengths of the forces, are in fact frozen accidents – haphazard values adopted in our particular inflation region (Susskind, 2006). Most inflation regions would possess biohostile values, either too big or too small; but very rarely, by accident, a "Goldilocks" region would emerge in which things are just right for life, and those are the regions (universes) that are observed and perhaps commented on.

That's fine and blindingly obvious, as far as it goes. But does it get us out of trouble? Does it provide an ultimate explanation for why things are as they are? No, it doesn't. As described, the multiverse remains a subset of all possible universes. For example, each universe comes equipped with a set of mathematical laws similar to those in our own universe. So a rule still divides what exists (universes with laws similar to ours) from what does not (universes with radically different laws, lawless universes). Tegmark (2003, 2004) proposes to extend the multiverse by including universes in which the laws might be described by very different mathematics, such as fractals or non-Cantorian sets. John Barrow (1991) considered how laws based on alternative mathematics might preclude the existence of information-processing systems rich enough to constitute observers. But even this retains an arbitrary aspect. Defining universes in terms of their mathematical properties is a prejudice that one might expect from a mathematical physicist, but it need not represent the ultimate categorization. After all, Tegmark proposes a rule that divides mathematical universes from non-mathematical ones. But one might consider all manner of non-mathematical criteria to label possible universes. How about universes constructed according to all possible aesthetic principles? Or all possible gods? Or all possible ethical principles?

To see the significance of this point, consider a multiverse that contains a subset of universes with all possible teleological principles. Now human beings find themselves living in a universe that in many respects behaves *as if* it is teleological.

---

[3] See, for example Linde (1990).   [4] See, for example, Greene (2000).

Darwinism is invoked in an attempt to eliminate teleology from the biological realm. But if one makes the *a priori* assumption that our universe is but a component of a multiverse, and that the said multiverse contains not merely copious examples of universes with traditional differential equation laws but also universes with explicitly teleological laws, then the job of the Darwinian is made considerably harder. Why? Well, it is now necessary to refute the claim that the universe appears teleological because it *is* teleological. After all, why explain what we observe as really a non-teleological universe cunningly masquerading as a teleological one, when plenty of *genuinely* teleological universes are in the multiverse that could do the job? To maintain this perverse stance, one would need to show that teleological universes are in some sense sparser in the multiverse than non-teleological Darwinian universes; otherwise, on Bayesian grounds, one would be justified in retaining the hypothesis of teleology. For all I know, this may be the case; but it has certainly not been demonstrated.

Similar reasoning may be applied to gods. All possible universes would certainly include some with traditional deities, others with alien deities, yet others with no deities. This might not bother scientists if the subset of deity-associated universes were very rare among all biophilic universes. But can we be sure of this? A few years ago, I was asked to debate Peter Atkins about the existence of God. As a humorous riposte to his invoking the now-standard multiverse explanations of cosmic biofriendliness, I pointed out the following consequence of even a mild (that is, non-Tegmarkian) version of the multiverse theory. In the multiverse, at least some universes must surely exist in which intelligent beings advance to the point of being able to simulate consciousness and virtual reality. Some so-called strong artificial intelligence (AI) specialists claim that even human beings will soon be able to create conscious machines. It is but a small step from that threshold to the point where such simulations reach a level of fidelity indistinguishable from what we (or at least I) now observe. This small step has already been taken by Hollywood (even if it is still awaited in the world of real science) in the guise of *The Matrix* movies, wherein real and virtual realities are so alike that they confuse the participants (and the viewers). I have serious reservations about such AI claims; but I am willing to concede, at least for the purposes of this argument, that such simulations are doable in principle and might in practice be achieved by building a quantum supercomputer. At any rate, let us accept that a multiverse rich enough to contain a limitless number of universes like ours will possess a subset in which simulated reality by technological civilizations is routine. What are the implications?

First, it is amusing to note that the simulated beings in these universes stand in the same relation to the simulating system (or its designer or operator) as human beings once stood in relation to the traditional biblical god. The simulating system is the creator and sustainer of the *in silico* (or perhaps *in quanto*) beings and can observe, interfere with, and if necessary pull the plug on the simulated creatures. Ironically, far from abolishing a god, the multiverse seems to offer a convincing proof of such a being's existence, at least in a subset of universes. (This lighthearted remark finds little favor among most theologians, who seek a god for *every* universe, not just the simulated ones.)

This line of speculation raises the obvious question of whether the universe that we humans observe is merely a simulation or is the Real Thing. Since a single simulating system may simulate an unlimited number of virtual worlds, the fake universes would soon proliferate and outnumber the real ones, once the technology had been mastered. It follows that a random observer would be more likely to inhabit a simulated world than a real world (Bostrom, 2003). To express it more formally, unless we invoke an arbitrary existence rule that divides "real" from "simulated" consciousness, and hence real from simulated universes, a multiverse rich enough to account for intelligent life is also a multiverse in which simulated beings are likely to greatly outnumber unsimulated beings. Conclusion: We are very probably simulated.

Could such a wild idea be tested? Indeed it could, according to Barrow (2003). He points out that the simulating system need not bother to render consistently every tiny feature of the simulated universes; it could achieve a great deal more economically by running a so-so simulation and then tweaking the parameters from time to time. For the simulated beings, this would appear as the cosmic equivalent of the scenery wobbling. Barrow cites an example of such a possible "glitch" in the laws of physics (2003).

All this is entertaining stuff; but my serious point is that once one goes down the slippery slope of ever-more elaborate multiverse models, the nature of reality becomes exceedingly murky. The multiverse is no longer simply a set of plausible universes with differing laws of physics and/or initial conditions. One must include in the grand inventory of universes all the fake ones along with the real ones. It gets worse. One may prove that a universal computer, or Turing machine, is capable of simulating another Turing machine, even one that is very different in structure. Thus, a PC may simulate a cellular automaton (pixels that wink on and off on a screen) that itself may simulate, say, a Macintosh computer. Translated into virtual realities, this says that virtual worlds may contain beings that can simulate their own virtual worlds, and so on (and on, *ad nauseam*, even if not *ad infinitum*, on account of the second law of thermodynamics). Any attempt to neatly decompose the multiverse into "real" and "virtual" components seems doomed to failure. Rather, minds and universes become entwined, fractal-like, in a stupendously intricate amalgam. There may even be no "basement" universe, no ultimate ground of reality, on which to build such a "tower of turtles." If so, the age-old dualism between mind and matter, creator and creature, real and imaginary, dissolves away. Most people would regard this conclusion as a *reductio ad absurdum* of the multiverse hypothesis, although some (for example, Rees, 2003) may embrace it fearlessly. My own position is that a little bit of multiverse is good for you, and probably unavoidable given the recent discoveries of physics and cosmology; but following the slippery multiverse slope to its ultimate conclusion is a descent into fantasy – quite literally!

If Henderson were alive today: more water wonders

When Henderson wrote *The Fitness of the Environment* (1913), the nature of life was not well understood. The enormous advances in biochemistry, molecular biology, earth sciences, and astrobiology lay decades ahead. If Henderson were alive today, how might he extend his analysis in the light of modern discoveries? One very obvious and dramatic development has a direct impact on his thesis, and that concerns the very word "environment." Today, we know that life can not only survive, but thrive, in environments that in Henderson's day would have been considered utterly extreme for biology. Microbes inhabit the searing water spewing from deep ocean volcanic vents and dwell in the sub-zero dry valleys of Antarctica. Other organisms inhabit pools so acidic that the fluid burns human flesh. Some microbes (called halophiles) can make a living in the extreme saline conditions of the mis-named Dead Sea and even in the radiation-drenched environment of nuclear waste pools (Postgate, 1996). The one condition that does seem to be indispensable is access to liquid water. Henderson, who stressed the importance of several peculiar properties of water for the success of life, would doubtless approve of the mission statement of NASA's Astrobiology Institute: "Follow the water!" Even the hardiest extremophile needs liquid water. This suggests extending Henderson's analysis to a much wider range of conditions in which liquid water exists, conditions that Henderson would never have dreamed might permit life.

What additional special properties of water might fascinate him?

Some of the richest and most studied ecosystems are the deep ocean volcanic vents, inhabited by *hyperthermophiles*: microbes that live happily in temperatures ranging from 90 °C to 121 °C. (At depth, water does not boil at the usual 100 °C.) Typical of such organisms is *Pyrodictium abysii*, with an optimum growth temperature of 110 °C. The current official record is held by *Pyrolobus fumerii*, which can reproduce at a temperature of 121 °C, although I have witnessed a species of archaea metabolizing for thirty minutes at 130 °C. This raises obvious questions about water's special properties: thermal conductivity and thermal capacity, solvent properties, interaction with cell membranes, viscosity, and surface tension. These properties, considered by Henderson in the temperature range associated with familiar life (0 °C–40 °C), need to be extended upward at least to 130 °C.

One may also consider the opposite extreme of the very low temperatures inhabited by psychrophiles. Because of the need for water to remain liquid, and the decline in the rate of metabolism with temperature, psychrophilic life is not as extensive, active, or diverse as thermophilic life, but it is not without its interesting features. Water will not freeze at 0 °C in saline conditions, and highly saline water will support halophilic life even at low temperatures. Another circumstance in which water can remain liquid at temperatures well below the normal freezing point is in thin films on the surface of dirt and ice crystals. Although these films may be only a few micrometers thick, reports exist of microbial communities inhabiting this niche. Psychrophiles have been found living in the Siberian permafrost, growing and reproducing at temperatures of −10 °C. At temperatures as low as −20 °C, they take in needed materials from their environment and appear to be capable of repairing damage (Shi *et al.*, 1997). Saline water will remain liquid at even lower temperatures than this. The chemical and physical properties of water in ultracold, micrometer-thick layers differ radically from those of liquid water in more familiar conditions, and an extension of Henderson's study into this regime would seem to be worthwhile. Similar "amazing water" analyses suggest themselves for conditions of high salinity, acidity, and alkalinity.

The earliest forms of life were almost certainly chemoautotrophs – organisms that derive energy and make biomass directly from inorganic substances – since these can thrive in the complete absence of pre-existing organic material. Plants are autotrophs, but they use the highly complex and sophisticated process of photosynthesis, which must have taken a long time to evolve. Chemotrophy, in which an organism uses chemical energy and inorganic raw materials, is simpler although less efficient. It also has the virtue that it can take place in the dark, thus opening up the subsurface zone for life (see the next section). Chemotrophs are known on earth today that can make a living directly from dissolved hydrogen and carbon dioxide, which are turned into methane (Chapelle *et al.*, 2002). It is likely that life began with such organisms (Davies, 2003). Their success depends crucially on three water-related factors. The first is the ability of water to dissolve hydrogen and carbon dioxide in sufficient concentration to provide the energy needed. The second is the dissociation of water into hydrogen and oxygen when it passes through hot rocks deep in the earth's crust. This provides the hydrogen "fuel" that the chemotrophs use, and results from a surprising and unusual process (Freund *et al.*, 2002). The third is the ability of water to circulate deep into the crust and convect back to the surface by percolating through the tiny rock pores. Without this continual cycling of fluid, the supply of raw materials needed to sustain chemotrophic life at or near the surface would soon be exhausted. The circulation rate, which will depend on such factors as the viscosity and coefficient of expansion of water, must be fast enough to "keep the food coming," but not so fast as to fail to dissolve the released hydrogen.

Water plays an indirect but crucial role in the story of life on earth through geophysical and astrophysical processes of which Henderson could have had no inkling. One of these is plate tectonics. Astrobiologists believe that a healthy planet must continually recycle material if equable conditions are to be maintained. For example, on earth, carbon becomes sequestered in carbonate rocks and is released again in the form of carbon dioxide when the rocks are subducted. Similarly, oxygen is prevented from building up to dangerous levels by tectonic activity, which continually exposes fresh material to be oxidized. Part of the reason Mars seems to be a dead planet is because its tectonic processes have ground to a halt. Water is a crucial ingredient in this story (Ragenauer-Lieb *et al.*, 2001). If the earth's crust were not hydrated, the basalt would be brittle. The water content gives the rock high plasticity that allows the plates to slide smoothly and material to flow steadily through the mantle.

On a larger scale, water has helped shape the solar system. The solar nebula that gave birth to the planets 4.6 billion years ago contained copious quantities of water. Because of the proximity of the hot protosun, the inner part of the nebula, where the earth formed, was largely desiccated and devolatilized. Water and volatile organics condensed near the periphery of the nebula (beyond the so-called snow line). The water formed small ice crystals that stuck together to make snow. Snowflakes aggregated and trapped dust particles to make "dirty snowballs." Over time, some snowballs became large enough to gravitate significantly. According to the standard model, the cores of the giant planets probably formed this way, with the smaller dirty snowballs left over as comets and icy planetesimals. And giant planets play a crucial role in maintaining earth's biofriendliness. Jupiter, in particular, is important to life on earth by sweeping up rogue comets that would otherwise menace earth. But enough comets were able to get through to deliver water to our parched planet, without which life would be impossible. A key step in the foregoing sequence is the aggregation of ice crystals. The stickiness of snow is a familiar property to all children who have engaged in a snowball fight. Without this stickiness, comets could not have delivered water to the early earth.

Water ice in the form of interstellar grains may have played an important role in prebiotic chemistry. These tiny particles permeate the interstellar medium, where they gather carbonaceous and silicate material and absorb ultraviolet radiation. Complex chemical processes on their surfaces produce a large variety of organic substances for possible later delivery to planets.

As a final example of new discoveries being made about water's remarkable properties, consider the work of Kolesnikov *et al.* (2004), who have experimented with water confined to carbon nanotubes and found some highly unusual behavior that might be important for plant osmosis and the transport of protons across cell membranes. If these expectations are confirmed, it would be worthwhile to elucidate the quantum mechanics (QM) of this configuration to determine how sensitively the said properties would depend on the electron and nuclear masses or the fine-structure constant.

Biogenesis: a damned close-run thing?

Another area of research that has received a great deal of attention since Henderson's book was published is biogenesis, the problem of how life started from non-life in the first place. In discussing the fitness of the environment for life, Henderson was largely silent on the process of how life began – that is, the pathway whereby lifeless chemicals assembled themselves into the first living cell. It is one thing for the environment to be surprisingly fit for life, quite another for it to be fit for the *emergence* of life. This "fitness of the incubator" is a largely unexplored area, for a very good reason: we have almost no idea how life *did* originate (Davies, 2003). If we knew the intricate chemical and physical pathway, we could determine just how critically the key steps depended on this or that property. We could theoretically vary some of the parameters (such as the mass of the electron) and calculate what effect it had on biogenesis. It is easy to imagine that the emergence of life depended on some felicitous combination of chemical properties that relied sensitively on the mass of the electron (say). But it is equally easy to imagine that the special type of organized complexity that represents life is robust enough to emerge under a wide range of conditions. In the absence of a plausible model for biogenesis, we simply do not know.

It is, however, possible to offer some general remarks on the matter of fine-tuning and biogenesis. As I have discussed in the previous section, life today is extremely resilient, able to exist across a wide range of conditions of pressure, temperature, pH, salinity, and even radiation exposure. The overriding restriction seems to be the presence of liquid water. However, life on earth has had 4 billion years to evolve specialized molecules and chemical procedures to optimize its performance and extend its environmental range. The first life forms are likely to have been much more at the mercy of a hostile environment, including a "fitness bottleneck" through which the system had to pass before its longevity was ensured. We do not know what this physical and chemical bottleneck was, but substantial fine-tuning may have been required for all the factors to work together. In itself, this is not a radical suggestion. The emergence of anything new requires a regime that is metastable, so that significant qualitative changes may take place. A physically and chemically stable system is, by definition, one in which nothing new is likely to happen. But a metastable system is also one that is likely to be compromised by changes in the fundamental physical parameters.

Let me give an example of what I have in mind. We now have good evidence that early life on earth was hyperthermophilic, consisting of microbes dwelling in high-temperature conditions (Davies, 2003). Sequence analysis of ribosomal RNA enables organisms to be located on the tree of life, and the trend seems to be that the deepest branches of the tree are occupied by hyperthermophiles, suggesting that these organisms are descendants of ancient life forms that evolved relatively little since the dawn of life (Brock and Goode, 1996). Furthermore, from what is known of the geological conditions of the early earth, our planet would have been subjected to a ferocious cosmic bombardment for about 700 million years after the formation of the solar system. This would have rendered the surface of the earth very hazardous for life: the largest impacting objects would have released enough energy to swathe the planet in incandescent rock vapor and boil the oceans dry (Sleep *et al.*, 1989; Maher and Stephenson, 1988). The safest location was a number of kilometers underground. Today, earth's crust is found to be teeming with microbes. Because temperature increases with depth, subsurface microbes are obligate thermophiles or hyperthermophiles. This too points to early life residing in a hot, deep location.

It is not possible to conclude from this, however, that life actually *started* in a hot, deep setting. Nevertheless, many astrobiologists believe that this is a plausible scenario. If they are right, then we may identify a temperature bottleneck for biogenesis. Above 100 °C, the stability of proteins and DNA is threatened with thermal disruption. Indeed, modern hyperthermophiles deploy customized heat shock enzymes that continually repair the thermal damage. So, on the one hand, the evidence points to a high-temperature setting for early life; on the other hand, modern life survives at high temperatures by using specialized molecules that have evolved over billions of years. If the first life forms had to get by without these specialized molecules, their viability would have been a damned close-run thing, to paraphrase the Duke of Wellington. So I am hypothesizing that life emerged from the edge of thermal disruption, using molecules and chemical processes that were only marginally stable in those conditions. Such a regime is likely to be sensitive to changes in parameters such as the fine-structure constant and particle masses, and once again a careful analysis would seem to be worthwhile.

A second, albeit even more speculative, possibility of fine-tuning in biogenesis concerns the role of quantum mechanics. Obviously, at some level life must receive a quantum mechanical description. However, most researchers think of biochemical processes in terms of classical ball-and-stick models. To be sure, quantum mechanics determines the shapes and chemical affinities of biological molecules; but additional quantum effects, such as superposition, entanglement, and tunneling, are not normally regarded as significant. But this point of view may be questioned (Davies, 2004). Some circumstantial evidence exists for non-trivial quantum effects in extant biology (Abbot, Davies and Pati, 2008). Regarding biogenesis, if life emerged from the molecular realm, then it perforce emerged from the quantum realm. Two possibilities then present themselves:

(1) that quantum uncertainty was a limiting factor and (2) that a prebiotic system exploited quantum effects in making the transition to life.

Consider each possibility in turn. If life is defined as a form of organized complexity, then quantum fluctuations place constraints on the fidelity of organizational processes. For example, complex molecular processes that demand fine-tuned choreography of the component parts are subject to the inherent limitations that apply to quantum timekeeping (Pešić, 1993). If the emergence of life was indeed a close-run thing, then the probability of biogenesis may be sensitive to any relative increase in quantum fluctuations. The second possibility is that the transition to life involved certain key quantum steps, such as resonant reactions or tunneling, that are highly sensitive to small changes in the heights and shapes of potential barriers, which in turn depend (to some extent) on the fine-structure constant and particle masses.

The role of catalysts in biogenesis is likely to have been absolutely critical. Biochemists have long suspected that catalytic surfaces, especially minerals such as clay crystals or the pores of ocean basalt, were crucial in assembling complex organic molecules. The building blocks of these molecules are adsorbed onto the surfaces and brought into conjunction, thereby facilitating their concatenation. Some researchers, such as Kauffman (1993) and Morowitz *et al.* (2000), envisage autocatalytic cycles (in three dimensions), in which the products of catalysis generate more of the catalyst, forming a self-organizing system leading eventually to life. The citric-acid cycle, which forms the basis of intermediary metabolism, is a possible candidate for a prebiotic self-organizing system. This cycle is already quite complicated, involving some dozens of different molecules, and the question arises as to how sensitive the stability of this cycle would be to changes in the fine-structure constant and particle masses. It is possible that this cycle of highly interdependent processes hangs together only as a result of some felicitous coincidences in the reaction and diffusion rates of different substances.

Henderson's legacy

Henderson could scarcely have imagined that a few decades after he wrote his book, "the environment" for life would consist (at least in some people's eyes) not just of the earth, or even the solar system, or even the universe, but of a vast assemblage of universes. Given that we may now entertain the possibility of alternative laws in neighboring universes, and anthropic selection of fitness "oases" within the multiverse, how may the overall significance of Henderson's fitness examples be assessed?

I should first like to introduce a distinction between two quite different forms of fitness that are often conflated in these discussions. The first concerns cases where, had the laws of physics differed slightly from their observed form, life would apparently have been impossible. I will not review here the many examples of this argument.[5] The second has to do with the felicitous conjunction of disparate biofriendly properties, of which the most famous example – made famous by Henderson – is water. It happens that water combines in one substance several key qualities (thermal, mechanical, chemical) that life exploits and indeed that are indispensable to life as we know it. What are we to make of this? Is it just a lucky fluke that the same stuff that has an anomalous expansion property when it freezes (enabling ice to float) also has superlative solvent properties or unusually high surface tension and/or efficient tectonic lubrication qualities, for example?

Viewed in a multiverse context, we may imagine a world in which, say, the mass of the electron is different. This would have a knock-on effect on all the above properties: N of them, say. Thus, as the mass changes, these properties cease to assume biofriendly values. So on the face of it, we may invoke a standard multiverse/anthropic explanation for why water is biofriendly. But this is a bit of a cheat. We still need to explain why *any* values for the electron mass exist in which *all N* of the key properties realize their biofriendly values *simultaneously*. My point is that the said N qualities are not independent, but interrelated through physics and chemistry. You can't change each of them in isolation without changing the others. So we are left with the mystery of why *any* universe in the multiverse contains *any* fluid that enjoys, in conjunction, all the properties needed for life.

Perhaps we can entertain a naturalistic explanation along the following lines. Parameters other than the mass of the electron may affect the properties of water. The mass of the proton and the neutron, hence the masses of hydrogen and oxygen, will affect the density and viscosity of water. The fine-structure constant will affect the strengths of chemical bonds. One could envisage these parameters being independently varied. Within the four-dimensional parameter space, one or more regions may then exist where most of the N key properties take biofriendly values. Barrow has pointed out (see Chapter 8, this volume) that simultaneously scaling the fine-structure constant and the product of the mass and charge of the electron leaves the non-relativistic Schrödinger equation invariant, implying no change in those properties of water that do not depend on relativistic effects. Therefore, one should envisage a large two-dimensional subspace of the parameter space as being biofriendly. Including the nuclear masses, however, breaks the invariance. A research program along these lines – to find out whether other regions exist of the larger-parameter space in which water possesses appropriate qualities – might be worthwhile.

This proposal is predicated on the assumption that one is free to vary the foregoing parameters independently. This may be unwarranted. A future unified theory

---

[5] See, for example, Barrow and Tipler (1986).

might well tie together quantities such as the masses of particles and the forces of interaction. In some envisaged unification schemes, all such parameters would be fixed by the final theory, leaving no freedom to vary them independently, apart from an overall scale factor (Greene, 2000). A good way to think about this point is to imagine a multidimensional parameter space, with each axis labeled by the masses of the fundamental particles, the coupling constants, and so forth. Suppose it has M dimensions, where M might be 30 or so. Each dimension will come with an associated biofriendly range of values. Obviously, in our universe all such biofriendly intervals intersect in a biofriendly M-dimensional volume, a region of parameter space where all life-permitting values occur simultaneously. Now suppose that the parameters are not all independent, but linked by an underlying theory. If the theory connects two parameters, it can be represented by a (generally curved) line in the two-dimensional subspace of the M-dimensional parameter space. This line had better intersect the biofriendly volume or the theory would be inconsistent with our existence. Continuing in this manner, a final theory that contained no free parameters would be a curve in the M-dimensional space that would pass through the biofriendly volume. This curve would define a particular multiverse model, being the ensemble of universes corresponding to the points on the curve. A different curve would define a different multiverse model characterized by a different final theory.

What would one then make of cosmic biofriendliness? One could appeal to the multiverse and anthropic selection to explain why our universe was positioned at a felicitous point *along the curve*, namely, a point that lies within the biofriendly volume. But it would still be a matter of some amazement that this curve obligingly intersected the key (and possibly very small) region of the M-dimensional parameter space in which all the physical parameters assumed their biofriendly values in conjunction. Note that a different final theory would lead to *a different curve*, one that in general would *not* intersect the biofriendly volume. So we seem to have merely shifted the problem up one level, changing the question from *Why this universe?* to *Why this multiverse?* Unless, of course, it could be demonstrated that only one final theory is logically possible (a most unlikely prospect, in my opinion). So, here we have the cosmic equivalent of the Henderson water problem: the happy and amazing conjunction of disparate but individually vital properties manifested in one system.

The project that Henderson began nearly a century ago suggested a deep link between life and the universe. Henderson's central idea – that the physical universe is intriguingly biofriendly – has endured through the revolutions in cosmology, physics, molecular biology, and astrobiology that followed in the decades after his work was published. Given these spectacular advances, one sees considerable scope for extending Henderson's ideas about water specifically, and the fitness of the universe more generally, to include the latest thinking in astrobiology and cosmology. The areas of biochemistry and molecular biology are almost entirely unexplored in this context, and incorporating them into this project could prove exceedingly rewarding.

References

Abbot, D., Davies, P. C. W. and Pati, A. (2008). *Quantum Aspects of Biology*. London: Imperial College Press.

Barrow, J. (1991). *Pi in the Sky*. Oxford: Oxford University Press.

Barrow, J. (2003). Glitch. *New Scientist* (7 June), p. 44.

Barrow, J. D. and Tipler, F. J. (1986). *The Anthropic Cosmological Principle*. Oxford: Oxford University Press.

Birrell, N. D. and Davies, P. C. W. (1978). Massless Thirring model in curved space: thermal states and conformal anomaly. *Physical Reviews* D**18**, 4408.

Bostrom, N. (2003). Are we living in a computer simulation? *Philosophical Quarterly*, **53** (211), 243.

Brock, G. R. and Goode, J. A., eds. (1996). *Evolution of Hydrothermal Ecosystems on Earth (and Mars?)*. New York, NY: Wiley.

Chaitin, G. (1988). Randomness in arithmetic. *Scientific American*, **259** (1), 80.

Chapelle, F. H. *et al.* (2002). A hydrogen-based subsurface microbial community dominated by methanogens. *Nature,* **415**, 312.

Davies, P. (1991). *The Mind of God*. London and New York: Simon and Schuster.

Davies, P. (2003). *The Origin of Life*. London: Penguin.

Davies, P. C. W. (2004). Does quantum mechanics play a non-trivial role in life? *BioSystems,* **78**, 69.

Davies, P. (2006). *The Goldilocks Enigma: Why the Universe Is Just Right for Life*. London: Penguin Books.

Davies, P. (2007). *Cosmic Jackpot*. London: Penguin Books.

Freund, F., Dickinson, J. T. and Cash, M. (2002). Hydrogen in rocks: an energy source for deep microbial communities. *Astrobiology*, **2**, 83.

Greene, B. (2000). *The Elegant Universe*. New York, NY: Vintage.

Haught, J. (1986). *What Is God?* New York, NY: Paulist Press.

Hawking, S. (1988). *A Brief History of Time*. New York, NY: Bantam, p. 174.

Henderson, L. J. (1913). *The Fitness of the Environment: An Inquiry into the Biological Significance of the Properties of Matter*. New York: Macmillan. Repr. (1958) Boston, MA: Beacon Press; (1970). Gloucester, MA: Peter Smith.

Kauffman, S. A. (1993). *The Origins of Order*. Oxford: Oxford University Press.

Kolesnikov, A. I., Zanotti, J., Loong, C. *et al.* (2004). Anomalously soft dynamics of water in a nanotube: a revelation of nanoscale confinement. *Physical Review Letters*, **93**, 035503.

Linde, A. (1990). *Inflation and Quantum Cosmology*. Boston, MA: Academic Press.

Maher, K. and Stephenson, D. (1988). Impact frustration of the origin of life. *Nature,* **331**, 612.

Morowitz, H. J. *et al.* (2000). The origin of intermediary metabolism. *Proceedings of the National Academy of Sciences*, USA, **97**, 7704.

Pešić, P. D. (1993). The smallest clock. *European Journal of Physics,* **14**, 90.

Postgate, J. (1996). *The Outer Reaches of Life*. Cambridge, UK: Cambridge University Press. See also Online Extreme Environments articles: www.lyon.edu/projects/marsbugs/extreme.html.

Ragenauer-Lieb, K., Yuen, D. and Branlund, J. (2001). The initiation of subduction: criticality by addition of water? *Science*, B**294**, 578.

Rees, M. (2001). *Our Cosmic Habitat*. Princeton, NJ: Princeton University Press.

Rees, M. (2003). Interview. *Edge* 116 (19 May).

Shi, T. *et al.* (1997). Characterization of viable bacteria from Siberian permafrost by 16S rDNA sequencing. *Microbial Ecology*, **33**, 169.

Sleep, N., Zahnle, K. and Kasting, J. (1989). Annihilation of ecosystems by large asteroid impacts on the early Earth. *Nature*, **342**, 139.

Susskind, L. (2005). *The Cosmic Landscape: String Theory and the Illusion of Intelligent Design*. New York: Little Brown.

Swinburne, R. (1993). *The Coherence of Theism*, rev. edn. Oxford: Clarendon Press.

Tegmark, M. (2003). Parallel universes. *Scientific American*, May, 40 (cover story).

Tegmark, M. (2004). Parallel universes. In *Science and Ultimate Reality: Quantum Theory, Cosmology and Complexity*, ed. J. D. Barrow, P. C. W. Davies and C. L. Harper, Jr. Cambridge, UK: Cambridge University Press, p. 459.

7

The interconnections between cosmology and life

Mario Livio

Four basic observations

Progress in cosmology in the past few decades has also led to new insights into the global question of the emergence of intelligent life in the universe. I am referring not to discoveries that are related to very localized regions, such as the detection of 200 extrasolar planetary systems (at the time of writing[1]), but rather to properties of the universe at large.

In order to set the stage properly for the topics that follow, I would like to start by presenting four observations with which essentially all astronomers agree. These four observations *define* the cosmological context of our universe and form the basis for any theoretical discussion.

1. Ever since the observations of Vesto Slipher in 1912–22 (Slipher, 1917) and Edwin Hubble (1929), we have known that the spectra of distant galaxies are red shifted.
2. Observations with the Cosmic Background Explorer (COBE) have shown that, to a precision of better than 10^{-4}, the cosmic microwave background (CMB) is *thermal*, with a temperature of 2.73 K (Mather *et al.*, 1994).
3. Light elements, such as deuterium and helium, have been synthesized in a high-temperature phase in the past (see, for example, Gamow, 1946; Alpher *et al.*, 1948; Hoyle and Tayler, 1964; Peebles, 1966; Wagoner *et al.*, 1967).
4. Deep observations, such as the Hubble Deep Field, and the Hubble Ultra Deep Field, have shown that galaxies in the distant universe look younger. Specifically, they are smaller (see, for example, Roche *et al.*, 1996; Ferguson *et al.*, 2004), and they have a higher fraction of irregular morphologies (see, for example, Abraham *et al.*, 1996). This is what one would expect from a higher rate of interactions and from the "building blocks" of today's galaxies.

---

[1] See http://en.wikipedia.org/wiki/Extrasolar_planet.

Fitness of the Cosmos for Life: Biochemistry and Fine-Tuning, ed. J. D. Barrow *et al.*
Published by Cambridge University Press. © Cambridge University Press 2007.

When the above four basic observational facts are combined and considered together, there is no escape from the conclusion that our universe is *expanding and cooling*. This conclusion is entirely *consistent* with the Hot Big Bang model. Sometimes, we hear the stronger statement that these observations "prove" that there was a Hot Big Bang. However, the scientific method does not truly produce "proofs" in the mathematical sense.

During the past decade, deep and/or special-purpose observations with a variety of ground-based and space-based observatories have advanced our understanding of the history of the universe far beyond the mere theory that a Big Bang occurred (see, for example, the determination of cosmological parameters by the Wilkinson Microwave Anisotropy Probe [WMAP]) (Spergel *et al.*, 2003). In particular, in spite of uncertainties that still exist, remarkable progress has been achieved in the understanding of cosmic star-formation history.

By using different observational tracers (e.g. the UV luminosity density) of star formation in high-redshift galaxies, tentative plots for the star formation rate (SFR) as a function of redshift have been produced (see, for example, Lilly *et al.*, 1996; Madau *et al.*, 1996; Steidel *et al.*, 1999; Giavalisco *et al.*, 2004; Stanway *et al.*, 2003). There is little doubt that the SFR rises from the present to about $z \approx 1$. What happens in the redshift range $z \approx 1$–6 is still somewhat controversial. Whereas some studies suggest that the SFR reaches a peak at $z \approx 1$–2 and then declines slightly toward higher redshifts (see, for example, Steidel *et al.*, 1999) or maybe even more than slightly (Stanway *et al.*, 2003) or stays fairly flat up to $z \approx 5$ (see, for example, Calzetti and Heckman, 1999; Pei *et al.*, 1999), others claim, more speculatively, that the SFR continues to rise to $z \approx 8$ (Lanzetta *et al.*, 2002). The last claim is based on the suggestion that previous studies had failed to account for the dimming effects of surface brightness, but it is not clear whether such a claim can be substantiated (see, for example, Giavalisco *et al.*, 2004). For my present purposes, however, it is sufficient that the history of the *global* SFR is on the verge of being determined (if it has not been determined already). A knowledge of the SFR as a function of redshift allows for meaningful constraints to be placed on the global emergence of carbon-based life for the first time.

Carbon-based life in the universe

The main contributors of carbon to the interstellar medium are intermediate-mass (1–8 $M_\odot$) stars (see, for example, Wood, 1981; Yungelson *et al.*, 1993; Timmes *et al.*, 1995) through the asymptotic giant branch and planetary nebulae phases. A knowledge of the cosmic SFR history, together with a knowledge of the initial stellar mass function (presently still uncertain for high redshift), therefore allows for an approximate calculation of the rate of carbon production as a function of redshift (see, for example, Livio, 1999). For a peaked SFR, of the type obtained by Madau *et al.* (1996), for instance, the peak in the carbon production rate is somewhat delayed (by $\leqslant 1$ billion years) with respect to the SFR peak. The decline in the carbon production rate is also shallower for $z \leqslant 1$ (than the decline in the SFR) because of the buildup of a stellar reservoir in the earlier epochs.

Assuming a "principle of mediocrity," one would expect the emergence of *most* carbon-based life in the universe to be perhaps not too far from the peak in the carbon production rate – around $z \approx 1$ (for a peak in the SFR at $z \approx 1$–2; other considerations, related to the production of radioactive elements, lead to similar conclusions [see, for example, Hogan, 2000]). As the timescale required to develop intelligent civilizations may be within a factor of 2 of the lifetime of F5 to mid-K stars (the ones possessing continuously habitable zones; see Kasting *et al.* [1993] and further discussion below), it can be expected that intelligent civilizations have emerged when the universe was $\geqslant 0$ Gyr old. A younger emergence age may be obtained if the SFR does not decline at redshifts $1.2 \leqslant z \leqslant 8$. The fact that statements about the time of emergence of life in the universe can even be made attests to the immense progress in observational cosmology.

Carbon features in most anthropic arguments. In particular, it is often argued that the existence of an excited state of the carbon nucleus (the 0_2^+ state) is a manifestation of fine-tuning of the constants of nature, which allowed for the appearance of carbon-based life.

Carbon is formed through the triple-α process in two steps. First, two α particles form the unstable (lifetime *c.* 10^{-16} s) ^{8}Be. Second, a third α particle is captured via ^{8}Be$(\alpha, \gamma)^{12}$C. Hoyle argued that, in order for the 3α reaction to proceed at a rate sufficient to produce the observed cosmic carbon, a resonant level must exist in ^{12}C, a few hundred keV above the ^{8}Be $+$ ^{4}He threshold; such a level was indeed found experimentally (Dunbar *et al.*, 1953; Hoyle *et al.*, 1953; Cook *et al.*, 1957).

The question of how fine-tuned this level needs to be for the existence of carbon-based life has been the subject of considerable research. The most recent work on this topic was done by Oberhummer and collaborators (see, for example, Oberhummer *et al.*, 2000; Csótó *et al.*, 2001; Schlattl *et al.*, 2004). These authors used a model that treats the ^{12}C nucleus as a system of 12 interacting nucleons, with the approximate resonant reaction rate

$$\gamma_{3\alpha} = 3^{3/2} N_\alpha^3 \left(\frac{2\pi\hbar^2}{M_\alpha k_B T} \right)^3 \frac{\Gamma_\gamma}{\hbar} \exp\left(-\frac{\varepsilon}{k_B T} \right) \tag{7.1}$$

Here M_α and N_α are the mass and number density of α particles, respectively, ε is the resonance energy (in the center-of-mass frame), Γ_γ is the relative width, and all other symbols have their usual meaning. These authors also introduced small variations in the strengths of the nucleon–nucleon interaction and in the fine-structure constant (affecting ε and Γ_γ) and calculated stellar models using the modified rates. In their initial work, Oberhummer *et al.* (2000) concluded that a change of more than 0.5% in the strength of the strong interaction or more than 4% in the strength of the electromagnetic interaction would result in essentially no production of carbon or oxygen (considering the $^{12}C(\alpha, \gamma)^{16}O$ and $^{16}O(\alpha, \gamma)^{20}Ne$ reactions) in any star. More specifically, a decrease in the strong-interaction strength by 0.5%, coupled with an increase in the fine-structure constant by 4%, resulted in a decrease in the carbon production by a factor of a few tens in 20 $M_\odot$ stars and by a factor of *c.* 100 in 1.3 $M_\odot$ stars. Taken at face value, this seemed to support anthropic claims for the extreme fine-tuning necessary for the emergence of carbon-based life.

Earlier calculations by Livio *et al.* (1989) indicated less impressive fine-tuning. They showed that shifting (artificially) the energy of the carbon resonant state by up to 0.06 MeV does not result in a significant reduction in the production of carbon. As this 0.06 MeV should be compared with the *difference* between the resonance energy in ^{12}C and the 3α threshold (calculated with the basic nucleon–nucleon interaction), it was not obvious that a particularly fantastic fine-tuning was required. Most recently, however, Schlattl *et al.* (2004) reinvestigated the dependence of carbon and oxygen production in stars on the 3α rate. These authors found that following the entire stellar evolution was crucial. They concluded that in massive stars C and O production strongly depends on the initial mass. In intermediate- and low-mass stars, Schlattl *et al.* found that the high carbon production during He shell flashes leads to a *lower* sensitivity of the C and O production to the 3α rate than inferred by Oberhummer *et al.* (2000). Schlattl *et al.* (2004) concluded by saying that "fine-tuning with respect to the obtained carbon and oxygen abundance is more complicated and far less spectacular" than that found by Oberhummer *et al.* (2000).

The nature of dark energy

In 1998, two teams of astronomers, working independently, presented evidence that the expansion of the universe is accelerating (Riess *et al.*, 1998; Perlmutter *et al.*, 1999). The evidence was based primarily on the unexpected faintness (by *c.* 0.25 mag) of distant ($z \approx 0.5$) Type Ia supernovas compared with their expected brightness in a universe decelerating under its own gravity. The results favored values of $\Omega_m \approx 0.3$ and $\Omega_\Lambda \approx 0.7$ for matter and "dark energy" density parameters, respectively. Subsequent observations of the supernova SN 1997ff, at the redshift of $z \simeq 1.7$, strengthened the conclusion of an accelerating universe (Riess *et al.*, 2001). This supernova appeared *brighter* relative to SNe in a coasting universe, as expected from the fact that at $z \approx 1.7$ a universe with $\Omega_m \approx 0.3$ and $\Omega_\Lambda \approx 0.7$ would still be in its decelerating phase. The observations of SN 1997ff do not support any alternative interpretation (such as dust extinction or evolutionary effects) in which supernovas are expected to dim monotonically with redshift. Measurements of the power spectrum of the CMB (see, for example, Abroe *et al.*, 2002; de Bernardis *et al.*, 2002; Netterfield *et al.*, 2002; and, most recently, the WMAP results, Bennett *et al.*, 2003) provide strong evidence for flatness ($\Omega_m + \Omega_\Lambda = 1$). When combined with estimates of Ω_m based on mass-to-light ratios, X-ray temperatures of intra-cluster gas, and dynamics of clusters (all of which give $\Omega_m \lesssim 0.3$; see, for example, Strauss and Willick, 1995; Carlberg *et al.*, 1996; Bahcall *et al.*, 2000), again a value of $\Omega_\Lambda \approx 0.7$ is obtained.

Arguably, the two greatest puzzles that physics faces today are:

1. What is the nature of dark energy, and why is its density, ρ_Λ, so small, but not zero? (Or, why does the vacuum energy gravitate so little?)
2. Why *now*? Namely, why do we find at present that $\Omega_\Lambda \approx \Omega_m$?

The first question reflects the fact that taking graviton energies up to the Planck scale, M_P, would produce a dark-energy density roughly of the order of

$$\rho_\Lambda \approx M_P^4 \approx (10^{18}\,\text{GeV})^4 \tag{7.2}$$

which misses the observed one, $\rho_\Lambda \approx (10^{-3}\,\text{eV})^4$, by more than 120 orders of magnitude. Even if the energy density in fluctuations in the gravitational field is taken only up to the supersymmetry-breaking scale, M_{SUSY}, we still miss the mark by a factor of 60 orders of magnitude because $\rho_\Lambda \approx M_{\text{SUSY}}^4 \approx (1\,\text{TeV})^4$. Interestingly, a scale $M_\Lambda \approx (M_{\text{SUSY}}/M_P)M_{\text{SUSY}}$ produces the right order of magnitude. However, although a few attempts in this direction have been made (see, for example, Arkani-Hamed *et al.*, 2000), no satisfactory model that naturally produces this scale has been developed.

The second question is related to the anti-Copernican fact that Ω_Λ may be associated with a cosmological constant, while Ω_m declines continuously (and, in any case, ρ_Λ may be expected to have a time behavior different from that of ρ_m), and yet the first time that we are able to measure both reliably we find that they are of the same order of magnitude.

The attempts to solve these problems fall into three general categories:

1. The behavior of "quintessence" fields
2. Alternative theories of gravity
3. Anthropic considerations

Attempts of the first type have concentrated in particular on "tracker" solutions (see, for example, Zlatev *et al.*, 1998; Albrecht and Skordis, 2002) in which the smallness of Ω_Λ is a direct consequence of the universe's old age. Generally, a uniform scalar field, ϕ, is taken to evolve according to

$$\ddot{\phi} + 3H\dot{\phi} + V'(\phi) = 0 \qquad (7.3)$$

where $V'(\phi) = \frac{dV}{d\phi}$ and H is the Hubble parameter. The energy density of the scalar field is given by

$$\rho_\phi = \frac{1}{2}\dot{\phi}^2 + V(\phi) \qquad (7.4)$$

and that of matter and radiation, ρ_m, by

$$\dot{\rho}_m = -3H(\rho_m + P_m) \qquad (7.5)$$

where P_m is the pressure. For a potential of the form

$$V(\phi) = \phi^{-\alpha} M^{4+\alpha} \qquad (7.6)$$

where $\alpha > 0$ and M is an adjustable constant ($M \ll M_P$), and a field that is initially much smaller than the Planck mass, one obtains a solution in which a transition occurs from an early ρ_m-dominance to a late ρ_ϕ-dominance (with no need to fine-tune the initial conditions). Nevertheless, for the condition $\rho_\phi \approx \rho_m$ to actually be satisfied at present requires (Weinberg, 2001) that the parameter M satisfy

$$M^{4+\alpha} \simeq (8\pi G)^{-1-\alpha/2} H_0^2 \qquad (7.7)$$

which is not easily explicable.

In order to overcome this problem, some quintessence models choose potentials in which the universe has periodically been accelerating in the past (see, for example, Dodelson *et al.*, 2000) so that dark energy's dominance today appears naturally.

A very different approach regards the accelerating expansion not as being propelled by dark energy, but rather as being the result of a modified gravity. For example, models have been developed (Deffayet *et al.*, 2002) in which ordinary particles are localized on a three-dimensional surface (3-brane) embedded in infinite-volume extra dimensions to which gravity can spread. The model is constructed in such a way that observers on the brane discover Newtonian gravity (four-dimensional) at distances that are shorter than a crossover scale, r_c, which can be of astronomical size. In one version, the Friedmann equation is replaced by

$$H^2 + \frac{k}{a^2} = \left(\sqrt{\frac{\rho}{3M_P^2} + \frac{1}{4r_c^2}} + e\frac{1}{2r_c^2} \right)^2 \qquad (7.8)$$

where ρ is the total energy density, a is the scale factor, and $e = \pm 1$.

In this case, the dynamics of gravity are governed by whether ρ/M_P^2 is larger or smaller than $1/r_c^2$. Choosing $r_c \approx H_0^{-1}$ preserves the usual cosmological results. At large cosmic distances, however, gravity spreads into extra dimensions (the force law becomes five-dimensional) and becomes weaker, directly affecting the cosmic expansion. Basically, at late times, the model has a self-accelerating cosmological branch with $H = 1/r_c$ (to leading order Equation 7.8 can be parameterized as $H^2 - H/r_c \simeq \rho/3M_P^2$). Interestingly, it has recently been suggested that the viability of these models can be tested by lunar ranging experiments (Dvali *et al.*, 2003). I should also note that the early WMAP results indicated an intriguing lack of correlated signal on angular scales greater than 60 degrees (Spergel *et al.*, 2003), reinforcing the low quadrupole seen already in COBE results. One possible, although at this stage very speculative, interpretation of these results is that they signal the breakdown of conventional gravity on large scales.

A third class of proposed solutions to the dark-energy problem relies on anthropic selection effects, and therefore on the *existence* of intelligent life in our universe. The basic premise of this approach is that some of the constants of nature are actually random variables, whose range of values and *a priori* probabilities are nevertheless determined by the laws of physics. The observed Big Bang, in this picture, is simply one member of an ensemble. It is further assumed that a "principle of mediocrity" applies; namely, we can expect to observe the most probable values (Vilenkin, 1995). Using this approach, Garriga *et al.* (2000; following the original idea of Weinberg, 1987) were able to show that, when the cosmological constant Λ is the only variable parameter, the order of magnitude coincidence $t_0 \approx t_\Lambda \approx t_G$ (where t_0 is the present time, t_Λ is the time at which Ω_Λ starts to dominate, and t_G is the time when giant galaxies were assembled) finds a natural explanation (see also Bludman, 2000).

Qualitatively, the argument works as follows.

In a geometrically flat universe with a cosmological constant, gravitational clustering can no longer occur after redshift $(1 + z_\Lambda) \approx (\rho_\Lambda/\rho_{m0})^{1/3}$ (where ρ_{m0} is the present matter density). Therefore, requiring that ρ_Λ does not dominate before redshift $z_{\max}$, at which the earliest galaxies formed, requires (see, for example, Weinberg, 1987)

$$\rho_\Lambda \lesssim (1 + z_{\max})^3 \rho_{m0} \tag{7.9}$$

One can expect the *a priori* (independent of observers) probability distribution $P(\rho_\Lambda)$ to vary on some characteristic scale, $\Delta\rho_\Lambda \approx \eta^4$, determined by the underlying physics. Irrespective of whether η is determined by the Planck scale (*c.* 10^{18} GeV), the grand unification scale (*c.* 10^{16} GeV) or the electroweak scale (*c.* 10^2 GeV), $\Delta\rho_\Lambda$ exceeds the anthropically allowed range of ρ_Λ (Equation 7.9) by so many orders of magnitude that it looks reasonable to assume that

$$P(\rho_\Lambda) = \text{const}, \tag{7.10}$$

over the range of interest. Garriga and Vilenkin (2001) and Weinberg (2001) have shown that this assumption is satisfied by a broad class of models, even though not automatically. With a flat distribution, a value of ρ_Λ picked randomly (and which may characterize a "pocket" universe) from an interval $|\rho_\Lambda| < \rho_\Lambda^{max}$ will, with a high probability, be on the order of ρ_Λ^{max}. The principle of mediocrity, however, means that we should observe a value of ρ_Λ that maximizes the number of galaxies. This suggests that we should observe the largest value of ρ_Λ that is still consistent with a substantial fraction of matter having collapsed into galaxies: in other words, $t_\Lambda \approx t_G$, as observed. Above, I argued that the appearance of carbon-based life may be associated roughly with the peak in the star formation rate, t_{SFR}. The "present time," t_0, is not much different (in that it takes only a fraction of a stellar lifetime to develop intelligent life), hence $t_0 \approx t_{SFR}$. Finally, hierarchical structure-formation models suggest that vigorous star formation is closely associated with the formation of galactic-size objects (see, for example, Baugh *et al.*, 1998; Fukugita *et al.*, 1998). Therefore, $t_G \approx t_{SFR}$, and we obtain $t_0 \approx t_G \approx t_\Lambda$.

Garriga *et al.* (2000) further expanded their discussion to treat not just Λ, but also the density contrast at recombination, σ_{rec}, as a random variable (see also Tegmark and Rees, 1998). The galaxy formation in this case is spread over a much wider time interval, and proper account has to be taken of the fact that the cooling of protogalactic clouds collapsing at very late times is too slow for efficient fragmentation and star formation (fragmentation occurs if the cooling timescale is shorter than the collapse timescale, $\tau_{cool} < \tau_{grav}$). Assuming an *a priori* probability distribution of the form

$$P(\sigma_{rec}) \sim \sigma_{rec}^{-\alpha} \tag{7.11}$$

Garriga *et al.* found that "mediocre" observers will detect $\sigma_{rec} \approx 10^{-4}$, $t_0 \approx t_G \approx t_\Lambda \approx t_{cb}$, as observed, *if* $\alpha > 3$ (here the "cooling boundary" t_{cb} is the time after which fragmentation is suppressed).

Other anthropic explanations for the value of the cosmological constant and the "why now?" problem have been suggested in the context of maximally extended ($N = 8$) supergravity (Kallosh and Linde, 2003; Linde, 2003). In particular, the former authors found that the universe can have a suffciently long lifetime only if the scalar field satisfies initially $|\phi| \lesssim M_P$ and if the value of the potential $V(0)$, which plays the role of the cosmological constant, does not exceed the critical density $\rho_0 \approx 10^{-120} M_P^4$.

Personally, I feel that anthropic explanations to the dark-energy problem should be regarded as the *last resort*, only after all attempts to find explanations based on first principles have been exhausted and failed. Nevertheless, the anthropic explanation may prove to be the correct one, if our understanding of what is truly *fundamental* is lacking. A historical example can help to clarify this last statement. Johannes Kepler (1571–1630) was obsessed by the following two questions:

1. Why were there precisely six planets? (Only Mercury, Venus, Earth, Mars, Jupiter, and Saturn were known at his time.)
2. What was it that determined that the planetary orbits would be spaced as they are?

The first thing to realize is that these "why" and "what" questions were a novelty in the astronomical vocabulary. Astronomers before Kepler were usually satisfied with simply recording the observed positions of the planets; Kepler was seeking a theoretical explanation. Kepler finally came up with preposterously fantastic (and absolutely wrong) answers to his two questions in *Mysterium cosmographicum*, published in 1597. He suggested that the reason for there being six planets is that there are precisely five Platonic solids. Taken as boundaries (with an outer spherical boundary corresponding to the fixed stars), the solids create six spacings. By choosing a particular order for the solids to be embedded in one another, with earth separating the solids that can stand upright (cube, tetrahedron, and dodecahedron) from those that "float" (octahedron and icosahedron), Kepler claimed to have explained the sizes of the orbits as well (the spacings agreed with observations to within 10%).

Today, we recognize the *main* problem with Kepler's model: Kepler did not understand that neither the number of planets nor their spacings are *fundamental* quantities that need to have an explanation from first principles. Rather, both are the result of historical accidents in the solar protoplanetary disk. Still, it is perfectly legitimate to give an anthropic "explanation" for earth's orbital radius. If that orbit were not in the continuously habitable zone around the sun (Kasting *et al.*, 1993), we would not be here to ask the question.

It is difficult to admit it, but our current model for the composition of the universe – *c.* 74% dark energy, *c.* 22% cold dark matter, *c.* 4% baryonic matter, and maybe *c.* 0.5% neutrinos – appears no less preposterous than Kepler's model. Although some version of string (or *M*-) theories may eventually provide a first-principles explanation for all of these values, it is also possible, in my opinion, that these individual values are in fact not fundamental, but accidental. Maybe the only fundamental property is the fact that *all the energy densities add up to produce a geometrically flat universe*, as predicted by inflation (Guth, 1981; Hawking, 1982; Steinhardt and Turner, 1984) and confirmed by WMAP (Spergel *et al.*, 2003). Clearly, for any anthropic explanation of the value of Ω_Λ to be meaningful at all, even in principle, one requires the existence of a large ensemble of universes, with different values of Ω_Λ. That this requirement may actually be fulfilled is precisely one of the consequences of the concept of "eternal inflation" (Steinhardt, 1983; Vilenkin, 1983; Linde, 1986; Goncharov *et al.*, 1987; Linde, 2003).

In most inflationary models, the timescale associated with the expansion is much shorter than the decay timescale of the false-vacuum phase, $T_{exp} \ll T_{dec}$. Consequently, the emergence of a fractal structure of "pocket universes" surrounded by false-vacuum material is almost inevitable (Garcia-Bellido and Linde, 1995; Guth, 2001; for a different view, see, for example, Bucher *et al.*, 1995; Turok, 2001). This ensemble of pocket universes may serve as the basis on which anthropic argumentation can be constructed (even though the definition of probabilities on this infinite set is non-trivial; see, for example, Linde *et al.*, 1995; Vilenkin, 1998). Furthermore, the construction of de Sitter vacua of supergravity in string theory has shown that the number of possible solutions is extraordinarily large (Kachru *et al.*, 2003), again indicating the possible existence of an ensemble of universes (a "landscape").

Is the fine-structure constant varying with time?

Another recent finding, which, *if confirmed*, may have implications for the emergence of life in the universe, is that of the cosmological evolution of the fine-structure constant $\alpha \equiv e^2/\hbar c$ (Webb *et al.*, 1999, 2001, and references therein). Needless to say, life as we know it places significant anthropic constraints on the range of values allowed for α. For example, the requirement that the lifetime of the proton would be longer than the main-sequence lifetime of stars results in an upper bound $\alpha \lesssim 1/80$ (Ellis and Nanopoulos, 1981; Barrow *et al.*, 2002a). The claimed detection of time variability was based on shifts in the rest wavelengths of redshifted UV resonance transitions observed in quasar absorption systems. Basically, the dependence of observed wave number at redshift z, ω_z, on α can be expressed as

$$\omega_z = \omega_0 + a_1\omega_1 + a_2\omega_2 \tag{7.12}$$

where a_1 and a_2 represent relativistic corrections for particular atomic masses and electron configurations, and

$$\omega_1 = \left(\frac{\alpha_z}{\alpha_0}\right)^2 - 1 \tag{7.13}$$

$$\omega_2 = \left(\frac{\alpha_z}{\alpha_0}\right)^4 - 1 \tag{7.14}$$

Here α_0 and α_z represent the present-day and redshift z values of α, respectively. By analyzing a multitude of absorption lines from many multiplets in different ions, such as Fe II and Mg II transitions in 28 absorption systems (in the redshift range $0.5 \lesssim z \lesssim 1.8$), and Ni II, Cr II, Zn II and Si IV transitions in some 40 absorption systems (in the redshift range $1.8 \lesssim z \lesssim 3.5$), Webb *et al.* (2001) concluded that α

was *smaller* in the past. Their data suggested a 4σ deviation

$$\frac{\Delta\alpha}{\alpha} = (-0.72 \pm 0.18) \times 10^{-5} \qquad (7.15)$$

over the redshift range $1.5 \lesssim z \lesssim 3.5$ (where $\Delta\alpha/\alpha = \frac{\alpha_z - \alpha_0}{\alpha_0}$). More recently, Murphy *et al.* (2003) analyzed a total of 128 absorption systems over the redshift range $0.2 < z < 3.7$ and obtained

$$\frac{\Delta\alpha}{\alpha} = (-0.574 \pm 0.102) \times 10^{-5} \qquad (7.16)$$

It should be noted, however, that the data are consistent with *no* variation for $z \lesssim 1$, in agreement with many previous studies (see, for example, Bahcall *et al.*, 1967; Wolfe *et al.*,1976; Cowie and Songaila, 1995).

Murphy *et al.* (2001, 2003) conducted a comprehensive search for systematic effects that could potentially be responsible for the result (e.g. laboratory wavelength errors, isotopic abundance effects, heliocentric corrections during the quasar integration, line blending, and atmospheric dispersion). Although they concluded that isotopic abundance evolution and atmospheric dispersion could have an effect, this was in the direction of actually amplifying the variation in α (to $\Delta\alpha/\alpha = (-1.19 \pm 0.17) \times 10^{-5}$). The most recent results of Webb *et al.* are not inconsistent with limits on α from the Oklo natural uranium fission reactor (which was active 1.8×10^9 years ago, corresponding to $z \approx 0.1$) and with constraints from experimental tests of the equivalence principle. The former suggests $\Delta\alpha/\alpha \simeq (-0.4 \pm 1.4) \times 10^{-8}$ (Fujii *et al.*, 2000), and the latter *allows* for a variation of the magnitude observed in the context of a general dynamical theory relating variations of α to the electromagnetic fraction of the mass density in the universe (Bekenstein, 1982; Livio and Stiavelli, 1998).

Before going any farther, I would like to note that what is desperately needed right now is an independent confirmation (or refutation) of the results of Webb *et al.* by other groups, through both additional (and preferably different) observations and independent analysis of the data. In this respect, it is important to realize that the reliability of the SNe Ia results (concerning the accelerating universe) was enormously enhanced by the fact that two separate teams (the Supernova Cosmology Project and the High-z Supernova Team) reached the same conclusion independently, using different samples and different data analysis techniques. A first small step in the direction of testing the variable α result came from measurements of the CMB. A likelihood analysis of BOOMERanG and MAXIMA data, allowing for the possibility of a time-varying α (which, in turn, affects the recombination time), found that in general the data may prefer a smaller α in the past (although the conclusion is not free of degeneracies) (Avelino *et al.*, 2000; Battye *et al.*, 2001). A second, much more important step came through an extensive analysis using the nebular emission lines of [O III] $\lambda\lambda$4959, 5007 Å (Bahcall *et al.*, 2004). Bahcall *et al.* found $\Delta\alpha/\alpha = (-2 \pm 1.2) \times 10^{-4}$ (corresponding to $\left|\alpha^{-1}\mathrm{d}\alpha/\mathrm{d}t\right| < 10^{-13}\,\mathrm{yr}^{-1}$, which they consider to be a null result, given the precision of their method) for quasars in the redshift range $0.16 < z < 0.8$.

Although this result is not formally inconsistent with the variation claimed by Webb *et al.*, the careful analysis of Bahcall *et al.* has cast some serious doubts on the ability of the "many-multiplet" method employed by Webb and his collaborators to actually reach the accuracy required to measure fractional variations in α at the 10^{-5} level. For example, Bahcall *et al.* have shown that, to achieve that precision, one needs to assume that the velocity profiles of different ions in different clouds are essentially the same to within 1 km s^{-1}. In the most recent work, Chand *et al.* (2004) observed 23 Mg II systems toward 18 QSOs in the redshift range $0.4 \leqslant z \leqslant 2.3$. The weighted mean value of variation in α they obtained was considerably smaller than that claimed by Murphy *et al.* (2003). Chand *et al.* found $\Delta\alpha/\alpha = (-0.06 \pm 0.06) \times 10^{-5}$. Furthermore, in a separate study, Ashenfelter *et al.* (2004) have shown that the synthesis of ^{25,26}Mg in low-metallicity asymptotic giant-branch stars produces isotopic ratios that could explain the data from $z < 1.8$ without invoking variations in the fine structure constant. However, even for extremely low metallicity (inconsistent with that observed in the $0.4 \leqslant z \leqslant 2$ redshift range), the expected effect produced by the different isotopic ratios is smaller than the one claimed by Murphy *et al.* (2003). Clearly, much more work on this topic is needed. I should also note right away that, in order not to be in conflict with the yield of ^{4}He, $|\Delta\alpha/\alpha|$ cannot exceed $c.\ 2 \times 10^{-2}$ at the time of nucleosynthesis (see, for example, Bergström *et al.*, 1999).

On the theoretical side, simple cosmological models with a varying fine structure constant have now been developed (see, for example, Sandvik *et al.*, 2002; Barrow *et al.*, 2002b). They share some properties with Kaluza–Klein-type models in which α varies at the same rate as the extra dimensions of space (see, for example, Damour and Polyakov, 1994) and with varying-speed-of-light theories (see, for example, Albrecht and Magueijo, 1999; Barrow and Magueijo, 2000).

The general equations describing a geometrically flat, homogeneous, isotropic, variable-α universe (Bekenstein, 1982; Livio and Stiavelli, 1998; Sandvik *et al.*, 2002) are the Friedmann equation (with $G = c \equiv 1$)

$$\left(\frac{\dot{a}}{a}\right)^2 = \frac{8\pi}{3}[\rho_m(1 + |\zeta_m|\,e^{-2\psi}) + \rho_r e^{-2\psi} + \rho\psi + \rho_\Lambda] \qquad (7.17)$$

the evolution of the scalar field varying α $(\alpha = \exp(2\psi)\,e_0^2/\hbar c)$

$$\ddot{\psi} + 3H\dot{\psi} = -\frac{2}{\omega}e^{-2\psi}\zeta_m\rho_m \qquad (7.18)$$

and the conservation equations for matter and radiation

$$\dot{\rho}_m + 3H\rho_m = 0 \qquad\qquad (7.19)$$
$$\dot{\rho}_r + 4H\rho_r = 2\psi\rho_r \qquad\qquad (7.20)$$

Here, ρ_m, ρ_r, ρ_ψ, ρ_Λ are the densities of matter, radiation, scalar field $\left(\frac{\omega}{2}\psi^2\right)$, and vacuum, respectively; $a(t)$ is the scale factor ($H \equiv \dot{a}/a$); $\omega = \hbar c / l^2$ is the coupling constant of the dynamic Langrangian (l is a length scale of the theory); and ξ_m is a dimensionless parameter that represents the fraction of mass in Coulomb energy of an average nucleon compared with the free proton mass.

Equations 7.17–7.20 were solved numerically by Sandvik *et al.* (2002) and Barrow *et al.* (2002b), assuming a negative value of the parameter ξ_m/ω, and the results are interesting both from a purely cosmological point of view and from the perspective of the emergence of life. First, the results are consistent both with the claims of a varying α of Webb *et al.* (2001) (which, as I noted, badly need further confirmation) and with the more secure, by now, observations of an accelerating universe (Riess *et al.*, 1998; Perlmutter *et al.*, 1999; Spergel *et al.*, 2003), while complying with the geological and nucleosynthetic constraints. Second, Barrow *et al.* (2002a,b) find that α remains almost constant in the radiation-dominated era and experiences a small logarithmic time increase during the matter-dominated era, but approaches a constant value again in the Λ-dominated era. This behavior has interesting anthropic consequences. The existence of a non-zero vacuum-energy contribution is now *required* in this picture to dynamically stabilize the fine structure constant. In a universe with zero Λ, α would continue to grow in the matter-dominated era to values that would make the emergence of life impossible (Barrow *et al.*, 2002a,b).

Clearly, the viability of all of the speculative ideas above relies at this point on the confirmation or refutation of time-varying constants of nature. Most recently, Barrow (2005) has shown that the isotropy of the microwave background imposes very stringent bounds on spatial variations of physical constants.

How rare is extraterrestrial intelligent life?

With the discovery of c. 200 massive extrasolar planets (see, for example, Mayor and Queloz, 1995; Marcy and Butler, 1996, 2000), the question of the potential existence of extraterrestrial, galactic, intelligent life has certainly become more intriguing than ever. This topic has attracted much attention and generated many speculative (by necessity) probability estimates. Nevertheless, in a quite remarkable paper, Carter (1983) concluded on the basis of the near-equality between the lifetime of the sun, $t_\odot$, and the timescale of biological evolution on earth, t_ℓ, that extraterrestrial intelligent civilizations are exceedingly rare in the galaxy. Most significantly, Carter's conclusion is supposed to hold even if the conditions optimal for the emergence of life are relatively common.

Let me reproduce here, very briefly, Carter's argument. The basic, and very crucial, assumption on which the argument is based is that the lifetime of a star, t_*, and the timescale of biological evolution on a planet around that star, t_ℓ (taken here, for definiteness, to be the timescale for the appearance of complex land life) are *a priori entirely independent*. In other words, the assumption is that land life appears at some *random* time with respect to the main-sequence lifetime of the star. Under this assumption, one expects that generally one of the two relations $t_\ell \gg t_*$ and $t_\ell \ll t_*$ applies (the set where $t_\ell \approx t_*$ is of negligible measure for two independent quantities). Let us examine each one of these possibilities. If *generally* $t_\ell \ll t_*$, it is very difficult to understand why, in the first system found to contain complex land life – the earth–sun system – the two timescales are nearly equal, $t_\ell \approx t_*$. If, on the other hand, *generally* $t_\ell \gg t_*$, then clearly the first system we find must exhibit $t_\ell \approx t_*$ (because for $t_\ell \gg t_*$ complex land life would not have developed). Therefore, one has to conclude that *typically* $t_\ell \gg t_*$, and that consequently, complex land life will generally not develop; earth is an extremely rare exception.

Carter's argument is quite powerful and not easily refutable. Its basic assumption (the independence of t_ℓ and t_*) appears on the face of it to be solid because t_* is determined primarily by nuclear burning reactions, whereas t_ℓ is determined by biochemical reactions and the evolution of species. Nevertheless, the fact that the star is the main energy source for biological evolution (light energy exceeds the other sources by 2–3 orders of magnitude; see, for example, Deamer, 1997) already implies that the two quantities are not completely independent.

Let me first take a purely mathematical approach and examine what it would take for the condition $t_\ell \approx t_*$ to be satisfied in the earth–sun system *without* implying that extraterrestrial intelligent life is extremely rare. Imagine that t_ℓ and t_* are not independent, but rather that

$$t_\ell / t_* = f(t_*) \tag{7.21}$$

where $f(t_*)$ is some *monotonically increasing* function in the narrow range $t_*^{\min} \lesssim t_* \lesssim t_*^{\max}$ that allows the emergence of complex land life through the existence of continuously habitable zones (corresponding to stellar spectral types F5 to mid-K; Kasting *et al.*, 1993). Note that for a Salpeter (1955) initial-mass function, the distribution of stellar lifetimes behaves as

$$\psi(t_*) \approx t_* \tag{7.22}$$

Consequently, if Relation 7.21 were to hold, it would in fact be *most probable* that where we first encountered an intelligent civilization we would find that $t_\ell / t_* \approx 1$, as in the earth–sun system. In other words, if we could identify some processes that are likely to produce a monotonically increasing t_*-t_ℓ/t_* relation, then the near equality of t_ℓ and t_* in the earth–sun system would find a natural explanation, with no implications whatsoever for the frequency of intelligent civilizations.

A few years ago, I proposed a simple toy model for how such a relation might arise (Livio, 1999). The toy model was based on the assumption that the appearance of land life has to await the buildup of a sufficient layer of protective ozone (Berkner and Marshall, 1965; Hart, 1978) and on the fact that oxygen in a planet's atmosphere is released in the first phase from the dissociation of water (Hart, 1978; Levine *et al.*, 1979). Given that the duration of this phase is inversely proportional to the intensity of radiation in the 1000–2000 Å range, a relation between t_ℓ and t_* can be established. In fact, a simple calculation gave

$$t_\ell/t_* \simeq 0.4(t_*/t_\odot)^{1.7} \tag{7.23}$$

precisely the type of monotonic relation needed.

I should be the first to point out that the toy model above is nothing more than that: a toy model. It does point out, however, that, at the very least, establishing a link between the biochemical and astrophysical timescales may not be impossible. Clearly, the emergence of complex life on earth required many factors operating together. These include processes that appear entirely accidental, such as the stabilization of the earth's tilt against chaotic evolution by the moon (see, for example, Laskar *et al.*, 1993). Nevertheless, we should not be so arrogant as to conclude everything from the one example we know. The discovery of many "hot Jupiters" (giant planets with orbital radii $\leqslant 0.05$ AU) has already demonstrated that the solar system may not be typical. In particular, Jupiter is a significant outlier (at the 2.3σ level) in the periastron distribution of all the extrasolar planets (Beer *et al.*, 2004). Although this could represent merely a selection effect (we should know within 5–10 years), the possibility exists that most of the observed extrasolar planetary systems have been formed in a way rather different from our own. We should none the less keep an open mind to the possibility that biological complexity may find other paths to emerge, making various "accidents," coincidences, and fine-tuning unnecessary. In any case, the final scientific assessment on life in the universe will probably come from biologists and observers – not from speculating theorists like myself.

Acknowledgments

This work has been supported by Grant 938-COS191 from the John Templeton Foundation. I am grateful to Andrei Linde, Heinz Oberhummer, and Keith Olive for their helpful comments.

References

Abraham, R. G., van den Bergh, S., Glazebrook, K. *et al.* (1996). *Astrophysical Journal Supplement*, **101**, 1.

Abroe, M. E., Balbi, A., Borrill, J. *et al.* (2002). *Monthly Notices of the Royal Astronomical Society*, **334**, 11.

Albrecht, A. and Magueijo, J. (1999). *Physical Review*, D**59**, 043516.

Albrecht, A. and Skordis, C. (2002). *Physical Review Letters*, **84**, 2076.

Alpher, R. A., Bethe, H. and Gamow, G. (1948). *Physical Review*, **73**, 803.

Arkani-Hamed, N., Hall, L. J., Colda, C. *et al.* (2000). *Physical Review Letters*, **85**, 4434.

Ashenfelter, T., Mathews, G. J. and Olive, K. A. (2004). Chemical evolution of Mg isotopes versus the time variation of the fine structure constant. *Physical Review Letters*, **92d**, 1102A.

Avelino, P. P., Martins, C. J. A. P., Rocha, G. *et al.* (2000). *Physical Review*, D**62**, 123508.

Bahcall, J. N., Sargent, W. L. W. and Schmidt, M. (1967). *Astrophysical Journal*, **149**, L11.

Bahcall, J. N., Steinhardt, C. L. and Schlegel, D. (2004). *Astrophysical Journal*, **600**, 520.

Bahcall, N. A., Cen, R., Davé, R. *et al.* (2000). *Astrophysical Journal*, **541**, 1.

Barrow, J. D. (2005). *Physical Review*, D**71**, 083520.

Barrow, J. D. and Magueijo, J. (2000). *Astrophysical Journal*, **532**, L87.

Barrow, J. D., Sandvik, H. B. and Magueijo, J. (2002a). *Physical Review*, D**65**, 123501.

Barrow, J. D., Sandvik, H. B. and Magueijo, J. (2002b). *Physical Review*, D**65**, 063504.

Battye, R. A., Crittenden, R. and Weller, J. (2001). *Physical Review*, D**63**, 043505.

Baugh, C. M., Cole, S., Frenk, C. S. *et al.* (1998). *Astrophysical Journal*, **498**, 504.

Beer, M. E., King, A. R., Livio, M. *et al.* (2004). *Monthly Notices of the Royal Astronomical Society*, **354**, 763.

Bekenstein, J. D. (1982). *Physical Review*, D**25**, 1527.

Bennett, C. L., Bay, M., Halpern, M. *et al.* (2003). *Astrophysical Journal*, **583**, 1.

Bergström, L., Iguri, S. and Rubinstein, H. (1999). *Physical Review*, D**60**, 045005.

Berkner, L. V. and Marshall, K. C. (1965). *Journal of Atmospheric Science*, **22**, 225.

Bludman, S. (2000). *Nuclear Physics*, A**663–4**, 865.

Bucher, M., Goldhaber, A. and Turok, N. (1995). *Physical Review*, D**52**, 3314.

Calzetti, D. and Heckman, T. M. (1999). *Astrophysical Journal*, **519**, 27.

Carlberg, R. G., Yee, H. K. C., Ellingson, E. *et al.* (1996). *Astrophysical Journal*, **462**, 32.

Carter, B. (1983). *Philosophical Transactions of the Royal Society of London*, A**310**, 347.

Chand, H., Srianand, R., Petitjean, P. *et al.* (2004). *Astronomy and Astrophysics*, **417**, 853.

Cook, C. W., Fowler, W. A. and Lauritsen, T. (1957). *Physical Review*, **107**, 508.

Cowie, L. L. and Songaila, A. (1995). *Astrophysical Journal*, **453**, 596.

Csótó, A., Oberhummer, H. and Schlattl, H. (2001). *Nuclear Physics*, A**688**, 560.

Damour, T. and Polyakov, A. M. (1994). *Nuclear Physics*, B**423**, 532.

Deamer, D. W. (1997). *Microbiology and Molecular Biology Reviews*, **61**, 239.

de Bernardis, P., Ade, P. A. R., Bock, J. J. *et al.* (2002). *Astrophysical Journal*, **564**, 559.

Deffayet, C., Dvali, G. and Gabadadze, G. (2002). *Physical Review*, D**65**, 044023.

Dodelson, S., Kaplinghat, M. and Stewart, E. (2000). *Physical Review Letters*, **85**, 5276.

Dunbar, D. N. F., Pixley, R. E., Wenzel, W. A. *et al.* (1953). *Physical Review*, **92**, 649.

Dvali, G., Gruzinov, A. and Zaldarriaga, M. (2003). *Physical Review*, D**68**, 024012.

Ellis, J. D. and Nanopoulos, D. V. (1981). *Nature*, **292**, 436.

Ferguson, H. C., Dickinson, M., Giavalisco, M. *et al.* (2004). *Astrophysical Journal*, **600**, L107.

Fujii, Y., Iwamoto, A., Fukahori, T. *et al.* (2000). *Nuclear Physics*, B**573**, 377.

Fukugita, M., Hogan, C. J. and Peebles, P. J. E. (1998). *Astrophysical Journal*, **503**, 518.

Gamow, G. (1946). *Physical Review,* **70**, 527.

Garcia-Bellido, J. and Linde, A. O. (1995). *Physical Review*, D**51**, 429.

Garriga, J. and Vilenkin, A. (2001). *Physical Review*, D**64**, 023517.

Garriga, J., Livio, M. and Vilenkin, A. (2000). *Physical Review*, D**61**, 023503.

Giavalisco, M., Dickinson, M., Ferguson, H. C. *et al.* (2004). *Astrophysical Journal*, **600**, L103.

Goncharov, A. S., Linde, A. D. and Mukhanov, V. F. (1987). *International Journal of Modern Physics*, A**2**, 561.

Guth, A. H. (1981). *Physical Review*, D**23**, 347.

Guth, A. H. (2001). In *Astrophysical Ages and Time Scales*, ed. T. von Hippel, C. Simpson and N. Mansit. San Francisco: Astronomical Society of the Pacific, p. 3.

Hart, M. H. (1978). *Icarus*, **33**, 23.

Hawking, S. W. (1982). *Physical Letters*, B**115**, 295.

Hogan, C. J. (2000). *Review of Modern Physics*, **72**, 1149.

Hoyle, F. and Tayler, R. J. (1964). *Nature*, **203**, 1108.

Hoyle, F., Dunbar, D. N. F. and Wenzel, W. A. (1953). *Physical Review*, **92**, 1095.

Hubble, E. (1929). *Proceedings of the National Academy of Sciences, USA*, **15**, 168.

Kachru, S., Kallosh, R., Linde, A. *et al.* (2003). *Physical Review*, D**68**, 046005.

Kallosh, R. and Linde, A. (2003). *Physical Review*, D**67**, 023510.

Kasting, J. F., Whitmore, D. P. and Reynolds, R. T. (1993). *Icarus,* **101**, 108.

Kepler, J. (1597). *Mysterium cosmographicum (The Secret of the Universe)*, transl. A. M. Duncan. Republished (1981), New York, NY: Abaris Books.

Lanzetta, K. M., Yahata, N., Pascarelle, S. *et al.* (2002). *Astrophysical Journal*, **570**, 492.

Laskar, J., Joutel, F. and Boudin, F. (1993). *Nature*, **361**, 615.

Levine, J. S., Hayes, P. B. and Walker, J. C. G. (1979). *Icarus*, **39**, 295.

Lilly, S. J., Le Fèvre, O., Hammer, F. *et al.* (1996). *Astrophysical Journal*, **460**, L1.

Linde, A. (2003). Inflation, quantum cosmology, and the anthropic principle. In *Science and Ultimate Reality: Quantum Theory, Cosmology and Complexity*, ed. J. D. Barrow, P. C. W. Davies and C. L. Harper, Jr. Cambridge, UK: Cambridge University Press, pp. 426–58.

Linde, A., Linde, D. and Mezhlumian, A. (1995). *Physical Letters*, B**345**, 203.

Linde, A. D. (1986). *Modern Physics Letters*, A**1**, 81.

Livio, M. (1999). *Astrophysical Journal*, **511**, 429.

Livio, M. and Stiavelli, M. (1998). *Astrophysical Journal*, **507**, L13.

Livio, M., Hollowell, D., Weiss, A. *et al.* (1989). *Nature*, **340**, 281.

Madau, P., Ferguson, H. C., Dickinson, M. *et al.* (1996). *Monthly Notices of the Royal Astronomical Society*, **283**, 1388.

Marcy, G. W. and Butler, R. P. (1996). *Astrophysical Journal*, **464**, L147.

Marcy, G. W. and Butler, R. P. (2000). *Publications of the Astronomical Society of the Pacific*, **112**, 137.

Mather, J. C., Cheng, E. S., Cottingham, D. A. *et al.* (1994). *Astrophysical Journal*, **420**, 439.

Mayor, M. and Queloz, D. (1995). *Nature*, **378**, 355.

Murphy, M. T., Webb, J. K., Flambaum, V. V. *et al.* (2001). *Monthly Notices of the Royal Astronomical Society*, **327**, 1223.

Murphy, M. T., Webb, J. K. and Flambaum, V. V. (2003). *Monthly Notices of the Royal Astronomical Society*, **345**, 609.

Netterfield, C. B., Ade, P. A. R., Bock, J. J. *et al.* (2002). *Astrophysical Journal*, **571**, 604.

Oberhummer, H., Csótó, A. and Schlattl, H. (2000). *Science*, **289**, 88.

Peebles, P. J. E. (1966). *Astrophysical Journal*, **146**, 542.

Pei, Y., Fall, S. M. and Hauser, M. G. (1999). *Astrophysical Journal*, **522**, 604.

Perlmutter, S., Aldering, G., Goldhaber, G. *et al.* (1999). *Astrophysical Journal*, **517**, 565.

Riess, A. G., Filippenko, A. V., Challis, P. *et al.* (1998). *Astronomical Journal*, **116**, 1009.

Riess, A. G., Nugent, P. E., Gilliland, R. L. *et al.* (2001). *Astrophysical Journal*, **560**, 49.

Roche, N., Ratnatunga, K. U., Griffths, R. E. *et al.* (1996). *Monthly Notices of the Royal Astronomical Society*, **282**, 1247.

Salpeter, E. E. (1955). *Astrophysical Journal*, **121**, 161.

Sandvik, H. B., Barrow, J. D. and Magueijo, J. (2002). *Physical Review Letters*, **88**, 031302.

Schlattl, H., Heger, A., Oberhummer, H. *et al.* (2004). *Astrophysics and Space Science*, **291**, 1.

Slipher, V. M. (1917). *Proceedings of the American Philosophical Society*, **56**, 403.

Spergel, D. N., Verde, L., Peiris, H. V. *et al.* (2003). *Astrophysical Journal* (suppl.) **148**, 175.

Stanway, E. R., Bunker, A. J. and McMahon, R. G. (2003). *Monthly Notices of the Royal Astronomical Society*, **342**, 439.

Steidel, C. C., Adelberger, K. L., Giavalisco, M. *et al.* (1999). *Astrophysical Journal*, **519**, 1.

Steinhardt, P. J. (1983). In *The Very Early Universe*, ed. G. W. Gibbons, S. Hawking and S. T. C. Siklos. Cambridge, UK: Cambridge University Press, p. 251.

Steinhardt, P. J. and Turner, M. S. (1984). *Physical Review*, D**29**, 2162.

Strauss, M. A. and Willick, J. A. (1995). *Physics Reports*, **261**, 271.

Tegmark, M. and Rees, M. J. (1998). *Astrophysical Journal*, **499**, 526.

Timmes, F. X., Woosley, S. E. and Weaver, T. A. (1995). *Astrophysical Journal* (suppl.) **98**, 617.

Turok, N. (2001). In *Birth and Evolution of the Universe*, ed. K. Sato and M. Kawasaki. Tokyo: Universal Academy Press, p. 1.

Vilenkin, A. (1983). *Physical Review*, D**27**, 2848.

Vilenkin, A. (1995). *Physical Review Letters*, **74**, 846.

Vilenkin, A. (1998). *Physical Review Letters*, **81**, 5501.

Wagoner, R. V., Fowler, W. A. and Hoyle, F. (1967). *Astrophysical Journal*, **148**, 3.

Webb, J. K., Flambaum, V. V., Churchill, C. W. *et al.* (1999). *Physical Review Letters*, **82**, 884.

Webb, J. K., Murphy, M. T., Flambaum, V. V. *et al.* (2001). *Physical Review Letters*, **87**, 091301.

Weinberg, S. (1987). *Physical Review Letters*, **59**, 2607.

Weinberg, S. (2001). In *Sources and Detection of Dark Matter and Energy in the Universe,* ed. D. B. Cline. Berlin: Springer, p. 18.

Wolfe, A. M., Brown, R. L. and Roberts, M. S. (1976). *Physical Review Letters*, **37**, 179.

Wood, P. R. (1981). In *Physical Processes in Red Giants*, ed. I. Iben, Jr. and A. Renzini. Dordrecht: Reidel, p. 205.

Yungelson, L., Tutukov, A. V. and Livio, M. (1993). *Astrophysical Journal*, **418**, 794.

Zlatev, I., Wang, L. and Steinhardt, P. J. (1998). *Physical Review Letters*, **82**, 896.

8

Chemistry and sensitivity

John D. Barrow

Introduction

In recent years, there has been great interest among some particle physicists and astronomers in assessing the dependence of the gross structure of the universe on the values of its defining constants of nature. This agenda has been partly motivated by the recognition that many of the most important structures in the universe, and its most crucial evolutionary pathways over billions of years of cosmic history, are surprisingly sensitive to the values of some of those constants. Since we lack any fundamental understanding of why any of the constants of nature take the values that they do, this state of affairs appears surprisingly fortuitous. It provokes cosmologists to consider our observed universe in the context of a wider ensemble of possible universes in which the laws of nature remain the same but the constants of nature, or the boundary conditions that specify the overall expansion dynamics of the universe, are allowed to change. In effect, a type of "stability" analysis is performed to ascertain how large such changes in the structure of the universe and its defining constants could be and still give rise to a recognizable universe.

These considerations have attracted renewed attention with the realization that our universe may have a significant non-uniform structure in space, in which the values of many of the quantities that we dub "constants" may in fact be variable values of spacetime-dependent fields. Their values in their ground state will be uniquely defined only if the ground state, or vacuum state, of the universe is unique. However, from what we know of string theories at present, it appears that there is a vast level of non-uniqueness about the vacuum state. String theory offers a vast landscape of different low-energy worlds in which the gallery of fundamental forces – their identities, multiplicity, and strengths – can fall out in many different ways. Some of these will allow complexity to develop in the universe, but most will

Fitness of the Cosmos for Life: Biochemistry and Fine-Tuning, ed. J. D. Barrow *et al.*
Published by Cambridge University Press. © Cambridge University Press 2007.

not. As a result, it is of considerable interest to study the extent to which observers are possible in different vacua [1, 2], whether the fact that we observe three large dimensions of space to exist in the universe is true everywhere, and whether the dimensionality of space has a dynamical or random origin. Thus the different "universes" either can be realized at different places in a very large, or spatially infinite, universe or can be different outcomes of a spontaneous symmetry-breaking process in the quantum gravitational era of the universe's evolution.

In assessing the likelihood that our universe arose from one of these string theoretic scenarios, we have to ask what the probability is of our type of universe arising from the subset of all possibilities that can permit living complexity to develop, not merely the unconditioned probability of our universe arising as a solution of the theory. Very similar considerations also arise in the context of the chaotic and eternal inflationary universe scenarios [3]. There, the final state of the fields that give rise to inflation can determine the values of some of the constants of nature and the low-energy symmetry group that determines the forces of nature. This final state can be different in different inflated regions of the universe, giving rise to an effective spatial variation of the "constants." Again, the challenge is to find a rigorous method to compute the probability distribution of outcomes and to determine the likelihood of "our" universe arising, as well as the probability of "our" universe arising in the subset of life-supporting universes. Taking a modified Copernican perspective, we ask whether we are typical among those universes in which intelligent life could evolve and persist.

These investigations in cosmology have been discussed in some detail elsewhere [4]. They have refocused attention on the sensitive aspects of the universe as a life-supporting environment. This type of examination of the sensitivity of the universe to the numerical values of its dimensionless constants and boundary conditions is very familiar to astrophysicists, but it appears to be almost unknown to chemists. Very few analyses of the sensitivity of life-supporting aspects of chemistry have been given by chemists in modern times (see, for example, [5]), although some extensive analyses have been made by physicists [6] under the title of the "anthropic principles." In addition, wider considerations of how local geophysical and astronomical "coincidences" have been exploited by the evolution of life are given by Ward and Brownlee under the banner of the "rare Earth hypothesis" [7] and by Gonzalez and Richards [8] in a recent survey that encompasses many aspects of solar system structure and geomorphology. In this chapter, and many of the others in this volume, we investigate ways in which crucial chemical and biochemical structures and mechanisms exhibit sensitivity to constants of nature and ambient conditions. We are interested in those pathways that are important for the origin and persistence of life. In addition to understanding the sensitivity of known pathways, we hope that these studies will provoke consideration of alternative biochemistries or the ways in which standard chemistry might be changed by unusual environments. For example, we have long appreciated that atomic and molecular structure is significantly changed in the presence of very strong magnetic fields and are beginning to appreciate that the unique properties of water can be significantly changed in thin films. In both these cases, atomic structures behave as though the world possessed just two accessible dimensions of space.

We begin with a historical summary of previous investigations of the special features of chemical structure by natural theologians of the eighteenth and nineteenth centuries, together with the important contributions of Henderson early in the twentieth century. We then describe some of the results of searching for fine-tunings in physics and astronomy before turning to introduce some points of possible current interest to chemists and biochemists.

Fine-tuning in history

Henderson

Lawrence J. Henderson, professor of chemistry at Harvard, wrote two books on the fitness of chemical elements, molecules, and compounds for the existence and functioning of the phenomenon that we call "life." The first, *The Fitness of the Environment* (1913) [9], is the inspiration for this volume. The other, less well known volume, entitled *The Order of Nature* (1917) [10], contains a summary of the arguments put forward in *Fitness*, together with a wider philosophical reflection on the conclusions to be drawn from them. The discussions are remarkable in that they predate quantum mechanics; however, they were by no means the first discussions of the fitness of chemistry for life. See, for instance, Prout's famous Bridgewater Treatise, *Chemistry, Meteorology and the Function of Digestion* (1834) [11]. Or see *Religion and Chemistry* (1880) by Josiah Cooke, a professor of chemistry and mineralogy at Harvard [12]. Cooke's work provoked further detailed discussions of the life-supporting properties of water, carbonic acid, C, N, and O by Chadbourne in his *Lectures on Natural Theology* (1870) [13]. And these properties were most magisterially addressed by Alfred Russel Wallace in his book *Man's Place in the Universe* (1903) [14]. Henderson's studies were the broadest and deepest, not superseded until Needham's *Order and Life* was published in 1936 [15]. Yet, despite their clear and compelling presentation, Henderson's books attracted little international attention at the time. J. S. Haldane briefly reviewed Henderson's second book for *Nature* [16], but it was only his son J. B. S. Haldane who grasped the significance of what Henderson was arguing and followed up with a number of short papers of his own about the reasons for the form of the laws of nature and the role of cosmological evolution in driving the pace of evolutionary change [17].

They probably appeared a little too speculative because they exploited non-standard cosmological models, with two timescales, advocated at the time by E. A. Milne.

Henderson's analysis of the wonderful appropriateness of hydrogen, oxygen, and carbon for the roles they play in biochemistry can be summarized by his claim that [9]

No other element or group of elements possesses properties which on any account can be compared with these. All such are deficient at many points, both qualitatively or quantitatively. . . . The unique properties of water, carbonic acid, and the three elements constitute, among the properties of matter, the fittest ensemble of characteristics for durable mechanism.

Henderson was well aware that he was following in the footsteps of those who had based an Argument from Design for the existence of God on the apparent fine-tuning of nature's structures and the forms of her laws. He even quotes from William Whewell's Bridgewater Treatise [18].

One suspects that the lack of contemporary interest in Henderson's work might have been because such ideas were closely associated with the teleological design arguments of the past, which had been discredited by the emergence of natural selection as a better explanation for biological fine-tuning and, to a far lesser extent, by the philosophical objections of Hume (1779) [19] and Kant (*c.* 1790) [20].

Henderson's contributions to this subject were not confined to his books. The philosopher Josiah Royce organized a private evening discussion group at Harvard that included Henderson (and even T. S. Eliot for a while) as a regular participant. During 1913–14, the group's discussions focused for three months on Henderson's *Fitness of the Environment* and its interpretations for science and philosophy [21, 22].

The legacy of design arguments

It is useful to distinguish two features of fine-tuning that have become confused by historians. On the one hand, design arguments have typically pointed to the fortuitous "fine-tuning" of different *outcomes* of the laws of nature: the way in which living things seemed tailor-made for their environments, or the fortuitous angle of tilt of the earth's rotation axis. Arguments of this sort were eloquently surveyed in William Paley's much-quoted *Natural Theology* (1802) [23], although it grew out of talks delivered from 1770 onwards. Many discussions of this type of fine-tuning of *outcomes* have taken place since ancient times; see [6] for an overview.

However, historians and theologians never seem to read the second half of Paley's book, where he turns to another type of argument (which he doesn't personally like so much because it is more removed from observation and less amenable to reasoning by analogy, which Paley especially liked) based on the fortuitous structure of the *laws and constants* of nature: the consequences of the inverse-square law of gravity or the strength of the gravitational force. Here Paley was greatly helped with the astronomical details by his friend John Brinkley, who was both Bishop of Cloyne in Cork, and Astronomer Royal for Ireland and Professor of Astronomy at Trinity College, Dublin. Fine-tuning arguments of this sort, based on the form of the laws of nature rather than their outcomes, began with Newton and appear first in print in Richard Bentley's *Boyle Lectures on Natural Theology* [24] for which he was strongly "coached" by Newton himself (this was the motivation for Newton's famous four "Letters to Bentley" in 1691 [25], although they were not published until 1756).

Design Arguments based on laws and invariants of nature were completely unaffected by natural selection, but few leading scientists appreciated the distinction. A notable exception was James Clerk Maxwell, who in 1873 emphasized the fact that "molecules" (what we would now call "atoms," in fact) were populations of identical particles whose properties were not acted on by natural selection, but that determined whether life could exist. Henderson was greatly influenced by Maxwell's ideas and quotes from him on many occasions. In fact, one of Maxwell's remarkable essays on the subject is reproduced in its entirety as an appendix to the *Order of Nature* [10]. This essay was a talk on "free will" delivered to the Apostles' conversazione group in Trinity College, Cambridge, in 1873, and also provides the first identification of the dynamical phenomenon that we now call "chaos" along with a discussion of its implications for determinism.

Fine-tuning in physics and cosmology

Motivations

The modern interest in fine-tuning exists primarily in particle physics, astrophysics, and cosmology, and its examples are frequently gathered together under the umbrella of "anthropic arguments" [6]. Cosmologists have been interested in the ways in which the observed values of dimensionless constants of nature are related to the ranges of values that permit the evolution of "life" or complexity, as we know it. This program has a variety of motivations, although its most interesting results generally arise as by-products of other investigations. In particular, it aims to:

(a) Understand which features of the universe and its constants are necessary conditions for the existence of complexity, chemistry, and life. As a by-product, we might consider whether alternative non-carbon-based biochemistries or different life-supporting environments to those found on earth might be possible.

(b) Determine whether radical new theories to understand coincidences between the values of different dimensionless constants of nature are warranted. Historically, Dicke [26] was the first to show that the "large number coincidences" of cosmology were necessary for observers (like ourselves) who rely on hydrogen-burning stars and do not require a time-variation in G, as Dirac had proposed in 1937 [27].

(c) Identify those aspects of the physical universe's structure that depend most sensitively on the actual values of fundamental constants and to discover whether these sensitivities offer new observational bounds on the possible time or space variation of the supposed "constants of nature" [4].

(d) Discover how many (if any) of the underlying constants of nature are fixed by the mathematical structure of the Theory of Everything and how many remain to be determined quasi-randomly by physical processes during the history of the universe.

Other worlds

The idea of counter-factual alternative "universes" or "many worlds" with different values of the constants of nature are considered in different contexts. These scenarios are the most common:

(a) Many worlds, in which all possible permutations and combinations of the constants of nature, laws, dimensions, and so forth, are considered. This ensemble of possible other worlds might be metaphysical, as in Leibniz' "all possible worlds," which was the motivation for Maupertuis's introduction of Least Action Principles in mechanics [6] in order to make mathematically precise what was meant by "possible worlds" and by "best world": the other worlds were the paths of non-minimal action, and the best world was the one whose dynamics followed the paths of least action! Or the ensemble might be physical (as might arise in a single spatially infinite universe). Some commentators see this as a way of avoiding any idea of special anthropocentric design in the structure of the universe because in an exhaustively random ensemble of possibilities a subset will necessarily exist that permits life (because we know one case certainly exists!), and we will necessarily be a member of it [28, 29]. No fine-tuning is required.

(b) Scenarios such as the chaotic or eternal inflationary universe, in which different quantities that we call "constants of nature" fall out differently in different parts of the universe as a result of symmetry-breaking or other random processes of quantum origin. It may be possible to determine the probability distribution of these outcomes in a prescribed set of inflationary theories (although this has not yet been calculated in a persuasive way). We would then be able to evaluate the probability of creating an expanding "habitable" domain old enough ($\approx 10^{10}$ years) and large enough to produce elements heavier than hydrogen and helium by stellar nucleosynthesis.

(c) Cosmologies in which some of the constants of nature are assumed to be slowly varying scalar fields, depending on time and position [4]. Such theories provide a fully self-consistent context in order to investigate the observational consequences of time-varying constants (as opposed to constants taking different – but still constant – numerical values).

(d) In any theory of the very early universe in which intrinsically random processes enter, for example through quantum uncertain initial conditions, quantum gravity [30], or symmetry-breaking, the ultimate prediction that such a theory makes will be probabilistic. In order to test these predictions, it is necessary to know what range of values for the predicted quantities could result in "observers" like ourselves. We need to know the conditional probability distributions of the predicted parameters given that observers can subsequently evolve, not the unconditioned probability distributions. If the peak of the probability distribution for some cosmological property predicted by quantum gravity does not permit any observers to exist, then we should not reject the theory.

(e) "Many worlds" interpretations of quantum mechanics in which the different worlds defined by measurement assume equal ontological status [31]. This interpretation of quantum mechanics is the one of choice for researchers in quantum cosmology, where no "observer" of the universe is required, as it is in the so-called Copenhagen Interpretation developed by Bohr.

The mystery of constants

The idea that some traditional constants might be varying has become of great interest because of recent observations consistent with variations in the fine-structure constant at high redshift [32, 33, 34]. However, the fact that string theories only appear to be consistent with many more dimensions of space than three means that the true constants of nature are defined in the total number of dimensions (nine or ten in current theories). The three dimensions we see are large. Any others must be assumed to be imperceptibly small. Our constants arise as shadows of the true higher-dimensional constants as a result of the unknown process by which the extra dimensions stay small while our three (why three?) have grown large. If the extra dimensions were to change in any way, then our three-dimensional "constants" would be observed to change at the same rate.

In all this work, it is good to remember that although we have made steady progress in measuring the constants of nature to ever greater numerical precision, we have no idea at all why any of them take the precise numerical values that they do. Physicists have never predicted or explained the value of any constant of nature. The most pressing problem in fundamental physics and cosmology at present is to understand why one apparent constant of nature – the cosmological constant – exists with a small ($\approx 10^{-121}$ in inverse Planck lengths), but non-zero, value – that is, sufficient to create the currently observed acceleration in the expansion of the universe.

Fine-tuning in stars and atoms

What depends on the value of α?

We are interested in those aspects of chemistry and biochemistry with special sensitivities to the values of the existing constants of nature, or to the local or global astronomical environment in which life evolves. The following examples will give the flavor of such considerations at a back-of-the-envelope level of detail; more examples can be found, with further detail, in references [35, 36, 37, 38].

Some examples

A variety of constraints are placed on the maximum value of the fine-structure constant, $\alpha = e^2/\hbar c \approx 1/137$, that is compatible with the existence of nucleons, nuclei, atoms, and stars under the assumption that the forms of the laws of nature remain the same.

(a) In simple grand unified gauge theories, the running of the fine-structure constant with energy due to vacuum polarization effects leads to an exponential sensitivity of the proton lifetime with respect to the low-energy value of α with $t_{pr} \sim \alpha^{-2} \exp[\alpha^{-1}] m_{pr}^{-1}\hbar c^{-2} \sim 10^{32}$ yrs. In order that the lifetime be less than the main sequence lifetime of stars, we have $t_{pr} < (Gm_{pr}^{-2})^{-1}m_{pr}^{-1}\hbar c^{-2}$, which implies that α is bounded above by $\alpha < 1/80$ approximately [39].

(b) The stability of nuclei is controlled by the balance between nuclear binding and electromagnetic surface forces. A nucleus (Z, A) will be stable if, roughly, $Z^2/A < 49(\alpha_s/0.1)^2(1/137\alpha)$. In order for carbon $(Z = 6)$ to be stable, we require $\alpha < 16\,(\alpha/0.1)^2$. Detailed investigations of the nucleosynthesis processes in stars have shown that a change in the value of α shifts the key resonance-level energies in the carbon and oxygen nuclei that are needed for the production of a mixture of carbon and oxygen from beryllium plus helium-4 and carbon-12 plus helium-4 reactions in stars. This unusual level structure was first noted by Hoyle [40] and has recently been explored in some detail by Oberhummer *et al.* [41]. Some of these upper bounds on α are model-independent in significant ways. However, sharper limits can be found by using our knowledge of the stability of matter derived from analysis of the Schrödinger equation.

(c) The value of α controls atomic stability. If α increases in value, then the innermost Bohr orbital contracts, and electrons will eventually fall into the nucleus when $\alpha > Z^{-1}m_{pr}/m_e$. Atoms all become relativistic and unstable to pair production. In order that the electromagnetic repulsion between protons does not exceed nuclear strong binding, $e^2/r_n < \alpha m_\pi c^2$, we require $\alpha < 1/20$. Lieb [42] has proved that atomic instability arises for the relativistic Schrödinger equation at $Z\alpha = \frac{2}{\pi}$. When the many-electron–many-nucleus problem is examined with the relativistic Schrödinger equation, there is a condition on α independent of Z for stability. If $\alpha < 1.94 = 0.0106$, stability occurs all the way up to the critical value $\alpha = \frac{2}{\pi Z} = 0.1061 \, (6/Z)$, whereas if $\alpha > 128/15\pi = 2.716$, the system is unstable for all values of Z. In the presence of arbitrarily large magnetic fields, matter composed of electrons and nuclei is known to become unstable if α or Z is too large.

(d) Using a series of non-linear inequalities that are more powerful and rigorous than conventional estimates using the Heisenberg Uncertainty Principle alone, Lieb and collaborators [42] analyzed the conditions needed for finite ground-state energies in atoms of any size. They proved that matter is stable if $\alpha < 0.06 = 1/16.67$ and $\alpha < 0.026 \, (6/Z)^{1/2}$. Thus the stability of matter with coulomb forces has been proved for non-relativistic dynamics, including arbitrarily large magnetic fields, and for relativistic dynamics without magnetic fields. In both cases stability requires that the fine-structure constant be not too large.

(e) If stars are to exist, it must be hot enough at their centers for thermonuclear reactions to occur. This requires α to be bounded above by $\alpha^2 < 20m_e/m_{pr}$. Carter [35] has also pointed out the existence of the very sensitive coincidence $\alpha^{12} \approx (m_e/m_{pr})^4 Gm_{pr}^{-2}\hbar^{-1}c^{-1}$ that must be met if stars are to undergo a convective phase, but this stringent condition no longer seems to be essential for planet formation.

These limits are the result of rather coarse-grained arguments. One imagines that if the form of life is specified more narrowly, then the limits arising from requirements for the persistence of stable planetary environments with atmospheres and on the biochemical functioning of that form of life would be rather stronger.

Fine-tuning in chemistry

Simultaneous variations of constants in chemistry

So far, as in almost all the existing literature, we have only mentioned limits arising when one constant (e.g. α) is changed. What happens if several are changed in value simultaneously? Interestingly, there is a simple case where precise conclusions are possible. The non-relativistic Schrödinger equation possesses a simple scaling invariance if we change the electron mass, m_e, and the fine-structure constant, α, simultaneously. The Hamiltonian for a non-relativistic atomic system is [6] (ignoring the kinetic energy of the nucleus, $\vec{x}_i$ is the position vector of the ith electron, $\vec{R}_j$ is the position vector of the jth nucleus):

$$H = \frac{-\hbar^2}{2m_e} \sum_i \frac{\partial_i}{\partial \vec{x}_i} \cdot \frac{\partial_i}{\partial \vec{x}_i} + e^2$$
$$\times \left[\sum_i \sum_j -\frac{Z_j}{|\vec{x}_i - \vec{R}_i|} + \frac{1}{2} \sum_{i<k} \frac{1}{|\vec{x}_i - \vec{x}_k|} - \frac{1}{2} \sum_{i<k} \frac{Z_j Z_l}{|\vec{R}_j - \vec{R}_l|} \right]$$

Now transform to new dimensionless position variables $(\vec{y}, \vec{S})$ by using the Bohr radius as the unit of length:

$$\vec{y} = \frac{\vec{x}}{a_0}, \quad \vec{S} = \frac{\vec{R}}{a_0}$$

where

$$a_0 = \frac{\hbar^2}{m_e c^2}$$

is the Bohr radius. The Hamiltonian transforms to

$$H = \alpha^2 m_e c^2 \left\{ -\frac{1}{2} \sum_i \frac{\partial_i}{\partial \vec{y}_i} \cdot \frac{\partial_i}{\partial \vec{y}_i} + \sum_i \sum_j - \frac{Z_i}{|\vec{y}_i - \vec{S}_j|} \right.$$
$$\left. + \frac{1}{2} \sum_{i<k} \frac{1}{|\vec{y}_i - \vec{y}_j|} + \frac{1}{2} \sum_{i<l} - \frac{Z_j Z_l}{|\vec{S}_j - \vec{S}_l|} \right\}$$

Thus, we see that if E was the original energy eigenstate and E' is the new energy eigenstate that results when α is changed to α', c is changed to c', and m_e is changed to m'_e, then

$$\frac{E'}{\alpha'^2 m'_e c'^2} = \frac{E}{\alpha^2 m_e c^2}$$

In effect, *in the new variables the Schrödinger equation can be written in a form in which the constants of nature do not appear.* All the unusual bond angles and properties of atomic structure that life exploits in non-relativistic atoms will arise in identical fashion in a world with new values of α and m_e. They come from the ubiquitous geometrical "factors of order 2π" that determine the eigenvalues of the Schrödinger equation. The double helices of its DNA molecules will differ only in size; its water molecules will display the same remarkable properties that flow from its special bond angles but will differ solely in overall scale.

The consequences of this invariance do not seem to have been addressed in the study of the sensitivity of biochemistry to the constants of nature. Note, however, that this scaling invariance does not hold when *relativistic* atomic structure effects are included or when other forces of nature intervene (as becomes inevitable for sufficiently large variations). Of course, one can hypothesize arbitrary changes in constants (including α and m_e) under the assumption that no Schrödinger equation governs them. We can conceive of these worlds with arbitrary atoms, but we cannot deduce any of the properties of those atoms without an underlying equation of which they are eigenstates. The scaling invariance does not govern them in this case, but the assumption is useless because without a Schrödinger (or Schrödinger-like) equation there is no way of deducing any atomic properties at all.

This scaling invariance also holds for Newtonian gravity (also an inverse square law force like the Coulomb force) and can be used to transform exact solutions of Newtonian gravity with a constant value of G into new ones in which G varies with time or in which the gravitating masses all vary with time [32].

Some effects of relativity

The scale invariance of atomic structure is broken when relativistic effects are introduced. They become important for heavy elements since relativistic effects can be gauged by the magnitude of orbital velocities $v \sim Z\alpha c$, where Z is the atomic number. They become significant when $Z\alpha$ becomes of order unity. Recently, there has been considerable interest in the sensitivity of relativistic contributions to atomic lines when very small changes (a few parts in 10^6) are made to the value of the fine-structure constant, α. For several years, John Webb, Michael Murphy, Victor Flambaum, Vladimir Dzuba, and I [33, 34] have been studying the distances between different absorption lines in the spectra of distant quasars and comparing them with line separations from identical atomic transitions in the laboratory. The observational program has completed detailed analyses of three separate quasar absorption line data sets taken at the Keck Telescope and finds persistent evidence that is consistent with the fine-structure constant, α, having been *smaller* in the past, at redshift z. The shift in the value of α for all the data sets is given provisionally by $\Delta\alpha/\alpha \equiv \{\alpha\{z\} - \alpha(0)\}/\alpha(0) = (-0.54 \pm 0.12) \times 10^{-5}$ for spectra originating in the redshift range $z = 0.2$–3.7, where $\alpha(0)$ is the value of the fine-structure constant here and now. This result is currently the subject of detailed analysis and reanalysis by the observers in order to search for possible systematic biases in the astrophysical environment or in the laboratory determinations of the spectral lines. This result is based on the analysis of about 950 spectra from 129 quasar systems in three separate samples, with both the observations and data analysis being done by three separate groups. A new VLT-UVES quasar sample of Mg II systems in the Southern sky has very recently been analyzed by Chand *et al.* [43], who report $\Delta\alpha/\alpha = (-0.06 \pm 0.06) \times 10^{-5}$ (assuming terrestrial isotope abundances[1]) from 18 quasars at $z = 0.4$–2.3 using the method of Webb *et al.*, who are also at present analyzing this high-quality dataset, but employing simpler data-analysis techniques. A significant difference between the results from the North and South sky subsamples has been recently noted in these studies with the Webb data, giving $\Delta\alpha/\alpha = (-0.66 \pm 0.12) \times 10^{-5}$ for the subsample of 96 quasars in the Northern sky and $\Delta\alpha/\alpha = (-0.36 \pm 0.19) \times 10^{-5}$ for the subsample of 32 quasars in the Southern sky [44].

---

[1] The Chand *et al.* data was reanalyzed to give $\Delta\alpha/\alpha = (-0.44 \pm 0.16) \times 10^{-5}$ in [51].

In order to carry out these astronomical studies of the constancy of α over ten billion years of light-travel time from the most distant quasars in the observational sample, Flambaum's group carried out detailed many-body computations of the energy levels of atomic species of relevance to the astronomical datasets. It was then possible to examine all the energy-level differences between pairs of lines in different multiplets in order to evaluate the sensitivity of these differences to very small changes in α. The power of this method is that the astronomical data are required to produce a simultaneous fit to many line separations with different contributions from two relativistic contributions to the energy levels. Although the energy differences between different relativistic lines can be systematically computed for a given value of α, it is not obvious ahead of computation what will be the effect on any particular energy level of a small change in the value of α. In fact, even the *sign* of the change in the energy level created by, say, a small increase in α is not obvious. Some line separations increase, whereas others decrease, for the same change in the value of α. This technique is therefore a powerful diagnostic of possible differences in the value of α at high redshift compared with its value here and now: about 50 times more powerful than direct laboratory determinations of the constancy of α by using atomic fountains. And so the reported variations have no consequences that are yet at an observable level in the laboratory. The interest of these studies for chemistry is that we have carried out a systematic examination of the sensitivity of different line separations to small changes in α, as shown in Figure 8.1.

Simulated spectra show how changing the value of the fine-structure constant affects the absorption of near-ultraviolet light by various atomic species. Each atom or ion has a unique pattern of lines. Changes in the fine-structure constant affect magnesium, silicon, and aluminum less than iron, zinc, nickel, and chromium. The consequences of different fractional changes (negative sign signifies a decrease) for the absorption wavelengths are shown.

Some heavy species such as Fe are fairly insensitive to changes in α, and so they can be used as "anchors" against which to measure the sensitivity of other lines by looking at the line separations between these species and Fe. In Figure 8.1, we see the effects of growing percentage changes in the value of the fine-structure constant on the energy levels of the species shown. As the relative shift in the value of α away from its usual value increases, we see the different responses of different atomic energy levels. The papers [33], [34] provide detailed tabulations of the changes to the energy levels in terms of a simple fitting formula that is quadratic in the value of α. This type of analysis allows us to identify the most sensitive diagnostics of variation in the value of α in the astronomical past when quasar absorption spectra were formed.

We note that investigations of the detailed consequences for atomic structure when constants of nature are varied by small amounts should start by evaluating

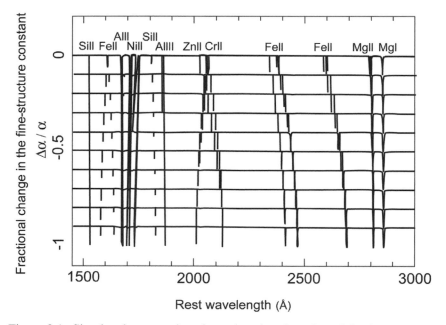

Figure 8.1. Simulated spectra show how changing the value of the fine-structure constant affects the absorption of near-ultra violet light by various atomic species. Each atom or ion has a unique pattern of lines. Changes in the fine-structure constant affect magnesium, silicon, and aluminum less than iron, zinc, nickel, and chromium. The consequences of different fraction changes (negative sign signifies a decrease) for the absorption wavelength are shown.

the effects due to shifts in relativistic fine structure. These give the most sensitive probes of any time variation in constants and also of their values in astronomically distant sites.

Sensitivity in networks

Chemists are extremely interested in networks of reactions that depend sensitively on local conditions of temperature and pressure. It is therefore of interest to see how a general analysis of such situations might be performed so as to identify those factors that play the leading role in sustaining a dynamical (which need not also be a thermodynamical) equilibrium. It is instructive to adapt an example from mathematical biology [45] that provides a very simple model example (for refinements and criticisms of its application to biological systems, see King and Pimm [46]). It can be generalized to other types of rate equation and different types of equilibria (for example, oscillatory limit-cycle attractors) as required.

Suppose that our chemical environment is controlled by n time-dependent variables x, and, for simplicity, assume that they satisfy a system of n linear ordinary differential equations

$$\dot{x} = Ax$$

where A is an $n \times n$ matrix. This system of equations has a solution $x = 0$ that we shall take to be the stable equilibrium solution for the situation under study. To make the stable equilibrium as simple as possible, we will assume that the eigenvalues of A are all equal to -1. Now, we are interested in the effects of small stochastic perturbations on this equilibrium. In particular, what does it take to destroy the equilibrium?

We add to A another matrix, S, whose elements are drawn from a normal probability distribution $N(0, \sigma^2)$ with zero mean and a variance denoted by σ^2. A fraction f of the elements of the matrix A will be populated in this way; the rest will be zero, so $0 \leq f \leq 1$. Thus, the parameter f is a measure of the complexity of the stochastic perturbations: as $f \to 1$, more and more of the n variables, x, are coupled together, whereas as $f \to 0$, the matrix S is very sparse, and different variables, x, are relatively uncoupled by the perturbations. Likewise, as n increases, so the system becomes larger, with more and more components of the vector x coming into play. Finally, as the variance of the stochastic perturbations, σ^2, increases, so the broader is the influence of random effects from one interaction pathway to others. Thus, the complex system is now turned into the stochastic differential equation

$$\dot{x} = [A + \{S\}]x$$

This system has a remarkably simple response to the stochastic perturbation. The equilibrium at $x = 0$ remains stable with probability 1 in the face of random perturbations so long as

$$n f \sigma^2 < 1$$

but if $n f \sigma^2 > 1$, it becomes unstable. The interpretation is simple. The instability condition requires the system to become very large ($n \to \infty$), very complex ($f \sim O(1)$), or highly connected ($\sigma^2 \to \infty$). In practice, living systems can avoid instability by evolving into a situation where the matrix A is effectively block diagonal, so that it behaves as though it is a smaller matrix with a far smaller value of n. Thus, in a complicated ecosystem, stability may be sustained because not all species interact with all others, only a small subset form a strongly interacting subset. More recently, a number of interesting mathematical studies have been made into the structure of connected networks (and the so-called small-world effect) following the famous discovery by Erdös and Rényi [47] that "almost any" random graph with more than

$n/2 \ln(n)$ links between pairs of its n points will be connected. These are likely to have implications for the stability of complex chemical or biochemical networks. It will also be important to specify carefully what is meant by stability. It may be enough for the effects of perturbations merely to be bounded rather than that they die to zero (i.e. "asymptotic stability" in mathematical parlance).

Sensitivity in chaotic systems

There has been much debate as to the sensitivity of the evolutionary process. Gould has argued that it is sensitive to small changes, so that if we replayed the tape of evolution again we would end up with a completely different result. Others, like Conway Morris [48], have argued that the same overall pattern of development would ensue, even though the details would vary. We can add just two things to this discussion. The first is that we must distinguish between "large" and "small" perturbations. Large perturbations that arise from outside the adaptive system, for example in the form of catastrophic impacts by comets or asteroids, clearly do have the power to change the whole course of evolution – by eliminating all life in the most extreme cases – and cannot be regarded as small perturbations. Second, it is important to understand more fully the nature of chaotic processes. Everyone is now familiar with the notion of a system that displays the propensities of chaos. The slightest perturbation will rapidly be amplified and produce significant effects. Hence it is tempting to draw Gould's conclusion that a chaotically unpredictable branching time series like the evolutionary process must end up giving a different outcome if its history is rerun with a slightly different starting state. However, the situation is more subtle than this. Consider the motions of molecules in a room. As Maxwell was the first to appreciate, these motions are chaotically unpredictable. Any small uncertainty in the trajectory of a given molecule will be exponentially amplified after a few collisions to become larger than the whole range of possibilities open to it. Yet, despite this unpredictability in the small, we have Boyle's Law in the large, linking the pressure, P, volume, V, and temperature, T, of the gas by the simple relation

$$PV/T = \text{constant}$$

Thus, although the individual molecular motions are chaotically unpredictable, their average behavior is entirely predictable and satisfies a particular probability distribution (the Maxwell–Boltzmann distribution). Quantities, such as the temperature that appear in Boyle's Law are measures of the average speed of the molecules. If we reran the "tape" of the history of our gas, we would find essentially the same average behavior, in accord with Boyle's Law, even though the individual trajectories of the molecules would be quite different.

This feature of probabilistic determinism, or well-defined average behavior, is a feature of most chaotic systems. It reveals why one should not conclude that the tape of evolution would have to come out differently if played again just because evolution is a chaotically unpredictable process in the mathematical sense.

The world is not enough

At the meeting that inspired this volume, it was clear that a sharp difference in "culture" exists between chemists and physicists. The latter were at home with the idea of "changing" the values of the constants of nature and evaluating the viability of the virtual realities so formed. They were used to the idea that other universes might exist, either in a Platonic sense or in the more mundane situation that constants are actually variables in our universe, and only in some universes or in some parts of our universe will life be possible. By contrast, chemists are generally loth to indulge in this type of perturbation of reality, and it is interesting to consider why this might be. Some of the reasons may be sociological, but other explanations may be consequences of the types of problem with which they are engaged.

Why are chemists generally reluctant to carry out analyses of the sensitivity of chemical and biochemical structures and pathways to small changes in the constants of nature? It is interesting that they will in general enter numerical values for constants at the outset of an analysis, and thus the sensitivity of a result to particular constants of nature will be hidden. By contrast, if constants are only evaluated numerically at the end of the calculation, their role is transparent. More important, I suspect, is the nature of the problems that chemists and biochemists address. In complex chemical networks, the environmental conditions of temperature and pressure can determine the chemical outcome, or in the situation of astrophysical chemistry, by the external cosmic ray flux. The effect of changing a constant of nature by a small amount can easily be totally dominated by the effects of small environmental changes. Also, in a complex network it is very hard to evaluate whether the effect of making a small change in a fundamental physical constant will destroy the possibility of a similar nearby equilibrium. Extremely non-linear systems are able to find new equilibria in a way that is not available to the situation, common in astrophysics, where a structure arises as a simple equilibrium between gravity and another force of nature. A classical series of examples are those displaying self-organized critical behavior, such as the sand-pile paradigm of Per Bak [49]. For such reason, along with the entirely sensible viewpoint that they have their hands full trying to understand this world without inventing others, chemists have, in the main, resisted the temptation to create virtual chemistries by changing dimensionless constants of nature. However, another type of study is available to them that gives definite information about the consequences of changing fundamental bonds and constants of nature. This is the process of element substitution. Thus, for example, the sensitivity of the properties of water (H_2O) can be studied by comparing it with the structure of heavy water (D_2O) formed by replacing hydrogen with its isotopes, deuterium or tritium, to make HDO, D_2O, and T_2O, or by changing the oxygen isotope from ^{16}O to ^{17}O or ^{18}O to form, say, $H_2{}^{17}O$, $H_2{}^{18}O$, or $D_2{}^{18}O$. The detailed properties of these hybrid forms of water can be found listed at the *Water Structure and Behavior* website maintained by Martin Chaplin [50]. Similar strategies can be observed in physics where muonic atoms (with muons substituted for electrons) or states of positronium show what happens if the proton-to-electron-mass ratio is changed in conventional atoms.

It is clear from these considerations that the complex problems faced in biochemistry require an approach different from the early-universe studies of cosmologists. The latter are used to clear-cut consequences of possible changes in the values of constants of nature or the dimensions of spacetime. The sciences of complexity, while open to investigations of this sort in many areas, also challenge us to understand the stability of complex networks and the couplings to their wider environments.

References

[1] L. Susskind. The anthropic landscape of string theory. hep-th/0302219.
[2] T. Banks, M. Dine and E. Gorbatov. Is there a string theory landscape? *Journal of High Energy Physics*, **08**, 058 (2004).
[3] A. Vilenkin. Unambiguous probabilities in an eternally inflating universe. *Physical Review Letters*, **81**, 5501 (1998); A. Linde. Quantum creation of an open inflationary universe. *Physical Review*, **D58**, 083514 (1998); J. Garriga, A. Linde and A. Vilenkin. Dark energy equation of state and anthropic selection. *Physical Review*, **D69**, 063521 (2004).
[4] J. D. Barrow. *The Constants of Nature*. London and New York: Random House (2002).
[5] H. J. Kreuzer, M. Gies, G. L. Malli *et al.* Has a possible change of the values of the physical constants a role in biological evolution? *Journal of Physics*, **A18**, 1571 (1985).
[6] J. D. Barrow and F. J. Tipler. *The Anthropic Cosmological Principle*. Oxford and New York: Oxford University Press (1986); see also J. D. Barrow. *The Artful Universe Expanded*. Oxford and New York: Oxford University Press (2005).
[7] P. Ward and D. Brownlee. *Rare Earth*. New York: Springer (1999).
[8] G. Gonzalez and J. W. Richards. *The Privileged Planet*. Washington, D.C. Regnery (2004).
[9] L. J. Henderson. *The Fitness of the Environment: An Inquiry into the Biological Significance of the Properties of Matter*. New York: Macmillan (1913); repr. Boston, MA: Beacon Press (1958); Gloucester, MA: Peter Smith (1970).
[10] L. J. Henderson. *The Order of Nature*. Cambridge, MA: Harvard University Press (1917).

[11] W. Prout. *Chemistry, Meteorology, and the Function of Digestion, Considered with Reference to Natural Theology*. Bridgewater Treatise VIII. London: The Royal Society (1834); 4 edns. by 1855.

[12] J. Cooke. *Religion and Chemistry*. New York: Scriven (1880). This work is discussed further, as is Darwin's theory of evolution by natural selection, by M. Valentine in *Natural Theology; or, Rational Theism*. Chicago: S. C. Griggs (1885). Cooke's book is quite rare but can be read online at http://etext.lib.virginia.edu/toc/modeng/public/CooReli.html.

[13] P. A. Chadbourne. *Lectures on Natural Theology*. New York: Putnam (1870).

[14] A. R. Wallace. *Man's Place in the Universe*. London: Chapman and Hall (1903).

[15] J. Needham. *Order and Life*. New Haven, CT: Yale University Press (1936).

[16] J. S. Haldane. *Nature*, **100**, 262 (1917).

[17] J. B. S. Haldane. Physical science and philosophy (contribution to debate). *Nature*, **139**, 1002 (1937); Radioactivity and the origin of life in Milne's cosmology. *Nature*, **158**, 555 (1944). See also Haldane's article (The origins of life) in the volume *New Biology 16*, ed. M. L. Johnson, M. Abercrombie and G. E. Fogg. London: Penguin (1955), p. 23.

[18] W. Whewell. *Astronomy and General Physics, Considered with Reference to Natural Theology*. Bridgewater Treatise III. London: The Royal Society (1833); 9 edns. by 1864.

[19] D. Hume. *Dialogues Concerning Natural Religion*, ed. N. Kemp Smith. Indiana: Bobbs Merrill (1977); first published in 1779, probably in Edinburgh.

[20] I. Kant. *Critique of Pure Reason*, ed. N. Kemp Smith. London: Macmillan (1968); first published in 1781.

[21] G. Smith, ed. *Josiah Royce's Seminar, 1913–1914: As Recorded in the Notebooks of Harry T. Costello*. New Brunswick, NJ: Rutgers University Press (1963).

[22] J. Parascandola. Organismic and holistic concepts in the thought of L. J. Henderson. *Journal of the History of Biology*, **4**, 64 (1971).

[23] W. Paley. *Natural Theology* (1802). In *The Works of William Paley*, ed. R. Lynam. Edinburgh: Baynes and Son (1825).

[24] R. Bentley. *A Confutation of Atheism from the Origin and Frame of the World*. London: Phoenix (1693).

[25] Newton's letters to Bentley can be found in I. B. Cohen. *Isaac Newton's Papers and Letters on Natural Philosophy and Related Documents*. Cambridge, MA: Harvard University Press (1958).

[26] R. H. Dicke. Principle of equivalence and weak interactions. *Reviews of Modern Physics*, **29**, 375 (1957); also, Dirac's cosmology and Mach's principle. *Nature*, **192**, 440 (1961).

[27] P. A. M. Dirac. The cosmological constants. *Nature*, **139**, 323 (1937).

[28] M. J. Rees. *Before the Beginning: Our Universe and Others*. New York: Simon and Schuster (1997).

[29] M. Tegmark. Is the "Theory of Everything" merely the Ultimate Ensemble Theory? *Annals of Physics (NY)*, **270**, 1 (1998).

[30] L. Smolin. *Three Roads to Quantum Gravity*. London: Weidenfeld (2000).

[31] For a range of views on the interpretations of quantum theory, see articles in J. D. Barrow, P. C. W. Davies and C. L. Harper, Jr., eds. *Science and Ultimate Reality: Quantum Theory, Cosmology and Complexity*. Cambridge, UK: Cambridge University Press (2004).

[32] J. D. Barrow. Time-varying *G*. *Monthly Notices of the Royal Astronomical Society*, **282**, 1397 (1996).

[33] M. Murphy, J. Webb, V. Flambaum *et al.* Possible evidence for a variable fine structure constant from QSO absorption lines – I. Motivations, analysis and results. *Monthly Notices of the Royal Astronomical Society*, **327**, 1208 (2001).

[34] J. K. Webb, V. V. Flambaum, C. W. Churchill *et al.* Evidence for time variation of the fine structure constant? *Physical Review Letters*, **82**, 884 (1999); J. K. Webb, M. T. Murphy, V. V. Flambaum *et al.* Further evidence for cosmological evolution of the fine structure constant. *Physical Review Letters*, **87**, 091301 (2001).

[35] B. Carter. Large number coincidences and the anthropic principle in cosmology. In *Confrontation of Cosmological Theories with Observational Data*, International Astronomical Union Symposium No. 63, ed. M. S. Longair. Dordrecht: Reidel (1974), pp. 291–8; also The anthropic principle and its implications for biological evolution. *Philosophical Transactions of the Royal Society of London*, A**310**, 347 (1983).

[36] B. J. Carr and M. Rees. The anthropic principle and the structure of the physical world. *Nature*, **278**, 611 (1979).

[37] P. C. W. Davies. *The Accidental Universe*. Cambridge, UK: Cambridge University Press (1982).

[38] C. J. Hogan. Why the universe is just so. *Reviews of Modern Physics*, **72**, 1149 (2000).

[39] J. Ellis and D. V. Nanopoulos. A refined estimate of the fine structure constant. *Nature*, **292**, 436 (1981).

[40] F. Hoyle, D. N. F. Dunbar, W. Wenzel *et al.* The 7.68 MeV state in C^{12}. *Physical Review*, **92**, 1095 (1953).

[41] H. Oberhummer, A. Csoto and H. Schlattl. Fine tuning the basic forces of nature through the triple alpha process in red giant stars. *Science*, **289**, 88 (2000).

[42] E. Lieb. The stability of matter. *Reviews of Modern Physics*, **48**, 553 (1976); J. Fröhlich, E. Lieb and M. Loss. Stability of Coulomb systems with magnetic fields. *Communications in Mathematical Physics*, **104**, 251 (1986); E. H. Lieb, M. Loss and J. P. Solovej. Stability of matter in magnetic fields. *Physical Review Letters*, **75**, 985 (1995).

[43] H. Chand, R. Srianand, P. Petitjean *et al.* Probing the cosmological variation of the fine structure constant: results based on VLT-UVES sample. *Astronomy and Astrophysics*, **417**, 853 (2004).

[44] C. Steinhardt. Constraints on field theoretical models for variation of the fine structure constant. *Physical Review*, D**71**, 043509 (2005).

[45] R. May. Will a large complex system be stable? *Nature*, **238**, 213 (1972); S. L. Pimm. The complexity and stability of ecosystems, *Nature,* **307**, 321 (1984).

[46] A. W. King and S. L. Pimm. Complexity, diversity, and stability: a reconciliation of theoretical and empirical results. *American Naturalist*, **122**, 145 (1983).

[47] P. Erdös and A. Rényi. On random graphs. *Publicationes Mathematicae*, Debrecen (Hungary), **6**, 290 (1959); for a discussion see N. Alon and J. H. Spencer. *The Probabilistic Method*. New York: Wiley (1992).

[48] S. Conway Morris. *Life's Solution: Inevitable Humans in a Lonely Universe*. Cambridge, UK: Cambridge University Press (2003).

[49] P. Bak. *How Nature Works*. Oxford: Oxford University Press (1997).

[50] M. Chaplin. *Water Structure and Behavior*. Available online at www.lsbu.ac.uk/water/data.html.

[51] M. T. Murphy, J. K. Webb and V. F. Flambaum. Revision of VLT/UVES constraints on a varying fine-structure constant. astro-ph/0612407.

9

Fitness of the cosmos for the origin and evolution of life: from biochemical fine-tuning to the Anthropic Principle

Julian Chela-Flores

The boundaries of philosophy, science, and theology

Some of the deeper questions that humans have raised are not always answerable within the boundaries of science. Instead, philosophers and theologians have approached such questions within their own domains of competence. One such example is provided by the question of purpose in evolution (see the discussion below). Indeed, the concept of purpose in a general sense may be understood as something that one sets before oneself as an object to be attained, an aim to be kept, a plan to be formulated. In attempting to give an answer to the question of "purpose in nature," we should discuss the main components of human knowledge in an integrated way, so as to ask the right questions in the right field of knowledge. This approach should encourage us to provide appropriate answers that are reasonable within philosophy, science, or theology. At this juncture, it may also be argued that the task of a scientist should be independent of those of the other areas of human culture (Russell, 1991, p. 13). On the other hand, it is surely useful to be aware that this view of the role of science that is "divorced" from both philosophy and natural theology can also be seen from a different point of view (Townes, 1995, p. 166): because science and religion are evolving and are similar in their search for truth, convergence of these independent searches for truth may occur in the future.

I wish to address some questions in philosophy and theology that are pertinent to the main subject of this chapter. In order not to go beyond the natural boundaries of either science or theology, I will discuss contemporary attempts to encompass Darwinian evolution in a natural theological context. To meet this objective, we need some definitions to make this chapter self-contained.

First, *teleology* may be considered in two separate ways: either it is a doctrine according to which everything in the cosmos has been designed with humans in

Fitness of the Cosmos for Life: Biochemistry and Fine-Tuning, ed. J. D. Barrow *et al.*
Published by Cambridge University Press. © Cambridge University Press 2007.

151

mind, or it can be interpreted as a theory of purposiveness in the cosmos (namely, phenomena that are to be explained in terms of its purpose, rather than by initial causes). This latter concept is intimately related to the Anthropic Principle, as employed either in physics or in biochemistry. Teleology is also related to various interpretations of fine-tuning in phenomena such as the nuclear reaction of beryllium atoms in the production of carbon, as originally pointed out by Fred Hoyle (Hoyle, 1975, pp. 401–2).

Second, the notion of *process* dominates the work of three philosophers of the last century: Henri-Louis Bergson (1859–1941), Samuel Alexander (1859–1938), and Alfred North Whitehead (1861–1947) (see Ayer, 1982, pp. 208–9). In passing, it is perhaps worth noting that process theology is based on the metaphysics of Whitehead, who rejected Divine Action in terms of causality, proposing that God acts persuasively in all events, but not necessarily in determining their character. Specifically, Whitehead pointed out the incoherence of belief in a lifeless universe.

Elsewhere in this book, Haught (Chapter 3) has emphasized God as the sole ground for the world's being (see also Haught, 1998). This approach to natural theology leads him to explain the world in terms of evolution, as understood within the Darwinian tradition (Russell, 1996). Russell focuses on features of process thought. This philosophical system is considered to be particularly helpful in the task of constructing an evolutionary theology that may throw some further light on Darwinism.

A far-reaching implication of the possibility of interpreting the evolutionary aspects of Darwinism within theology is that the evolutionary process begins at the molecular level of biochemistry. In fact, such "chemical" evolution is a time-honored discipline that has been studied extensively in the past, particularly during the last decade of the twentieth century (see, for example, Ponnamperuma and Chela-Flores, 1993; Chela-Flores *et al.*, 2001). In this chapter, I endeavor to show that fine-tuning in biochemistry is a well-defined problem. Its evolutionary aspects should, in principle, be able to be integrated into a framework of natural theology, for instance in approaches such as kenotic process theology.

Process philosophy, or "process thought," attempted to provide a common metaphysical basis for discussions of science and religion. Some criticisms have been raised in the past (Polkinghorne, 1996, p. 28): if physics is to be appropriate for process thought, this school of philosophy has to face an ongoing debate. In particular, continuity seems to be intrinsic to quantum mechanics (for example, the Schrödinger equation is a differential equation). For Polkinghorne, at least, the mathematics of process thought should be that of "difference" equations, instead of "differential" equations (with their implied underlying continuity).

On the other hand, we should keep in mind deeper issues that are currently under debate as quantum gravity aims to provide a coherent theory of spacetime.

Spacetime is a dynamic entity, and as such it would have quantum properties (Rovelli, 2000). Both current and future developments in theoretical physics have to investigate the concept of *discrete* excitations of space itself. Thus, process thought, as a philosophical system, cannot be ruled out at present, because of an unfinished debate in theoretical physics.

To complete our discussion of contemporary attempts to encompass Darwinism in a natural theological context, we ask whether there is evidence of purpose in the cosmos. If we allow the simultaneous approaches of philosophy, theology, and science, it seems possible to reconsider the question that has been raised in the past: Is there evidence of purpose in the cosmos? And, in particular, is there any evidence of purpose in biochemistry?

We can consider how the cosmos itself is well fitted for the origin and evolution of life. I will discuss how the combined approaches of philosophy, science, and ultimately natural theology can help us to begin to discuss the intelligibility of the universe in a rational way. Care is needed in addressing the right questions within their corresponding cultural domains.

First, we consider the origin and evolution of intelligent behavior in the cosmos, examining the case of life on earth. As already pointed out above, investigations of the evolution of life on earth (Darwinism) can be incorporated in natural theology (process theology), at least in principle. The subsequent arguments in this chapter, within the boundaries of science, should, I hope, be useful for their interpretation in terms of theological issues.

Biochemical fine-tuning and fitness of the cosmos for life

Our starting point for studying the fitness of the cosmos for the origin and evolution of life is Lawrence J. Henderson's influential *The Fitness of the Environment* (1913). As a graduate of and professor at Harvard University, Henderson's main interests ranged widely, and he became a physiologist, chemist, biologist, philosopher, and sociologist. He discussed the question of teleology in biochemistry to give some rationale to the question of the fitness of the environment for the evolution of life. For many chemical compounds, he discussed the difficulties that the evolution of life would have encountered had these compounds not been freely available in the environment. One obvious example used by Henderson was water, the search for which even today is a main objective of our explorations of the solar system, especially Mars, Europa, Titan, and Enceladus (the tiny Saturn moon).

Today, we need to search the roots of Henderson's biocentrism at the molecular level. In fact, fine-tuning in biochemistry is represented by the strength of the chemical bonds that makes the universal genetic code possible. Neither transcription nor translation of the messages encoded in RNA and DNA would be possible if the strength of the bonds had different values. Hence, life, as we understand it today, would not have arisen.

In this chapter, I will argue in favor of the fitness of the cosmos for the origin and evolution of life without touching on the question of teleology. Instead, I approach the subject by considering biological evolution in the universe, as well as the evolution of the structure of the cosmos itself. I will touch on the evolution of solar systems, interstellar matter, and finally various aspects of the cosmos – all in relation to the emergence of life. However, I hasten to point out that arguments based on science can nevertheless be a source of inspiration for reconsidering the bases of natural theology. I will argue that the fitness of the universe for the origin and evolution of life can be best understood not only through convergence in biochemistry, but also through a range of convergences based on observations of phenomena in the space sciences.

The Weak Anthropic Principle in cosmology and biochemistry

In cosmology or biology, we may inquire whether general mechanisms (for instance, natural selection and adaptive radiation), as well as special values of some physical constants, could be interpreted together in natural theology as indications of purpose. The example of "fine-tuning" in physics has led to a weaker and a stronger version of the Anthropic Principle, which is concerned with the question of the bases of life, particularly intelligent life, in the cosmos. I have no difficulty in accepting what has come to be known as the "Weak Anthropic Principle" in physics:

Change the laws (and constants of nature), and the universe that would emerge most likely would not be compatible with life.

Biochemistry offers a clearly analogous statement:

First omit the observed cosmic abundance of the biogenic elements that are favorable to life. Then omit the environments (earth-like planets or Europa-like satellites) that favor evolution and adaptive radiation. The consequence of omitting both factors is that life most likely would not arise.

However, difficulties certainly would arise, both in cosmology as well as in biology, if we allowed some degree of teleology to be brought into the argument. Here, of course, I am referring to formulation of the "Strong Anthropic Principle":

- The laws of nature and the physical constants were established so that human beings would arise in the universe.
- The distribution of earth-like environments and Europa-like satellites was laid out so that not only life, but at least in certain circumstances human beings, would also arise in the universe.

The general mechanisms of nature, according to the evidence that we can infer from the biota of earth, which by now is at least 3 billion years old, imply that the evolution of intelligent behavior seems inevitable. What is not evident is the inevitability of the emergence of human beings.

The intimately related concepts of the Anthropic Principle and fine-tuning in living systems (Carr and Rees, 2003) are topics that would be simpler to understand with knowledge of more than a single instance of emergence of life on earth. On the other hand, the West's religious traditions go back to Jewish theology: a sole omnipotent God created heaven and earth, and subsequently life on earth. This view of our origins has traditionally been referred to as the "First Genesis." With the emergence of astrobiology (Chela-Flores, 2001), we can start to explore the possibility of the occurrence of a "Second Genesis" – namely, whether the evolution of intelligent behavior is inevitable in an evolving cosmos, given the present laws of cosmology (general relativity) and the general mechanisms of biological evolution (natural selection and adaptive radiation). If we were to change these laws and mechanisms, the arguments supporting the inevitability of the evolution of intelligent behavior would not stand, and thus the evolution of intelligent beings would not necessarily take place. This aspect of evolution has clear implications for the Weak Anthropic Principle.

A first aspect of convergence: cultural convergence

In the search for answers that go beyond the boundaries of a given area of human culture, we should first consider whether the various approaches would ever converge. The concept of convergence enters our discussion in three different contexts. In this section, I consider convergence in different cultural areas (Townes, 1995, p. 166). The other two aspects that are discussed in the following two sections are (1) convergence at a cosmic level, a subject that is essentially based in the space sciences, and (2) convergence in biology, a topic that is central to understanding Darwinian evolution.

Both science and religion are concerned with the common understanding of life in the universe. Because they largely address the same questions, we would expect that both aspects of human culture should at some point converge. With subsequent progress in philosophy, science, and theology, convergence seems unavoidable, although human culture does not seem to show any evidence of convergence at present. The status of the relationship between these three disciplines – philosophy, science, and theology – has been discussed in the past (John Paul II, 1992). In this chapter, an integrated approach to the questions regarding the fitness of the universe for the origin and evolution of life aims to avoid a splintered culture.

A second aspect of convergence: convergence at the cosmic level

In this section, I will discuss five cases of convergence at the cosmic level. First, from organic chemistry we know that nuclear synthesis is relevant for the generation of the elements of the periodic table beyond hydrogen and helium and, eventually, for the first appearance of life in solar systems. The elements synthesized in stellar interiors are required for making the organic compounds that have been observed in the circumstellar, as well as the interstellar, medium in comets and other small bodies. The same biogenic elements are also needed for synthesis of the biomolecules of life. Moreover, the spontaneous generation of amino acids in the interstellar medium is suggested by general arguments based on biochemical experimentation: the study of amino acids in the room-temperature residue of an interstellar ice analog has yielded 16 amino acids, some of which are also found in meteorites (Muñoz Caro *et al.*, 2002; see also Bernstein *et al.*, 2002). These factors help us to understand the first steps in the eventual habitability of planets.

On the other hand, the concept of cosmic convergence has a second aspect that may be inferred from what we know about the small bodies, such as the Murchison meteorite. These bodies may even play a role in the origin of life: according to chemical analyses in this particular meteorite, we find basic molecules that are needed for the origin of life such as lipids, nucleotides, and more than 70 amino acids (Cronin and Chang, 1993). Most of the amino acids are not relevant to life on earth and may be unique to meteorites.

This demonstrates that those amino acids present in the Murchison meteorite, which also play the role of protein monomers, are indeed of extraterrestrial origin. In addition, chemical analysis has demonstrated the presence of a variety of amino acids in the Ivuna and Orgueil meteorites (Ehrenfreund *et al.*, 2001). If the presence of biomolecules on the early earth is due in part to the bombardment of interplanetary dust particles, comets, and meteorites, then the same phenomenon could be taking place in any other solar system.

Interstellar gas provides yet another illustration of the convergent phenomenon that occurs at a cosmic level. Indeed, solar systems, many of which are now known, originate from interstellar dust that is constituted mainly of the fundamental elements of life, such as C, N, O, S, P, and a few others. When a star explodes into a supernova, all the elements that have originated in its interior as a result of thermonuclear reactions are expelled, thus contributing to the interstellar dust. The star itself collapses under its own gravity, compressing its matter to a degenerate state; the laws of microscopic physics eventually stabilize its collapse into a white dwarf. Stellar evolution of stars more massive than the sun is far more interesting. After the star has burnt out its nuclear fuel, a catastrophic explosion follows in which an enormous amount of energy and matter is released. These supernova explosions are the source of enrichment of the chemical composition of the interstellar medium. This chemical phenomenon, in turn, provides new raw material for subsequent generations of star formation, which leads to the formation of planets. Late in their evolution, stars are still poor in some of the heavier biogenic elements (for instance, magnesium and phosphorus). Such elements are the product of nucleosynthesis triggered in the extreme physical conditions that occur in the supernova event itself. By this means, the newly synthesized elements are disseminated into interstellar space, becoming dust particles after a few generations of star births and deaths.

An additional case that argues in favor of convergence at a cosmic level is emerging from what we are beginning to learn about the origin of planetary systems around stars. Our solar system formed in the midst of a dense interstellar cloud of dust and gas, essentially a circumstellar disk around the early sun. Some evidence suggests that this event was triggered by the shock wave of a nearby supernova explosion more than five billion years ago. Indeed, some evidence indicates the presence of silicon carbide (carborundum, SiC) grains in the Murchison meteorite, a fact demonstrating that they are matter from a type II supernova (Hoppe *et al.*, 1997). We may now be observing an extrasolar circumstellar disk around a young, three-million-year-old, sun-like star in the constellation Monoceros (Kerr, 2002). Several earlier examples of circumstellar disks are known, including a significantly narrow one around an eight-million-year-old star. The narrowness of this disk suggests the presence of planets constraining the disk (Schneider *et al.*, 1999). The following additional information further supports the arguments in favor of universal mechanisms of convergence in the formation of solar systems. The matter of the original collapsing interstellar cloud does not coalesce into the star itself, but collapses into the spinning circumstellar disk, where planets are thought to be formed by a process of accretion. Some planetesimals collide and stay together because of the gravitational force. In addition, a variety of small bodies are formed in the disk, prominent among which are comets, asteroids, and meteorites, completing the components that make up a solar system.

Finally, the fifth example of what I have called "cosmic convergence" is provided by the convergent origin of hydrospheres and atmospheres. The earliest preserved geologic period (the lower Archaean) may be considered as representing the tail end of the "heavy bombardment period." During that time, various small bodies, including comets, collided frequently with the early precursors of the biomolecules that eventually ignited the evolutionary process on earth and in its oceans. In addition, comets may be the source of other volatile substances significant to the biosphere, as well as the biochemical elements that were precursors of the biomolecules. Collisions with comets, therefore, are thought to have played a significant role in the formation of the hydrosphere and atmosphere of habitable planets, including earth. The sources of comets are the Oört cloud and Kuiper belt. These two components of the outer solar system seem to be common in other solar systems. Hence, in this cosmic sense, we recognize evolutionary convergence.

A third aspect of convergence: the case for convergence in biology

The question of evolutionary convergence in the context of the life sciences has been discussed extensively (Conway Morris, 2003; Chapter 11, this volume; Chela-Flores, 2001, pp. 149–62; 2003; 2006; Akindahunsi and Chela-Flores, 2004). We are assuming that natural selection is the main driving force of evolution in the universe, a hypothesis made earlier elsewhere (Dawkins, 1983). For these reasons, it is relevant to question whether local environments that were favorable for the emergence of life on the early earth were at all unique, occurring exclusively in our own solar system. Another view on the universal validity of biology in the cosmos has been advanced in the context of the basic building blocks (Pace, 2001): it seems likely that the basic building blocks of life anywhere will be similar to our own. Amino acids are readily formed from simple organic compounds and occur in extraterrestrial bodies, such as meteorites. Functions that are suggested as being common to life elsewhere in the cosmos serve to capture adequate energy from physical and chemical processes to conduct the chemical transformations that are necessary for life: lithotropy, photosynthesis, and chemosynthesis. Other factors that argue in favor of the universality of biochemistry are physical (temperature, pressure, and volume) and genetic constraints (see below).

In general, we may say that features that become more, rather than less, similar through independent evolution will be called "convergent." In fact, convergence in biology is often associated with similarity of function, as in the evolution of wings in birds and bats. New World cacti and the African spurge family provide an example. Some other examples are the euphorbs, such as *Euphorbia stapfii*, and some members of the Madagascar Didieraceae *(Didiera madagascariensis)*. These plants are similar in appearance, being succulent, spiny, water-storing, and adapted to desert conditions (Tudge, 1991, p. 67; Nigel-Hepper, 1982, p. 81). However, they are classified in separate and distinct families, sharing characteristics that have evolved independently in response to similar environmental challenges. Hence, we may say that this is a typical case of convergence.

When we look at convergence at the biochemical level, we can further document the general question of evolution in the life sciences. Convergent evolution is manifest at the active sites of enzymes and in whole proteins, as well as in the genome itself, as the following examples show.

- The northern sea cod *(Boreogadus saida)* is an economically important marine fish of the family Gadidae found on both sides of the North Atlantic. The distantly related order Perciformes with its suborder Percoidei contains the sea basses, sunfishes, perches, and, more relevant to our interest, the notothenioid fishes from the Antarctic *(Dissotichus mawsoni)*. In spite of their distant relationship with cods, they have evolved the same type of antifreeze proteins, in which the amino acids threonine, alanine, and proline repeat (Chen *et al.*, 1997). These proteins are active in the fish's blood and avoid freezing by preventing the ice crystals from growing. The Antarctic fish protein arose over seven million years ago, whereas the Arctic cod first appeared about three million years ago (both species arose in different episodes of genetic shuffling).
- The blind cavefish *Astyanax fasciatus* are sensitive in two long-wavelength visual pigments. In humans, the long-wavelength green and red visual pigments diverged about thirty million years ago. The mammalian lineage diverged from fishes about four hundred million years ago, but a recent episode in evolution has provided fish multiple-wavelength-sensitive green and red pigments. Genetic analysis demonstrates that the red pigment in humans and fish evolved independently from the green pigment by a few identical amino acid substitutions (Yokoyama and Yokoyama, 1990), a clear case of evolutionary convergence at the molecular level.
- Convergence may also occur when the sequence and structure of molecules are very different, but the mechanisms by which they act are similar. Serine proteases have evolved independently in bacteria (e.g. subtilisin) and vertebrates (e.g. trypsin). Despite their very different sequences and three-dimensional structures, in each the same set of three amino acids form the active site. The catalytic triads are His57, Asp102, and Ser195 (trypsin) and Asp32, His64, and Ser221 (subtilisin) (Doolittle, 1994; A. Tramontano, personal communication).

Evolutionary convergence in biology has been best documented at the level of animals. The evolutionary biology of the Bivalvia, at the level of both zoology and paleontology, provides multiple examples of convergence and parallel evolution, a fact that makes the interpretation of their evolutionary history difficult (Harper *et al.*, 2000). Specific examples of convergence in mollusks have been pointed out in various families of the gastropods (camaenid, helminthoglyptid, and helcid snails). The shells of the camaenid snails from the Philippines and the helminthoglyptid snails from Central America resemble each other and also members of European helcid snails. These distant species, in spite of having quite different internal anatomies, have grown to resemble each other morphologically in response to their environment. In other words, in spite of considerable anatomical diversity, mollusks from these distant families have come to resemble one another in terms of their external calcareous shell (Tucker Abbott, 1989, pp. 7–8).

In addition, we should recall that among birds the Passeriformes (including swallows) may be confused with Apodiformes (including swifts), but are not related to them. Swallows and swifts provide a classical example of evolutionary convergence.

Although unrelated, swallows are generally similar to swifts in size, proportion, and aerial habits (Clench and Austin, 1983). Members of these two orders differ widely in anatomy, and their similarities are the result of convergent evolution on different stocks that have become adapted to the same lifestyles in similar ecosystems for both species.

Can convergent pathways of evolution in the cosmos be foreseen?

Above I have argued that fine-tuning in biochemistry is represented in molecular biological terms by the strength of chemical bonds that make the universal genetic code possible. The messages coded in RNA and DNA would not be possible if the strengths of the bonds had different values. Hence, life, as we understand it today, would not have arisen. Subsequent evolutionary stages beyond molecular evolution in biochemistry (i.e. beyond chemical evolution of the building blocks of life) will depend on certain factors that can be documented with further research in the geologic record of hydrothermal vent communities and with the exploration of the solar system. I will review some of them, beginning with the geologic record. Some evidence indicates that once life originates, provided sufficient (geologic) time is available, evolution will provide living organisms with the opportunity to occupy every conceivable environment. This notion further favors the hypothesis that once life appears at a microscopic level on a given planet or satellite, the eventual evolution of intelligent behavior is just a matter of time.

The inevitability of some of the earliest stages of the evolution of life on earth can be illustrated with careful analysis of the geologic record. For instance, Cambrian fauna, such as lamp-shells (inarticulate brachiopods) and primitive mollusks (Monoplacophora), were maintained during Silurian times by microorganisms that lived in hydrothermal vents (Little *et al.*, 1997). Many examples of such fossils have been retrieved from the Silurian Yaman Kasy sulfide deposit. This volcanogenic site is located in the Orenburg district (southern Urals, Russia). In modern vent communities, monoplacophorans have been recovered at the Mid-Atlantic Ridge (37° 50′ N), and brachiopods from mid-ocean ridges are also recorded. However, taxonomic analysis of Cenozoic fossils suggests that shelly vent taxa are not ancestors of modern vent mollusks or brachiopods (Little *et al.*, 1998). We may conclude that modern vent taxa support the hypothesis that the vent environment is not a refuge for evolution.

In fact, evidence exists that since the Paleozoic and through the Mesozoic era, taxonomic groups have moved in and out of vent ecosystems through time: no single taxon has been able to escape evolutionary pressures. Some independent support for deep-water extinction has also been presented (Jacobs and Lindberg, 1998). These findings rule out the possibility that deep-sea environments are refuges against evolutionary pressures. In other words, the evidence so far does not support the idea that there could be environments where ecosystems might escape biological evolution, even in the apparently unassailable depths of the oceans. This gives considerable support to the hypothesis that any microorganism, in whatever environment on earth or elsewhere, would be inexorably subject to evolutionary pressures.

As I have shown above, fossils from Silurian hydrothermal vent fauna demonstrate that species have become extinct in locations that at first sight seem to be far removed from the pressures of evolutionary forces. Given that there are no refuges against evolution, we can raise the question whether over geologic time it was inevitable that the most primitive cellular blueprint bloomed into full eukaryogenesis and beyond, along convergent evolutionary pathways, ultimately to organisms displaying intelligent behavior (Chela-Flores, 1998).

To investigate beyond the geologic record whether subsequent evolutionary stages lead to the evolution of intelligent behavior, we turn to the exploration of our solar system. In order to investigate, in the short term, whether the evolution of life is subject to convergence throughout the cosmos, we have at least two possibilities. First, we can directly test whether evolution of intelligent behavior has followed a convergent evolutionary pathway elsewhere in the universe by means of the Search for Extraterrestrial Intelligence (SETI) project (Ekers *et al.*, 2002). Unfortunately, no signal that could definitely be interpreted as originating from an advanced civilization has ever been detected.

A second alternative, although much more restricted in scope, is currently in progress. We can test for the possible existence of the lowest stages of the evolutionary pathway within the solar system, namely at the level of microorganisms. One approach is currently being carried out in terms of the search for life on Mars. Another approach, still within the solar system, is to search in due course for life on the Jovian satellite Europa (Chela-Flores, 2003, 2006). Even beyond our solar system, scientific research may help us to decide whether environments exist that fulfill conditions favorable to life's origin and evolution. This is due in part to the fact that we are aware of multiple examples of solar systems. In addition, we suppose that stable conditions persist in extrasolar planets. By stable conditions, it should be understood that the planet (or satellite), where life may evolve, is bound to a long-lived star. In other words, the time available for the origin and evolution of life should be sufficient to allow life itself to evolve, before the solar system of the host planet or satellite reaches the final stages of stellar evolution, such as at the red-giant and supernova phases.

It is also assumed that major collisions of large meteorites with a habitable world are infrequent after the solar system has passed through its early period of formation. Under such stable conditions, the gradual action of natural selection would be expected to be the dominant mechanism in evolution. Fortunately, the existence of stable earth-like planetary conditions is an empirical question for which we will be able to give partial answers in the foreseeable future. Reliable observational techniques are currently being provided to image Jupiter-like planets orbiting at several astronomical units from their corresponding stars. Hence, we may conclude that in the not-too-distant future we will be able to address the following question on the evolution of intelligent behavior: is the evolution of the cosmos "fine-tuned" for the inevitable emergence of intelligent behavior throughout the cosmos? The assumed universality of biological evolution suggests a positive answer to this question, provided that stable planetary conditions are maintained in a given planet, or satellite, over geologic time.

A third factor in favor of the inevitability of the evolution of intelligent behavior in the cosmos is natural selection, which seems to be powerful enough to shape terrestrial organisms to similar ends, independent of historical contingency. Likewise, in view of the assumed universality of biology, we would expect evolutionary processes to take place in the cosmos that are mechanistically similar to those that have driven the evolution of life on earth. I will discuss some examples that support this view. Before approaching the question of convergent evolution, however, we should first recall that the set of factors influencing the relative degree to which earth's biota has been shaped is still a debatable topic. According to the hypothesis of universal Darwinism, life on earth, and possibly elsewhere, may have been shaped either by contingency or by the gradual action of natural selection. It may be possible to document convincingly whether, independent of historical contingency, natural selection is powerful enough for organisms living in similar environments to be shaped to similar ends. For this reason, I highlight the following examples, which suggest that, to a certain extent and in certain conditions, natural selection may be stronger than chance.

- Black European fruit flies *(Drosophila subobscura)* were transported to California more than twenty years ago. This event has provided the possibility of testing the role of natural selection in two different continental environments. Pacific coast *D. subobscura* (from Santa Barbara to Vancouver) were compared in wing-length with European specimens (from Southern Spain to the middle of Denmark). After half a dozen generations living in similar conditions, the increase in wing length was almost identical (four percent). This is a compelling case in favor of the key role played by natural selection in evolution (Huey *et al.*, 2000).
- Anole lizards from some Caribbean islands (*Anolis* spp.) provide another example of evolutionary convergence. In Cuba, Hispaniola (shared by Haiti and the Dominican Republic), Jamaica, and Puerto Rico (the so-called Greater Antilles), the observed phenomenon suggests that in similar environments adaptive radiation can overcome historical contingencies to produce strikingly similar evolutionary outcomes. We could even say that *replicated adaptive radiation* has occurred in the various islands. In fact, it has been

shown that although many species were known to thrive on these islands, some groups of lizards from different islands living in similar environments also look similar (Losos *et al.*, 1998). Genetic analysis has shown that similar traits have evolved in distantly related species for coping with similar environments (such as living in treetops or on the ground): anoles that live on the ground have long, strong hind legs, whereas those living in treetops have large toe-pads and short legs. Repeated evolution of similar groups of species (both morphologically and ecologically) suggests that adaptation is responsible for the predictable evolutionary responses of the anole lizards of the Caribbean. Indeed, we can speak in this case of evolutionary history repeating itself (Vogel, 1998).

Finally, in order to decide whether the standard laws of physics and biology imply the evolution of intelligent behavior, it is instructive to appreciate the implications of the existence of several *constraints on chance*. These constraints are relevant to the question of whether life elsewhere might follow pathways analogous to the ones it has already followed in terrestrial evolution. Christian de Duve has enumerated various examples of constraints on chance (de Duve, 1995, pp. 296–7; 2002; Chapter 10, this volume).

- Not all genes are equally significant targets for evolution. The genes involved in significant evolutionary steps are few in number; these are the so-called regulatory genes. In these cases, mutations may be deleterious and consequently are not fixed.
- Once a given evolutionary change has been retained by natural selection, future changes are severely constrained; for example, once a multicellular body plan has been introduced, future changes are not totally random, as the viability of the organisms narrows down the possibilities. For instance, once the body plan of mammals has been adopted, mutations such as those that are observed in *Drosophila*, which exchange major parts of their body, are excluded. Such fruit-fly mutations are impossible in the more advanced, mammalian body plan.
- Not every genetic change retained by natural selection is equally decisive. Some may tend to increase biodiversity rather than contribute to a significant change in the course of evolution.

Implicit in Darwin's work is chance represented by the randomness of mutations in the genetic patrimony and their necessary filtering by natural selection. However, the novel point of view that astrobiology forces on us is to accept that randomness is built into the fabric of the living process. Yet, contingency, which is represented by the large number of possibilities for evolutionary pathways, is limited by a series of constraints.

Natural selection necessarily seeks solutions for the adaptation of evolving organisms to a relatively limited number of possible environments. From cosmochemistry we know that the elements used by the macromolecules of life are ubiquitous in the cosmos.

To sum up, a finite number of environments force a limited number of options on natural selection for the evolution of organisms. We expect convergent evolution to occur repeatedly, wherever life arises. Consequently, it makes sense to search for the analogs of the attributes that we have learned to recognize on earth, especially the evolution of intelligent behavior.

Conclusion

Data from the current fleet of space probes that are capable of searching for signs of life in the solar system suggest that extraterrestrial life could be identified sometime in the near future. Thus, if we can settle the question of the occurrence of a Second Genesis elsewhere in the universe, additional information would be available to discuss the question of fine-tuning in biochemistry. Earlier, I extended the meaning of convergence from biology to the space sciences in an effort to provide a solid scientific basis for the concept that the universe is fit for the origin and evolution of life.

Closely related to the issue of extending convergence from the life to the space sciences is the subject of the intelligibility of life in the universe, a significant topic that requires an explanation. In this chapter, two aspects of the intelligibility of life in the universe were discussed in scientific terms, namely, the origin and the evolution of intelligent behavior in the universe. The arguments were centered on whether evolution is dominated by contingency, or by the gradual action of natural selection. Random gene changes accumulating over time may imply that the course of evolution is generally unpredictable, but constraints on chance, as argued above, put some powerful bounds on the degree of uncertainty.

Other contingent factors are the extinction of species due to asteroid collisions or other calamities, although these uncertainties may affect only the evolution of single lineages. However, such questions are of lesser interest to the larger issues that are relevant either to natural theology, or to science. This is particularly true when the question of fine-tuning is raised in physics, or biochemistry – that is, whether the appearance of biological features such as vision, locomotion, nervous systems, brains, and intelligent behavior is inevitable, rather than the preservation of a given species. I have further argued throughout this chapter that contingency does not contradict a certain repetition of natural history. Evolution allows a certain degree of predictability of the eventual biological properties that are likely to evolve, mainly because of convergence in the life sciences, but especially, as we have stressed repeatedly, because of the useful new concept of convergence in the space sciences.

References

Akindahunsi, A. A. and Chela-Flores, J. (2004). On the question of convergent evolution in biochemistry. In *Life in the Universe*, ed. J. Seckbach, J. Chela-Flores, T. Owen, *et al.* Dordrecht: Kluwer Academic Publishers, pp. 135–8.

Ayer, A. J. (1982). *Philosophy in the Twentieth Century*. London: Unwin Paperbacks.

Bernstein, M. P., Dworkin, J. P., Sandford, S. A. *et al.* (2002). Racemic amino acids from the ultraviolet photolysis of interstellar ice analogues. *Nature*, **416**, 401–3.

Carr, B. J. and Rees, M. J. (2003). Fine-tuning in living systems. *International Journal of Astrobiology*, **2**(2), 1–8.

Chela-Flores, J. (1998). The phenomenon of the eukaryotic cell. In *Evolutionary and Molecular Biology: Scientific Perspectives on Divine Action*, ed. R. J. Russell, W. R. Stoeger and F. J. Ayala. Vatican City State/Berkeley, CA: Vatican Observatory and the Center for Theology and the Natural Sciences (CTNS), pp. 79–99.

Chela-Flores, J. (2001). *The New Science of Astrobiology: From Genesis of the Living Cell to Evolution of Intelligent Behavior in the Universe*. Dordrecht: Kluwer Academic Publishers.

Chela-Flores, J. (2003). Testing evolutionary convergence on Europa. *International Journal of Astrobiology*, **2**(4), 307–12.

Chela-Flores, J. (2006). The sulphur dilemma: are there biosignatures on Europa's icy and patchy surface? *International Journal of Astrobiology*, **5**(1), 17–22.

Chela-Flores, J., Owen, T. and Raulin, F., eds. (2001). *The First Steps of Life in the Universe*. Dordrecht: Kluwer Academic Publishers.

Chen, L., DeVries, A. L. and Cheng, C.-H. C. (1997). Convergent evolution of antifreeze glycoproteins in Antarctic notothenioid fish and Arctic cod. *Proceedings of the National Academy of Sciences, USA*, **94**, 3817–22.

Clench, M. H. and Austin, O. L. (1983). Apodiform. In *Macropaedia*. Chicago: *The Encyclopaedia Britannica*, vol. 1, pp. 1052–66.

Conway Morris, S. (2003). *Life's Solution: Inevitable Humans in a Lonely Universe*. Cambridge, UK: Cambridge University Press.

Cronin, J. R. and Chang, S. (1993). Organic matter in meteorites: molecular and isotopic analyses of the Murchison meteorite. In *The Chemistry of Life's Origins*, ed. J. M. Greenberg, C. X. Mendoza-Gomez and V. Pirronello. Dordrecht: Kluwer Academic Publishers, pp. 209–58.

Dawkins, R. (1983). Universal Darwinism. In *Evolution from Molecules to Men*, ed. D. S. Bendall. Cambridge, UK: Cambridge University Press, pp. 403–25.

de Duve, C. (1995). *Vital Dust: Life as a Cosmic Imperative*. New York: Basic Books (HarperCollins).

de Duve, C. (2002). *Life Evolving: Molecules, Mind and Meaning*. New York: Oxford University Press.

Doolittle, R. F. (1994). Convergent evolution: the need to be explicit. *Trends in Biochemical Science*, **19**, 15–18.

Ehrenfreund, P., Glavin, D. P., Botta, O. *et al.* (2001). Extraterrestrial amino acids in Orgueil and Ivuna: tracing the parent body of CI type carbonaceous chondrites. *Proceedings of the National Academy of Sciences, USA*, **98**, 2138–41.

Ekers, R. D., Kent Cullers, D., Billingham, J. *et al.*, eds. (2002). *SETI 2020*. Mountain View, CA: SETI Press.

Harper, E. M., Taylor, J. D. and Crame, J. A., eds. (2000). The evolutionary biology of the Bivalvia. *Geological Society Special Publication*, **177**, 1–494.

Haught, J. F. (1998). Darwin's gift to theology. In *Evolutionary and Molecular Biology: Scientific Perspectives on Divine Action*, ed. R. J. Russell, W. R. Stoeger and F. J. Ayala. Vatican City State/Berkeley, CA: Vatican Observatory and the Center for Theology and the Natural Sciences (CTNS), pp. 393–418.

Henderson, L. J. (1913). *The Fitness of the Environment: An Inquiry into the Biological Significance of the Properties of Matter*. New York: Macmillan. Repr. (1958) Boston, MA: Beacon Press; (1970) Gloucester, MA: Peter Smith.

Hoppe, P., Strebel, R., Eberhadt, P. *et al.* (1997). Type II supernova matter in a silicon carbide grain from the Murchison meteorite. *Science*, **272**, 1314–17.

Hoyle, F. (1975). *Astronomy and Cosmology: A Modern Course*. San Francisco: W. H. Freeman and Company.

Huey, R., Gilchrist, G., Carlson, M. *et al.* (2000). Rapid evolution of a geographic cline in size in an introduced fly. *Science*, **287**, 308–9.

Jacobs, D. K. and Lindberg, D. R. (1998). Oxygen and evolutionary patterns in the sea: onshore/offshore trends and recent recruitment of deep-sea faunas. *Proceedings of the National Academy of Sciences, USA*, **95**, 9396–401.

John Paul II (1992). Discorso di Giovanni Paolo II alla Pontificia Accademia delle Scienze, *L'Osservatore Romano*, 1 November, p.1.

Kerr, R. A. (2002). Winking star unveils planetary birthplace. *Science*, **296**, 2312–13.

Little, C. T. S., Herrington, R. J., Maslennikov, V. V. *et al.* (1997). Silurian hydrothermal-vent community from the southern Urals, Russia. *Nature*, **385**, 146–8.

Little, C. T. S., Herrington, R. J., Maslennikov, V. V. *et al.* (1998). The fossil record of hydrothermal vent communities. In *Modern Ocean Floor Processes and the Geologic Record*, ed. R. A. Mills and K. Harrison. London: Geological Society, pp. 259–70.

Losos, J. B., Jackman, T. R., Larson, A. *et al.* (1998). Contingency and determinism in replicated adaptive radiations of island lizards. *Science*, **279**, 2115–18.

Muñoz Caro, G. M., Meierhenrich, U. J., Schutte, W. A. *et al.* (2002). Amino acids from ultraviolet irradiation of interstellar ice analogues. *Nature*, **416**, 403–6.

Nigel-Hepper, F. (1982). *Kew: Gardens for Science and Pleasure*. London: Her Majesty's Stationary Office.

Pace, N. R. (2001). The universal nature of biochemistry. *Proceedings of the National Academy of Sciences, USA*, **98**, 805–8.

Polkinghorne, J. (1996). *Scientists as Theologians*. London: SPCK.

Ponnamperuma, C. and Chela-Flores, J., eds. (1993). *Chemical Evolution: Origin of Life*. Hampton, VA: A. Deepak Publishing.

Rovelli, C. (2000). The century of the incomplete revolution: searching for a general relativistic quantum field theory. *Journal of Mathematical Physics*, **41**, 3776–800.

Russell, B. (1991). *History of Western Philosophy and Its Connection with Political and Social Circumstances from the Earliest Times to the Present Day*. London: Routledge.

Russell, R. J. (1996). Introduction. In *Quantum Cosmology and the Laws of Nature: Scientific Perspectives on Divine Action*, 2nd edn., ed. R. J. Russell, N. C. Murphy and C. J. Isham. Vatican City State: Vatican Observatory, pp. 1–31.

Schneider, G., Smith, B. A., Becklin, E. E. *et al.* (1999). NICMOS imaging of the HR 4796A circumstellar disk. *Astrophysical Journal*, **513**, L1217–30.

Townes, C. H. (1995). *Making Waves*. Woodbury, NY: American Institute of Physics.

Tucker Abbott, R. (1989). *Compendium of Landshells*. Melbourne, FL: American Malacologists, Inc.

Tudge, C. (1991). *Global Ecology*. London: Natural History Museum Publications.

Vogel, G. (1998). For island lizards, history repeats itself. *Science*, **279**, 2043.

Yokoyama, R. and Yokoyama, S. (1990). Convergent evolution of the red- and green-like visual pigment genes in fish, *Astyanax fasciatus*, and humans. *Proceedings of the National Academy of Sciences, USA*, **87**, 9315–18.

Part III

The fitness of the terrestrial environment

10

How biofriendly is the universe?

Christian de Duve

Introduction

We live in a biofriendly world. Were it otherwise, we wouldn't be around. The question is, therefore, how biofriendly is it? Physicists have addressed this question and have come to the conclusion that if any of the fundamental physical constants were a little smaller or a little larger than they are, the universe would be very different from what it is and unable to produce or harbor living organisms. Not everyone, however, subscribes to the concept of "fine-tuning" embodied in the so-called Anthropic Principle, some preferring instead the notion of a "multiverse," in which our universe is only one in trillions of trillions, perhaps the only one that, by mere chance, happened to have the right combination of constants to enable it to serve as our birthplace and abode.

In contrast, biologists and other scientists interested in biology generally take the universe for granted and ask instead to what extent the manifestations of life, including humankind, fit within the existing physical and cosmic framework. Nothing could better illustrate the depth of their ignorance on this subject than the diversity of answers they have given, which cover virtually the whole array of possibilities. Many agree with the late Jacques Monod, who, in his best-seller *Chance and Necessity* (1971), expressed his skepticism in the oft-quoted sentence: "The Universe was not pregnant with life, nor the biosphere with man." In my *Vital Dust* (1995), I defended the diametrically opposed view, also shared by many scientists, that life and mind are "cosmic imperatives," likely to exist in many areas of the universe. The recent interest in astrobiology and the SETI (Search for Extra-Terrestrial Intelligence) project rest on this assumption. Some scientists make a distinction between life and mind, but once again in two diametrically opposed versions. In *Rare Earth* (2000), the geologist Peter Ward and the astronomer Donald Brownlee

Fitness of the Cosmos for Life: Biochemistry and Fine-Tuning, ed. J. D. Barrow *et al.*
Published by Cambridge University Press. © Cambridge University Press 2007.

have claimed that life is likely to develop very frequently, but that the probability of its giving rise to mind through evolution is exceedingly small, so that we may well be the only conscious, intelligent beings in the universe, even though simpler living organisms may abound. In contrast, the paleontologist Simon Conway Morris, one of the editors of this book, argues in *Life's Solution* (2003) that life may be very rare, perhaps unique, but that, once arisen, it was bound to give rise to human beings. Humans, he states, are "inevitable in a lonely universe." Both views make us exceptional, but for totally different reasons.

In this chapter, I propose to look at the problem through the eyes of a biochemist. I will first consider the main features that define life as we know it and then ask how such features could have arisen and what environmental conditions would have been necessary for them to appear. Finally, in trying to put these notions within a cosmic perspective, I hope to derive some assessment of the probability of life's origin, a sort of quantitative estimate of the universe's "biofriendliness."

The hallmarks of life

What Is Life? This question, the title of a famous book by Erwin Schrödinger (1944), has been answered in many different ways, most often stressing the ability of living organisms to maintain improbable structures with the help of outside energy, reproduce themselves, and undergo evolution. My own answer to the question is simple – many would call it simplistic, even tautological: *Life is all that is common to all known living organisms*. This definition, based on concrete facts that are beginning to be well understood, allows a number of key properties to be singled out as characteristic of life. I will briefly review these properties in the following pages. For additional information, the reader can refer to my previously published books (1984, 1991, 1995, 2002, 2005) and to standard biochemistry textbooks.

Water and minerals

All living organisms contain abundant water and a number of mineral constituents, mostly Na, K, Cl, Ca, and Mg, together with various trace elements, such as Fe, Cu, Mn, Zn, Ni, Co, Mo, Si, F, I, and Se, which, although present in very small amounts, often play functional roles of central importance.

Organic building blocks

The bulk of all living matter is constructed with little more than fifty different organic building blocks, mostly sugars, amino acids, nitrogenous bases, fatty acids, and a few more specialized substances, made with only six elements, represented, not very euphonically, by the acronym CHNOPS (carbon, hydrogen, nitrogen, oxygen, phosphorus, sulfur). These building blocks are joined together in different ways to form a number of metabolic intermediates, coenzymes, and other relatively small molecules that are found in many different organisms, sometimes in all. What accounts for the main differences between the great variety of life forms – whether microbes, plants, fungi, animals, or humans – is the ways in which the building blocks are assembled into macromolecules, such as polysaccharides, lipids, and, especially, nucleic acids and proteins.

Catalysis

Of the thousands of chemical reactions that make up metabolism, virtually none would occur without an appropriate catalyst, or enzyme. Even a reaction such as the combination of carbon dioxide with water to form carbonic acid, which has the rare distinction of taking place spontaneously at a significant rate, is catalyzed by an enzyme (carbonic anhydrase), which accelerates the reaction more than a million-fold.

Protein enzymes

Most enzymes are proteins, which owe their unique functional versatility to the number and diversity of their building blocks. The distinctive side chains that characterize the various proteinogenic amino acids include aliphatic and aromatic molecules, which either are unsubstituted and hydrophobic or bear one of a number of different hydrophilic groups exhibiting a great diversity of physical properties, thus allowing a wide variety of possible interactions. In proteins, the amino acids are linked into long polypeptide chains that usually contain several hundred amino acids.

With the exception of structural proteins, which often remain linear, the polypeptide chains of proteins usually fold into complex, three-dimensional conformations. In enzymes, these conformations bring together certain functional groups of amino acids that are distantly located in the chain, so as to delimit specific binding sites and catalytic centers, disposed in such a manner that reaction substrates are fished out from the surrounding medium, immobilized in a particular configuration, and caused to interact. In addition, many enzymes also bear separate sites, called allosteric, which bind regulatory substances that modify the enzyme's functional properties.

More than ninety percent of the chemical reactions catalyzed by enzymes are either group transfers (including hydrolyses, in which water serves as acceptor of the transferred group) or electron transfers. These two types of reaction, as will be seen, both play major roles in energy transactions. Other enzymatic reactions are mostly isomerizations or reversible molecular splittings.

Cofactors

Protein enzymes frequently operate with the help of metal ions and with that of organic cofactors, which either are covalently bound to the enzyme molecule (prosthetic groups) or act in free, soluble form (coenzymes). As is to be expected, these cofactors serve mostly as carriers in transfer reactions. Several are nucleotide derivatives or have a nucleotide-like structure, thus being related to nucleic acid components. Many contain a vitamin in their molecule.

Ribozymes

A few biological catalysts are RNA molecules. Known as ribozymes, these catalytic RNAs are involved in RNA splicing and other forms of RNA processing. A special ribozyme of central importance is the ribosomal peptidyl transferase, the catalyst in ribosomes responsible for peptidyl transfer in protein synthesis.

Energy

The central role of ATP

Throughout the living world, energy is derived from the environment and used to support life by mechanisms that almost invariably end up depending, directly or indirectly, on the hydrolysis of the terminal pyrophosphate bond of ATP, giving ADP and inorganic phosphate (P_i):

$$ATP + H_2O \longrightarrow ADP + P_i \tag{10.1}$$

This reaction, in which ATP is sometimes replaced by another NTP, which in turn is subsequently regenerated at the expense of ATP, is exergonic. It releases free energy (that is, energy convertible into work, henceforth to be referred to simply as energy) in an amount that, under physiological conditions, is on the order of 14 kcal, or 59 kJ, per gram molecule of ATP split. This value defines the size of the energy packages used in biological energy transactions, what may be seen as the standard bioenergetic currency unit. If ATP hydrolysis takes place freely, the energy released by the reaction is dissipated as heat. Living systems contain a number of transducers that convert this energy into some form of work, including the mechanical work performed in animal motility; the osmotic work of forcing substances into cells, or out of them, against a concentration gradient; the electric work dependent on the active transport of ions and the resulting membrane potentials; the emission of light in bioluminescence; and, especially, the many forms of chemical work involved in the synthesis of natural compounds.

Biosynthetic processes are as varied as their products, but share some key features. One property they have in common is that they rely, in most cases, on the linking together of small molecules with removal of water (dehydrating condensations). Thus, sugars join into polysaccharides; amino acids into proteins; fatty acids and alcohols into lipid esters; bases, pentoses, and phosphate into nucleic acids; and so on. In an aqueous medium, such reactions are always strongly endergonic. They are almost invariably supported by ATP, or by some other ATP-related NTP, by way of sequential group transfer mechanisms. In these mechanisms, one of the reactants accepts a piece of the ATP molecule – inorganic phosphate, AMP, or inorganic pyrophosphate – forming a complex from which the reactant thus activated is transferred to the other reactant. I have detailed these mechanisms elsewhere (1984, 1991, 2000, 2005).

Besides fueling biosynthetic processes, ATP hydrolysis also serves to spark many metabolic reactions. Thus, phosphorylation of substrates – sugars, for example – often initiates their metabolic transformation. In addition, a number of receptor-mediated, regulatory mechanisms depend on the splitting of ATP (or GTP).

Living cells can function only a few seconds on their content of ATP and other NTPs, which need to be continually regenerated at the expense of their split products for life to be maintained. Exceptionally, this regeneration may take place at the expense of stored, high-energy phosphate compounds, including the phosphagens of animal cells and, especially, a phosphate polymer, or polyphosphate, found in some bacterial cells. These reservoirs, however, can serve only as a buffer and need to be replenished with the help of ATP. Therefore, all the energy expenditures of a cell are ultimately dependent on the regeneration of ATP by energy-yielding metabolic processes. These processes consist almost exclusively of electron transfers.

Electron transfers

In these transactions, electrons – often combined with protons into hydrogen atoms – are transferred from a donor to an acceptor down a difference in energy level, which is measured by the difference between the redox potentials of the two reactants. A very large number of such reactions participate in metabolism. To be able to serve for ATP regeneration, electron transfers must satisfy two conditions: (1) they must release enough energy to power the reversal of Reaction 10.1; and (2) they must be coupled to the reversal of this reaction – that is, constrained in such a way that they can take place only if ADP and inorganic phosphate condense to ATP at the same time.

The energy condition requires the donor and acceptor to be separated by a difference in redox potential of at least $(600/n)$ mV, n being the number of electrons transferred. This value represents the minimum needed to provide 14 kcal, or 59 kJ, per gram molecule, which we have seen is the energy released by Reaction 10.1 and, therefore, the minimum amount of energy required for its reversal. There are essentially two ways in which this difference is created, depending on whether or not light energy is involved. When it is, an electron in the chlorophyll molecule is displaced to a higher energy level with the help of light and allowed to fall back to a lower energy level through a phosphorylating system, to be either similarly recycled (cyclic photophosphorylation) or used for biosynthetic reductions (see below). In all other cases, electrons are provided from the outside at an appropriately high energy level, most often by organic foodstuffs, as in all heterotrophic organisms, but sometimes also by mineral donors, as in chemotrophic prokaryotes. After powering ATP assembly, the electrons are taken up at a lower level of energy by an outside acceptor. Most often (oxidative phosphorylation), this acceptor is molecular oxygen, which combines with four electrons and four protons to give two molecules of water. In certain prokaryotes, the acceptor is a mineral oxidized substance, for example nitrate or sulfate. Exceptionally, no outside electron acceptor is required; the final acceptor arises metabolically. Such is the case in alcoholic or lactic fermentation. In addition, all organisms can, when deprived of their external energy supply (either light or electron donors), use their stored reserves, of fat or carbohydrate, for example, or even their own substance as a source of high-energy electrons.

Coupling mechanisms

Throughout the living world, the coupling between electron transfer and ATP assembly is mediated predominantly by electrochemical mechanisms dependent on protonmotive force. In these mechanisms, known as carrier-level phosphorylations, electrons travel along complex chains of membrane-embedded carriers organized in such a way that the electrons can move only if protons are translocated at the same time from one side of the membrane to the other, from which they can return only by way of a proton-dependent, ATP-assembling system. In some cases, this machinery may function in reverse, lifting electrons from a lower to a higher energy level with the help of ATP generated at some other site.

A second coupling mechanism, of minor quantitative but immense qualitative importance, involves soluble enzyme systems and intermediates. It concerns the oxidation of a few selected aldehydes and α-keto-acids, with NAD^+ serving as electron acceptor. The substrates are oxidized jointly with a thiol (R'-SH) in such a manner that a thioester of the acid is formed:

$$R-\overset{\overset{\displaystyle O}{\|}}{C}-(COO)H + NAD^+ + R'-SH \Leftrightarrow R-\overset{\overset{\displaystyle O}{\|}}{C}-S-R' + NADH + H^+(+CO_2)$$

$$(10.2)$$

This type of reaction is almost unique in that it couples the donation of electrons to the closure of a chemical bond, the thioester bond. This bond, which is energetically equivalent to the pyrophosphate bonds of ATP, can be used to power ATP assembly by sequential group transfer:

$$R{-}\overset{\overset{\displaystyle O}{\|}}{C}{-}S{-}R' + ADP + P_i \Leftrightarrow R{-}\overset{\overset{\displaystyle O}{\|}}{C}{-}OH + R'{-}SH + ATP \qquad (10.3)$$

Adding up the two reactions, one finds that the aldehyde is oxidized to the corresponding acid, or that the α-keto-acid is oxidatively decarboxylated to the corresponding acid and CO_2, while ATP is assembled, with the thiol playing a catalytic role. This thiol may be a cysteine residue in the enzyme protein; or an enzyme-linked phosphopantetheine molecule; or coenzyme A, a soluble coenzyme in which phosphopantetheine is linked to 3'-phospho-AMP; or lipoic acid, also known as thioctic acid, a molecule that can serve at the same time as electron carrier and as group carrier, a unique distinction.

Reactions of this kind are called substrate-level phosphorylations, as opposed to the protonmotive force-dependent carrier-level phosphorylations mentioned earlier. Substrate-level phosphorylations are involved in only a few processes; but these belong to the most central metabolic systems, the glycolytic chain and the Krebs cycle, making substrate-level phosphorylations both universal and indispensable. In addition, the reactions are readily reversible (especially with aldehydes as substrates). They can serve in the reduction of a carboxyl group ($-$COOH) to a carbonyl group ($=$CO) with the help of energy provided by the splitting of ATP. As we will see, this process is of key importance in biosynthetic reductions.

It should be further noted that thioesters, with coenzyme A as carrier, play a key role as donors of acyl groups in the synthesis of lipids and other esters from free acids and alcohols. In such cases, the acid is activated to the corresponding thioester by Reaction 10.3, running from right to left or, more frequently, by the analogous reaction in which ATP is split to AMP and inorganic pyrophosphate.

Biosynthetic reductions

All living organisms carry out some reductive reactions. The conversion of sugars to fatty acids and the formation of deoxyribose from ribose are important examples of such reactions. In heterotrophic organisms, these processes are supplied with electrons by foodstuffs and, if needed, with energy by ATP. The importance of biosynthetic reductions is much greater in autotrophic organisms because these organisms manufacture their substance with very simple, oxidized building blocks, mostly H_2O, CO_2, NO_3^-, and SO_4^{2-}. They thus need a considerable additional supply of electrons, as well as of energy, as environmental electrons are usually provided at a low energy level.

Chemotrophic autotrophs derive the required electrons from mineral donors, for example hydrogen sulfide (H_2S). They lift these electrons to the high energy level needed for the reductions with the help of ATP (or protonmotive force) generated by the downhill transfer of electrons from the same donors to an outside acceptor (most often oxygen). Primitive phototrophic bacteria likewise obtain electrons from mineral donors, but depend largely on light energy for lifting the electrons to a higher energy level. This process is catalyzed by a chlorophyll-dependent system known as photosystem I. The more advanced phototrophs, which include cyanobacteria, eukaryotic algae, and plants, have a second chlorophyll-dependent system, photosystem II, which withdraws electrons from water, releasing molecular oxygen, and lifts them to an intermediate energy level, with the help of light energy. From this level, the electrons are transferred to photosystem I for further energization and utilization for biosynthetic reductions. In their passage from photosystem II to photosystem I, the electrons suffer a partial energy fall coupled with ATP assembly (non-cyclic photophosphorylation).

Interestingly, in the majority of autotrophic organisms, whether chemotrophic or phototrophic, the top energy boost whereby electrons are lifted to the level from which they can serve to reduce carboxyl to carbonyl groups – mostly 3-phosphoglycerate to 3-phosphoglyceraldehyde, as in reverse glycolysis – is accomplished by a thioester-mediated mechanism (reversal of Reactions 10.3 and 10.2), confirming the central importance of this mechanism.

Information

Biological information circulates universally by means of three processes: replication of DNA, transcription of DNA into RNA, and translation of RNA into protein. In addition, viruses with an RNA genome possess an enzyme that catalyzes one of two processes that are not carried out by cells: replication of RNA and reverse transcription of RNA into DNA. These five processes are shown schematically in Figure 10.1.

These processes have two aspects, one involving energy, the other information. I will treat these two aspects separately.

Biosynthetic mechanisms

In essence, biological information-transfer processes all rely on syntheses and, as such, operate by the same kind of ATP-fueled mechanisms that carry out biosyntheses in general. The synthesis of nucleic acids, which underlies all forms of replication and transcription, is special in that a single transfer reaction completes the process. Mononucleotide groups (either dNMPs or NMPs) are transferred to the

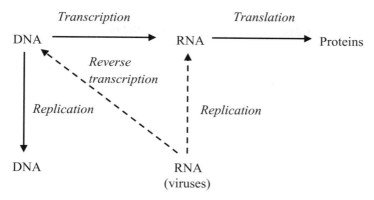

Figure 10.1. Biological information transfers.

3′ end of the growing polynucleotide chain from the corresponding dNTPs or NTPs, with release of inorganic pyrophosphate, which is hydrolyzed, thereby ensuring the irreversibility of the process.

Protein assembly involves reactants carried by transfer RNAs (tRNAs). The bond linking amino acids to tRNAs is created by a sequential group-transfer process in which ATP is split into AMP and inorganic pyrophosphate, which is further hydrolyzed, driving the reaction, as above. The two transfer steps of this process are catalyzed by a single enzyme, called aminoacyl-tRNA synthetase. The peptide bond itself is formed at the expense of the ATP-derived aminoacyl-tRNA bond by transfer of the growing peptide chain from its tRNA carrier to the amino group of a tRNA-borne amino acid, yielding a peptide chain lengthened by one amino-acid unit and attached to the tRNA bearing the new amino acid. This reaction takes place on the surface of ribosomes, which are small particles consisting of about fifty percent protein and fifty percent RNA (rRNAs). Interestingly, the peptidyl transferase that catalyzes the reaction is not a protein, but an rRNA. This is one of the rare known instances of involvement of a catalytic RNA, the only one in which the substrate is not an RNA molecule.

Templates

The biosynthetic mechanisms mentioned differ from all others in being guided by a template molecule that dictates which building block is to be added at each step. This template molecule is invariably a nucleic acid, and the mechanism by which it exerts its control relies universally on base pairing, the phenomenon whereby adenine joins specifically with thymine or with uracil, and guanine joins likewise with cytosine. These relationships are summarized by the following cardinal relationships:

$$A \Leftrightarrow T \text{ or } U \quad G \Leftrightarrow C \tag{10.4}$$

In the synthesis of nucleic acids, the template is a nucleic acid – of the same kind in replication, of opposite kind in transcription (see Figure 10.1, above) – which imposes, by base pairing, the nature of the mononucleotide to be added at each step. In protein assembly, the template is a messenger RNA molecule (mRNA), derived by transcription (eventually followed by splicing) from the DNA gene in which the sequence of the protein is encoded. The mRNA dictates, by successive base triplets, or codons, the choice of the amino acid to be added at each step. This phenomenon is mediated by base pairing between the mRNA codons and complementary base triplets, or anticodons, present in the amino-acid-bearing tRNAs. Translation itself is carried out earlier by the aminoacyl-tRNA synthetases, which ensure, by their specificities, that amino acids and tRNAs are correctly matched. Exactly twenty varieties of such enzymes exist, one for each proteinogenic amino acid.

Base pairing also has an important structural function. It is responsible for the joining of two complementary DNA chains (or, exceptionally, RNA, in certain viruses) into the celebrated double helix. It also causes single polynucleotide chains (mostly RNA) to bundle up into complex, three-dimensional arrangements. This property accounts for the many functions accomplished by RNA molecules, as opposed to DNA, which is essentially inert chemically and has as its sole functions to serve as a repository of genetic information and as a template for the transfer of this information in replication and transcription.

The genetic code

The genetic code is the set of correspondences between amino acids and codons. In it, 61 of the 64 codons represent amino acids, with up to six codons standing for the same amino acid; the remaining three codons serve as stop signals. With very rare exceptions (in some mitochondria, for example, and in a few microorganisms), which arose late in evolution, the genetic code is the same for all living organisms. Workers have long puzzled about the relationships that underlie the code, which have been attributed to anything between a "frozen accident" and a set of strict, chemical complementarities between amino acids and codons or anticodons. The latest answer is that the genetic code is most likely the product of selective optimization. The structure of the code is such that it minimizes the adverse consequences of mutations leading to the replacement of one base by another in a codon. As has long been pointed out, such changes often either do not alter the nature of the coded amino acid or replace it with an amino acid of similar physical properties so that the mutant protein remains functional. Modeling experiments have shown the code to be close to optimal in this respect (Freeland *et al.*, 2003).

Homochirality

The amino acids used for protein synthesis are all, with the exception of glycine, which is not optically active, chiral molecules of L configuration. Similarly, DNA and RNA are made exclusively with the D isomers of their constitutive pentoses, deoxyribose and ribose, which happen to be levorotatory, like the L-amino acids. This peculiarity has been the object of much discussion and speculation. It is viewed by many as one of the most mysterious hallmarks of life.

It should be noted first that the fact that only enantiomers of the *same* optical rotation sign are found in proteins and nucleic acids – the "homo" part of homochirality – may be viewed as a necessity, most probably a consequence of natural selection. Protein molecules containing both D- and L-amino acids, nucleic acid constructed with the two enantiomers of the pentoses, would lack some of their key biological properties. Life as we know it could not exist with heterochiral proteins or nucleic acids. The alleged mystery thus concerns the chirality part, the actual nature of the enantiomers used: L- rather than D-amino acids, and D- rather than L-pentoses.

Much has been written on this topic. Relevant to the subject of this chapter is the fact that there seems to be no compelling cosmic reason – there may be a local one – why life should use one or the other enantiomer. The possibility of life forms similar to those we know, but functioning with proteins made of D-amino acids and with nucleic acids constructed with L-pentoses, is not ruled out in the present state of our knowledge. Whether D-amino acids could be used with D-pentoses, or the opposite, seems less likely if, as may well have been the case, steric interactions between the two kinds of molecule played an important role in the origin of life.

Membranes

Membranes are indispensable biological constituents. No living cell can exist without at least a peripheral membrane. In addition, many cells contain internal, membrane-bounded structures. Biological membranes consist universally of a lipid fabric within which a number of proteins are embedded.

The fabric of membranes

The basic fabric of all membranes is the lipid bilayer, a bimolecular layer consisting of amphiphilic molecules typically made of a three-carbon core or skeleton to which are attached, on one side, two long, hydrophobic chains made exclusively of carbon and hydrogen and, on the other, a hydrophilic head containing a variety of chemical substances that share the property of being electrically charged or polarized:

$$CH_3\diagdown\diagdown\diagdown\diagdown\diagdown\diagdown\diagdown\diagdown\diagdown\diagdown\diagdown\diagdown\diagdown$$

$$CH_3\diagdown\diagdown\diagdown\diagdown\diagdown\diagdown\diagdown\diagdown\diagdown\diagdown\diagdown\diagdown\diagdown$$

(10.5)

The three-carbon core of such molecules most often consists of glycerol, with the hydrophobic tails being made of long-chain fatty acids linked to glycerol by ester bonds – except in archaebacteria, where they are formed by isoprenoid alcohols linked to glycerol (in the chiral configuration opposite to that shown in Scheme 10.5 by ether bonds. Sphingolipids differ from this general scheme in that the long-chain aminoalcohol sphingosine provides both the core and the top hydrophobic chain, the second hydrophobic chain belonging to a long-chain fatty acid attached to sphingosine by an amide bond.

In phospholipids, which are the most important membrane lipids, the hydrophilic head consists of a negatively charged phosphate ester group, often esterified by some positively charged nitrogenous, hydroxylated substance, such as choline, ethanolamine, or serine (which also carries an extra negative group). In glycolipids, a variety of carbohydrate compounds, occasionally associated with a negatively charged acidic group (sialate or sulfate), make up the hydrophilic head.

When mixed with water under certain conditions, molecules with the above configuration spontaneously organize into closed lipid bilayers. In this structure, two monomolecular layers join by their hydrophobic faces, creating an inner, oily film, lined on both sides by the hydrophilic heads in contact with water inside and outside the vesicle:

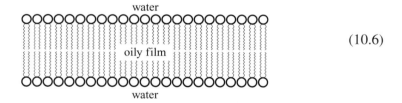

(10.6)

Biological membranes owe a number of distinctive properties to the physical characteristics of the lipid bilayers with which they are constructed. One such property is flexibility. Because their constituent molecules can readily slide along one another in semi-fluid fashion, the films can adopt a wide variety of shapes, thus allowing membranes to fold and mold themselves around almost any object.

Another important property is self-sealing. Lipid bilayers and the membranes arising from them invariably exist in the form of closed sacs, or vesicles, with a strong, inherent tendency to adopt this shape. This ability accounts for the spontaneous repair of pierced or injured membranes and for the various reorganization phenomena involved in the fission and fusion of vesicles that underlie many fundamental cellular processes, such as endocytosis, exocytosis, and vesicular transport.

In most membranes, the basic lipid bilayer fabric is enriched in various ways. The inner, fatty film of the bilayer often provides shelter for a number of hydrophobic molecules, for example cholesterol, which is an obligatory constituent of the outer membrane of all eukaryotic cells. Most importantly, all biological membranes contain a variety of specialized proteins of crucial importance.

Membrane proteins

Proteins are able to associate with lipid bilayers because several of the amino acids used for protein synthesis have hydrophobic groups. When a large enough number of such amino acids are present close together in a given stretch of the protein chain, they form a short, rod-shaped segment sufficiently hydrophobic to be accommodated by the bilayer. If such a segment ends the chain, it serves as anchor, with the bulk of the molecule hanging from one or the other face of the bilayer. If a hydrophobic segment is situated inside a protein chain, it can form a transmembrane bridge, with the two parts of the molecule separated by the segment protruding on opposite faces of the bilayer. Finally, if, as is often the case, other such segments follow one another along the protein chain, the molecule will snake in and out across the bilayer, with both ends coming out on the same face or on opposite faces of the bilayer, depending on whether the number of hydrophobic segments is even or odd.

Membrane proteins have multiple functions. They comprise enzymes acting on locally concentrated substrates; transport systems of various kinds, including the ionic pumps responsible for bioelectric manifestations; translocators serving in the specific delivery of proteins and other macromolecules across membranes; electron-transfer chains and other components of protonmotive machineries; photosystems and their ancillary cofactors; and a variety of receptors involved in transmembrane signaling.

Cells

Life cannot exist without cells. The minimal cell contains, within a surrounding membrane, a DNA genome, a complete machinery for replicating this genome and expressing it by way of RNA and protein molecules, and a set of enzymes and cofactors capable of building the cell's constituents from surrounding materials and of deriving energy from the outside in a form suitable for the support of all these activities.

Present in the simplest prokaryotes, these attributes are complemented by a variety of additional properties in higher prokaryotes and, especially, in eukaryotic cells, which are distinguished by a fenced-off nucleus and by complex cytomembrane structures, cytoskeletal and motor elements, and endosymbiont-derived organelles.

A key feature of cells is the ability to multiply by division. Initiated by DNA replication – and held in check as long as this process has not taken place – cell division involves a number of steps, different in prokaryotic and eukaryotic cells, but leading in both to the splitting of the surrounding membrane into two distinct, closed membranes, each enveloping a full copy of the genome and enough of all other cell components to endow each daughter cell with autonomy. Relatively unchecked in prokaryotes, which tend to multiply exponentially as long as enough usable matter and energy are available, cell division is subject to complex regulations in eukaryotic cells, especially those of multicellular organisms, in keeping with the individual requirements and mutual relationships of tissues and organs. Cancer, as is known, is a major harmful consequence of a defect in this control.

The requisites of life

Having reviewed the key features of life on earth, we are now ready to examine the universe's fitness for life. In addressing this question, I will limit myself to life as we know it, ignoring the frequently evoked possibility of other life forms, differing from extant ones by the nature of their organic constituents and/or by the mechanisms whereby they build their own substance, exploit environmental energy sources, or handle information – and perhaps even constructed from elements other than CHNOPS. Such open-minded conjectures, which satisfy the commendable principle of "leaving no stone unturned," have stimulated many valuable, theoretical reflections on the nature of life and the kind of chemistry needed to support it. In practice, however, these conjectures have drawn attention mostly to the unique fitness of present-day biochemistry as underpinning the main properties of life. I will not consider them further.

Regarding fitness, an important distinction must be made between fitness to harbor life and fitness to generate life. The former question is to some extent trivial. We know from experience that life, once it has taken hold, can adapt to almost any kind of milieu, from freezing polar waters to superheated volcanic jets; from swamps, lakes, and oceans to the deepest of rocks and the driest of deserts; from almost salt-free freshwaters to saturated brine; from pristine springs to the most polluted canals and other sites created by human technology; and across a wide pH range, from biting acids to caustic alkalis. Considering the diversity and, in some cases, simplicity of the environments in which certain forms of life manage to thrive, one is tempted to believe that the universe has countless spots where some kind of living organism would be able to survive.

Fitness to generate life, however, raises an entirely different question, which could be answered only if we knew how life actually originated. Although we don't know this, I will nevertheless attempt to address the question on the basis of what is known of the nature of life and what is suspected of its origin. My account will be brief, and I again refer the reader to my previous publications for additional information and bibliography (1991, 1995, 2002, 2005). In discussing these requisites of life, I will follow the same order as in the first part of this chapter.

Water and minerals

Liquid water is indispensable for life and must have been so for its origin as well. As far as is known, the universe contains plenty of water. The liquid state carries with it a more stringent condition in that its temperature range, although it varies with pressure, remains relatively narrow. Considering the number and diversity of planets, moons, and other celestial bodies that presumably exist in the universe, this condition can hardly be forbidding. It seems most likely that many sites *capable* of holding liquid water exist, in our galaxy and in others, and that at least a fraction of those actually *do* contain it. Even in our solar system, earth is probably not the only such site. Liquid water may lie under the permafrost believed to cover parts of Mars and under the ice plates detected on the surface of Jupiter's moon Europa.

As to the various minerals associated with life, they are likely to be present in sufficient amounts in many of the water-containing celestial bodies. It is significant, in this respect, that no element of atomic number higher than 34 is found in living organisms, even in trace amounts. Up to that limit, elements are all relatively abundant in many stars.

Organic building blocks

When, in 1953, Stanley Miller announced that he had observed the formation of amino acids and other biological molecules "under possible primitive Earth conditions," the scientific world (including the media) was thrilled by the possibility, evoked by Miller's historic experiments, that some of the building blocks of life *could* have arisen spontaneously on our planet. Surprisingly, nothing like the excitement generated by Miller's results or by those of his co-workers and followers greeted the much more astonishing discovery that the building blocks of life actually *do* arise spontaneously in many parts of the cosmos.

This is the startlingly significant message that extraterrestrial objects of various kinds – interstellar dust particles, neighboring planets and their moons, comets, and, especially, meteorites – have divulged to high-resolution spectroscopic explorations and to direct analyses using instruments borne by spacecraft or applied to materials that have fallen to earth. It is now known that a great variety of organic radicals and molecules exist in many extraterrestrial sites never visited by any living organism (for reviews, see Botta and Bada, 2002; Ehrenfreund *et al.*, 2002)

Because of the extremely rarefied and cold conditions under which it operates, cosmic chemistry is nothing like the kind of chemistry we know. But it obeys the same laws and, like organic chemistry on earth, exploits the unique associative properties of the carbon atom. Dominant among its products are small molecules and radicals consisting of only a few atoms, short aliphatic hydrocarbons believed, for example, to form the seas observed on the surface of the Saturn moon Titan, the ubiquitous polyaromatic hydrocarbons (PAHs), and fullerene-like polycarbon associations. In addition, cosmic chemistry seems indiscriminately to make, in amounts that decrease with molecular size, most of the combinations that can arise from the radicals. These combinations include amino acids, nitrogenous bases, sugars, and other moderately sized molecules that are found in living organisms, but with no obvious bias in favor of those that serve as biological building blocks over those that do not. Chirality could be an exception, revealed by slight excesses of L- over D-amino acids in some meteoritic material. This matter is, however, still the object of much debate. The bias, to the extent that it exists, is most likely secondary and due to asymmetric destruction rather than to synthesis.

Much is still expected from this fruitful research field. Particularly encouraging is the possibility of reproducing in the laboratory some of the events that take place in outer space. Two distinct groups, one American (Bernstein *et al.*, 2002) and the other European (Muñoz Caro *et al.*, 2002), have investigated the effects of ultraviolet irradiation, under about the lowest temperature and highest vacuum conditions attainable, on simulated interstellar ice analogs containing water, methanol, and ammonia as the main components with the addition of cyanide in the first research project and carbon monoxide and carbon dioxide in the other. Both groups have observed the formation of a variety of organic compounds, including a number of amino acids.

It is generally accepted that products of cosmic chemistry were showered on the nascent earth and could likewise fall on many other celestial bodies, brought down by comets, meteorites, and cosmic dust. But there is no agreement on the contribution of such products to the origin of life. Some researchers believe that the bulk of the building blocks of life came from outer space, others that they arose in the primitive terrestrial atmosphere by the kinds of process Miller and others have tried to reproduce in the laboratory. Most likely, both sources contributed, but in proportions that remain to be evaluated.

Faced with these facts, one is obviously tempted to assume that the products of cosmic/terrestrial chemistry did, in reality, serve as the building blocks from which life first arose. This view is not unanimously accepted. At least two prominent investigators, the German Günter Wächtershäuser (1998) and the American

Harold Morowitz (1999; Chapter 18, this volume), have vigorously defended the theory that life started from scratch in autotrophic fashion. Their proposed schemes differ, but have in common that they attribute a central role to a reverse, reductive Krebs cycle, from which the first amino acids would have arisen by amination of certain intermediates. A recent cladistic analysis has, however, led to the opposite conclusion that Krebs-cycle intermediates arose from amino acids by deamination (Cunchillos and Lecointre, 2002).

On the whole, it would be very surprising if cosmic chemistry had nothing to do with the development of biochemistry. Attributing the many similarities between the two chemistries to a meaningless coincidence strains the boundaries of credibility. If we accept the more plausible view, then the main message of all these discoveries, from the point of view of our discussions, is *that the chemical "seeds" of life are present throughout the cosmos.* This, as far as I know, is the strongest argument yet uncovered in favor of a "biofriendly" universe.

Catalysis

Chemical "seeds" are a necessary, but not a sufficient, condition for the development of life. For the seeds to "germinate," that is, for the building blocks to join into proteins, nucleic acids, and other key biological constituents, suitable catalysts are indispensable. What is true of present-day life no doubt also applies to emerging life, which, however, could not have had available the enzymes, ribozymes, and coenzymes that are at work today.

Many investigators have searched the mineral world for possible catalysts of biogenic reactions. Suggestive results have been obtained with metal ions, zinc for example (Lohrmann *et al.*, 1980), clays (Cairns-Smith, 1982; Ferris, 1998), double-layer metal hydroxide minerals (Pitsch *et al.*, 1995), pyrite (Wächtershäuser, 1998), and iron–sulfur complexes (Cammack, 1983; Wächtershäuser, 1998). To what extent such materials may have been crucial to the origin of life on earth is, however, not known. In addition, doubts have often been expressed that mineral catalysts could, alone, have sufficed to launch life.

Much attention has therefore been given to the suggestion that ribozymes may have played a key role in the origin of life in the so-called RNA world (Gilbert, 1986). There are, indeed, strong reasons that favor this possibility. Protein synthesis, for example, almost certainly was developed by interacting RNA molecules. But RNA itself can obviously not have catalyzed its own synthesis nor all the protometabolic reactions that came before it.

For a number of years, I have advocated the view that the first catalysts were short peptides and related compounds that I have called "multimers" (1991). Different from proteins, which they are supposed to have preceded, these substances could have contained D- as well as L-amino acids, some not found in proteins, together with other components, such as hydroxyacids, resembling in this respect a vast group of natural substances of bacterial and fungal origin. Known as non-ribosomal peptides and polyketides, these substances include a number of antibiotics and other biologically active molecules and have in common that they are synthesized by way of the thioesters of their precursor acids (Cane, 1997; Walsh, 2004).

As I have pointed out, substances of this kind would be most likely, because of chemical similarities, to mimic, be it only in rudimentary fashion, some of the catalytic properties of protein enzymes. Such properties have not been described in natural substances, but they could be looked for, both in natural and in artificial compounds (de Duve, 2003). Furthermore, the possibility that such substances could have arisen spontaneously in a prebiotic setting is plausible. Made of building blocks that are abundant among the primary products of cosmic chemistry, the posited multimers could have formed by fairly simple assembly mechanisms, which could even, like their natural counterparts, have involved thioester precursors. As discovered by Wieland even before the biological role of thioesters came to be known, peptides can form spontaneously from the thioesters of amino acids (see survey by Wieland, 1988). Several other investigators have described the synthesis of peptides under plausible prebiotic conditions (see, for example, Huber and Wächtershäuser, 1998). Particularly interesting are the experiments by a group of Japanese workers that have observed amino acid oligomerization in a hot–cold flow reactor designed to simulate a hydrothermal vent (Imai *et al.*, 1999; Ogata *et al.*, 2000; Yokoyama *et al.*, 2003). This point, together with the hypothetical involvement of thioesters, is of interest with respect to the possibility that life may have started in a volcanic setting (see below).

Substances resembling my multimers could even be formed by cosmic chemistry. There is evidence of the presence of peptides in meteorites, although apparently limited to the simple dipeptide glycyl–glycine (Shimoyama and Ogasawara, 2002). In the simulation experiments reported above (Bernstein *et al.*, 2002; Muñoz Caro *et al.*, 2002), free amino acids were obtained in substantial amounts in the products of UV-irradiated interstellar ice analogs only if the material had first been subjected to acid hydrolysis. The possibility that the amino acids may be present in the form of peptides is explicitly raised by the investigators. It is thus conceivable that peptides and analogous substances accompanied basic building blocks in the cometary showers that are suspected of having delivered chemical seeds of life to the prebiotic earth.

Energy

Any consideration of the energetic underpinning of emerging life must allow a crucial role to ATP and the other NTPs, especially since these compounds also serve to make RNA, most likely the first biological information-bearing substance (see below). It is interesting and possibly significant that inorganic pyrophosphate substitutes for ATP in a few metabolic reactions today. These reactions have been studied in particular detail by the Swedish couple Margaret and Herrick Baltscheffsky (1992), who believe them to go back to a "pyrophosphate world," a hypothetical early stage in the origin of life in which inorganic pyrophosphate was the first bearer of high-energy phosphate bonds. As mentioned above, another inorganic pyrophosphate derivative, polyphosphate, serves as a reservoir of such bonds in certain bacteria.

If inorganic pyrophosphates were essential to the development of life, this fact would put a severe constraint on the type of site where life could start, as pyrophosphates are very rare in the mineral world. Almost the only place where they might be expected to be present is a hot, volcanic setting (*pyr* means fire in Greek), where they could arise from inorganic phosphates and have indeed been detected in one instance (Yamagata *et al.*, 1991).

Whatever the role of inorganic pyrophosphate, an early development of the organic moieties of NTPs is mandated by the participation of these molecules in RNA synthesis. This problem is still largely unsolved, even though it has engaged a considerable amount of research. Ribose and the various purine and pyrimidine bases that make up natural nucleosides can all arise by relatively simple reactions. Some of these substances could even be among the products of cosmic chemistry, having been detected in trace amounts in meteoritic materials. What remains unknown, however, is how these molecules joined into nucleosides, were phosphorylated, and, especially, emerged to occupy the central position they hold in the scheme of life. It is possible that subsequent selection, at the RNA level, rather than specific chemistry, was responsible for this emergence (de Duve, 2002, 2005).

Thioesters are another group of substances that could have played a key role in primitive energy transactions. We have seen how these substances are uniquely involved in both group and electron transfers and in phosphorylation processes that are associated with some of the most primordial metabolic reactions and do not, unlike those dependent on protonmotive force, require complex membrane-embedded systems. We have also seen how my hypothetical multimers could have arisen from the thioesters of amino acids (see above). These facts and others have led to my proposal of a "thioester world" (1998).

Thioesters form by condensation between organic acids, which are typical products of cosmic chemistry, and thiols, which are organic derivatives of hydrogen sulfide (H_2S), a characteristic component of volcanic fumes. Here, therefore, is yet another hint of a volcanic cradle of life.

These considerations raise the question of the actual source of energy whereby nascent life was fueled. Volcanic heat, by way of pyrophosphates and, perhaps, thioesters, is a possible source, although not likely to have, by itself, supported the entire development of life. Light, which supports most of the biosphere today, has often been envisaged, but does not, in my opinion, qualify as a primitive source of energy. Its utilization depends on complex membrane-linked systems and, especially, requires the ability to continue functioning in periods of darkness. Today, unilluminated phototrophs subsist on their reserves and, if need be, on their own substance, by coupled electron-transfer mechanisms similar to those used by heterotrophs. Most likely, therefore, some primitive electron-transfer reactions were already involved at an early stage, perhaps connecting thioesters and pyrophosphates in a manner presaging today's substrate-level phosphorylations. If so, what were the participants in these early transactions and how do they relate to the nature of the relevant environment?

This question is unresolved. When Miller designed his celebrated experiments, he was much influenced by the theory of his mentor, Harold Urey, who believed the atmosphere of the early earth to have been strongly reducing. This theory has, however, since fallen into disfavor, throwing doubt on the validity of Miller's results, especially since similar experiments performed with a neutral atmosphere yielded negligible quantities of organic material. The controversy is not yet resolved, but it may become irrelevant if cosmic synthesis, rather than atmospheric formation, should turn out to be the main source of life's building blocks. In such an event, oxidative conditions, rather than reducing ones, would be called for to support the first electron-transfer reactions, with preformed building blocks serving as electron donors.

Although molecular oxygen was most likely absent on the prebiotic earth, there was probably no dearth of potential electron acceptors in the mineral world at the time. Sulfate and its reduction product, sulfite, which serve as electron acceptors for some bacteria, are possible acceptors. This is an attractive possibility because sulfite is a common component of volcanic fumes, and its reduction could have yielded hydrogen sulfide.

The matter must remain open, however, since, as mentioned above, some investigators do not believe that early life was fed by products of cosmic chemistry, but propose instead an autotrophic origin of life, thus reinstating the hypothesis of a reducing milieu. In the scheme devised by Wächtershäuser (1998), the necessary electrons are assumed to be supplied by sulfide ions (S^{2-}) condensing into disulfide (S_2^{2-}). Ferrous iron is taken to drive this reaction thermodynamically by precipitating the disulfide formed into the highly insoluble FeS_2 constituent of the mineral pyrite, which is also supposed to provide a catalytic surface for the reaction. Experiments have indeed shown that this process can support certain reductions. Interestingly, this "iron–sulfur world," which combines two major participants of biological electron transfers in a suggestive fashion, also presupposes a volcanic setting.

In recent years, the "hot-cradle" theory has been boosted by the discovery of deep-sea hydrothermal vents and of the many strange forms of life that inhabit those dark and seemingly inhospitable recesses. It has been argued against this theory that the fragility of many of the constituents of life precludes a hot environment and even mandates freezing temperatures (Miller and Bada, 1988). The fact remains that some organisms thrive at temperatures as high as 110 °C. Also, one would expect primitive forms of life to be more robust than their more sophisticated descendants. Certainly, the traces pointing to a volcanic birthplace apparently left in life's most fundamental processes cannot be lightly disregarded.

Information

Information most probably entered nascent life with the appearance of RNA. One need only look at Figure 10.1 to realize that DNA can be dispensed with provided that RNA can be replicated. In such a DNA-free "RNA world," RNA would have served both as the repository of genetic information and as the agent of the expression of this information, first by itself and later by way of its protein translation products. On the other hand, the key functions carried out by RNA molecules in protein synthesis indicate strongly that proteins – defined as special polypeptides made with a distinct set of twenty amino acids – are an "invention" of RNA. It has been proposed, because of the molecular complexity of RNA, that this substance may itself have been preceded by some simpler information-bearing molecule. This hypothesis, however, is unsupported by any evidence.

The problem of RNA formation is a *chemical* problem. This point is of cardinal importance. The first RNA molecules were the products, probably together with a number of similar compounds, of the early chemistry whereby life was launched; they were not "intended" to serve as information carriers. It was their unique base composition that, by lending itself to pairing, probably allowed their specific replication and amplification, which, in turn, caused them to emerge and to be molded by selection into the first bearers of genetic information and into the first catalysts of protein synthesis, as well as, perhaps, of other processes (de Duve, 2005).

In light of these considerations, it would seem that the main requirements for the appearance of RNA were the conditions, already briefly referred to, that allowed the formation of its NTP precursors, presumably in connection with primitive energy transfers. Two restrictions may be added. First, there is the problem of assembly, which probably required a special catalyst (possibly later supplemented by RNA molecules). Furthermore, there is the need for sufficient stability of the assembled molecules. RNA is a fragile substance, especially in an alkaline medium. Thus, a certain degree of acidity may be another quality to be added to our reconstruction of the cradle of life. This point is of interest as adaptation to media of increasing acidity may have played a role in the development of protonmotive force (de Duve, 1991).

Membranes

Biological membranes, as we have seen, are constructed on lipid bilayers. What were the first amphiphilic molecules capable of associating into bilayers, and how did they arise?

There are several possibilities. As found by the American investigator David Deamer (1998), small amounts of molecules capable of assembling into bilayers, probably medium-length fatty acids, can be extracted from meteorites, suggesting a possible cosmic origin of the first amphiphilic molecules.

Another possibility is that the first membranes arose from phospholipid-like molecules produced by primitive metabolism. Phosphate, with or without additional components, could have provided the hydrophilic heads of the molecules. As to their hydrophobic tails, the French chemist Guy Ourisson has built a strong case in favor of polyisoprenoid molecules, the building blocks of ether lipids, as opposed to long-chain fatty acids, the main constituents of ester lipids (see Ourisson and Nakatani, 1994; Chapter 19, this volume). In nature, isoprenoid chains arise from isopentenyl pyrophosphate, a relatively simple precursor.

Finally, the possible intervention of hydrophobic peptides or other multimers (de Duve, 1991) may deserve consideration, as they might have provided the pores needed for exchanges of matter across the membranes, to be replaced later by proteins.

Amphiphilic molecules with sufficiently bulky tails readily assemble into vesicular lipid bilayers. A whole industry has been built around the production of artificial vesicles of this kind, or liposomes, which are widely used in the cosmetic and drug industries. A condition for the formation of liposomes is vigorous mechanical agitation, which is generally provided by ultrasound. Thus, we may add agitation to our reconstruction of the cradle of life. Pressured jets, such as those that make up hydrothermal vents, could possibly create the necessary turbulence.

Cells

Whether life started with the formation of vesicular structures or became encapsulated at a later stage is still a matter of much debate. What seems likely is that some kind of protocells already existed by the time RNA appeared. This contention is based on the assumption, which looks reasonable, that the RNA molecules that launched the "RNA world" and developed protein synthesis must have emerged by Darwinian selection. A direct selection process, based on the stability and replicability of the molecules themselves, could conceivably have accounted for the early stages of this evolution. But soon, an indirect process, based on the usefulness of the RNA molecules or of their protein products, must perforce have been at work. Competing protocells capable of multiplication are an indispensable condition of such a selective process.

As seen in the preceding subsection, some kind of vesicular structure could have formed spontaneously by agitation as soon as the appropriate amphiphilic molecules were present. But vesicles are not necessarily cells. Two supplementary conditions, at least, had to be fulfilled. First, the membranes bounding the protocells must have had permeability properties such as to keep functional components inside while allowing incoming foodstuffs and exiting waste products to pass through. In addition, the protocells must have been capable of growth and division if they were to participate in a process of Darwinian selection.

How such conditions were fulfilled is not known. In terms of environment, they imply the existence of a milieu capable of accommodating a large number of protocells, of providing them with necessary nutrients, and of disposing of their waste. The calmer surroundings of hydrothermal vents could conceivably have provided such a milieu.

Discussion

In dealing with the question "How biofriendly is the universe?" I have intentionally restricted my analysis to life and the universe as we know them. Conjecturing on possible forms of life obeying a different kind of chemistry seems to me absurd in the absence of any sign that they exist or any hint of the kind of chemistry on which they could be based. Similarly, speculating on other possible universes as background for an appreciation of the "fine-tuning" and "anthropic" qualities of the one we inhabit is beyond my competence and must be left to physicists, who have given it abundant attention. Within these self-imposed limits, I have addressed the problem mainly in the light of biochemistry, which is now understood in considerable detail, and of what is known or suspected about the origin of life, a topic still in its infancy but subject to intensive research.

The chemical seeds of life

What is probably the most significant and revealing lesson emerging from the preceding survey comes from outer space: the chemistry of life is written into that of the cosmos. It has long been recognized that living organisms are built with the most abundant elements. Recent findings have now shown that this biofriendliness extends from atoms to molecules. Organic chemistry, far from being an exclusive prerogative of living organisms, as its name was intended to signify, is universal. Moreover, some of the products of cosmic chemistry are identical or closely similar to the most central building blocks of life. Chemical seeds of life arise spontaneously in outer space and on many celestial objects.

It is difficult to attribute this fact to a meaningless coincidence. This, however, is what one would have to believe if, as some researchers maintain, life developed in autotrophic fashion from carbon dioxide and very small building blocks, and not from preformed organic molecules (see above). The problem remains open because autotrophy obviously did develop at some stage by mechanisms that must be accounted for. On the other hand, the evidence is very strong that our young planet – or whatever site harbored the origin of earth life – must have been abundantly supplied with amino acids and other biogenic substances delivered by comets, meteorites, and cosmic dust. It seems highly likely that these substances participated significantly in the development of life, whatever the contribution of local syntheses.

The cradle of life

Seeds can germinate only in fertile soil. Thus, if, as appears likely, cosmic chemistry did play a role in the origin of life on earth, the key question facing us is: how frequent, or infrequent, in the universe are the conditions that allowed the products of cosmic chemistry to become fruitfully involved in the generation of living cells?

Our knowledge on this topic is still very scanty. All we have are a few indications that life may have originated in a volcanic milieu, perhaps one resembling present-day hydrothermal vents. What seems to be needed, therefore, according to this reconstruction of the cradle of life, is a rocky body containing a magmatic core, pockets of liquid water, and, in between, a fissured crust allowing water to seep through the cracks and resurge in the form of pressured, turbulent, overheated, acidic, sulfurous, metal-laden jets. If this reconstruction is correct, the question raised reduces to estimating the probability of celestial objects elsewhere in the universe that are likely to harbor hydrothermal vents physically and chemically similar to those found on earth.

This question is obviously unanswerable in the present state of our knowledge. In any case, it is not for the biologist to answer. All that can be said is that the present direction of astronomical research favors multiplicity rather than uniqueness. It is already known that planetary systems are far from rare. It is true that only large planets orbiting close to their sun have been detected so far, but this is because of technical limitations that may well be overcome by the advances of tomorrow. The existence of earth-like planets is not excluded; it is seen as likely by many experts.

The very real possibility remains, however, that our reconstruction of the cradle of life is incomplete and that some special condition not included in our assessment has to be fulfilled for life to arise. Perhaps the magnetic field must be just right. Or the planet's orbit must have the right degree of ellipticity, its rotation axis the right tilt. Or there may be a need for a moon of just the right mass circling at just the right distance. With enough rare conditions to be satisfied, planet earth could indeed turn out to be unique in the entire universe, a "cosmic fluke" (Conway Morris, 2003).

On the whole, sheer numbers would seem to argue against rarity. Considering the number of sunlike stars believed to exist in our galaxy (on the order of thirty billion) and the estimated number of galaxies in the universe (about one hundred billion), the odds seem to favor the existence of multiple earth-like planets capable of giving rise to life. Even some defenders of a "rare earth" share this opinion (Ward and Brownlee, 2000).

The probability of life

Sticking to our metaphor, what we have seen so far is that the chemical seeds of life are universally available and that the kind of soil in which the seeds germinated to give rise to present-day life probably exists in other sites in the universe. One question remains: how probable is it that germination will succeed when the seeds fall on the right soil? In other words, given the building blocks and the appropriate environmental conditions, what is the probability that viable cells will actually emerge?

In considering this question, elsewhere I have vigorously defended the view that life was virtually bound to arise under the conditions that prevailed at the site of its birth (1995, 2002, 2005). This opinion rests mainly on the fact that life must have arisen by chemical processes and that chemistry deals with highly deterministic, reproducible mechanisms. To be sure, contingency became added to chemical determinism when replicable molecules, most likely RNA, began to be made. This phenomenon introduced continuity into the course of events, but, also, the risk of its accidental loss by mutation. Many researchers have emphasized the chancy nature of such developments.

As I have pointed out on a number of occasions, the intervention of chance does not necessarily exclude inevitability. What counts is the number of opportunities available for the occurrence of a given event, as compared with the probability of that event. Many instances indicate that, when faced with a large number of distinct chance opportunities, nascent life enjoyed the freedom of exploring the range of possibilities extensively enough to approach selective optimization. Many of the "hallmarks" singled out in the first part of this chapter may have been the outcome of such a process (de Duve, 2005).

Admittedly, the possibility cannot be ruled out that some rare substance or some improbable chance occurrence played a decisive, non-reproducible role in the appearance of life. Scientific caution commands such reservation, but common sense argues against it. One doesn't readily see how a complex network of interconnected chemical reactions, such as those that must have initiated life, could have rested critically on a single, improbable chance factor. Such a caveat looks more like an *ad hoc* hypothesis than like a reasonable conjecture.

Nevertheless, the fact remains that attempts at reproducing key events leading to the development of life in the laboratory have met with little success so far. We still have no inkling, for example, of how such central compounds as ATP and the other NTPs may have arisen. Achieving the synthesis of these substances under plausible prebiotic conditions would go far toward accrediting the deterministic theory. Until this happens, the question stays open. In this connection, the indications gathered on the possible conditions under which life arose may be useful guides in the design of experiments. We have seen that this approach has already yielded some interesting results. Success could also depend critically on finding the right catalysts among possible mineral components of the prebiotic environment or, more likely, among the organic products of primitive chemistry (de Duve, 2003).

All these uncertainties would be dispelled if clear evidence of extraterrestrial life could be obtained. Even then, kinship with earth life would have to be ruled out. Discovering life on Mars, for example, might not be decisive in itself. It is not considered impossible that Martian life could originate from earth, or terrestrial life from Mars, or both from some third site in the solar system. The two forms of life would have to differ in a significant way, for example in the chirality of some key constituent, for their independent origin to be incontrovertibly established.

The concerted effort that is going on at present in the new discipline of astrobiology, sometimes also called exobiology or bioastronomy – a discipline with three names but no known object! – will perhaps lead to the discovery of extraterrestrial life some time in the future. If it does not, however, all one will be able to state is that life is not extremely frequent. Finding no sign of life in the minuscule part of the cosmos accessible to our explorations in no way can serve as proof of the rarity of life, let alone its uniqueness.

Conclusion

In conclusion, we live in a biofriendly universe; but just how biofriendly is not precisely known. Therefore, the only scientifically valid attitude with respect to the question is prudent agnosticism. If, however, some hypothesis is to be favored, as a guide for experimentation, for example, or as a justification for research funding, or, perhaps more importantly, as an inspiration for one's world view, available clues support the assumption that our universe is such that generation of life was obligatory, probably in many sites and at many times. Turning around Monod's famous saying (see above), the universe was, and presumably still is, "pregnant with life."

References

Baltscheffsky, M. and Baltscheffsky, H. (1992). Inorganic pyrophosphate and inorganic pyrophosphatases. In *Molecular Mechanisms in Bioenergetics*, ed. L. Ernster. Amsterdam: Elsevier, pp. 331–48.

Bernstein, M. P., Dworkin, J. P., Sandford, S. A. *et al.* (2002). Racemic amino acids from the ultraviolet photolysis of interstellar ice analogues. *Nature*, **416**, 401–3.

Botta, O. and Bada, J. L. (2002). Extraterrestrial organic compounds in meteorites. *Surveys in Geophysics*, **23**, 411–67.

Cairns-Smith, A. G. (1982). *Genetic Takeover and the Mineral Origins of Life*. Cambridge, UK: Cambridge University Press.

Cammack, R. (1983). Evolution and diversity in the iron-sulfur proteins. *Chemica Scripta*, **21**, 87–95.

Cane, D. E., ed. (1997). Polyketide and nonribosomal polypeptide biosynthesis. *Chemical Reviews*, **97**, 2463–706 (includes 13 papers on the topic).

Conway Morris, S. (2003). *Life's Solution: Inevitable Humans in a Lonely Universe*. Cambridge, UK: Cambridge University Press.

Cunchillos, C. and Lecointre, G. (2002). Early steps of metabolism evolution inferred by cladistic analysis of amino acid catabolic pathways. *Comptes Rendus Biologies*, **325**, 119–29.

Deamer, D. W. (1998). Membrane compartments in prebiotic evolution. In *The Molecular Origins of Life*, ed. A. Brack. Cambridge, UK: Cambridge University Press, pp. 189–205.

de Duve, C. (1984). *A Guided Tour of the Living Cell*. New York, NY: Scientific American Books.

de Duve, C. (1991). *Blueprint for a Cell*. Burlington, NC: Neil Patterson Publishers, Carolina Biological Supply Company.

de Duve, C. (1995). *Vital Dust: Life as a Cosmic Imperative*. New York, NY: Basic Books.

de Duve, C. (1998). Clues from present-day biology: the thioester world. In *The Molecular Origins of Life*, ed. A. Brack. Cambridge, UK: Cambridge University Press, pp. 219–36.

de Duve, C. (2000). *The Origin of Life: Energy.* Vol. 1, *Frontiers of Life*. San Diego, CA: Academic Press.

de Duve, C. (2002). *Life Evolving: Molecules, Mind, and Meaning*. New York, NY: Oxford University Press.

de Duve, C. (2003). A research proposal on the origin of life. *Origins of Life and Evolution of Biospheres* (formerly *Origins of Life and Evolution of the Biosphere*), **33**, 559–74.

de Duve, C. (2005). *Singularities: Landmarks on the Pathways of Life*. New York, NY: Cambridge University Press.

Ehrenfreund, P., Irvine, W., Becker, L. *et al.* (an International Space Science Institute Team) (2002). Astrophysical and astrochemical insights into the origin of life. *Reports on Progress in Physics*, **65**, 1427–87.

Ferris, J. P. (1998). Catalyzed RNA synthesis for the RNA world. In *The Molecular Origins of Life*, ed. A. Brack. Cambridge, UK: Cambridge University Press, pp. 255–68.

Freeland, S. J., Wu, T. and Keulmann, N. (2003). The case for an error minimizing standard genetic code. *Origins of Life and Evolution of Biospheres* (formerly *Origins of Life and Evolution of the Biosphere*), **33**, 457–77.

Gilbert, W. (1986). The RNA world. *Nature*, **319**, 618.

Huber, C. and Wächtershäuser, G. (1998). Peptides by activation of amino acids by CO on (Ni.Fe) surfaces: implications for the origin of life. *Science*, **281**, 670–2.

Imai, E., Honda, H., Hatori, K. *et al.* (1999). Elongation of oligopeptides in a simulated submarine hydrothermal system. *Science*, **283**, 831–3.

Lohrmann, L., Bridson, P. K. and Orgel, L. E. (1980). Efficient metal-ion catalyzed template-directed oligonucleotide synthesis. *Science*, **208**, 1464–65.

Miller, S. L. (1953). A production of amino acids under possible primitive earth conditions. *Science*, **117**, 528–9.

Miller, S. L. and Bada, J. L. (1988). Submarine hot springs and the origin of life. *Nature*, **334**, 609–11.

Monod, J. (1971). *Chance and Necessity*, transl. A. Wainhouse. New York, NY: Knopf.

Morowitz, H. J. (1999). A theory of biochemical organization, metabolic pathways, and evolution. *Complexity*, **4**(6), 39–53.

Muñoz Caro, G. M., Meierhenrich, U. J., Schutte, W. A. *et al.* (2002). Amino acids from ultraviolet irradiation of interstellar ice analogues. *Nature*, **416**, 403–6.

Ogata, Y., Imai, E., Honda, H. *et al.* (2000). Hydrothermal circulation of sea water through hot vents and contribution of interface chemistry to prebiotic synthesis. *Origins of Life and Evolution of Biospheres* (formerly *Origins of Life and Evolution of the Biosphere*), **30**, 527–37.

Ourisson, G. and Nakatani, T. (1994). The terpenoid theory of the origin of cellular life: the evolution of terpenoids to cholesterol. *Chemistry and Biology*, **1**, 11–23.

Pitsch, S., Eschenmoser, A., Gedulin, B. *et al.* (1995). Mineral induced formation of sugar phosphates. *Origins of Life and Evolution of Biospheres* (formerly *Origins of Life and Evolution of the Biosphere*), **25**, 297–334.

Schrödinger, E. (1944). *What Is Life?* Repr. Cambridge, UK: Cambridge University Press.

Shimoyama, A. and Ogasawara, R. (2002). Peptides and diketopiperazines in the Yamato-791198 and Murchison carbonaceous chondrites. *Origins of Life and Evolution of Biospheres* (formerly *Origins of Life and Evolution of the Biosphere*), **32**, 165–79.

Wächtershäuser, G. (1998). Origin of life in an iron-sulfur world. In *The Molecular Origins of Life*, ed. A. Brack. Cambridge, UK: Cambridge University Press, pp. 206–18.

Walsh, C. T. (2004). Polyketide and nonribosomal peptide antibiotics: modularity and versatility. *Science*, **303**, 1805–10.

Ward, P. D. and Brownlee, D. (2000). *Rare Earth*. New York, NY: Springer-Verlag.

Wieland, T. (1988). Sulfur in biomimetic peptide syntheses. In *The Roots of Modern Biochemistry: Fritz Lipmann's Squiggle and Its Consequences*, ed. H. Kleinkauf, H. von Döhren and L. Jaenicke. Berlin and New York: Walter de Gruyter, pp. 213–21.

Yamagata, Y., Watanabe, H., Saitoh, M. *et al.* (1991). Volcanic production of polyphosphates and its relevance to prebiotic evolution. *Nature*, **352**, 516–19.

Yokoyama, S., Koyama, A. Nemoto, A. *et al.* (2003). Amplification of diverse catalytic properties of evolving molecules in a simulated hydrothermal environment. *Origins of Life and Evolution of Biospheres* (formerly *Origins of Life and Evolution of the Biosphere*), **33**, 589–95.

11

Tuning into the frequencies of life: a roar of static or a precise signal?

Simon Conway Morris

Introduction

A glance at a bacterium, and a humpback whale, will reveal not only two immensely complex organisms, but also two very different life forms. Each is, in its respective way, constrained by a whole series of physical and chemical factors. One of the most obvious aspects is the fluid environment in which they live, albeit at scales that in being separated by about eight orders of magnitude are determinative of radically different behaviors. The bacterium's world is submillimetric, and accordingly it is dominated by constraints of viscosity. Motion involves the remarkable method of flagellar propulsion (the nearest thing to a wheel in biology; see, for example, Berg, 2003). When its flagella stop beating, the bacterium ceases its movement in a distance equivalent to the diameter of a hydrogen atom. The humpback whale, by contrast, occupies a liquid environment with which we are somewhat more familiar, although our swimming ability is feeble compared with the whale's oceanic travel range of thousands of kilometers. Fully aquatic, the humpback occupies a world that to us is both alien, with its complex system of echolocation, and familiar, with its ability to communicate – which includes singing.

Despite such wide divergences, the basic point of commonality is that both bacteria and whales live in environments where the physical controls imposed by the physico-chemical properties of water – be they viscosity or acoustic transmission – predetermine what is biologically possible. To this simple example could be added many other physical and chemical constraints. In the case of water – echoing Lawrence Henderson's (1913) prescient remarks on the way in which the physico-chemical glove matches the hand of life – one could list such factors as its power as a solvent, dielectric properties, transparency, latent heat of evaporation, and decrease of density when frozen. These properties are all central in various ways to life

Fitness of the Cosmos for Life: Biochemistry and Fine-Tuning, ed. J. D. Barrow *et al.*
Published by Cambridge University Press. © Cambridge University Press 2007.

itself, as well as more generally necessary to the habitability of the planet. Other examples, such as the strength of the carbon–carbon bond or the central role of phosphorus in biochemistry (Westheimer, 1987), will be familiar as properties that are essential to life. Putative alternatives are difficult to imagine.

Physicochemical factors in various ways necessarily predetermine what organisms can and cannot achieve, and they therefore delimit the habitable "box" of all biological possibilities. Biologists, of course, freely acknowledge that form and function are intimately connected to the prevailing physicochemical constraints of the real world. Yet typically they would exhibit restraint from making anything but the broadest predictions about how life might turn out on some other, earth-like planet – or indeed how the history of life might unfold if, by some miraculous intervention, the clock were reset to the time of the Cambrian "explosion" or earlier. In the common view, we might have flight, but not insects; maybe heterotrophs, but not fungi (let alone mushrooms); maybe animals, but not intelligent arborealists. Is this view of generic contingency warranted? If physico-chemical factors substantially constrain possibilities, might they not also constrain and predetermine evolutionary tendencies and trajectories?

General patterns?

The purpose of this review is two-fold. First, it is to question the prevailing belief (or, if the reader prefers, paradigm) that evolution has effectively no predictabilities, let alone destinations. This is an issue reviewed at some length in my book *Life's Solution* (Conway Morris, 2003). Here I particularly want to explore issues I dealt with there only cursorily, or not at all, not least because new information has since become available. A second purpose of direct relevance to the theme of this volume is to ask (if only in outline) what some of the basic and necessary physicochemical underpinnings of life might be. To reiterate Henderson's phrase, the question at hand is what physico-chemical factors might contribute to the fitness of life itself.

These two questions are interrelated. Ultimately we would like to be able to determine, on the basis of some common framework, the probability that sentient life forms would emerge that could formulate the very questions posed here. To begin to address this area, we must acknowledge that while the organization of life is hierarchical (protein chemists and evolutionary sociologists seldom attend one another's meetings, perhaps regrettably), we have little understanding of (but plenty of speculation about) how organic complexities emerge and what their key predeterminants are. So, too, as addressed later in this chapter, we need to ask whether a particular evolutionary pathway critically depends on a given innovation. This is not to say that a particular biological property (say ion channels, elastic proteins, or multicellularity) cannot be understood in an evolutionary context. Far from it. The point to note is that our descriptions are specific to the instance in question. We are not yet capable of addressing general principles of constraint and directionality within a coherent scientific framework.

Another largely unacknowledged tension within evolutionary biology is between historical perspectives and the desire to establish general principles, if not laws. The former aspect often remains implicit, not least in much of the biochemical and physiological literature. Evolutionary divergences (or, for that matter, convergences) by definition must follow historical pathways. Unsurprisingly, this historical dimension has encouraged interest in contingent events, most famously in the example of treating the Cretaceous–Tertiary (K/T) impact as a roll of the dice, setting evolution in a new direction. Such a focus of attention, arguing for a diversion or derailing of evolutionary trajectories, naturally undermines the search for general principles in biology. In this regard, the discipline remains notoriously refractory. Indeed, its lack of a law-like framework has led to the cliché of distinguishing physics (and chemistry) as "hard" and biology (and geology) as "soft." The search for general principles remains, therefore, a perennial challenge. In this context, important recent contributions toward the possibility of general "laws" include discussions of such topics as: (i) allometric scaling in organisms (see, for example, West *et al.*, 1999; Savage *et al.*, 2004; but see Kozlowski and Konarzewski, 2004); (ii) ontogenetic growth (West *et al.*, 2001); (iii) movement and locomotion (Marden and Allen, 2002; Taylor *et al.*, 2003; Hedenström, 2004; Linden and Turner, 2004); and, from a different perspective, (iv) latitudinal diversity gradients (Hillebrand, 2004).

Stumbling blocks in evolution

One indication of the potential difficulties in comprehending the fundamental dynamics of evolution is those instances where research areas seemingly have failed. Such conclusions may seem harsh and in no way are meant to denigrate or dismiss the enormous intellectual and scientific effort given to such topics. Nevertheless, the list is interesting, and among those that come most straightforwardly to mind are the origin(s) of life, artificial intelligence, and artificial life.

Each in its own way thus far has failed to match its initial sense of scientific promise. Notably, in the cases of the origin of life (see, for example, Dose, 1988) and in the quest to generate artificial intelligence (see Horgan, 1997), for decades the respective communities have been claiming to be on the edge of significant breakthroughs. Problems, however, have turned out to be far more difficult than initially expected. In the case of the origin of life, it is not clear what physico-chemical system now holds the greatest promise for further progress. Hydrothermal vent analogs (see, for example, Cody *et al.*, 2000; Lebrun *et al.*, 2003) seem potentially very interesting, not least because of their possible occurrence on early Mars and

present-day Europa. But at a basic scientific level of insight, it is far from clear why
any particular set of experiments, synthesizing one set of simple organic molecules
and defining one set of protometabolic pathways, should be any more likely to
lead to deep illumination compared with other sets of experiments that follow different hunches and hopes. Perhaps more progress has taken place in the case of
computational artificial intelligence, but ambitions toward the emergence of artificial machine sentience seem to be as remote as ever. To be sure, another *in silico*
venture, that of artificial life, where a computer program allows entities in a "virtual world" to "evolve" from a starting point of unremarkable conditions, certainly
seems an important and fruitful area of investigation (see, for example, Lenski
et al., 2003). However, I am struck by how uninteresting the end results tend to be
when compared with the concrete products of real evolution, constructed as they
are out of the full toolbox of nature and not just purely mathematical procedures
digitally instantiated.

   A simple reply would be that these model systems were never meant to represent
the real world. This point certainly has force in the case of artificial life, although
the question remains how successful the approach will be in helping to decide
whether the biological world we know is largely contingently accidental or, as I
argue *ab initio*, is more strongly predictable. The point, in summary, is that we have
continuing debates over the possible existence of general theoretical perspectives
such as might address questions about degrees of contingency and directionality, but
at present we do not yet have what might form a fundamental theory of evolution.

   In the remainder of this chapter, one of my purposes is to explore a series of
related points concerning the emergence of biologically complex forms. Note that
when discussing these points the underlying themes of "fine-tuning" and "fitness
of the environment" are not neglected in considering the emergence, continuation,
diversification, and irreversible increase of complexity of biological organizations
over protracted intervals of geological time. This view of life raises two additional questions: (1) Do ultimate limits to the process of evolutionary diversification
exist? and (2) Insofar as physicochemical fine-tuning and fitness of the environment are effectively universal properties (see also de Duve, 1995), to what degree
would the evolution of life on earth carry specific expectations and implications for
exobiology?

Chemical underpinnings

That life is chemically based is beyond dispute, and any attempt to revivify a
vitalistic program in terms of an *élan vital* is doomed to failure. Williams and Fraústo
da Silva (2003) have gone on to argue that thermodynamics and the rules of chemical
assembly impart a strong directionality to evolution. From the point of view of the

organism, they argue that the principal constraints revolve around the nature of intracellular reductive chemistry, the challenge of oxygen, and the advantages of cooperative interactions in the context of ecosystems. They further suggest that these constraints led to "changes [that] were in an inevitable progression, and were not just due to blind chance" (p. 323). This claim surely implies that even to a background of mutational or other Darwinian change there remained "a fixed overall route." In the context of "fitness of the environment," the Williams–Fraústo da Silva paper is, therefore, exceptionally important in its claim that the "overall activity of the [blind] watchmaker was constrained by the nature of changing chemicals and the thermodynamic equilibrium conditions of the environment" so that "[l]ife was in a physical tunnel and there was only one way to go" (p. 335). Hence, Williams and Fraústo da Silva suggest that not only was the emergence of eukaryotes inevitable, but so also was the evolution of plants and animals. Their paper does not, however, make claims for particular organismal specificities. Although it seems to be implicit in their analysis that humans are also a logical outcome of this process, nowhere do they state that such an evolutionary outcome is inevitable. Nor, I hasten to add, would I necessarily expect them to address this topic. To reiterate, their arguments are effectively chemical and thermodynamic.

What, then, might be the basis for a road map of evolution? I will start with the assumption that all life is carbaquist, thus doubly dependent on the properties of water (Henderson, 1913) and on the ability of carbon to form both flexible chains and chemical bonds. Liquid ethane as an alternative solvent and silicon-based life forms are among the alternatives, but appear to suffer significant drawbacks. So life, perhaps all life, originates as carbaquist; but for the time being, locations on any map involving the origin of life might as well be labeled "Here be dragons" (see also Conway Morris, 2003). My hunch is that the chemical pathway to life will turn out to be extraordinarily specific, involving the customary "happy co-incidences." At present, however, the diversity of experimental approaches is a clear indication that both the process and place of assembly remain highly speculative. Difficulties in understanding the origin of life are all the more puzzling because of evidence that the basic parameters presumed to be essential in the initiation of this process are probably universal. In part, the idea of such "universality" follows because of the ease of synthesis of many of the principal building blocks of life. Detection of amino acids, nucleic acids (or precursors), and hydrocarbons in carbonaceous meteorites also suggests that carbon chemistry probably is the molecular substrate on which all life sits. So far as the amino acids are concerned, one estimate is that approximately three-quarters of those found in terrestrial life would have extraterrestrial counterparts (Weber and Miller, 1981).

So too in his research program looking at the so-called etiology of DNA, the team led by Albert Eschenmoser (1999) shows that few alternatives match the effectiveness of DNA itself. Nevertheless, it is worth remembering three implications of this research program: (i) as Eschenmoser stresses, the principle of molecular choice revolves around optimization, rather than maximization (for example, of such properties as conformational flexibility); (ii) even if alternative DNAs match, or even exceed, the optimized properties of "real" DNA, this does not necessarily mean this advantage extends to the operational milieu of chromatin, DNA proteins such as the histones, chromosomes, and the cell; and (iii) although various alternatives to the sugar and nucleic acid building blocks are also feasible, in at least many cases their synthesis in any prebiotic situation seems improbable. This is not to say that DNA could not have been preceded by a viable and more primitive precursor, such as the TNA α-threofuranosyl oligonucleotide (Schöning *et al.*, 2000); yet here too questions of possible chemical pathways and ease of prebiotic synthesis cannot be ignored.

Universal biochemistry?

What of the next stage, the integration of basic biomolecules into a functioning biochemistry? Pace (2001) has made a strong argument for the universality of such a biochemistry. This view is relatively uncontroversial, and it echoes Wald's famous remark that candidates for biochemistry exams would be at equal advantage whether they sat for them here or many light-years away on a planet orbiting Arcturus. In a different vein, Denton *et al.* (2002) have argued that the basic protein folds are an axiomatic product of amino acid assembly, constrained by constructional rules that impart a law-like behavior reflecting innate and inevitable tendencies toward constructing particular types of biological structure. The extent to which such constraints apply to specific types, such as seven-helical, membrane-spanning proteins, or globins, which are often associated with particular functions (in these two cases, respectively, signal transduction and oxygen transport), remains to be seen. The evidence for functional convergences in these and other proteins (Conway Morris, 2003; Zakon, 2002; see also Beuth *et al.*, 2003; Charnock *et al.*, 2002; Cheng *et al.*, 2003; Dupuy *et al.*, 2002; Litvak and Selinger, 2003; Hamburger *et al.*, 1999; Boffelli *et al.*, 2004; Kryukov *et al.*, 2002; Johns and Somero, 2004) suggests that here, too, there may be important rules of assembly.

The extent to which biochemical systems are genuinely universal is, of course, an unresolved question. Many biologists, I suspect, would not be overly surprised to find close analogs to the citric acid cycle operating in extraterrestrial biospheres. On the other hand, most would be surprised, possibly astonished, if those similarities extended, for example, to signal transduction and respiratory proteins. So how do we proceed? Even the resolution of life on Mars is probably many years in the future, and the likelihood of visiting exoplanetary biospheres is still centuries away.

At the moment, the only realistic ways to test such suppositions seem to fall into two categories. The first involves the engineering of a biological system, either to reconstruct a putative primitive state or to define alternative organizational possibilities. Steven Benner (2000) notes that the former approach is one of considerable significance in that it opens the door to a possible rapprochement of the biological and physical sciences. In other words, by moving biology toward the establishment of long-sought-after general principles, paradoxically it might help to explain the extraordinary specificity and precision of many biological systems. Less often, as Benner also reminds us, is the gap between the efficacy of natural systems and the clumsiness of artificial imitations given its proper emphasis.

The second line of inquiry is to identify alternatives (or rivals) in order to explore degrees of similarity purely in terms of comparative operational terms and relative efficacy. The engineering of biological systems to explore alternatives is, unsurprisingly, driven by available genetic technologies. In a stimulating overview, Bennett (2003) discusses how biological novelty can be laboratory-generated with a view to testing the question of whether a diversity of adaptive pathways exists. He also explores the corollary of whether evolution has preferred directions and inevitabilities of outcome, concluding that outcomes may well be predictable, but need not be exclusive. As Bennett stresses, such experimentation may lead to the emergence of unanticipated features. His review is important because it helps to define a general research program to explore predictability in an adaptive context, although, as will become apparent (see also Conway Morris, 2003), it would be a mistake to assume that the natural world itself cannot supply the necessary information about "alternative adaptive solutions" (Bennett, 2003, p. 9) – that is, by examples of evolutionary convergence.

A related theme of inquiry is the attempt to infer the evolution of a biological system, especially in the context of proteins. For proteins, the ancestral states and historical pathways can be established on the basis of functionalities, typically dependent on amino acids at key sites and the adaptive shifts from precursor arrangements (see, for example, Malcolm *et al.*, 1990; Benner, 2002; Zhang and Rosenberg, 2002; Gaucher *et al.*, 2003). Such an approach is important not only because of its implications for the so-called neutral theory of protein evolution, but also because it opens the possibility of linking these changes to environmental circumstances. Such a line of enquiry also offers the chance to consider whether the evolved system is in any way optimized.

The second approach in determining the boundaries of life, and thereby the constraints that may point to universal principles, is to consider alternative solutions to a common problem. Examples from a number of molecular systems are available. Here I will address briefly the case of the respiratory proteins, which employ either iron (hemoglobin/myoglobin and hemerythrin/myoerythrin) or copper (hemocyanin). Although it is possible that all are ultimately derived from an ancestral metalloprotein (see, for example, Volbeda and Hol, 1989), the general consensus is that they represent distinct solutions to problems of providing respiratory oxygen. I review elsewhere (Conway Morris, 2003) evidence that hemoglobin and myoglobin are both convergent and, in the latter case, the noteworthy convergence with an enzyme (indoleamine 2,3-dioxygenase) in some mollusks (Conway Morris, 2003, pp. 288–9). So too the hemocyanins of mollusks and arthropods are most likely convergent solutions (Burmester, 2002; see also Immesberger and Burmester, 2004).

The case of hemerythrin (see, for example, Stenkamp, 1994; Kurtz, 1999) is of particular interest for several reasons. It is clearly distinct from the other hemorespiratory proteins and also occurs as analogous blood and muscle types (see, for example, Takagi and Cox, 1991). Intriguingly, its occurrence in animals (specifically the brachiopods, sipunculans, priapulids, and a number of annelids) makes no phylogenetic sense in terms of our current understanding of metazoan relationships. This is most obvious in terms of the placement of the priapulids in the ecdysozoans (i.e. related to arthropods), whereas the other three phyla are all lophotrochozoans. Even within this latter superclade, the distribution of hemerythrin does not seem to be phylogenetically informative.

Thus, even among the polychaete annelids, the occurrences of hemerythrin (see Rouse and Pleijel, 2001) do not appear to map in an evolutionarily coherent manner. In addition, many other annelids employ either hemoglobin or the closely related blood protein chlorocruorin (see Weber, 1978). Identifying the antecedents to hemerythrin is also problematic. The molecule evidently belongs to a larger class of proteins with non-heme carboxylated-bridge diiron sites (see, for example, Coufal *et al.*, 2000). Proteins comparable to hemerythrin are known in various bacteria (see, for example, Herrmann *et al.*, 1980; Beeumen *et al.*, 1991). Although convergence cannot be ruled out, it is possible that the sporadic occurrence of hemerythrins in the Metazoa is due to lateral gene transfer (see, for example, Beeumen *et al.*, 1991) in a way that is analogous to the cellulose gene of the tunicates (Matthysse *et al.*, 2004; Nakashima *et al.*, 2004).

This case study of hemerythrin raises a number of issues relevant to the arguments for (or against) universal biochemistries. First, with this molecule, can we rule out the possibility of molecular convergences (see Herrmann *et al.*, 1980)? One might note, for example, that even within the lophotrochozoans, specifically brachiopods and sipunculans, the sequence divergence of their hemerythrins is considerable (Joshi and Sullivan, 1973). Irrespective of origin, the hemerythrin in animals is most likely the product of lateral gene transfer; but is there only one source or several? Could, for example, different non-heme carboxylated-bridge diiron site proteins "donate" hemerythrin in separate circumstances? What are those circumstances? If, as seems likely, hemerythrin is polyphyletic, then why is its occurrence so rare? Given that hemoglobin and hemocyanin are much more widespread, should we judge, in the words of Stenkamp (1994, p. 724), hemerythrin as "an evolutionary development that failed to provide the best means of solving the reversible oxygen-binding problem"? Is hemerythrin one of life's "solutions," but effectively a curiosity? Could we go so far as to speculate that all biospheres will certainly have hemoglobin, very likely have a copper-based respiratory protein analogous to the hemocyanins, and also have much rarer respiratory proteins, which on occasion might be hemerythrin?

What hemerythrin does share with hemoglobin and hemocyanin, however, is its adaptation of dealing with dioxygen chemistry. In Kurtz's (1999) evocative words, this is "Nature's [way of] tiptoeing along the edges of the energetic barrier separating reversible O_2 binding from O—O bond cleavage without crossing it" (p. 97). Such "tiptoeing" is, of course, a striking example of evolutionary "navigation" (Conway Morris, 2003). Finally, as Kurtz notes, this triad of solutions begs the question of whether other oxygen-carrying proteins used for respiration still await discovery. If they do, we can now be confident that in certain critical ways they must conform to some general rules.

Bifurcation points and critical "choices"

Arguments, therefore, are made that at the molecular level strong constraints certainly exist, and perhaps we should even look to inevitabilities. In terms of their molecular architecture, alien biospheres will be similar – perhaps very similar – to the only one we now know. They will not be identical, and interesting alternatives must exist. Yet the example of hemerythrin and the questions of its origin(s), possible co-option(s), and relative efficacy (especially in comparison with the near-ubiquitous hemoglobin [Kurtz, 1999] suggest that we can begin to tackle the problem of how to define alternative trajectories in evolution. This perspective is clear even if we are (necessarily) limited to the one planet we happen to inhabit.

Let us then accept that any biosphere will have much the same molecular substrate. The objection could still be raised that this base level of biomolecular commonality has little bearing on the diversification of life. To return to hemerythrin, this protein occurs in animals as diverse as annelids, brachiopods, priapulids, and sipunculans; but (so far as we know) it has little, if any, bearing on the diversification of these three sorts of worm and a bivalved animal. Rather, one needs now to shift perspective to ask to what extent the nexus of evolution is controlled or governed by predeterminants, possible pathways, and incumbencies. Do these undermine, if not negate, any possibility of evolutionary inevitabilities? Recall that, from a historical perspective, evolution is faced with repeated "decisions" based on predetermined realities. To date, it is not clear to what extent a pre-existing substrate, say a particular protein family, is absolutely necessary for the emergence of a given complexity such as, to take an extreme example, intelligence. Nor is it clear whether evolutionary incumbency – the preoccupation of a metaphorical high ground – can in principle permanently frustrate the emergence of a particular complex system. These are unresolved questions, but the evidence (Conway Morris, 2003) seems very much to point in the other direction; that is, toward inevitabilities of outcome. In the case of the absolute necessity of a given molecular substrate, the emergence of highly complex systems (notably in animals in the form of social systems, tool use, cultural transmission, and vocalizations) from markedly different beginnings suggests that a given molecular substratum cannot be an absolute predeterminant.

Incumbency is a more difficult question. However, the clear evidence for the emergence of recurrent levels of complexity during geological time (see, for example, Knoll and Bambach, 2000) suggests that no pre-existing system can impose a complete impediment to further diversification and integration. That is evidently what happened on this planet, but it begs the question, of course, of whether radically alternative biospheres exist. All one can say is that if we combine the insights in the emergence of successive complexities (Knoll and Bambach, 2000) with the evidence from evolutionary convergence, then so far as earth-like planets are concerned it would seem that what we see around us on earth is a fair guide to what we would see anywhere else in the galaxy (Conway Morris, 2003).

Evolutionary inherency

Much of the preceding discussion hinges on the question of when in evolutionary history a particular molecule or developmental system appeared. These are generally difficult to date accurately; yet, despite this relative uncertainty, it is clear that a number of molecules and molecular mechanisms essential for the function of a complex system had evolved at a much earlier stage and had then been elaborated, often in distinct ways, in different lineages. Such is the case, for example, with glumate-based cell signaling (Dennison and Spalding, 2000). It also is clear in the instance of the repeat protein known as *armadillo* (Coates, 2003). On the basis of these and other examples, it seems reasonable to conclude that much of the complexity of animals and plants is evidently inherent at a microbial level. Self-evidently, animals possess intermediate-level structures defined by such properties as nervous systems, sensory organs, and contractile tissues. These structures, however, utilize such key molecules as acetylcholine (Wessler *et al.*, 1999), crystallins (see, for example, Piatigorsky, 1992), and myosin (Berg *et al.*, 2001), all molecules that had evolved hundreds of millions, if not billions, of years before animals arose.

Molecular convergence

In this section, I consider how the fitness of the environment not only applies to biochemistry, but, as shown by several examples, also has implications for the inevitable emergence of complex systems. The first topic concerns remarkable and emerging new evidence for various sorts of molecular convergence. At the outset, this would be expected to be surprising for the simple reason that, with an astronomically large number of alternatives, a random process should not generate coincidences of even short identical sequences. That is the principle. But, in fact, considerably more examples exist than might otherwise be imagined (see Conway Morris, 2003; Zakon, 2002; see also Beuth *et al.*, 2003; Charnock *et al.*, 2002; Cheng *et al.*, 2003; Dupuy *et al.*, 2002; Litvak and Selinger, 2003; Hamburger *et al.*, 1999; Boffelli *et al.*, 2004; Kryukov *et al.*, 2002; Johns and Somero, 2004). Such examples are less surprising when functional constraints are considered, such as those associated with active enzymatic sites (for recent examples see Beuth *et al.*, 2003; Johns and Somero, 2004).

A particularly interesting case concerns an example of convergent evolution in a virus, specifically the bacteriophage ϕX174 (Bull *et al.*, 1997). Admittedly, this example is somewhat artificial inasmuch as it is laboratory-based and involved in an experiment designed to study the evolutionary response to a change in a rather general environmental property, in this case an increase in temperature. Despite the constraints of the protocol, the results were important: the study of five lineages associated with two bacterial hosts (*Escherichia coli* and *Salmonella typhimurium*) showed striking convergence at a few key sites in the genome. The response to heat, therefore, was evidently selective. As Bull and colleagues also point out, despite the generality of the imposed environmental change, the experiments are strongly constrained both by the similar selective environment (specifically, an increase from 38 °C to 43.5 °C) and by the fact that the starting point of each viral lineage had a practically identical genome. On the other hand, it is important to note that, without the prior knowledge of the convergence, the true evolutionary tree of divergence would have been unrecoverable. Moreover, although Bull and colleagues stress the unusual nature of this case, they also note the possibility that molecular convergences may be more prevalent than has generally been expected. Given the widely acknowledged reality of extreme sensitivities of certain molecular sites (although remarkably unexplored [but see Axe, 2000]), such convergences are actually unsurprising. This is despite the average scientist's reaction – almost always precisely the opposite, with exclamations of astonishment and the use of such adjectives as "stunning" (Conway Morris, 2003, p. 128).

Evidence for molecular convergence is emerging in other quarters. Examples include both transcriptional machinery and developmental processes. In the context of the former, particularly important is the analysis by Conant and Wagner (2003) of gene circuits in *E. coli* and *Saccharomyces cerevisiae* (yeast). Of the four circuit types, for example those known as bi-fan and feed-forward in yeast (and two equivalents in *E. coli*), the evidence strongly points to the great majority having independent origins. Conant and Wagner remark:

Our results also suggest that convergent evolution . . . may have an important role in the higher organizational level of gene circuits. Stephen Jay Gould famously asked what would be conserved if life's tape, its evolutionary history, was replayed . . . Transcriptional regulation circuits, it seems, might come out just about the same.   *(p. 265)*

Although these examples of molecular convergence can always be put in a particular context, the emerging evidence for various sorts of convergence in developmental biology is attracting considerable attention, principally because of continuing discussions of the nature of biological homology and the roles of constraint. Put briefly, would we not expect the same structure to have the same genetic basis? And, if this basis is wanting, will the structure then never evolve? Possibly not. Evidence for possible convergence in developmental mechanisms falls into three broad categories. The first concerns the repeated and independent emergence of a particular phenotype, but on the basis of the same regulatory mechanism. Examples may be quite specific; for instance, the loss of trichomes in a larval fly (Sucena *et al.*, 2003) or melanism in birds (Hoekstra and Price, 2004; Mundy *et al.*, 2004). Alternatively, they may be more general, as in the evolution of lecithotropic, direct-developing sea urchin larvae (Wray, 2002; see also Nielsen *et al.*, 2003). The focus of interest is the reiteration of a phenotype, presumably in an adaptive context, and the unsurprising observation that a particular developmental pathway will be recruited "on demand." The homoplastic spanner this throws into the painfully literal cladistic machinery will be self-evident. Although these convergences involve examples from animals, a comparable result has also been documented in bacteria. This involved a study of directed evolution (for 20,000 generations) in twelve populations of *E. coli*. Rather remarkably, fifty-nine genes showed parallel changes in expression patterns – all in the same direction (Cooper *et al.*, 2003).

A second category is potentially more significant. In this case, a very similar phenotype emerges from a different developmental basis. Here, too, one can find well-documented cases in fly development (see, for example, Gompel and Carroll, 2003; Wittkopp *et al.*, 2003), as well as melanism in the cat family (Eizirik *et al.*, 2003). This category is arguably much more important because it is central to the identification of convergence in phenotypes. The evolutionary route is navigated in very different ways, but leads to the same "solution" (Conway Morris, 2003).

A third category is where complex structures are "built" by recruitment of similar genetic modules; although they are nominally the "same," they are patently not homologous. The almost identical construction of the insect wing and the vertebrate limb is a striking example (see Tabin *et al.*, 1999).

Closely linked to the question of convergence (and constraint) in developmental systems is the clear evidence for repeated co-option of developmental genes for new functions. So widespread is this phenomenon that major difficulties arise in determining the "primitive" function of a given gene. In this context, it also is difficult to decide what, if any, are the general rules of engagement: are certain genes recruited "come what may"? In specific cases, it does look as though the organism in question had "no choice." This is evident, for example, in the repeated recruitment of the *otx* gene to tube-foot development in direct-developing sea urchins (Nielsen *et al.*, 2003). Other possible examples of apparent redeployment are cases such as genes essential for the development of the eye being recruited for the expression of spermatocyte (Fabrizio *et al.*, 2003), muscle (Heanue *et al.*, 1999; see also Relaix and Buckingham, 1999), and salivary duct tissue (Jones *et al.*, 1998). These presumably are fortuitous; but, alternatively, deeper constraints may make the co-option of these genes (and their protein products) effectively inevitable. Too little is known at present to clarify this uncertainty.

Social microbes, intelligent plants?

The emergence of complex systems has another interesting dimension concerning the evolution of analogous systems. In one way, such similarities are hardly surprising given that they call on a common molecular repertoire. They remain important, however, because they give us some sense of the range of alternatives and a healthy reminder that, although animals may be the only group with the potential to understand creation, other players reinforce our sense of its richness, diversity, and even strangeness. Thus, Crespi (2001; see also Rainey and Rainey, 2003; Velicer and Yu, 2003) notes how it is that "all the hallmarks of a complex and coordinated social life" (p. 178) are identifiable in microbial communities, with analogs of cooperation, division of labor, communication, and sociality being identifiable. In fact, Crespi also identifies seven social phenomena that represent behavioral convergences with "higher" organisms.

In an even more ambitious vein, Trewavas (2003) explores how aspects of communication, computation, and intentionality in plants reflect not only their adaptive plasticity, but also a form of biological capacity not inaccurately described as "intelligence." At first sight, his analysis seems to be decidedly heterodox. Yet his fundamental point seems clear: that the many analogs of animal intelligence, such as stimulus response and memory, are predicated on a broadly similar chemical basis. The analogs are just that. Trewavas offers interesting speculations on future agendas for describing plant communication. He also notes a way in which the multiple networks offered by separate ramets allow the potential for a sort of multiple channel form of integrated information processing (see also Firn, 2004, and reply by Trewavas, 2004).

Higher-level convergences?

Evolution spans a series of levels. A typical order would read: molecules (or genes), cells, individuals (or ramets), societies, and communities. Questions of where the main drivers might lie in this hierarchy, and what interactions are between these levels, continue to be areas of active debate (see, for example, Gregory, 2004). In the context of convergence, it is generally agreed that even if societal structures (e.g. fission–fusion, monogamy) show convergence, at yet higher levels of community structure convergence generally is not observed. Such a conclusion is not surprising. Communities are seen largely as *ad hoc* associations. At least in some cases, such as Pleistocene forests, the reassembly of a community following the retreat of the ice sheets and recolonization seem mostly haphazard. Some counter-exceptions, however, emerge; as with other examples of convergence at lower levels of the evolutionary hierarchy, it is possible that general principles will be discovered. Here I touch on four disparate examples. These suggest that, in certain circumstances, operational rules apply beyond the species level. The central theme is the recognition of so-called ecomorphs. This term refers to unrelated species that evolve similar morphologies in response to equivalent functional demands within a given environment. As with other examples of convergence, the degrees of similarity are seldom precise, but can still be striking.

Particular attention has been paid to ecomorphs in fish (Conway Morris, 2003, p. 133), among which the African cichlids are particularly instructive (see, for example, Allender *et al.*, 2003; Kassam *et al.*, 2003; Koblmüller *et al.*, 2004), although other groups also provide relevant examples (see, for example, Brosset, 2003; Knouft, 2003). Perhaps more celebrated are the instances of anolid ecomorphs in the Caribbean islands (Conway Morris, 2003, p. 125; see also Losos *et al.*, 2003). More recently, Gillespie (2004) produced an absorbing survey of community assembly of Hawaiian spiders. She identified four distinct ecomorphs that have emerged independently multiple times. She writes: ". . . within any community, similar sets of ecomorphs arise . . . suggest[ing that] universal principles underlie community assembly" (p. 356). In the case of the mammalian guilds, the similarity of community structure has long been appreciated (Conway Morris 2003, p. 132). Nevertheless, these ecomorphs continue to attract interest (see, for example, Meehan and Martin, 2003). And the examples of plant ecomorphs, most famously the xerophytes, are almost too well-known to require comment. However, comparisons may also be made at the community level involving very disparate plants, notably bryophytes versus higher plants (Steel *et al.*, 2004). Such examples again point towards possible general rules.

Since the addition of molecular data to phylogenetic studies (hitherto largely dependent on morphological information) it has become clear that homoplasy is rampant. It is also clear that in numerous cases the earlier reliance on morphological similarity in establishing a given phylogeny was seriously misplaced. Again and again, morphological features turn out to be convergent. Many such examples could be given. From the recent literature, let me note the cases of the pulmonate snails and the multiple invention of a lung (Grande *et al.*, 2004); tropical scleractinian corals (Fukami *et al.*, 2004); mushrooms (O'Donnell *et al.*, 2001; see also Hibbett *et al.*, 1997a,b); and ferns (Ranker *et al.*, 2004). In reality, this is hardly surprising. If one has a given biological "form," it is to be expected that evolution not only will "explore" and populate ecological and functional space, but will do so repeatedly. Thus, given various constraints, groups inevitably will navigate again and again toward particular solutions.

Although this evidently applies to particular clades, the notion that a general principle exists so that any form will be convergent is likely to be greeted with skepticism. The reasons for this are complex. In outline, I suggest the source of this skepticism is actually anti-evolutionary, revolving around typological and essentialist views of body plans, especially at the level of the phylum. Thus, whereas one might concede that convergence is prevalent in the pulmonate snails, scleractinians, fungi, or ferns, it will be widely assumed that the anthozoans (and encompassing cnidarians), gastropods (and encompassing mollusks), fungi (or at least basidiomycetes), and tracheophytes (and so ferns) are monophyletic. This may be correct. But it is difficult to test, in terms of both the fossil record and the molecular data. The point to appreciate is that if the processes of evolution giving rise to biological forms are effectively the same at whatever taxonomic level, then one has no reason to suppose that the process of evolutionary "exploration" does not occur at taxonomic levels, which, in hindsight, we would identify as equivalent to class or even phylum. Thus, during the Cambrian "explosion," it would be as likely that various groups were "trying" to become cnidarians or mollusks. If correct, this is an important point. Not only will homoplasy be universal, but at any given stage of the evolution of life we should not be surprised to see a number of groups all attempting to run along similar evolutionary trajectories, not only in terms of body plans, but in such general developments as biomineralization, the invasion of the land, or the rise of advanced intelligence.

To appreciate this in more detail, consider the following list of connected examples: photosynthesis, chlorophyll, chloroplasts, water-conducting tissue (xylem), flowers, and a rose garden. Most evolutionary biologists would, I suspect, see this list as one of ever-decreasing evolutionary probabilities. As far as planetary life is concerned, photosynthesis may well be universal (see also Wolstencroft and Raven, 2002). Such may also be true of chlorophyll (Wald, 1974). (In passing, we need to note that although proteorhodopsin can also act as a phototrophic molecule [Beja *et al.*, 2001], its use is restricted to particular bacteria and was co-opted from its prior use as a proton pump. Moreover, although it may well provide an energetic advantage to the bacteria, apparently no evidence exists that it can actually assist in the fixation of carbon.) Beyond photosynthesis and possibly chlorophyll, the rest of my list would be regarded as a series of fortuitous evolutionary innovations of only terrestrial significance. Such a view, however, might be premature. Thus, although they are generally regarded as monophyletic, Stiller *et al.* (2003) present evidence that chloroplasts may have arisen independently several times. So, too, xylem has evolved twice (Ligrane *et al.*, 2002), as have flowers (see Conway Morris, 2003, pp. 135–8). On this basis, only the rose garden is unique, although here we need to remind ourselves that flowers also show recurrent homoplasies (see, for example, Hufford, 1997).

Evolutionary convergence, therefore, gives an interesting and largely unappreciated slant on life. Nevertheless, does it really provide the sought-after generalities that might also mesh with Henderson's concept of the "fitness of the environment"? Discussion is hindered, not only because of the inevitable specializations in biology – watching a biochemist, an ecologist, and a paleontologist in animated conversation is an all-too-rare sight – but also because we are uncertain what depends on what. Thus, the gene-centered view, popularized by Richard Dawkins, may be seen to be not only over-reductionist, but in some cases possibly seriously misleading. Consider, for example, the classic case of haplodiploidy as an explanation of the highly organized colonies observed in the eusocial insects. Despite the elegance of this hypothesis and its neat explanation of how non-reproductive individuals retain a genetic benefit even though they cannot directly pass on their genes, an alternative explanation for kin selection is the benefit conferred by a balance between individual risk and colony viability (a sort of life insurance). This seems, at least in some circumstances, to be a more convincing explanation for the origin and maintenance of eusociality (see, for example, Queller and Strassman, 1998; Field *et al.*, 2000; Landi *et al.*, 2003). The point here is not to dispute the reality of kin selection, but to stress that the life-insurance hypothesis, together with other features such as worker policing and individuality (in terms of recognition and apparent choice of activity), removes the reductionist stamp to eusociality with the implicit assumption of blind and robotic forces.

Yet finding a common ground is difficult. For example, the debate of "molecules versus morphology" remains inconclusive, despite the remarkable advances in developmental biology. The truth surely is that our growing understanding of adaptive complexes only serves to reinforce our (or at least my) wonder at the robustness, integration, and sophistication of biological systems. Molecules obviously play a part, but they are very far from being the whole story. No wonder these systems are so hard to dissect. Moreover, although the adaptive explanation (and penalty for failure) is self-evident, it is worth reminding ourselves of the immense difficulties that confront any *ab initio* imitation of these systems, perhaps most obviously in the arena of artificial intelligence.

Paradox of stability and change

Organisms appear on the whole as though they had been precisely engineered. They operate as though aware of their environment, and invariably they are able to modify it – albeit on a variety of scales. In addition, they are capable of sophisticated computational exercises operating across a variety of timescales. Not only are organisms remarkably robust in their developmental pathways and environmental responses, but they also represent extraordinary machines in their ability to engage in sophisticated internal "conversations" whereby feedback loops ensure long-term homeostasis. No matter how familiar we are with these factors, we too easily take them for granted. Yet, the fine-tuning of organisms is entirely extraordinary, as even a glance at a living cell and its biochemical intricacies will confirm: machine-like, but unlike any machine we can build. These remarks apply as much to inevitable inefficiencies, and here the Rubisco enzyme comes to mind. A critical future test with this enzyme will be to see how successful any "improvement" that accompanies technological intervention will actually be. If Rubisco has not been improved during billions of years of evolution, it is not clear that we will do much better. Rubisco is far from perfect, but it may still be the best.

The integration and complexity of the organism, not to mention the emergence of form from a fertilized zygote in multicellular groups, depend, of course, on a molecular substrate. This is classically the area of developmental biology and genetic intervention. Yet, despite the many remarkable successes (of which the result of *Pax-6* expression in terms of unrestricted eye development across the body of a fly is perhaps the most iconic), this area of evolution contains unresolved paradoxes. The central dilemma revolves around the place of genetic conservation versus innovation. The former is most famously expressed in the role of *Hox* genes in axial patterning, which to the first approximation are identical in fly and mouse. Similar arguments have been applied to key features of animal organization, such as eye and heart development and definition of dorsoventrality, or in plants, such as development of flowers (see, for example, Theissen and Becker, 2004). Such a program has considerable evolutionary implications. First, the recognition of such genetic systems in more primitive groups invites identification of homologous structures. Second, the existence of such gene-control systems may provide a possible framework to discuss evolutionary convergent directionalities and constraints.

The Cnidaria (a phylum familiar to the non-expert as the group containing jelly-fish and sea anemones) exemplifies this possibility. Thus, model cnidarians, notably *Hydra* and *Nematostella*, provide a molecular template from which it is reasonably supposed that more advanced metazoans emerged. Finnerty *et al.* (2004) argue for a fundamental equivalence of body axes (anterior–posterior and dorsoventral defining bilaterality) in cnidarians and higher metazoans. Similarly, an important developmental gene family known as COE (an acronym based on three important genes: Col/Olf-1/EBF) found in vertebrates and invertebrates is proposed, on the basis of its expression patterns in cnidarian development, to have played an ancestral role in chemoreception (Pang *et al.*, 2004). Such examples are powerful metaphors for both the molecular basis of evolution and the deep-seated unity of biological organization, in this case of animals. Yet, such views at least require qualification. Bosch and Khalturin (2002), for example, emphasize the limits to developmental conservation and the relatively unacknowledged importance of novel genes that have equivalent functions, but not molecular homology, with other systems. So too Primus and Freeman (2004) caution against over-simplistic identification of similar function, for example in the β-catenin signaling pathway, in widely differ-ent groups. Indeed, as the net of genomic investigation widens, the versatility and flexibility of these genetic systems will become apparent. For example, Locascio *et al.* (2002) explore the history of the gene family Snail, a zinc-finger protein, and specifically products of a gene-duplication event known as *snail* and *slug*. These are associated with many important developmental features, including mesoderm and neural formation. The Snail family of genes is widespread. Yet in the vertebrates, Locascio and colleagues have shown that for three features (lens, premigratory neural crest tissue, and tailbud mesenchyme), *snail* and *slug* show a "much higher degree of plasticity and complexity than expected" (p. 16845).

Such qualifications deserve to be widely appreciated, even though the extent of recruitment and redeployment of developmental genes in certain examples is well attested. The echinoderms provide an excellent example (see, for example, Wray and Lowe, 2000). At first sight, the straightforward conclusion would be that the molecular components of evolution are no different from any other and will be co-opted when necessary. So, too, it would seem reasonable to assume that such co-option is governed by local rules, if not opportunity. This, however, may not entirely correspond with emerging evidence for convergence in developmental pathways. Thus, in the case of the echinoderms, Wray and Lowe point to several instances of possible convergence (see also Nielsen *et al.*, 2003). Other examples suggest that such convergences may be more widespread than generally realized. Some examples were given earlier in this chapter in the general discussion of molecular convergence. The same phenotype may emerge by the repeated employment of the same genes, albeit independently of each other (see, for example, Mundy *et al.*, 2004; Sucena *et al.*, 2003). This is probably the most frequent situation and also the most unsurprising. Much more significant is a similar phenotype arising from different genotypes (see, for example, Gompel and Carroll, 2003; Reeves and Olmstead, 2003; Wittkopp *et al.*, 2003). This dichotomy – of similar phenotypes arising from the same genotype as opposed to the same phenotype emerging from divergent genotypes – will need considerable refinement as the details of the pathways are elucidated. For example, in a study of evolutionary parallelism in plants, the multiple origins of a particular trait depend on a single developmental program. Despite this, in each case the specific genetic changes are different (Yoon and Baum, 2004).

The potential importance of this area of evolutionary convergence has not been overlooked. Rudel and Sommer (2003) write:

Our understanding of the role of parallel evolution to produce similar structures in closely related taxa and of the causes and mechanisms for the occurrence of convergent evolution is a black box, yet an understanding of how and why disparate species within groups of related organisms independently generate the same solution *de novo* and obtain an analogous morphology may be the most important pursuit of all . . . The phenomena of convergence, parallelism, and homoplasy . . . will be one of the largest challenges for the future study of evolution of developmental mechanisms.   *(p. 32)*

The eternal return

If divergence and convergence are considered antithetical, then most evolutionary biologists typically will ascribe the greater role to the former. This accords well with the vast diversity of the biosphere, as well as with the pervasive influence of the body plan concept whereby the organic world can be divided into a finite number of "designs." Yet, within any clade, the phenomenon of convergence is ubiquitous, and, as already noted, careful analysis of morphological features in the elucidation of phylogenies typically reveals rampant homoplasy (see, for example, Ranker *et al.*, 2004). And, interestingly, in successive diversifications within a clade convergence becomes increasingly prevalent (Wagner, 2000); in essence, the clade "runs out of things to do." In such cases, it is generally the case that the character states that turn out to be homoplastic are often relatively simple. It is important to stress, however, that in terms of the emergence of complex biological properties, such as those associated with social structure and intelligence, convergences again are found to be widespread.

This divergence of approaches that look at evolution in terms of divergence versus convergence doubtless will continue to generate fruitful dialog. Each, taken alone, will lead to remarkable insights. However, a wider aim is to delineate the basic "landscape" of evolution and the extent to which navigation to stable points (perhaps even attractors) is achieved along "superhighways" compared with "country lanes." Will such discoveries help to map aspects of Henderson's notion of the biocentric "fitness of the environment"?

The best test, of course, would be the discovery and documentation of one or more extraterrestrial biospheres, assuming any exist. Even the exercise of trying to envisage biospheres on planets with physical parameters very far removed from those of the earth, in terms of, perhaps, size (and gravity) or atmospheric density, might refine our appreciation of the evolutionary possibilities. A terrestrial context, however, might also serve to give some guidelines. Consider, for example, the surprising fact that in various plant biomes a particular form has never evolved within the indigenous flora. Thus, as Mack (2003) points out, it is rather remarkable that succulents have not evolved in Yunnan, or (with few exceptions) annuals in New Zealand, or (for that matter) sand-binding grasses (marram) in the Pacific Northwest. So too in a number of cases, such as the spread of alien prickly acacia in Australia, it is surprising that none of the native acacias have evolved in similar structures. The importance of these "gaps" becomes clear on the arrival of those preadapted aliens that aggressively colonize the given biome. Not only can this lead to radical adjustments of the flora, but it also can lead to the definition of new habitats that may trigger further ecological change. This has been particularly evident in Hawaii since the introduction of the red mangrove. Note also the potential disaster awaiting the temperate forests of the Pacific Northwest in North America if bamboo is allowed to naturalize.

Mack's (2003) analysis is a compelling instance of the importance of phylogenetic constraints, even though in many cases the absence of an "expected" life form remains rather puzzling. When, however, these observations are set in the wider context of convergence in plants (see Conway Morris, 2003), then one can begin to resolve the dichotomy between divergence (and phylogenetic constraint) and convergence (and evolutionary inevitability). Local histories are not only important, but fascinating. They are also a reminder of the plenitude of the world. On the other hand, wider principles are at play. Were we ever to be able to wander in an extraterrestrial forest, we likely would still struggle to reconcile the expected strangeness with the unexpected familiarity.

Futures

Such inherencies and analogs suggest that the motors of adaptation and ecological diversification make the emergence of complex biological systems, say an eye seeing a rose garden, probable, and perhaps even inevitable. The basic similarity of these analogs indicates that radical, alien alternatives may be much less likely than is often thought. Such arguments would, I suggest, apply to the emergence of all biological complexities, including intelligence (Conway Morris, 2003) and, as Grammer *et al.* (2003) have recently argued, also to the appreciation of beauty. Consider the evolutionary inevitability of intelligence. There now appears to be specific support from experiments designed to establish whether cognitive functions are uniquely propositional and so in principle unable to function in the absence of symbolic-based language with implicit semantics and syntax. The alternative concept is that cognitive functions might be based fundamentally on an analog-coding process whereby a continuum between cognitive abilities and lower-level sensory mechanisms exists. One way to test this is to examine numerosity (the ability of animals to count) insofar as this property presumably involves both the sensory perception ("I see two dogs") and abstraction ("I see two extraterrestrials, not two dogs"). Such experiments (see, for example, Nieder and Miller, 2003) indicate that the analog-coding hypothesis is correct. At the moment, such results make the evolutionary emergence of human cognition unremarkable. But they also fail to explain why in certain important respects it is unique.

Another intriguing question is whether the fitness of the environment confers any predictability on the future of the evolutionary process. Certainly, the likelihood of a universal biochemistry and the inevitability of organismic convergence at many levels (Conway Morris, 2003) provide a necessary framework. But what of the future evolution of intelligence? Hofman (2001) has convincingly argued that the potential for further encephalization in the hominids is limited. The stock response, of course, is to postulate computer-based extensions of our cognitive ability. A more radical alternative is to suggest that in the evolution of (any) biosphere the rise of bird- and mammal-like intelligences with a potential for developing technologies is geologically short-lived and is replaced by other systems, such as insect eusocial colonies. Such a view will seem far-fetched. It may be that the complexities achievable by insects and their eusocial societies have gone as far as they are able. Nevertheless, some degree of caution might be advisable. This is for at least three reasons. First, extrapolating evolutionary trends into the future has been generally regarded as a futile exercise, although in principle a proper understanding of the evolutionary "landscape" might confer a degree of predictability. Even so, from our present perspective, estimating the potential of a complex system is fraught with difficulties. Suppose we encountered a pre-australopithecine and set it in the context of a late Miocene African ecosystem. Would we, *ab initio*, predict ourselves?

The second reason for caution is that although we know quite a lot about the increasing complexity of the biosphere through geological time, we have little understanding about how the complex systems interact and reciprocate. A popular suggestion, for example, concerns the co-evolution of angiosperms and insects. Similarly, one can speculate about the evolution of sonar (bat, cetacean) and the prey response. Clearly, interactions occur; but what are the evolutionary dynamics? So, too, with the independent emergence of many complex insect societies – how are they driving the rest of the biosphere, and on what timescales?

Finally, consider the sometimes underappreciated complexity of the social insects themselves. These include not only complex societies of varying types, but also sophisticated communication (e.g. bee "dance," quorum sensing), co-ordinated activity (e.g. wing beating), memory, sleep, ability to distinguish same from different, metabolic control and resource appropriation, and, most famously, the agriculture in attine ants and an analogous system in termites. As noted earlier, it is possible that the systems are incapable of further elaboration. To my mind, however, this may well be incorrect, and in any event our views of intelligence are still too narrow. Indeed, as our knowledge of complex organic systems capable of memory, communication, and computation grows, we may find ourselves considering systems even more arcane than that of the social insects.

In conclusion, I wish simply to comment that, although the notion of the "fitness of the environment" is most obvious and direct at the level of biochemistry, the interconnections of life via evolution effectively predetermine the process. This situation evidently imparts both directionality and, more controversially, inevitability.

In summary, wherever life exists, there will, in due course, evolve mind. Whether it is always our type of mind is altogether another question.

Acknowledgments

I acknowledge support from the John Templeton Foundation, and I warmly thank Sandra Last and Vivien Brown for typing numerous drafts. Critical remarks by Charles Harper greatly improved the paper.

References

Allender, C. J. *et al.* (2003). Divergent selection during speciation of Lake Malawi cichlid fishes inferred from parallel radiations in nuptial coloration. *Proceedings of the National Academy of Sciences, USA*, **100**, 14074–9.

Axe, D. D. (2000). Extreme functional sensitivity to conservative amino acid changes on enzyme exteriors. *Journal of Molecular Biology*, **301**, 585–96.

Beeumen, J. J. *et al.* (1991). The primary structure of ruberythrin, a protein with inorganic pyrophosphate activity from *Desulfovibrio vulgaris*. *Journal of Biological Chemistry*, **266**, 20645–53.

Beja, O. *et al.* (2001). Proteorhodopsin phototrophy in the ocean. *Nature*, **411**, 786–9.

Benner, S. A. (2000). Unite efforts and conquer mysteries of artificial genetics. *Science*, **290**, 1506.

Benner, S. A. (2002). The past as the key to the present: resurrection of ancient proteins from eosinophils. *Proceedings of the National Academy of Sciences, USA*, **99**, 4760–1.

Bennett, A. F. (2003). Experimental evolution and the Krogh principle: generating biological novelty for functional and genetic analyses. *Physiological and Biochemical Zoology*, **76**, 1–11.

Berg, H. C. (2003). The rotary motor of bacterial flagella. *Annual Review of Biochemistry*, **72**, 19–54.

Berg, J. S., Powell, B. C. and Cheney, R. E. (2001). A millennial myosin census. *Molecular Biology of the Cell*, **12**, 780–94.

Beuth, B., Niefind, K. and Schomburg, D. (2003). Crystal structure of creatinase from *Pseudomonas putida*: a novel fold and a case of convergent evolution. *Journal of Molecular Biology*, **332**, 287–301.

Boffelli, D., Cheng, J.-F. and Rubin, E. M. (2004). Convergent evolution in primates and an insectivore. *Genomics*, **83**, 19–23.

Bosch, T. C. G. and Khalturin, K. (2002). Patterning and cell differentiation in *Hydra*: novel genes and the limits of conservation. *Canadian Journal of Zoology*, **80**, 1670–7.

Brosset, A. (2003). Convergent and divergent evolution in rain-forest populations and communities of cyprinodontiform fishes (*Aphyosemion* and *Rivulus*) in Africa and South America. *Canadian Journal of Zoology*, **81**, 1848–93.

Bull, J. J. *et al.* (1997). Exceptional convergent evolution in a virus. *Genetics*, **147**, 1497–1507.

Burmester, T. (2002). Origin and evolution of arthropod haemocyanins and related proteins. *Journal of Comparative Physiology*, B**172**, 95–107.

Charnock, S.-J. *et al.* (2002). Convergent evolution sheds light on the anti-β-elimination mechanism common to family 1 and 10 polysaccharide lyases. *Proceedings of the National Academy of Science, USA*, **99**, 12067–72.

Cheng, Z. *et al.* (2003). Highly divergent methyltransferases catalyze a conserved reaction in tocophenol and plastoquinone synthesis in cyanobacteria and photosynthetic eukaryotes. *Plant Cell*, **15**, 2343–56.

Coates, J. C. (2003). Armadillo repeat proteins: beyond the animal kingdom. *Trends in Cell Biology*, **13**, 463–71.

Cody, G. D. *et al.* (2000). Primordial carbonylated iron-sulphur compounds and the synthesis of pyruvate. *Science*, **289**, 1337–40.

Conant, G. C. and Wagner, A. (2003). Convergent evolution of gene circuits. *Nature Genetics*, **34**, 264–6.

Conway Morris, S. (2003.) *Life's Solution: Inevitable Humans in a Lonely Universe.* Cambridge, UK: Cambridge University Press.

Cooper, T. F., Rozen, D. E. and Lenski, R. E. (2003). Parallel changes in gene expression after 20,000 generations of evolution in *Escherichia coli. Proceedings of the National Academy of Sciences, USA*, **100**, 1072–7.

Coufal, D. E. *et al.* (2000). Sequencing and analysis of the *Methylococcus capsulatus* (Bath) soluble methane monooxygenase genes. *European Journal of Biochemistry*, **267**, 2174–85.

Crespi, B. J. (2001). The evolution of social behavior in microorganisms. *Trends in Ecology and Evolution*, **16**, 178–83.

de Duve, C. (1995). *Vital Dust: Life as a Cosmic Imperative*. New York, NY: Basic Books (HarperCollins).

Dennison, K. L. and Spalding, E. P. (2000). Glutamate gated Ca^{2+} fluxes in *Arabidopsis*. *Plant Physiology*, **124**, 1511–14.

Denton, M. J., Marshall, C. J. and Legge, M. (2002). The protein folds as Platonic forms: new support for the pre-Darwinian conception of evolution by natural law. *Journal of Theoretical Biology*, **219**, 325–42.

Dose, K. (1988). The origin of life: more questions than answers. *Interdisciplinary Science Reviews*, **13**, 348–56.

Dupuy, F. *et al.* (2002). α1,4-fucosyltransferase activity: a significant function in the primate lineage has appeared twice independently. *Molecular Biology and Evolution*, **19**, 815–24.

Eizirik, E. *et al.* (2003). Molecular genetics and evolution of melanism in the cat family. *Current Biology*, **13**, 448–53.

Eschenmoser, A. (1999). Chemical etiology of nucleic acid structure. *Science*, **284**, 2118–24.

Fabrizio, J. J., Boyle, M. and DiNardo, S. (2003). A somatic role for *eyes absent* (*eya*) and *sine oculis* (*so*) in *Drosophila* spermatocyte development. *Developmental Biology*, **258**, 117–28.

Field, J. *et al.* (2000). Insurance-based advantage to helpers in a tropical hover wasp. *Nature*, **404**, 869–71.

Finnerty, J. R. *et al.* (2004). Origins of bilateral symmetry: *Hox* and *Dpp* expression in a sea anemone. *Science*, **304**, 1335–7.

Firn, E. (2004). Plant intelligence: an alternative point of view. *Annals of Botany*, **93**, 345–51.

Fukami, H. *et al.* (2004). Conventional taxonomy obscures deep divergence between Pacific and Atlantic corals. *Nature*, **427**, 832–5.

Gaucher, E. A. *et al.* (2003). Inferring the palaeoenvironment of ancient bacteria on the basis of resurrected proteins. *Nature*, **425**, 285–8.

Gillespie, R. (2004). Community assembly through adaptive radiation in Hawaiian spiders. *Science*, **303**, 356–9.

Gompel, N. and Carroll, S. B. (2003). Genetic mechanisms and constraints governing the evolution of correlated traits in drosophilid flies. *Nature*, **424**, 931–5.

Grammer, K. *et al.* (2003). Darwinian aesthetics: sexual selection and the biology of beauty. *Biological Reviews*, **78**, 385–407.

Grande, C. *et al.* (2004). Molecular phylogeny of Euthyneura (Mollusca: Gastropoda). *Molecular Biology and Evolution*, **21**, 303–13.

Gregory, T. R. (2004). Macroevolution, hierarchy theory, and the C-value enigma. *Paleobiology*, **3**, 179–202.

Hamburger, Z. A. *et al.* (1999). Crystal structure of invasin: a bacterial integrin-binding protein. *Science*, **286**, 291–5.

Heanue, T. A. *et al.* (1999). Synergistic regulation of vertebrate muscle development by *Dach2, Eya2*, and *Six1*, homologs of genes required for *Drosophila* eye formation. *Genes and Development*, **13**, 3231–43.

Hedenström, A. (2004). A general law for animal locomotion. *Trends in Ecology and Evolution*, **19**, 217–19.

Henderson, L. J. (1913). *The Fitness of the Environment: An Inquiry into the Biological Significance of the Properties of Matter*. New York, NY: Macmillan. Repr. (1958) Boston, MA: Beacon Press; (1970) Gloucester, MA: Peter Smith.

Herrmann, K. M., Schultz, J. and Hermodson, M. A. (1980). Sequence homology between the tyrosine-sensitive 3-deoxy-D-*arabino*-heptulosonate 7-phosphate synthase from *Escherichia coli* and hemerythrin from Sipunculida. *Journal of Biological Chemistry*, **255**, 7079–81.

Hibbett, D. S., Grimaldi, D. and Donoghue, M. J. (1997a). Fossil mushrooms from Miocene and Cretaceous ambers and the evolution of the Homobasidiomycetes. *American Journal of Botany*, **84**, 981–91.

Hibbett, D. S. *et al.* (1997b). Evolution of gilled mushrooms and puffballs inferred from ribosomal DNA sequences. *Proceedings of the National Academy of Sciences, USA*, **94**, 12002–6.

Hillebrand, H. (2004). On the generality of the latitudinal diversity gradient. *American Naturalist*, **163**, 192–211.

Hoekstra, H. E. and Price, T. (2004). Parallel evolution is in the genes. *Science*, **303**, 1779–81.

Hofman, M. A. (2001). Brain evolution in hominids: are we at the end of the road? In *Evolutionary Anatomy of the Primate Cerebral Cortex,* ed. D. Falk and K. R. Gibson. Cambridge, UK: Cambridge University Press, pp. 113–27.

Horgan, J. (1997). *The End of Science: Facing the Limits of Knowledge in the Twilight of the Scientific Age*. London: Little, Brown.

Hufford, L. (1997). The roles of ontogenetic evolution in the origins of floral homoplasies. *International Journal of Plant Sciences*, **158** (suppl. 6), 565–80.

Immesberger, A. and Burmester, T. (2004). Putative phenoloxidases in the tunicate *Ciona intestinalis* and the origin of the arthropod hemocyanin superfamily. *Journal of Comparative Physiology*, B**174**, 169–80.

Johns, G. C. and Somero, G. N. (2004). Evolutionary convergence in adaptation of proteins to temperature: A_4-lactate dehydrogenase of Pacific damselfishes (*Chromis* spp.). *Molecular Biology and Evolution*, **21**, 314–20.

Jones, N. A. *et al.* (1998). The *Drosophila Pax* gene *eye gone* is required for embryonic salivary duct development. *Development*, **125**, 4163–74.

Joshi, J. G. and Sullivan, B. (1973). Isolation and preliminary characterization of hemerythrin from *Lingula unguis*. *Comparative Biochemistry and Physiology*, B**44**, 857–67.

Kassam, D. D. *et al.* (2003). Morphometric analysis on ecomorphologically equivalent cichlid species from Lake Malawi and Tanganyika. *Journal of Zoology, London*, **260**, 153–7.

Knoll, A. H. and Bambach, R. K. (2000). Directionality in the history of life: diffusion from the left wall or repeated scaling of the right? *Paleobiology*, **26** (suppl.: Deep time, paleobiology's perspective), 1–14.

Knouft, J. H. (2003). Convergence, divergence, and the effect of congeners on bodysize ratio in stream fishes. *Evolution*, **57**, 2374–82.

Koblmüller, S., Salzburger, W. and Sturmbauer, C. (2004). Evolutionary relationships in the sand-dwelling cichlid lineage of Lake Tanganyika suggests multiple colonization of rocky habitats and convergent origin of biparental mouthbrooding. *Journal of Molecular Evolution*, **58**, 79–96.

Kozlowski, J. and Konarzewski, M. (2004). Is West, Brown and Enquist's model of allometric scaling mathematically correct and biologically relevant? *Functional Ecology*, **18**, 283–9.

Kryukov, G. V. *et al.* (2002). Selenoprotein R is a zinc-containing stereo-specific methionine sulphoxide reductase. *Proceedings of the National Academy of Sciences, USA*, **99**, 4245–50.

Kurtz, D. M. (1999). Oxygen-carrying proteins: three solutions to a common problem. *Essays in Biochemistry*, **34**, 85–100.

Landi, M. *et al.* (2003). Low relatedness and frequent queen turnover in the stenogastrine wasp *Eastenogaster fraterna* favor the life insurance over the haplodiploid hypothesis for the origin of eusociality. *Insectes Sociaux*, **50**, 262–7.

Lebrun, E. *et al.* (2003). Arsenite oxidase, an ancient bioenergetic enzyme. *Molecular Biology and Evolution*, **20**, 686–93.

Lenski, R. E. *et al.* (2003). The evolutionary origin of complex features. *Nature*, **423**, 139–44.

Ligrane, R. *et al.* (2002). Diversity in the distribution of polysaccharide and glycoprotein epitopes in the cell walls of bryophytes: new evidence for the multiple evolution of water-conducting cells. *New Phytologist*, **156**, 491–508.

Linden, P. F. and Turner, J. S. (2004). "Optimal" vortex rings and aquatic propulsion mechanisms. *Proceedings of the Royal Society*, B**271**, 647–53.

Litvak, Y. and Selinger, Z. (2003). Bacterial mimics of eukaryotic GTPase-activating proteins (GAPs). *Trends in Biochemical Sciences*, **28**, 628–31.

Locascio, A. *et al.* (2002). Modularity and reshuffling of *Snail* and *Slug* expression during vertebrate evolution. *Proceedings of the National Academy of Sciences, USA*, **99**, 16841–6.

Losos, J. B. *et al.* (2003). Niche lability in the evolution of a Caribbean lizard community. *Nature*, **424**, 542–5.

Mack, R. M. (2003). Phylogenetic constraint, absent life forms, and preadapted alien plants: a prescription for biological invasions. *International Journal of Plant Sciences*, **164** (suppl. 3), S185–96.

Malcolm, B. A. *et al.* (1990). Ancestral lysozymes reconstructed, neutrality tested, and thermostability linked to hydrocarbon packing. *Nature*, **345**, 86–9.

Marden, J. H. and Allen, L. R. (2002). Molecules, muscles, and machines: universal performance characteristics of motors. *Proceedings of the National Academy of Sciences, USA*, **99**, 4161–6.

Matthysse, A. G. *et al.* (2004). A functional cellulose synthase from ascidian epidermis. *Proceedings of the National Academy of Sciences, USA*, **101**, 986–91.

Meehan, T. J. and Martin, L. D. (2003). Extinction and re-evolution of similar adaptive types (ecomorphs) in Cenozoic North American ungulates and carnivores reflect van der Hammen's cycles. *Naturwissenschaften*, **90**, 131–5.

Mundy, N. I. *et al.* (2004). Conserved genetic basis of a quantitative plumage trait involved in mate choice. *Science*, **303**, 1870–3.

Nakashima, K. *et al.* (2004). The evolutionary origin of animal cellulose synthase. *Development, Genes and Evolution*, **214**, 81–8.

Nieder, A. and Miller, E. K. (2003). Coding of cognitive magnitude: compressed scaling of numerical information in the primate prefrontal cortex. *Neuron*, **37**, 149–57.

Nielsen, M. G. *et al.* (2003). Evolutionary convergence in *otx* expression in the pentameral adult rudiment in direct-developing sea urchins. *Development, Genes and Evolution*, **213**, 73–82.

O'Donnell, K. *et al.* (2001). Evolutionary relationships among mucoralean fungi (Zygomycota): evidence for family polyphyly on a large scale. *Mycologia*, **93**, 286–96.

Pace, N. R. (2001). The universal nature of biochemistry. *Proceedings of the National Academy of Sciences, USA*, **98**, 805–8.

Pang, K., Matus, D. Q. and Martindale, M. Q. (2004). The ancestral role of COE genes may have been in chemoreception: evidence from the development of the sea anemone, *Nematostella vectensis* (Phylum Cnidaria: Class Anthozoa). *Development, Genes and Evolution*, **214**, 134–8.

Piatigorsky, J. (1992). Lens crystallins. Innovation associated with changes in gene regulation. *Journal of Biological Chemistry*, **267**, 4277–80.

Primus, A. and Freeman, G. (2004). The cnidarian and the canon: the role of Wnt/β-catenin signalling in the evolution of metazoan embryos. *BioEssays*, **26**, 474–8.

Queller, D. C. and Strassman, J. E. (1998). Kin selection and social insects. *BioScience* **48**, 65–175.

Rainey, P. B. and Rainey, K. (2003). Evolution of cooperation and conflict in experimental bacterial populations. *Nature*, **425**, 72–4.

Ranker, T. A. *et al.* (2004). Phylogeny and evolution of grammitid ferns (Grammitidaceae): a case of rampant morphological homoplasy. *Taxon*, **53**, 415–28.

Reeves, P. A. and Olmstead, R. G. (2003). Evolution of the TCP gene family in Asteridae: cladistic network approaches to understanding regulatory gene family diversification and its impact on morphological evolution. *Molecular Biology and Evolution*, **20**, 1997–2009.

Relaix, F. and Buckingham, M. (1999). From insect eye to vertebrate muscle: redeployment of a regulatory network. *Genes and Development*, **13**, 3171–8.

Rouse, G. W. and Pleijel, F. (2001). *Polychaetes*. Oxford: Oxford University Press.

Rudel, D. and Sommer, R. J. (2003). The evolution of developmental mechanisms. *Developmental Biology*, **264**, 15–37.

Savage, V. M. *et al.* (2004). The predominance of quarter-power scaling in biology. *Functional Ecology*, **18**, 257–82.

Schöning, K.-U. *et al.* (2000). Chemical etiology of nucleic acid structure: the α-threofuranosyl-(3′-2′) oligonucleotide system. *Science*, **290**, 1347–51.

Steel, J. B. *et al.* (2004). Are bryophyte communities different from higher-plant communities? Abundance relations. *Oikos*, **104**, 479–86.

Stenkamp, R. E. (1994). Dioxygen and hemerythrin. *Chemical Reviews*, **94**, 715–26.

Stiller, J. W., Reel, D. C. and Johnson, J. C. (2003). A single origin of plastids revisited: convergent evolution in organellar genome content. *Journal of Phycology*, **39**, 95–105.

Sucena, E. *et al.* (2003). Regulatory evolution of *shavenbaby/ovo* underlies multiple cases of morphological parallelism. *Nature*, **424**, 935–8.

Tabin, C. J., Carroll, S. B. and Panganiban, G. (1999). Out on a limb: parallels in vertebrate and invertebrate limb patterning and the origin of appendages. *American Zoologist*, **39**, 650–63.

Takagi, T. and Cox, J. A. (1991). Primary structure of myohemerythrin from the annelid *Nereis diversicolor*. *FEBS Letters*, **285**, 25–7.

Taylor, G. K., Nudds, R. L. and Thomas, A. L. R. (2003). Flying and swimming animals cruise at a Strouhal number tuned for high power efficiency. *Nature*, **425**, 707–11.

Theissen, G. and Becker, A. (2004). Gymnosperm orthologues of Class B floral homeotic genes and their impact on understanding flower origin. *Critical Reviews in Plant Sciences*, **23**, 129–48.

Trewavas, A. (2003). Aspects of plant intelligence. *Annals of Botany*, **92**, 1–20.

Trewavas, A. (2004). Aspects of plant intelligence: an answer to Firn. *Annals of Botany*, **93**, 353–7.

Velicer, G. J. and Yu, Y.-T. N. (2003). Evolution and novel cooperative swarming in the bacterium *Myxococcus xanthus*. *Nature*, **425**, 75–8.

Volbeda, A. and Hol, W. G. J. (1989). Pseudo 2-fold symmetry in the copper-binding domain of arthropodan haemocyanins: possible implications for the evolution of oxygen transport proteins. *Journal of Molecular Biology*, **206**, 531–46.

Wagner, P. J. (2000). Exhaustion of morphologic character states among fossil taxa. *Evolution*, **54**, 365–86.

Wald, G. (1974). Fitness in the universe: choices and necessities. *Origins of Life and Evolution of Biospheres*, **5**, 7–27.

Weber, A. L. and Miller, S. L. (1981). Reasons for the occurrence of the twenty coded protein amino acids. *Journal of Molecular Evolution*, **17**, 273–84.

Weber, R. E. (1978). Respiratory pigments. In *Physiology of Annelids*, ed. P. J. Mill. London: Academic Press, pp. 393–446.

Wessler, I., Kirkpatrick, C. J. and Racke, K. (1999). The cholinergic "pitfall": acetylcholine, a universal cell molecule in biological systems, including humans. *Clinical and Experimental Pharmacology and Physiology*, **26**, 198–205.

West, G. B., Brown, J. H. and Enquist, B. J. (1999). The fourth dimension of life: fractal geometry and allometric scaling of organisms. *Science*, **284**, 1677–9.

West, G. B., Brown, J. H. and Enquist, B. J. (2001). A general model for ontogenetic growth. *Nature*, **413**, 628–31.

Westheimer, F. H. (1987). Why nature chose phosphates. *Science*, **235**, 1173–8.

Williams, R. J. P. and Fraústo da Silva, J. J. R. (2003). Evolution was chemically constrained. *Journal of Theoretical Biology*, **220**, 323–43.

Wittkopp, P. J. *et al.* (2003). *Drosophila* pigmentation evolution: divergent genotypes underlying convergent phenotypes. *Proceedings of the National Academy of Sciences, USA*, **100**, 1808–13.

Wolstencroft, R. D. and Raven, J. A. (2002). Photosynthesis: likelihood of occurrence and possibility of detection on Earth-like planets. *Icarus*, **157**, 535–48.

Wray, G. A. (2002). Do convergent developmental mechanisms underlie convergent phenotypes? *Brain, Behavior and Evolution*, **59**, 327–36.

Wray, G. A. and Lowe, C. J. (2000). Developmental regulatory genes and echinoderm evolution. *Systematic Biology*, **49**, 28–51.

Yoon, H. S. and Baum, D. A. (2004). Transgenic study of parallelism in plant morphological evolution. *Proceedings of the National Academy of Sciences, USA*, **101**, 6524–9.

Zakon, H. H. (2002). Convergent evolution on the molecular level. *Brain, Behavior and Evolution*, **59**, 250–61.

Zhang, J.-Z. and Rosenberg, H. F. (2002). Complementary advantageous substitutions in the evolution of an antiviral RNase of higher primates. *Proceedings of the National Academy of Sciences, USA*, **99**, 5486–91.

12

Life on earth: the role of proteins

Jayanth R. Banavar and Amos Maritan

Introduction

It is now believed that our universe was created around 13.8 billion years ago. Our planet earth came into existence around 4.5 billion years ago. For nearly a billion years or so after it was formed, the earth was stark and bereft of life. The matter contained on earth was inorganic with relatively small molecules. There were endless rock formations, oceans, and an atmosphere.

And then there was life.

The problem of how life was created is a fascinating one. Our focus is on looking at life on earth and asking how it works. The lessons we learn provide hints to the answer to the deep and fundamental question pondered by our ancients: Was life on earth inevitable? Then there are the questions posed by Henderson [1]: Is the nature of our physical world biocentric? Is there a need for fine-tuning in biochemistry to provide for the fitness of life in the cosmos – or, even less ambitiously, for life here on earth? Surprisingly, as we will show, a physics approach turns out to be valuable for thinking about these questions.

All living organisms have a genetic map consisting of a one-dimensional string of information encoded in the DNA molecule. An essential question that one seeks to answer is how an organism converts that information into a three-dimensional living being.

Life has many common patterns. All living cells follow certain simple "universal" themes. As eloquently described by Hoagland and Dodson [2] in their classic book *The Way Life Works*, these patterns include the following properties.

- Life builds from the bottom up.
- Life uses a few themes to generate many variations.
- Life organizes with information.

Fitness of the Cosmos for Life: Biochemistry and Fine-Tuning, ed. J. D. Barrow *et al.*
Published by Cambridge University Press. © Cambridge University Press 2007.

- Life tends to optimize rather than maximize.
- Life is opportunistic.
- Life competes within a cooperative framework.
- Life is interconnected and interdependent.

The revolution in molecular biology [3] sparked by the discovery [4] of the structure of the DNA molecule 50 years ago has led to a breathtakingly beautiful description of life. Life employs well-tailored chain molecules to store and replicate information, carry out a dizzying array of functionalities, and provide a molecular basis for natural selection. The complementary base-pairing mechanism in DNA combined with its double-helix structure serves as a repository of information and provides an elegant mechanism for replication [4]. The replication is prone to errors or mutations; these errors, which are the basis of evolution, are in turn copied in future generations [5]. Using the RNA molecule as an intermediary, the information contained in the DNA genes is translated into proteins, which are linear chains of amino acids. Unlike the DNA molecule, which has one structure, protein molecules [6, 7] fold into thousands of native-state structures under physiological conditions. For proteins, form determines functionality, and the rich variety of observed forms underscores the versatility of proteins. There then follows a complex orchestrated dance in which proteins catalyze reactions, sometimes speeding up reactions by more than a factor of 10 billion, interact with one another, and finally feed back into the genes to regulate the synthesis of other proteins [3]. Thus, life originates from the dynamics of large networks. Living cells are complex systems whose constituents interact selectively, non-linearly, and in a temporally orchestrated manner to yield coherent and robust behavior.

Our focus in this chapter is on understanding a key component of this network of life: the globular proteins that act as enzymes. In his book *For the Love of Enzymes* published in 1989 [8], the brilliant biochemist Arthur Kornberg wrote eloquently:

What chemical feature most clearly enables the living cell and organism to function, grow and reproduce? Not the carbohydrate stored as starch in plants or glycogen in animals, nor the depots of fat. It is not the structural proteins that form muscle, elastic tissue, and the skeletal fabric. Nor is it DNA, the genetic material. Despite its glamor, DNA is simply the construction manual that directs the assembly of the cell's proteins. The DNA is itself lifeless, its language cold and austere. What gives the cell its life and personality are enzymes. They govern all body processes; malfunction of even one enzyme can be fatal. Nothing in nature is so tangible and vital to our lives as proteins, and yet so poorly understood and appreciated by all but a few scientists.

A protein molecule is large and has many atoms. In addition, the water molecules surrounding the protein play a crucial role in its behavior. At the microscopic level, the laws of quantum mechanics can be used to deduce the interactions, but the degrees of freedom are far too many for the system to be studied in all its detail. When one attempts to look at the problem in a coarse-grained manner [9] with what one hopes are the essential degrees of freedom, it is very hard to determine what the effective potential energies of interaction are. This situation makes the protein problem particularly daunting, and no solution has yet been found.

Over many decades, many experimental data have been accumulated; yet theoretical progress has been somewhat limited. The problem is highly interdisciplinary and touches on biology, chemistry, and physics; it is often hard to distill the essential features of each of the multiple aspects of the problem. The great successes of quantum chemistry in the determination of the structure of the DNA molecule [4] and in the spectacular prediction that helices and sheets [10, 11, 12] are the building blocks of protein structures have spurred much work using detailed chemistry for understanding the protein problem. Such work has been very insightful in providing useful hints on how proteins behave at the atomic scale in performing their tasks. The missing feature in such a theoretical approach, of course, is that it treats each protein as a special entity with all the attendant details of the sequence of amino acids, their intricate side-chain atoms, and the water molecules. Such an approach, while quite valuable, neither has as a goal nor can lend itself to a unified way of understanding seemingly disparate phenomena pertaining to proteins. Reinforcing this, experiments, which are very challenging, are carried out on one protein at a time and cry out for an understanding of the behavior of an individual class of protein.

The lessons we have learned from physics are of a different nature. The history of physics is replete with examples of the elucidation of connections between what seem to be distinct phenomena and the development of a unifying framework, which, in turn, leads to new observable consequences [13]. Indeed, strong evidence suggests that globular proteins share many common characteristics: their ability to fold rapidly and reproducibly in order to create a hydrophobic core, the fact that there seem to be a relatively small number (on the order of a few thousand) of distinct modular folds made up of helices and almost planar sheets, the fact that protein folds are flexible and versatile in order to accomplish the dizzying array of functionalities that these proteins perform, and the unfortunate tendency of proteins to aggregate and form amyloids, which are implicated in human diseases.

Researchers have made many attempts at using physics-based approaches (such as all-atom molecular simulations, simpler coarse-grained models that retain what one hopes are the essential degrees of freedom, and statistical approaches for determining the effective interactions between them) for understanding proteins [9]. These methods have provided valuable insights on how one might think about the problem and have served as a means of understanding experimental data. Yet, no simple unification has been achieved in a deeper understanding of the key principles at work in proteins.

Our goal here is to present a different approach to the protein problem that we have developed with many collaborators. We restrict ourselves to globular proteins, which display the rich variety of native-state structures. Other interesting and important classes of proteins [14] include membrane proteins and fibrous proteins, but we do not consider them here. Our focus is on understanding the origin of protein structures and how they form the basis for both functionality and natural selection. Our work points to a unification of the various aspects of all proteins: symmetry and geometry determine the limited menu of folded conformations that a protein can choose from for its native-state structure. Such structures are in a marginally compact phase in the vicinity of a phase transition and are therefore eminently suited for biological function. These structures are the molecular target for the powerful forces of evolution. Proteins are well-designed sequences of amino acids that fit well into one of these predetermined folds and are prone to misfolding and aggregation, leading to the formation of amyloids, which are implicated in debilitating human diseases [15, 16] such as Alzheimer's syndrome, light-chain amyloidosis, and spongiform encephalopathies. We will show how our framework touches on the issue of fine-tuning for life on earth. Michael Denton, in Chapter 13 of this book, discusses further the fascinating consequences of our principal result that the menu of protein folds is determined by physical law.

Phases of matter: from spheres to tubes

The fluid and crystalline phases of matter can be readily understood [17] in terms of the behavior of a simple system of hard spheres. The standard way of ensuring the self-avoidance of a system of uniform hard spheres is to consider all pairs of spheres and require that their centers are no closer than their diameter. Studies of hard spheres have a venerable history [18], including early work by Kepler on the packing of cannonballs in a ship's hold. Each hard sphere can be thought of as a point-particle or a zero-dimensional object with its own private space of spatial extent equal to its radius. Generalizing to a one-dimensional object, one must consider a line or a string, with private space associated with each point along the line, leading to a uniform tube of radius of cross-section or thickness, Δ, with its axis defined by the line. (Likewise, one could consider a collection of interacting tubes.) The new feature, with respect to the hard-sphere system, is that a flexible tube has self-interaction. The generalization of the hard-sphere constraint to the description of the self-avoidance of a tube of non-zero thickness is as follows [19]. Consider all triplets of points along the axis of the tube. Draw circles through each

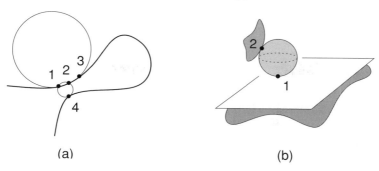

(a)   (b)

Figure 12.1. Interpretation of multi point distances for a curve and a surface [19]. (a) Three-point distance for curve. Given any three distinct points, r is the radius of the unique circle that contains the points. When the points are from the same neighborhood on a curve, such as points 1, 2, and 3, r is close to the local radius of curvature. When points, such as 1, 2, and 4, are taken from two different neighborhoods of the curve that are close to intersection, r approximates (half of) the distance of closest approach of the curve to itself. (b) Tangent-point distance for a surface. Given two distinct points 1 and 2 on a surface, ρ is the radius of the unique sphere that contains both points and is tangent to the surface at point 1. When the points are neighbors on the surface, ρ approximates the absolute value of the local normal radius of curvature in the direction defined by the two points (not illustrated). When points are taken from different neighborhoods that are close to intersection, ρ approximates (half of) the distance of closest approach.

of the triplets, and ensure that none of the radii are less than the tube thickness [20]. This prescription surprisingly entails discarding pairwise interactions and working with effective many-body interactions for the description of the self-avoidance of a tube or sheet of non-zero thickness, as shown in Figure 12.1.

One may visualize a tube as the continuum limit of a discrete chain of tethered disks or coins [21] of fixed radius separated from one another by a distance a in the limit of $a \to 0$. The inherent anisotropy associated with a coin (the heads-to-tails direction being different from the other two) reflects the fact that a special local direction at each position is defined by the locations of the adjacent objects along the chain. An alternative description of a discrete chain molecule is a string-and-beads model in which the tethered objects are spheres. The key difference between these two descriptions is the different symmetry of the tethered objects. On compaction, spheres tend to surround themselves isotropically with other spheres, unlike the tube situation in which nearby tube segments need to be placed parallel to one another. Even for unconstrained particles, deviations from spherical symmetry (replacing a system of hard spheres with one of hard rods, for example) lead to rich new liquid crystal phases [22, 23]. Likewise, the tube and a chain of tethered spheres exhibit quite distinct behaviors.

Marginally compact tubes

Consider a tube of non-zero thickness undergoing compaction to expel the water away from the interior of its structure in the folded state. (In order to make connection with proteins, it is useful to note that the backbone of all amino acids contains a carbon atom called a C_α atom. In a coarse-grained description, this atom may be chosen as the representative of the amino acid. Our tube can be thought of as a discrete chain of C_α atoms of the protein backbone.) As we have discussed, the notion of a tube thickness is captured by ensuring that none of the three-body radii is smaller than a threshold value equal to the radius of the tube. Let us postulate that the attractive interactions promoting compaction are pairwise and have a given range. (Because we are considering a discrete situation, it is quite valid to have pairwise interactions, which need to be discarded in a continuum description of a tube.) We will need to specify one dimensionless quantity, which we will call X, that is the ratio of the thickness of the tube to the range of the attractive interactions.

When X is very large compared with 1, the tube is so fat that it is unable to benefit from the attractive interactions. The constraints of the three-body interaction dominate (the pairwise interaction plays no role), and one then obtains a swollen phase consisting of all self-avoiding conformations that satisfy the three-body radius constraint associated with the non-zero tube thickness. A vast majority of these conformations are ineffective in expelling the water from the interior of the structure. The non-zero thickness is loosely analogous to restricted space that others are not allowed to trespass on. Imagine that your friend sits in the center of his/her room and requests that no one enter the room. The thickness then is proportional to the width of the room. If the range of attractive interactions is very small compared with this size, your ability to benefit from interactions with him/her is compromised by the fact that you cannot enter the room, and, for all practical purposes, it is as though your interactions with him/her were turned off. At the other extreme, for a tube with a very small X compared with 1, one also obtains many, many conformations. This is because, in the room analogy, your interaction with your friend is sufficiently long-range that you have a lot of flexibility in where you position yourself. From a dynamical point of view, the structures obtained when $X \ll 1$ are somewhat inaccessible because the energy landscape is studded with numerous multiple minima. This situation is one in which the pairwise attractive interactions dominate and the three-body radii constraints do not matter.

On varying X, we find two regimes: the phase with an effective long-range attraction and the swollen phase, both with tremendous degeneracies, as shown in Figure 12.2. There is a "twilight zone" between these two phases, viewed as day and night, when X is just shy of 1. (We alert experts in the protein field that this

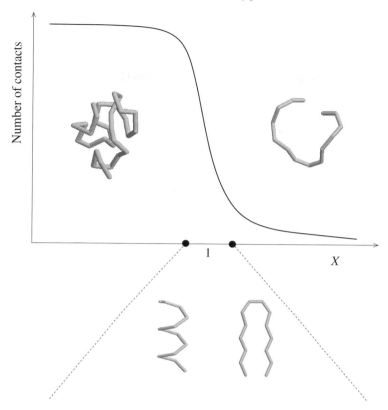

Figure 12.2. Sketch of the maximal number of contacts that a short, compact tube can make as a function of X, the dimensionless ratio of the tube thickness to the range of the attractive interaction. When X is large compared with 1, one obtains a swollen phase. At the other extreme, when $X \ll 1$, one finds a highly degenerate compact phase. The twilight zone between these two phases occurs in the vicinity of $X \sim 1$ and is characterized by marginally compact structures. The figure shows typical tube conformations in each of the phases.

crossover that we characterize colloquially as a "twilight zone" has no relationship to, and should not be confused with, the same terminology sometimes used in the studies of sequence similarity.) In this twilight zone is a rich interplay of the pairwise attractive interactions and the constraints imposed by the three-body interaction. This is a situation in which you are able to interact with your friend, but can only do so by positioning yourself right outside his/her room.

In the twilight zone, a tube is barely able to avail itself of the attractive interactions promoting compaction. In this region of parameter space, the forces promoting compaction just set in, and one would expect to obtain marginally compact structures that have the ability to expel the water from the interior. In addition, because the scale of the interaction strength is relatively small, one would expect a low ordering crossover temperature with entropic effects not being too important. Furthermore, the physical picture of a tube (recall that a tube can be thought of as many anisotropic coins tethered together) leads to a strongly anisotropic interaction between nearby tube segments: it is better to position them parallel rather than perpendicular to one another. Thus, in the twilight zone, one has a relatively weak and strongly anisotropic interaction. Because the tube segments have to position themselves next to one another and with the right relative orientation in order to avail themselves of the attractive self-interaction, one would expect a cooperative transition with few intermediates – the tube will need to snap into its correctly folded configuration. Also, because of the loss of flexibility regarding the relative positioning and orientation of nearby tube segments, one would expect a large decrease in the degeneracy.

In this marginally compact state, the number of candidate tube structures is somewhat limited. The ground-state structures of a short tube in the marginally compact phase subject to compaction are shown in the bottom panel of Figure 12.3 below.

Tubes and proteins

There is a truly remarkable coincidence between the structures one obtains in the marginally compact physical state of matter of short tubes and the building blocks of protein "native-state" structures, as shown in Figure 12.3. Proteins [14] are linear chains of amino acids, of which there are twenty naturally occurring types with distinct side chains. The backbone and several of the side chains are hydrophobic; under physiological conditions, globular proteins fold rapidly and reproducibly to somewhat compact conformations called their "native-state structures." In their native state, a hydrophobic core is created that is space-filling, as shown in Figure 12.4, and water is expelled from the interior. Even though human cells have hundreds of thousands of proteins, the total number of distinct folds that they adopt in their native states is only on the order of a few thousand [24, 25, 26]. Furthermore, these structures seem to be evolutionarily conserved [27, 28]. Proteins are relatively short-chain molecules; indeed, longer globular proteins form domains that fold autonomously [29]. The building blocks of protein structures are helices, hairpins, and almost planar sheets (see Figure 12.3). Strikingly, short tubes, with no heterogeneity, in the marginally compact phase form helices with the same pitch to radius ratio as in real proteins [30] (see Figure 12.4) and almost planar sheets made up of zig-zag strands. It is interesting to note that the helix is a very natural conformation for a tube and occurs without any explicit introduction of hydrogen bonding. As in the tube case, small globular proteins show a two-state behavior [31, 32, 33, 34, 35] in their folding pattern.

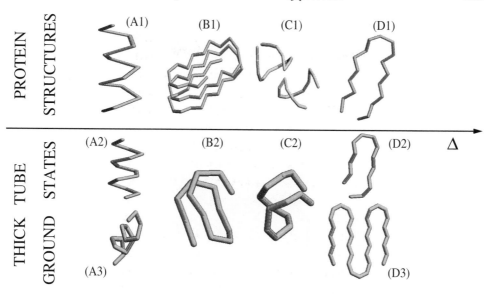

PROTEIN STRUCTURES

THICK TUBE GROUND STATES

Figure 12.3. Building blocks of biomolecules and ground-state structures associated with the marginally compact phase of a short tube. The axis in the middle indicates the direction along which the tube thickness Δ increases. The top row shows some of the building blocks of biomolecules; the bottom row depicts the corresponding structures obtained as the ground-state conformations of a short tube. (A1) is an α-helix of a naturally occurring protein, whereas (A2) and (A3) are the helices obtained in our calculations: (A2) has a regular contact map, whereas (A3) is a distorted helix in which the distance between successive atoms along the helical axis is not constant but has period 2. (B1) A helix of strands in the alkaline protease of *Pseudomonas aeruginosa*; (B2) the corresponding structure obtained in our computer simulations. (C1) The "kissing" hairpins of RNA and (C2) the corresponding conformation obtained in our simulations. Finally, (D1) and (D2) are two instances of quasi-planar hairpins. The first structure is from the same protein as before (the alkaline protease of *Pseudomonas aeruginosa*); the second is a typical conformation found in our simulations. The sheet-like structure (D3) is obtained for a longer tube (see [21] for more details about the simulation).

Let us make the *constructive hypothesis* that the extraordinary similarity between the structures adopted by short tubes in the marginally compact phase and the building blocks of protein native-state structures is not a mere coincidence. We postulate instead that the tube picture presented above is a paradigm for understanding protein structures. Quite generally, such postulates are of limited utility unless one is able to unify seemingly unrelated aspects of the problem and make new predictions amenable to experimental verification. In our case, although the tube idea is theoretical, a wealth of experimental data is already available on proteins. Before we proceed to explore the consequences of our hypothesis, we

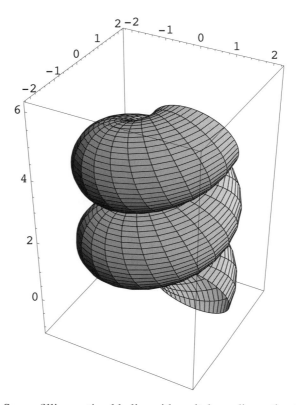

Figure 12.4. Space-filling optimal helix, with a pitch : radius ratio of 2.512. This value is determined by requiring that the local radius of curvature of the axis of the helix is equal to half the minimum distance of closest approach between different turns of the helix. The corresponding tube (which can be thought of as being inflated uniformly around the axis) is space-filling. Strikingly, the same geometry is found, within 3%, for α-helices in the native-state structures of proteins [30].

will first link the tube picture with the protein problem, using experiments as a guide.

Let us begin by asking whether the backbone of a protein can be described as a tube. Figure 12.5 indeed shows that, in its native state, the protein backbone can be thought of as the axis of a tube of approximate radius of cross-section (Δ) equal to 2.7 Å. Interestingly, the tube radius shows small variations, especially in the vicinity of backward bends [36].

The marginally compact phase in tubes occurred for a finely tuned ratio, X, around 1, of the tube thickness to the range of attractive interactions. Strikingly, for proteins, this quantity is self-tuned around this value. Steric interactions lead to a vast thinning of the phase space that protein structures can explore [37, 38]. Physically, the notion of a thick chain or a tube follows directly from steric interactions in a protein: one

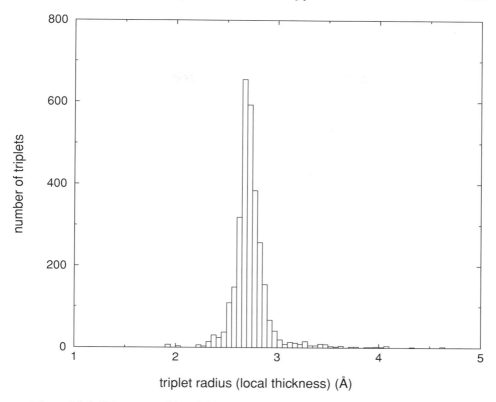

Figure 12.5. Histogram of local thicknesses computed for all residues of different protein native structures, when the virtual chain formed by the backbone C_α atoms is viewed as a discretized thick tube. At a given residue, the local thickness is simply the minimum triplet radius over all triplets containing that residue.

needs room around the backbone to house the amino-acid side chains without any overlap. The same side chains that determine the tube thickness also control the range of attraction: the outer atoms of the side chain interact through a short-range interaction screened by the water. This self-tuning is a quite remarkable feature of proteins.

The system, being poised in this twilight zone, has several significant advantages between large and small values of X with the limited number of marginally compact structures as the candidate native-state conformations. The thermodynamic limit of a tube of infinite length has a first-order transition, on decreasing the tube thickness, between a swollen phase and a compact phase. This phase transition is characterized, nevertheless, by a diverging length scale: the propensity for nearby tube segments to be aligned just right with respect to one another leads to a diverging persistence length, defined as the characteristic length over which memory of the tube orientation is preserved.

Let us briefly review the well-studied subject of phase transitions and critical phenomena [39]. Examples of critical points include a magnet at the onset of ordering, a liquid–vapor system at the critical temperature and pressure, and a binary liquid system that is about to phase-separate. The key point is that the fluctuations in a system at its critical point occur at all scales, and the system is exquisitely sensitive to tiny perturbations. Even though sharp phase transitions can occur only in infinitely large systems, behavior akin to that at a phase transition is observed for systems of finite size as well. Indeed, for a system near a critical point, the largest scale over which fluctuations occur is determined either by how far away one is from the critical point or by the finite size of the system.

A magnet at low temperatures compared with its critical temperature is well magnetized and is not very sensitive to a tiny external field. After all, when the magnetization is large, small perturbations do not lead to major consequences. Similarly, a magnet at very high temperatures is not very sensitive to a tiny external field because the strong thermal fluctuations dominate and the ordering tendencies are rather small. However, at the critical point, where an onset of the magnetization is about to occur, the system is very sensitive to an applied magnetic field, and indeed the magnetic susceptibility for an infinite system diverges.

Nature, in her desire to design proteins to serve as smart and versatile machines, has used a system poised near a phase transition to exploit this sensitivity. Indeed, it is well-known that proteins utilize conformational flexibility [40] to achieve optimal catalytic properties [6]. That protein structures are poised near a phase transition provides the versatility and the flexibility needed for the amazing range of functions that proteins perform.

The rapid folding of small proteins can be understood in terms of the inherent anisotropy of a tube and the self-tuning of the two key length scales, the tube thickness and the range of the attractive interactions. In the marginally compact phase, in order to avail themselves of the attractive interactions, nearby segments of the tube have to snap into place parallel to and right up against one another. As stated before, both in the tube picture and in proteins, the helix and the sheet are characterized by such parallel space-filling alignment of nearby tube segments. In proteins, such an arrangement serves to expel the water from the protein core. As shown by Linus Pauling and co-workers [10, 11], hydrogen bonds provide the scaffolding for both helices and sheets and place strong geometrical constraints stemming from quantum chemistry.

Beyond the tube archetype: a refined tube model informed by protein data

We turn now to a marriage of the tube idea and the wealth of information available from a variety of experimental probes [6, 41] in preparation for the task of exploring the consequences of our hypothesis. Recall that three-body local and non-local radius constraints describe the self-avoidance of a tube [19]. Unlike unconstrained matter, for which pairwise interactions suffice, for a chain molecule it is necessary to define the context of the object that is part of the chain. This is most easily carried out by defining a local Cartesian co-ordinate system whose three axes are defined by the tangent to the chain at that point, the normal, and the binormal that is perpendicular to both the other two vectors. A study [42] of the experimentally determined native-state structures of proteins from the Protein Data Bank [43] reveals that clear amino-acid-aspecific geometrical constraints are placed on the relative orientation of the local co-ordinate systems because of constraints due to sterics and the chemistry of backbone hydrogen bonds.

Recently, we [42] have carried out Monte Carlo simulations of short *homopolymers*, chains made up of just one type of amino acid, subject to these geometrical constraints and physically motivated interaction energies, a local-bending energy penalty, e_R, an overall hydrophobicity, e_W, and effective hydrogen-bond energies. The resulting phase diagram and the associated structures are depicted in Figure 12.6. In keeping with the behavior of the archetype tube discussed earlier, in the vicinity of the swollen phase one obtains distinct assembled tertiary structures, quite akin to real protein structures, on making small changes in the interaction parameters.

The marginally compact phase has distinct structures, including a single helix, a bundle of two helices, a helix formed by β-strands, a β-hairpin, three-stranded β-sheets with two distinct topologies, and a β-barrel-like conformation. These structures are the stable ground states in different parts of the phase diagram. Furthermore, conformations such as the β–α–β motif are found to be competitive local minima. The specific structure depends on the precise values of the local radius of curvature penalty (a large penalty forbids tight turns associated with helices, resulting in an advantage for sheet formation) and the strength of the hydrophobic interactions (a stronger overall attraction leads to somewhat more compact well-assembled tertiary structures). The topology of the phase diagram allows for the possibility of conformational switching, leading to the conversion of an α-helix to a β-topology on changing the hydrophobicity parameter, analogous to the influence of denaturants or alcohol in experiments [45].

Free-energy landscape of proteins

Many previous studies of proteins have been done from a physics point of view. The standard approach is to assume an overall attractive short-range potential that serves to lead to a compact conformation of the chain in its ground state. In the absence of amino-acid specificity or when dealing with a homopolymer, the ground

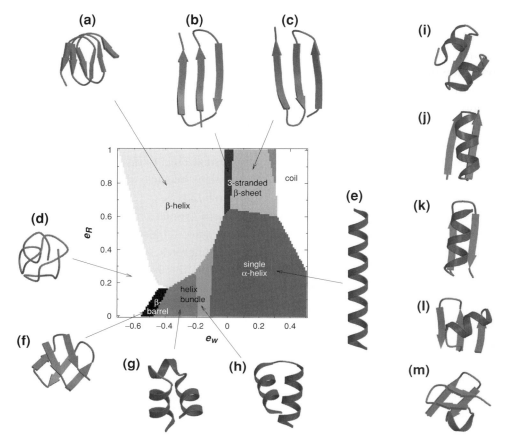

Figure 12.6. Phase diagram of ground-state conformations. The ground-state conformations were obtained by means of Monte Carlo simulations of chains of 24 C_α atoms; e_R and e_W denote the local radius of curvature energy penalty and the solvent-mediated interaction energy, respectively. More than 600 distinct local minima were obtained in different parts of parameter space by running simulations starting from a randomly generated initial conformation. The temperature is set initially at a high value and then decreased gradually to zero. Parts (a), (b), (c), (e), (f), (g), (h) are the Molscript representation of the ground-state conformations that are found in different parts of the parameter space, as indicated by the arrows. Conformations (i), (j), (k), (l), (m) are competitive local minima. In the phase at topmost right, the ground state is a 2-stranded β-hairpin (not shown). Two distinct topologies of a 3-stranded β-sheet (dark and light phases) are found, corresponding to conformations shown in Conformations (b) and (c), respectively. The white region on the left of the phase diagram has large attractive values of e_W; the ground-state conformations are compact globular structures with a crystalline order induced by hard-sphere-packing considerations [44] and not by hydrogen bonding (Conformation (d)).

state is highly degenerate and comprises all maximally compact conformations, as shown in Figure 12.7(a). This ground-state degeneracy grows exponentially with the length of the homopolymer. The role played by sequence heterogeneity is to break this degeneracy, leading to a very rugged landscape with a specific native-state conformation, which, of course, depends on the amino-acid sequence, as shown in Figure 12.7(b). The motion of a protein in a rugged landscape can be subject to trapping in local minima, and a model protein may not be able to fold rapidly. Glassy behavior may ensue because of such trapping. Bryngelson and Wolynes [46] suggested that a principle of minimal frustration is at work in which there is a nice fit between a given sequence and its native-state structure, carving out a funnel-like landscape [47] that promotes rapid folding and avoids the glassy behavior, as shown in Figure 12.7(c).

Indeed, the common belief in the field of proteins is that given a sequence of amino acids, with all the attendant details of the side chains and the surrounding water, one obtains a funnel-like landscape with the minimum corresponding to its native-state structure. Each protein is characterized by its own landscape. In this scenario, the protein sequence is all-important, and the protein-folding problem, besides becoming tremendously complex, needs to be attacked on a protein-by-protein basis.

In contrast, our model calculations show that the large number of common attributes of globular proteins [48, 49] reflect a deeper underlying unity in their behavior. At odds with conventional belief, a consequence of our hypothesis is that the gross features of the energy landscape of proteins result from the amino-acid-aspecific common features of all proteins. This landscape is *(pre)sculpted* by general considerations of geometry and symmetry (Figure 12.7(b)). Our unified framework suggests that the protein-energy landscape should have thousands of broad minima corresponding to putative native-state structures.

The key point is that for each of these minima the desirable funnel-like behavior is already achieved at the homopolymer level *in the marginally compact part of the phase diagram*. The self-tuning of two key-length scales – the thickness of the tube and the interaction range – so that they are comparable to each other and the interplay of the three energy scales – hydrophobic, hydrogen bond, and bending – in such a way as to stabilize marginally compact structures also provide the close cooperation between energy gain and entropy loss needed for the sculpting of a funneled energy landscape.

Consequences of a presculpted free-energy landscape

Recent work has shown that the rate of protein folding is substantially the same [50], even with large changes in the amino-acid sequence [50, 51], as long as the overall

Homopolymer (maximally compact)

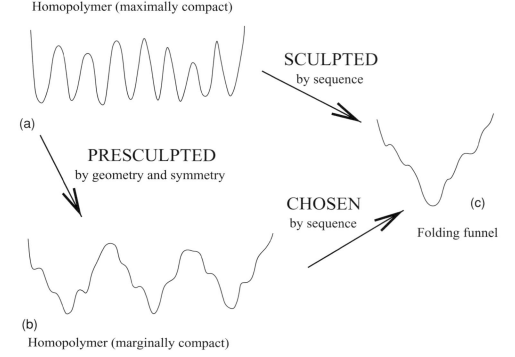

(a)

SCULPTED
by sequence

PRESCULPTED
by geometry and symmetry

CHOSEN
by sequence

(c)

Folding funnel

(b)

Homopolymer (marginally compact)

Figure 12.7. Simplified one-dimensional sketches of energy landscape. The quantity plotted on the horizontal axis schematically represents a distance between different conformations in the phase space; the barriers in the plots indicate the energy needed by the chain to travel between two neighboring local minima. (a) Rugged energy landscape for a homopolymer chain with an attractive potential promoting compaction as, for, example, in a string-and-beads model. Many distinct, maximally compact ground-state conformations have roughly the same energy, separated by high-energy barriers (the degeneracy of ground-state energies would be exact in the case of both lattice models and off-lattice models with discontinuous square-well potentials). (b) Presculpted energy landscape for a homopolymer chain in the marginally compact phase. The number of minima is greatly reduced and the width of their basin increased by the introduction of geometrical constraints. (c) Funnel-energy landscape for a protein sequence. As folding proceeds from the top to the bottom of the funnel, its width, a measure of the entropy of the chain, decreases cooperatively with the energy gain. Such a distinctive feature, crucial for fast and reproducible folding, arises from careful sequence design in models whose homopolymer energy landscape is similar to (a). In contrast, funnel-like properties already result from considerations of geometry and symmetry in the marginally compact phase (b), thereby making the goals of the design procedure the relatively easy task of stabilizing one of the presculpted funnels followed by the more refined task of fine-tuning the putative interactions of the protein with other proteins and ligands.

topology of the folded structure is the same. Furthermore, mutational studies [52, 53] have shown that the structures of the transition states are also similar and in complete accord with our picture of a pre-sculpted free-energy landscape.

Sequence design [54] would favor the appropriate native-state structure over the other putative ground states, leading to a free-energy landscape conducive to rapid and reproducible folding of that particular protein. Nature has a choice of twenty amino acids for the design of protein sequences. A pre-sculpted landscape greatly facilitates the design process. Indeed, within our model, we find that a crude design scheme, which takes into account the hydrophobic (propensity to be buried) and polar (desire to be exposed to the water) character of the amino acids is sufficient to carry out a successful design of sequences with one or the other of the structures shown in Figure 12.6. The matching of the hydrophobic profile of the designed sequence to the burial profile [55] (as measured by the number of neighbors within the range of the hydrophobic interaction) leads to the correct fold in a Monte Carlo simulation.

Also, as is seen experimentally, many protein sequences adopt the same native-state conformation [56]. Once a sequence has selected its native-state structure, it is able to tolerate a significant degree of mutability except at certain key locations [54]. Furthermore, multiple protein functionalities can arise within the context of a single fold [57].

One successful method of predicting protein structure is based on threading [58]. The basic idea is entirely consistent with our findings. Using pieces of native-state structures of longer proteins as possible candidate structures of a shorter protein, the technique is simple because, instead of determining the structure from *ab initio* calculations, one merely has to select from among the putative native-state structures. The documented success of the threading method confirms that each protein does not fashion its own native-state structure, but merely selects from the menu of predetermined folds.

Amyloid phase of proteins

A range of human diseases such as Alzheimer's, spongiform encephalopathies, and light-chain amyloidosis lead to degenerative conditions and involve the deposition in tissue of plaque-like material arising from the aggregation of proteins [15, 16, 59, 60]. In prions [60], one observes a transition from α- to β-rich structures, which favors aggregation and causes the disease bovine spongiform encephalopathy (BSE). It has been argued [61] that the formation of amyloid fibrils occurs in a hierarchical way starting from a chiral β-strand. The resulting structures arise from a competition between the free-energy gain from the aggregation and the elastic-energy cost of the distortion. A variety of proteins not involved in these diseases also form aggregates very similar to those implicated in the diseased state [16, 59]. This suggests [16] that the tendency for proteins to aggregate is a generic property of polypeptide chains with the specific sequence of amino acids playing, at best, a secondary role.

Let us consider the phase obtained when the tube is sufficiently long (or when there are many interacting tubes) and is subject to attractive interactions leading to compaction. In this phase, the tube is stretched out locally with nearby sections parallel to one another (or the tubes are stacked parallel to one another in a periodic arrangement) and does not have the richness we associate with protein native-state structures. Returning to the protein, one may ask whether some structures are the analogs of those found in this so-called semicrystalline phase.

In collaboration with Hoang, Seno, and Trovato, we have found that a β-sheet structure is a significant competitor with a large basin of attraction in a region where the stable phase is a helix, reinforcing the possibility that the interaction between several proteins could stabilize the formation of extended hydrogen-bonded β-sheets via the aggregation of individual chains. These kinds of structures, which resemble the basic structures associated with amyloid fibrils, thus seem to belong to the general class of predetermined folds, but this time for multiple proteins, and should be seen ubiquitously in generic proteins [16, 59]. This suggests that the key to preventing such aggregates is to stabilize helices in such proteins and evolutionary mechanisms such as proteasomes, molecular chaperones [62], and ubiquitination enzymes [16, 59].

Our protein-tube hypothesis shows that long chains of amino acids have a tendency to form amyloids rather than maintain their protein-like shape. Indeed, nature has on suitable occasions thwarted this tendency by dividing the protein into substantially independent domains that fold autonomously and are then assembled together. This suggests that the variety of protein folds increases with length up to a certain point at which they are supplanted by the formation of domains or amyloids.

In a recent paper, Fandrich and Dobson [63] suggested that

amyloid formation and protein folding represent two fundamentally different ways of organizing polypeptides into ordered conformations. Protein folding depends critically on the presence of distinctive side chain sequences and produces a unique globular fold. By contrast . . . amyloid formation arises primarily from main chain interactions that are, in some environments, overruled by specific side chain contacts.

Our results are in complete accord with the suggestion that amyloid structures may arise from the generic properties of the proteins, with the details of the amino-acid side chains playing a secondary role. However, our work suggests that instead of an "inverse side chain effect in amyloid structure formation [63]," there is a unifying theme in the behavior of proteins. Just as the class of cross-linked β-structures are determined from geometrical considerations, the menu of protein native-state structures is also determined by the common attributes of globular proteins: the inherent anisotropy associated with a tube and the geometrical constraints imposed by hydrogen bonds and steric considerations.

Natural selection and protein interactions

Natural selection

Traditionally, the framework of evolution in life works through two aspects of organization, the genotype and the phenotype. The genotype is the heritable information encoded in the DNA, which is translated through the RNA molecules into proteins. The phenotype is valuable for adaptation and at the molecular level plays a key role in natural selection. One conventionally assumes that a selection of phenotypes leads to an enhancement in the numbers of the genotype. Furthermore, mutations of the genotype lead to the possibility of new phenotypes.

Let us consider the situation at two levels: the sequence level (which is the genotype because it is a direct translation from the evolving DNA molecules) and the structure level (which we can think of as the phenotype). As pointed out by Maynard Smith [64], as the sequence undergoes mutation, the mutated sequences must traverse a continuous network without passing through any intermediaries that are non-functioning. Thus, one seeks a connected network in sequence space for evolution by natural selection to occur. Considerable evidence accumulated since the pioneering suggestion of Kimura [65] and King and Jukes [66] shows that much of evolution is neutral. The experimental data strongly support the view that the "random fixation of selectively neutral or very slightly deleterious mutants occurs far more frequently in evolution than selective substitution of definitely advantageous mutants" [67]. Also "those mutant substitutions that disrupt less the existing structure and function of a molecule (conservative substitutions) occur more frequently in evolution than more disruptive ones" [67]. Thus, although one has a "random walk" in sequence space that forms a connected network, there is no similar continuous variation in structure space.

These facts are in accord with our result of a pre-sculpted free-energy landscape that is shared by all proteins and has thousands of local minima corresponding to putative native-state structures – not too few because that would not lead to sufficient diversity, and not too many because that would lead to too rugged a landscape with little hope that a protein could fold reproducibly and rapidly into its native-state structure. Indeed, many proteins share the same native-state fold, and often the substitution of one amino acid for another does not lead to radical changes in the native-state structure, underscoring the fact that it is not the details of the amino acid side chains that sculpt the free-energy landscape, but rather some overarching features of symmetry and geometry that are common to all proteins. In this respect, the phase of matter that forms the native-state structures is one that is determined by physical law rather than by the plethora of microscopic details analogous to the limited menu of possible crystal structures.

Protein interactions and functionality

Anfinsen [32] wrote in 1973, "Biological function appears to be more a correlate of macromolecular geometry than of chemical detail." Much recent progress has been made in extracting information on biological function and protein interactions [68] from the structure of proteins and the complexes they form [69]. A protein structure chosen from the predetermined menu of folds contains information on the topology of the folded state. In addition, one can glean information on the nature of the exposed surface and crystal packing and the existence of clefts or other geometrical features (which are often the active sites of enzymes). The picture is completed by knowledge of the sequence of amino acids that folds into the structure, using which one can infer the amino-acid composition of the exposed surfaces, the location of mutants, and conserved residues and evolutionary relationships. For some structural families, function is highly conserved, whereas for others one can use the types of information described above to guess the function [70].

There is increasing evidence that evolution along with natural selection allows nature to use variations on the same theme facilitated by the rich repertory of amino acids to create enzymes that are able of catalyzing a remarkable array of diverse and complex tasks in the living cell. The key point, of course, is that a constant backdrop of folds not shaped by sequence but determined by physical law is necessary for molecular evolution to work in this manner. Were the folds not immutable and themselves subject to Darwinian evolution, the possibility of creating many subtle and wonderful variations on the same theme would not exist. The pre-sculpted landscape is the crucial feature that leads to a predetermined menu of immutable folds.

Other consequences

The picture we have developed based on the tube-protein hypothesis has several attractive features. First, as noted before, protein structures lie in the vicinity of a phase transition to the swollen phase, which confers on them exquisite sensitivity to the effects of other proteins and ligands. The flexibility of different parts of the protein depends on the amount of constraints placed on them from the rest of the protein [40]. From this point of view, it is easy to understand how loops, which are not often stabilized by backbone hydrogen bonds, can play a key role in protein functionality.

It is useful to reconsider how nature uses the variety of amino acids for sequence design. The existence of a pre-sculpted free-energy landscape with broad minima corresponding to the putative native-state structures and the existence of neutral evolution demonstrate that the design of sequences that fit a given structure is relatively easy, leading to many sequences that can fold into a given structure. This freedom facilitates the accomplishment of the next-level task of evolution through natural selection: the design of optimal sequences, which not only fold into the desired native-state structure, but also are fit in the environment of other proteins. A useful protein is one that can interact with other proteins in a synergistic manner and at the same time is not subject to the tendency to aggregate into the harmful amyloid form. This suggests that protein-engineering studies aimed at improving enzymatic function should focus on two aspects in a sequential manner. First, the family of sequences that fold into a desired target structure need to be selected, and then a finer design needs to be carried out in the context of the substrates and the other proteins with which the target protein interacts. Unlike the generality of geometry and symmetry that leads to the menu of native-state folds, what we have here is a problem of chemistry acting within the fixed background of the physically determined structures. These considerations suggest that, when the information becomes available, protein–protein interaction networks [71] can be fruitfully viewed not only as the interactions between proteins, but also as the interactions between the structures that house them.

The two characteristics required for protein native-state structures to be targets of an evolutionary process are stability and diversity. Stability is needed because one would not want to mutate away a DNA molecule able to code for a useful protein, and diversity is needed to allow evolution to build complex and versatile forms. The mechanism for natural selection arises naturally in this context: DNA molecules that code for amino-acid sequences that fit well into one of these predetermined folds and have useful functionality thrive at the expense of molecules that create sequences that are not useful. Indeed, in this picture, sequences and functionality evolve in order to fit within the constraints of these folds, which, in turn, are immutable and determined by physical law.

The situation is somewhat reminiscent of a content-addressable memory [72], in which partial information is converted by the brain to recover the complete information. Such content-addressable memories [72], as well as the energy landscape [73] suitable for prebiotic evolution [74], have been modeled through spin glasses [75]. The energy landscape of spin glasses is also characterized by diversity and stability arising from randomness and frustration, which is quite distinct from the the physical mechanisms of short tubes in the marginally compact phase.

In conventional spin glasses, randomness, which plays a role somewhat similar to amino-acid-specific interactions in proteins [76], sculpts, through frustration, an energy landscape with many local minima. Indeed, a non-random exchange interaction between spins would lead to periodic order with much simpler behavior. In spin glasses, starting from a random spin configuration, it is hard to reach a specific local minimum unless the exchange constants are tuned in a clever way, as in a content-addressable memory. The landscape is not invariant on changing the exchange interactions and can be fashioned at will. For proteins, on the other hand, our analysis shows that a rich landscape is obtained even in the absence of any sequence heterogeneity, and the nature of the ground states is determined by geometry and symmetry and is therefore immutable [28].

Summary and perspective

Symmetry and geometry place strong constraints on the types of infinite-sized crystal structures, and there are exactly 230 distinct space groups in three dimensions. Proteins are finite-sized objects. Our analysis demonstrates that the same kind of symmetry and geometrical considerations lead to a finite number of protein folds. This number grows with the size of the protein, but is limited by the fact that proteins beyond a characteristic length form either autonomous domains or amyloids. Unlike the crystalline state of matter, proteins are characterized by an inherent anisotropy because of their tube-like character. A given crystalline structure transcends the material that is housed in it: common salt adopts the face-centered-cubic lattice structure, as do the well-packed cannonballs of Kepler [18]. Likewise, different sequences of proteins can be housed in the same protein fold and yet be able to perform different functions [57]. Protein structures are modular in form, being simple assemblages of helices and strands connected by tight turns.

The unified picture leads to a single free-energy landscape with two distinct classes of structure. The amyloid phase is dominated by β-strands linked to one another in a variety of forms, whereas the native-state structure menu is an assembly of α-helices and β-structures. Nature has exploited these native-state structures in the context of the workhorse molecules of life. The selection mechanism for genetic evolution at the molecular level lies in the ability of the protein encoded by the gene to fold well into one of the predetermined folds and have a useful function. Unfortunately, however, the proximity of this beautiful phase to the generic amyloid phase underscores how life can easily malfunction as soon as aggregational tendencies of proteins come to the fore. One cannot but marvel at the robustness of life.

The protein problem, which lies at the intersection of many disciplines, is highly complex. Evolution complicates the situation even further. Human design allows for an engineer to devise entirely new ways of accomplishing certain tasks – a classic example is the replacement of vacuum tubes with semiconductor transistors. Nature does not have this luxury in evolutionary design. Nature takes what she has, tinkers with it, and builds on it. Thus, the notion of optimal design is not particularly relevant, and the future is very strongly correlated with the present and the past. A slightly different turn of events could have led to conspicuously different life forms. This picture of nature muddling along through evolution combined with the inherent complexity of proteins makes the problem very daunting. Yet, within this complexity, there is a stunning simplicity provided by the fixed backdrop of the protein folds determined by physical law, in the context of which sequences and functionalities are shaped by evolution.

Let us revisit the classic theoretical work, shown in Figure 12.8, of Pauling [10, 11] and Ramachandran [37]. Both of them considered the protein backbone that is the common part of all proteins. Pauling and his co-workers explored the types of structures that are consistent with both the backbone geometry and the formation of hydrogen bonds. They predicted that helices and sheets are the structures of choice in this regard. Ramachandran and his co-workers carried out their pioneering work more than a decade after Pauling. They considered the role of excluded volume or steric interactions between nearby amino acids along the sequence in reducing the available conformational phase space. Astonishingly, the two significantly populated regions of the Ramachandran plot correspond to the α-helix and the β-strand. Even though hydrogen bonds and steric constraints are not related to each other, they are both promoters of helices and sheets. Is this concurrence of events a mere accident? The marginally compact phase of short tubes has helices and sheets as its preferred structures. In order for nature to take advantage of this phase of matter, proteins, which obey physical law, may have been selected to conform to the tube geometry. Hydrogen bonds serve to enforce the parallelism of nearby tube segments, a feature of both helices and sheets, and steric constraints emphasize the non-zero thickness of the tube and serve to position it in the marginally compact phase. Because the marginally compact phase is a finite-size effect, proteins tend to be relatively short compared with conventional macromolecules, including DNA. Indeed, proteins seem to be a vivid example of the adaptation of nature to her own laws.

Nature has used several classes of chain molecule – DNA, RNA, and proteins – to create living matter. (Small, unconstrained molecules are not very useful for this because one cannot associate a natural context for them.) The DNA molecule carries information and is able to replicate itself, albeit with occasional mistakes. Importantly, these mistakes, which form the basis of molecular evolution, are also replicated in future generations. The RNA molecule plays a key role in translating the information contained in the DNA for making proteins. Proteins, the workhorse

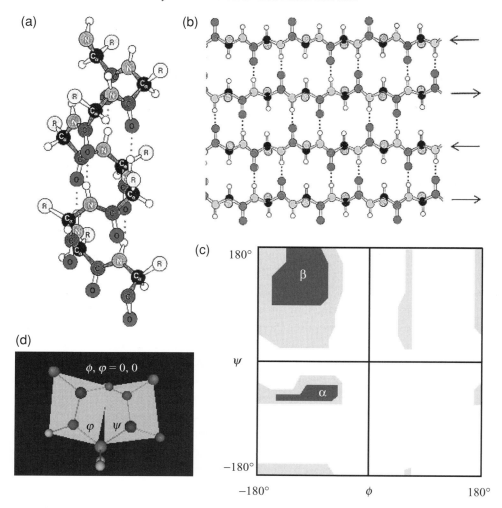

Figure 12.8. Pauling and Ramachandran revisited. The top row depicts the "classic" structures of an α-helix (a) and a pleated β-sheet (b). The main-chain backbone atoms and the C^β atoms of the side-chain groups are shown (shading is different for (a) and (b)). Hydrogen bonds, which stabilize the structures, are shown as dotted lines. In the bottom row we show the Ramachandran plot (c) describing how the torsional degrees of freedom (ψ, φ), the backbone dihedral angles within an all-atom representation, are constrained by steric effects. The shaded areas in the plot correspond to allowed regions in conformational space. The structures (a) and (b) stabilized by hydrogen bonding indeed lie squarely within the sterically accessible regions. An example of a dipeptide conformation disallowed because of steric hindrance is shown in (d).

molecules of life, are wonderful molecular machines that carry out a variety of functions and speed up chemical reactions by orders of magnitude. A single protein may have a variety of capabilities, but the work it does, although efficient, is monotonous. The situation changes dramatically when one has a collection of proteins organized in a network. These proteins interact with one another, catalyze chemical reactions, turn the gene on or off, and lead to the robust and coherent behavior that we associate with life.

The structure of the DNA molecule provides a beautiful explanation of how it is able to encode information and the mechanism underlying its replication. Proteins, on the other hand, are less well understood. One could ask what kind of a phase of matter one would choose to house protein structures in order to accommodate the important roles that these molecules of life play. Our work suggests that a very special, previously unstudied phase of matter is associated with the marginally compact phase of short tubes with a thickness specially tuned to be comparable to the range of attractive interactions promoting the compaction. This phase is a finite-size effect and exists only for relatively short tubes; it is poised near a phase transition of a new kind that lends itself to flexibility in the structure; the structures that one finds in the marginally compact phase are space-filling and modular in construction, being made up of two principal building blocks – helices and sheets; the total number of distinct folds is relatively small and only on the order of a few thousand or so, and proteins are able to fold rapidly and reproducibly into them. The price that nature pays for utilizing this novel phase of matter is the relative ease with which aggregation of multiple tubes can occur, leading to amyloid formation.

In his insightful book, *The Fitness of the Environment* [1], Henderson extended the notion of Darwinian fitness to argue that "the fitness of [the] environment is quite as essential a component as the fitness which arises in the process of organic evolution." Strikingly, the chemistry of proteins ensures that they are self-tuned to occupy the marginally compact phase of short tubes. One cannot but marvel at how several factors – the steric interactions; hydrogen bonds, which provide the scaffolding for protein structures; the constraints placed by quantum chemistry on the relative lengths of the hydrogen and covalent bonds; the near planarity of the peptide bonds; and the key role played by water – all reinforce and conspire with one another to place proteins in this novel phase of matter.

Proteins have proved to be difficult to understand because of (1) their inherent complexity with twenty types of amino acids and the role played by water; (2) their relatively short length compared with generic human-made polymers, which means they are therefore likely to be characterized by "non-universal" behavior; and (3) the complexities associated with the random process of evolution. Nevertheless, our work suggests an underlying stunning simplicity. Although sequences and functionalities of proteins evolve, the folds that they adopted, which in turn determine function, seem to be determined by physical law and are not subject to Darwinian evolution. In that regard, these folds may be thought of as immutable or Platonic. Protein folds do not evolve: rather, the menu of possible folds is determined by physical law. In that sense, it is as if evolution acts in the theater of life and shapes sequences and functionalities, but does so within the fixed backdrop of the Platonic folds.

Henderson wrote [1]:

The properties of matter and the course of cosmic evolution are now seen to be intimately related to the structure of the living being and to its activities; they become, therefore, far more important in biology than has been previously suspected. For the whole evolutionary process, both cosmic and organic, is one, and the biologist may now rightly regard the universe in its very essence as biocentric.

His intriguing ideas continue to provoke thought even as we strive to understand the connections between life and the laws of nature.

Acknowledgments

We are indebted to our collaborators Trinh Hoang, Flavio Seno, and Antonio Trovato for their invaluable contributions to the ideas and work presented here; to Michael Denton for numerous stimulating discussions; and to John Barrow and Pamela Contractor for many helpful suggestions. This work was supported by the John Templeton Foundation.

References

[1] L. J. Henderson. *The Fitness of the Environment: An Inquiry into the Biological Significance of the Properties of Matter*. New York: Macmillan (1913). Repr. Boston, MA: Beacon Press (1958); Gloucester, MA: Peter Smith (1970).

[2] M. Hoagland and B. Dodson. *The Way Life Works*. New York, NY: Random House (1995).

[3] B. Alberts *et al. Molecular Biology of the Cell*. New York, NY: Garland Science Publishing (2002); D. Voet and J. G. Voet. *Biochemistry*. New York, NY: John Wiley and Sons (2003).

[4] J. D. Watson and F. H. C. Crick. Molecular structure of nucleic acids – a structure for deoxyribose nucleic acid. *Nature*, **171** (1953), 737.

[5] M. Kimura. *Population Genetics, Molecular Evolution and the Neutral Theory*. Chicago, IL: University of Chicago Press (1994).

[6] A. Fersht. *Structure and Mechanism in Protein Science: A Guide to Enzyme Catalysis and Protein Folding*. New York, NY: W. H. Freeman and Company (1999).

[7] A. V. Finkelstein and O. Ptistyn. *Protein Physics: A Course of Lectures*. New York, NY: Academic Press (2002).

[8] A. Kornberg. *For the Love of Enzymes*. Cambridge, MA: Harvard University Press (1989).
[9] For a brief review of computational approaches to the protein-folding problem, see J. R. Banavar and A. Maritan. Computational approach to the protein-folding problem. *Proteins*, **42** (2001), 433.
[10] L. Pauling, R. B. Corey and H. R. Branson. The structure of proteins – 2 hydrogen-bonded helical configurations of the polypeptide chain. *Proceedings of the National Academy of Sciences, USA*, **37** (1951), 205.
[11] L. Pauling and R. B. Corey. Configurations of polypeptide chains with favored orientations around single bonds – 2 new pleated sheets. *Proceedings of the National Academy of Sciences, USA*, **37** (1951), 729.
[12] D. Eisenberg. The discovery of the α-helix and β-sheet, the principal structural features of proteins. *Proceedings of the National Academy of Sciences, USA*, **100** (2003), 11207.
[13] A. P. Lightman. *Great Ideas in Physics*. New York, NY: McGraw-Hill (2000).
[14] T. E. Creighton. *Proteins: Structures and Molecular Properties*. New York, NY: W. H. Freeman and Company (1993).
[15] J. W. Kelly. The alternative conformations of amyloidogenic proteins and their multi-step assembly pathways. *Current Opinions in Structural Biology*, **8** (1998), 101.
[16] C. M. Dobson. Protein folding and disease: a view from the first Horizon Symposium. *National Review of Drug Discoveries*, **2** (2003), 154.
[17] P. M. Chaikin and T. C. Lubensky. *Principles of Condensed Matter Physics*. Cambridge, UK: Cambridge University Press (2000).
[18] G. G. Szpiro. *Kepler's Conjecture*. New York, NY: John Wiley (2003).
[19] J. R. Banavar, O. Gonzalez, J. H. Maddocks *et al.* Self-interactions of strands and sheets. *Journal of Statistical Physics*, **110** (2003), 35.
[20] A theorem proved by O. Gonzalez and J. Maddocks (Global curvature, thickness, and the ideal shapes of knots. *Proceedings of the National Academy of Sciences, USA*, **96** [1999], 4769) in the context of knot topologies shows the remarkable connection between the three-point recipe and the tube thickness.
[21] J. R. Banavar, A. Flammini, D. Marenduzzo *et al.* Geometry of compact tubes and protein structures. *ComPlexUs*, **1** (2003), 4.
[22] P. G. de Gennes and J. Prost. *The Physics of Liquid Crystals*. Oxford: Oxford University Press (1995).
[23] S. Chandrasekhar. *Liquid Crystals*. Cambridge, UK: Cambridge University Press (1977).
[24] C. Chothia and A. V. Finkelstein. The classifications and origins of protein folding patterns. *Annual Review of Biochemistry*, **59** (1990), 1007.
[25] C. Chothia. Proteins – 1000 families for the molecular biologist. *Nature*, **357** (1992), 543.
[26] T. Przytycka, R. Srinivasan and G. D. Rose. Recursive domains in proteins. *Protein Science*, **2** (2002), 409.
[27] C. Chothia, J. Gough, C. Vogel *et al.* Evolution of the protein repertoire. *Science*, **300** (2003), 1701.
[28] M. Denton and C. Marshall. Laws of form revisited. *Nature*, **410** (2001), 417.
[29] P. L. Privalov. Stability of proteins: proteins which do not present a single cooperative system. *Advances in Protein Chemistry*, **35** (1982), 1.
[30] A. Maritan, C. Micheletti, A. Trovato *et al.* Optimal shapes of compact strings. *Nature*, **406** (2000), 287.

[31] A. Ginsburg and W. R. Carroll. Some specific ion effects on the conformation and thermal stability of ribonuclease. *Biochemistry-US*, **4** (1965), 2159.

[32] C. B. Anfinsen. Principles that govern folding of protein chains. *Science*, **181** (1973), 223.

[33] S. E. Jackson. How do small single-domain proteins fold? *Folding and Design*, **3** (1998), R81.

[34] R. L. Baldwin and G. D. Rose. Is protein folding hierarchic? I. Local structure and peptide folding. *Trends in Biochemical Science*, **24** (1999), 26.

[35] D. Baker. A surprising simplicity to protein folding. *Nature*, **405** (2002), 39.

[36] J. R. Banavar, A. Maritan, C. Micheletti *et al.* Geometry and physics of proteins. *Proteins*, **47** (2002), 315.

[37] G. N. Ramachandran and V. Sasisekharan. Conformation of polypeptides and proteins. *Advances in Protein Chemistry*, **23** (1968), 283.

[38] R. Srinivasan and G. D. Rose. LINUS: a hierarchic procedure to predict the fold of a protein. *Proteins*, **22** (1995), 81.

[39] H. E. Stanley. Scaling, universality, and renormalization: three pillars of modern critical phenomena. *Reviews of Modern Physics*, **71** (1999), S358.

[40] D. J. Jacobs, A. J. Rader, L. A. Kuhn *et al.* Protein flexibility predictions using graph theory. *Proteins*, **44** (2001), 150; A. J. Rader, B. M. Hespenheide, L. A. Kuhn *et al.* Protein unfolding: rigidity lost. *Proceedings of the National Academy of Sciences, USA*, **99** (2002), 3540.

[41] S. W. Englander, N. W. Downer and H. Teitelbaum. Hydrogen exchange. *Annual Review of Biochemistry*, **41** (1972), 903; C. K. Woodward and B. D. Hilton. Hydrogen exchange kinetics and internal motions in proteins and nucleic acids. *Annual Review of Biophysics and Biomolecular Structure*, **8** (1979), 99; J. B. Udgaonkar and R. L. Baldwin. NMR evidence for an early framework intermediate on the folding pathway of ribonuclease A. *Nature*, **335** (1988), 694; A. Matouschek, J. T. Kellis Jr., L. Serrano *et al.* Transient folding intermediates characterized by protein engineering. *Nature*, **346** (1990), 440; S. E. Radford and C. M. Dobson. Insights into protein folding using physical techniques: studies of lysozyme and alpha-lactalbumin. *Philosophical Transactions of the Royal Society of London*, B **348** (1995), 17; A. R. Fersht. Characterizing transition states in protein folding: an essential step in the puzzle. *Current Opinion in Structural Biology*, **5** (1995), 79.

[42] T. X. Hoang, A. Trovato, F. Seno *et al.* Geometry and symmetry presculpt the free-energy landscape of proteins. *Proceedings of the National Academy of Sciences, USA*, **101** (2004), 7960.

[43] H. M. Berman *et al.* The Protein Data Bank. *Nucleic Acids Research*, **28** (2000), 235.

[44] Y. Zhou, C. K. Hall and M. Karplus. First-order disorder-to-order transition in an isolated homopolymer model. *Physical Review Letters*, **77** (1996), 2822.

[45] K. D. Wilkinson and A. N. Meyer. Alcohol-induced conformational changes of ubiquitin. *Archives of Biochemistry and Biophysics*, **250** (1986), 390; E. Dufour and T. Haertle. Alcohol-induced changes of beta-lactoglobulin-retinol-binding stoichiometry. *Protein Engineering*, **4** (1990), 185; M. M. Harding, D. H. Williams and D. N. Woolfson. Characterization of a partially denatured state of a protein by two-dimensional NMR: reduction of the hydrophobic interactions in ubiquitin. *Biochemistry-US*, **30** (1991), 3120; P. Fan, C. Bracken and J. Baum. Structural characterization of monellin in the alcohol-denatured state by NMR: evidence for beta-sheet to alpha-helix conversion. *Biochemistry-US*, **32** (1993), 1573; K. Shiraki, K. Nishikawa and Y. Goto. Trifluoroethanol-induced stabilization of the

alpha-helical structure of beta-lactoglobulin: implication for non-hierarchical protein folding. *Journal of Molecular Biology*, **245** (1995), 180.

[46] J. D. Bryngelson and P. G. Wolynes. Spin-glasses and the statistical-mechanics of protein folding. *Proceedings of the National Academy of Sciences, USA*, **84** (1987), 7524.

[47] P. E. Leopold, M. Montal and J. N. Onuchic. Protein folding funnels: a kinetic approach to the sequence-structure relationship. *Proceedings of the National Academy of Sciences, USA*, **89** (1992), 8271; P. G. Wolynes, J. N. Onuchic and D. Thirumalai. Navigating the folding routes. *Science*, **267** (1995), 1619; K. A. Dill and H. S. Chan. From Levinthal to pathways to funnels. *Nature Structural Biology*, **4** (1997), 10.

[48] J. R. Banavar and A. Maritan. Colloquium: geometrical approach to protein folding – a tube picture. *Reviews of Modern Physics*, **75** (2003), 23.

[49] J. D. Bernal. Structure of proteins. *Nature*, **143** (1939), 663.

[50] D. Perl, C. Welker, T. Schindler *et al.* Conservation of rapid two-state folding in mesophilic, thermophilic and hyperthermophilic cold shock proteins. *Nature Structural Biology*, **5** (1998), 229.

[51] D. S. Riddle, J. V. Santiago, S. T. Bray *et al.* Functional rapidly folding proteins from simplified amino acid sequences. *Nature Structural Biology*, **4** (1997), 805; D. E. Kim, H. Gu and D. Baker. The sequences of small proteins are not extensively optimized for rapid folding by natural selection. *Proceedings of the National Academy of Sciences USA*, **95** (1998), 4982.

[52] V. Villegas, J. C. Martinez, F. X. Aviles *et al.* Structure of the transition state in the folding process of human procarboxypeptidase A2 activation domain. *Journal of Molecular Biology*, **283** (1998), 1027; F. Chiti, N. Taddei, P. M. White *et al.* Mutational analysis of acylphosphatase suggests the importance of topology and contact order in protein folding. *Nature Structural Biology*, **6** (1999), 1005.

[53] J. C. Martinez and L. Serrano. Obligatory steps in protein folding and the conformational diversity of the transition state. *Nature Structural Biology*, **6** (1999), 1010; D. S. Riddle, V. P. Grantcharova, J. V. Santiago *et al.* Experiment and theory highlight role of native state topology in SH3 folding. *Nature Structural Biology*, **6** (1999), 1016.

[54] J. S. Richardson and D. C. Richardson. The de novo design of protein structures. *Trends in Biochemical Science*, **14** (1989), 304; W. F. DeGrado, Z. R. Wasserman and J. D. Lear. Protein design, a minimalist approach. *Science*, **243** (1989), 622; M. H. Hecht, J. S. Richardson, D. C. Richardson *et al.* De novo design, expression, and characterization of felix – a 4-helix bundle protein of native-like sequence. *Science*, **249** (1990), 884; C. P. Hill, D. H. Anderson, L. Wesson *et al.* Crystal-structure of alpha-1 – implications for protein design. *Science*, **249** (1990), 343; C. Sander and R. Schneider. Database of homology-derived protein structures and the structural meaning of sequence alignment. *Proteins*, **9** (1991), 56; S. Kamtekar, J. M. Schiffer, H. Y. Xiong *et al.* Protein design by binary patterning of polar and nonpolar amino-acids. *Science*, **262** (1993), 1680; A. P. Brunet, E. S. Huang, M. E. Huffine *et al.* The role of turns in the structure of an alpha-helical protein. *Nature*, **364** (1993), 355; A. R. Davidson and R. T. Sauer. Folded proteins occur frequently in libraries of random amino-acid-sequences. *Proceedings of the National Academy of Sciences USA*, **91** (1994), 2146; M. W. West, M. X. Wang, J. Patterson *et al.* De novo amyloid proteins from designed combinatorial libraries. *Proceedings of the National Academy of Sciences, USA*, **96** (1999), 11211; Y. Wei, S. Kim, D. Fela *et al.* Solution structure of a de novo protein from a designed

combinatorial library. *Proceedings of the National Academy of Sciences, USA*, **100** (2003), 13270.

[55] G. D. Rose, A. R. Geselowitz, G. J. Lesser *et al.* Hydrophobicity of amino acid residues in globular proteins. *Science*, **229** (1985), 834; S. Miller, J. Janin, A. M. Lesk *et al.* Interior and surface of monomeric proteins. *Journal of Molecular Biology*, **196** (1987), 641; C. Lawrence, I. Auger and C. Mannella. Distribution of accessible surfaces of amino acids in globular proteins. *Proteins*, **2** (1987), 153.

[56] J. U. Bowie, J. F. Reidhaar-Olson, W. A. Lim *et al.* Deciphering the message in protein sequences – tolerance to amino-acid substitutions. *Science*, **247** (1990), 1306; W. A. Lim and R. T. Sauer. The role of internal packing interactions in determining the structure and stability of a protein. *Journal of Molecular Biology*, **219** (1991), 359; D. W. Heinz, W. A. Baase and B. W. Matthews. Folding and function of a t4 lysozyme containing 10 consecutive alanines illustrate the redundancy of information in an amino-acid-sequence. *Proceedings of the National Academy of Sciences, USA*, **89** (1992), 3751; B. W. Matthews. Structural and genetic analysis of protein stability. *Annual Review of Biochemistry*, **62** (1993), 139.

[57] L. Holm and C. Sander. An evolutionary treasure: unification of a broad set of amidohydrolases related to urease. *Proteins*, **28** (1997), 72.

[58] D. T. Jones, W. R. Taylor and J. M. Thornton. A new approach to protein fold recognition. *Nature*, **358** (1992), 86.

[59] S. E. Radford and C. M. Dobson. From computer simulations to human disease: emerging themes in protein folding. *Cell*, **97** (1999), 291; M. Bucciantini, E. Giannoni, F. Chiti *et al.* Inherent toxicity of aggregates implies a common mechanism for protein misfolding diseases. *Nature*, **416** (2002), 507; M. Dumoulin, A. M. Last, A. Desmyter *et al.* A camelid antibody fragment inhibits the formation of amyloid fibrils by human lysozyme. *Nature*, **424** (2003), 783; F. Chiti, M.Stefani, N. Taddei *et al.* Rationalization of the effects of mutations on peptide and protein aggregation rates. *Nature*, **424** (2003), 805.

[60] S. B. Prusiner. Prions. *Proceedings of the National Academy of Sciences, USA*, **95** (1998), 13363.

[61] A. Aggeli, I. A. Nyrkova, M. Bell *et al.* Hierarchical self-assembly of chiral rod-like molecules as a model for peptide beta-sheet tapes, ribbons, fibrils, and fibers. *Proceedings of the National Academy of Sciences, USA*, **98** (2001), 11857.

[62] A. L. Horwich, E. U. Weber-Ban and D. Finley. Chaperone rings in protein folding and degradation. *Proceedings of the National Academy of Sciences, USA*, **96** (1999), 11033.

[63] M. Fandrich and C. M. Dobson. The behaviour of polyamino acids reveals an inverse side chain effect in amyloid structure formation. *EMBO Journal*, **21** (2002), 5682.

[64] J. Maynard Smith. Natural selection and concept of a protein space. *Nature*, **225** (1970), 563.

[65] M. Kimura. Evolutionary rate at the molecular level. *Nature*, **217** (1968), 624.

[66] J. L. King and T. H. Jukes. Non-Darwinian evolution. *Science*, **164** (1969), 788.

[67] M. Kimura. How genes evolve: a population geneticist's view. *Annales de Génétique–Paris*, **19** (1976), 153.

[68] C. von Mering, R. Krause, B. Snel *et al.* Comparative assessment of large-scale data sets of protein–protein interactions. *Nature*, **417** (2002), 399.

[69] J. M. Thornton, A. E. Todd, D. Milburn *et al.* From structure to function: approaches and limitations. *Nature Structural Biology*, **7** (2000), 991 (suppl. S).

[70] C. A. Orengo, A. E. Todd and J. M. Thornton. From protein structure to function. *Current Opinion in Structural Biology*, **9** (1999), 374.

[71] A. C. Gavin, M. Bösche, R. Krause *et al.* Functional organization of the yeast proteome by systematic analysis of protein complexes. *Nature*, **415** (2002), 141; Y. Ho, A. Gruhler, A. Heilbut *et al.* Systematic identification of protein complexes in *Saccharomyces cerevisiae* by mass spectrometry. *Nature*, **415** (2002), 180.

[72] J. J. Hopfield. Neural networks and physical systems with emergent collective computational abilities. *Proceedings of the National Academy of Sciences, USA*, **79** (1982), 2554.

[73] P. W. Anderson. Suggested model for prebiotic evolution – the use of chaos. *Proceedings of the National Academy of Sciences, USA*, **80** (1983), 3386.

[74] M. Eigen, P. Schuster, W. Gardiner *et al.* The origin of genetic information. *Scientific American*, **244** (1981), 78.

[75] K. Binder and A. P. Young. Spin glasses: experimental facts, theoretical concepts, and open questions. *Reviews of Modern Physics*, **58** (1986), 801; M. Mezard, G. Parisi and M. Virasoro, *Spin Glass Theory and Beyond*. Singapore: World Scientific (1987).

[76] J. D. Bryngelson, J. N. Onuchic, N. D. Socci *et al.* Funnels, pathways, and the energy landscape of protein-folding – a synthesis. *Proteins*, **21** (1995), 167; D. K. Klimov and D. Thirumalai. Criterion that determines the foldability of proteins. *Physical Review Letters*, **76** (1996), 4070; H. S. Chan and K. A. Dill. Protein folding in the landscape perspective: chevron plots and non-arrhenius kinetics. *Proteins*, **30** (1998), 2.

13

Protein-based life as an emergent property of matter: the nature and biological fitness of the protein folds

Michael J. Denton

The thesis I shall present in this book is that the biosphere does not contain a predictable class of events but is a particular event, certainly compatible indeed with first principles, but not deducible from those principles and therefore essentially unpredictable.

Jacques Monod (1972), *Chance and Necessity*

The process of crystallization in inorganic nature . . . is . . . the nearest analogue to the formation of cells . . . should we not therefore be justified in putting forward the proposition that the formation of the elementary parts of organisms is nothing but a crystallization and the organism nothing but an aggregate of such crystals?

Theodore Schwann (1847), *Microscopical Researches*

The laws of light as of gravitation being the same [on other planets] . . . the inference as to the possibility of the vertebrate type being the basis of organization of some of the inhabitants of other planets will not appear so hazardous.

Richard Owen (1849), *On the Nature of Limbs*

Introduction

In his great classic *The Fitness of the Environment* (1913), Lawrence J. Henderson examined the fitness of the basic chemical constituents and chemical processes used by living organisms on earth and of the general environment, including the hydrosphere and atmosphere of the earth, and argued that the laws of nature and the properties of matter appear uniquely and maximally fit for life as it exists on earth. Towards the end of *Fitness*, he summarized his findings:

Fitness of the Cosmos for Life: Biochemistry and Fine-Tuning, ed. J. D. Barrow *et al.*
Published by Cambridge University Press. © Cambridge University Press 2007.

The fitness of the environment results from characteristics which constitute a series of maxima – unique or nearly unique properties of water, carbonic acid, the compounds of carbon, hydrogen, and oxygen and the ocean – so numerous, so varied, so nearly complete among all things which are concerned in the problem, that together they form certainly the greatest possible fitness. No other environment consisting of primary constituents made up of other known elements, or lacking water and carbonic acid, could possess a like number of fit characteristics . . . to promote . . . the organic mechanism we call life.   *(p. 272)*

And, in the last sentence of *Fitness*, Henderson concludes with the challenging claim that "the biologist may now rightly regard the universe in its very essence as biocentric."

Henderson's core point – that the existence of carbon-based life depends on a remarkably fit "ensemble of natural constituents" and that carbon-based life as it exists on earth is integral to nature and in a sense *preordained in the properties of matter* – can hardly be contested. After ninety years, no evidence has come to light of any alternative ensemble of natural or synthetic ingredients that could promote the "organic mechanism we call life." However, whether the ensemble is "uniquely and maximally fit," as Henderson claims, exhibiting "not a single disability" (p. 267), which might be required to defend a teleological conclusion, remains a question for debate.

Henderson accepted natural selection as a cause of biological fitness; but, as he points out, the basic chemical constituents that he examines are features of the natural environment that existed *before* life emerged on earth. Natural selection may mold the organism, but not the *preceding* natural environmental constituents, such as water and carbon dioxide. In his words: "This latter component of fitness, antecedent to adaptations [is] a natural result of the properties of matter and the characteristics of energy in the course of cosmic evolution" (p. 275). In *The Order of Nature* (1917, p. 206), Henderson makes it clear that his argument depends on "a harmonious unity among the abstract *changeless* characteristics of the universe" (emphasis added). And he stresses that the properties of the constituents he is considering are "perfectly changeless in time" (p. 201) and "antedate" the process of evolution (p. 191).

The logical structure of Henderson's argument, which he lays out at the end of chapter 2 of *Fitness* (pp. 67–71), is quite straightforward. He identifies a key natural constituent or process of life and then he examines exhaustively its various chemical and physical properties that contribute to its biological fitness, arguing that these appear to be maximally fit. He then considers what other *natural* alternatives might exist that could perform the same role, concluding that no alternatives are remotely as fit and that the constituent of interest (water, carbon dioxide, the carbon atom, etc.) exhibits a unique and maximal fitness for its biological role. He then inquires whether they constitute "a unique ensemble of fitness, among all possible chemical substances, for a living organism" (p. 71). Henderson's method would appear to be a general approach or procedure for assessing the "fitness" of any *natural* biochemical constituent of life for its biological role. (Whether his treatment is as exhaustive as he claims, and whether the properties of the individual constituents such as water, carbonic acid, etc. are as maximally fit as he claims, remain, as mentioned above, matters for debate.)

Henderson's claim may be controversial, but his thesis is easy to refute. The discovery of "alternatives" as fit as water or carbon dioxide for carbon-based life, the existence of only one single obvious "disability" in any of the constituents he examines (water, carbon dioxide, carbon compounds, etc.), or the construction of any comparable alternative self-replicating system with cells built out of a set of different materials (silica, or a different suite of carbon compounds, for example) would immediately undermine the claim that the cosmos is uniquely biocentric: that the properties of matter are uniquely fit for carbon-based life as it exists on earth. If a vast number of different types of biology are compatible with the laws of physics, obviously the claim of a unique biocentricity in nature for life *as it exists on earth* is fatally undermined. For example, the discovery of life on Mars, based on a fundamentally different chemistry, would immediately negate the concept of a unique fitness in the order of things for life on earth. In being easy to refute, Henderson's thesis is a robust scientific theory in terms of Karl Popper's philosophy of science (Popper, 1965).

Since Henderson

The idea that the cosmos is in some sense biocentric has been supported over the past several decades by the discovery of biocentric fine-tuning of the fundamental physical constants (see also the contributions of other authors in this volume), the so-called cosmic coincidences (Car and Rees, 1979; Davies, 1982; Barrow and Tipler, 1986). One such coincidence is the "lucky" fact that the nuclear resonances of C^{12} and O^{16} are exactly what they need to be if carbon is to be synthesized and accumulate in any quantity in the interior of stars. The energy levels of these resonances ensure that C^{12} is first synthesized in stellar interiors from collisions between Be^8 and helium nuclei and that the carbon synthesized is not depleted later. This discovery was made by Hoyle in 1953 while working at Caltech with William Fowler (Hoyle, 1964). An intriguing aspect of the discovery is, as Hoyle later pointed out (1994, p. 256), that it was a prediction from the Anthropic Principle. From the cosmic abundance of carbon, Hoyle inferred probable coincidences in the nuclear resonances that facilitated and promoted the synthesis of carbon (Barrow and Tipler, 1986, pp. 250–5). Hoyle's discovery was widely acclaimed, not only as a major scientific discovery, but also as evidence for the biocentricity of nature.

Hoyle himself commented on the energy levels of the resonances: "If you wanted to produce carbon and oxygen in roughly equal quantities by stellar nucleosynthesis, these are the two levels you would have to fix, and your fixing would have to be just about where these levels are found to be" (Davies, 1982, p. 118).

Additional support for Henderson's thesis has also come from advances in bio-chemical knowledge. Very little was known in 1913 about the critical role of metals in life (Williams and Fraústo da Silva, 2003) or of the remarkable fitness of some of the more complex material constituents of the cell, including the lipid bilayer membrane (Trinkaus, 1984, pp. 51–3) and the nucleic acids (Eschenmoser, 1999) (see also other chapters in this volume). In my view, nearly everything we have learned since Henderson's day about the material constituents of life, especially since the molecular biological revolution in the 1950s, is at least *consistent with* Henderson's claim that the properties of matter are uniquely fit for life as it exists on earth (Denton, 1998).

Proteins

The proteins have been long invested with a mystical aura. Even the word "protein" – proposed in 1838 by the Swedish chemist Berzelius in a letter to his student Gerrit Jan Mulder (Fruton, 1999, p. 171), derived from the Greek meaning "holding first place" – conveys the idea that these molecules play a unique and vital role in biology. In his *Molecular Biology of the Gene*, Jim Watson, writing well over a hundred years later, remarked:

Through the first quarter of this century . . . the most important group of macromolecules was believed to be the proteins . . . [the discovery by Sumner that enzymes were proteins] did not dispel the general aura of mystery about proteins . . . it was still possible as late as 1940, for some scientists to believe that these molecules would eventually be shown to have features unique to life . . . 　　　　*(Watson, 1976, pp. 25–6)*

Berzelius's choice of the word "protein" turned out to be remarkably prophetic. Proteins are quite literally the stuff of life. Apart from the transmission of genetic information, they carry out individually or in groups nearly all the essential activities on which the life of the cell depends and most of the catalytic functions in the cell. In addition, the core of nearly all of the cell's supramolecular assemblies, such as microtubules and microfilaments, is composed of proteins. Indeed, complex cellular life would be inconceivable without what Jacques Monod called their "demonical" catalytic abilities (1972, p. 64). The importance of proteins is such that much of the cell's functioning, including the role of DNA and RNA, the genetic code, and the basic logic of the genetic system, is organized primarily for their manufacture. Life on earth – life as we know it – is in essence a protein-based phenomenon.

This centrality of the proteins leads to an obvious test of the "Henderson hypothesis." If the cosmos is uniquely fit for life as it exists on earth, then, given their primal significance, the proteins should be another *uniquely fit ensemble of natural forms* (analogous to, but more complex than, the ensemble of simple natural forms discussed by Henderson in *Fitness*). To show that this "anthropic prediction" holds first necessitates finding convincing evidence (1) that the protein folds are genuine natural forms determined by physical law that, like water and carbon dioxide or other natural forms such as atoms or crystals, are universal "givens" of the order of nature whose basic architectural and core functional properties are thus "antecedent" to the appearance of life; and (2) that they are uniquely fit for their biological role as the universal "toolkit" for carbon-based life. If either of these predictions fails, whatever else may be true of nature, we can conclude definitively that the order of nature is *not uniquely fit* for the special form of modern cellular protein-based life as it exists on earth today.

The 1000 protein folds

The 1000 protein folds are the basic stable units of the protein universe and are therefore, given the centrality of proteins in the modern cell system, the very building blocks of all life on earth. The folds are complex molecules made up of several thousand atoms and consisting of a linear chain of approximately 80–200 amino acids folded into a complex three-dimensional configuration. They have a hierarchic structure, being built up mainly out of two well-known secondary structural motifs – alpha (α) helices and beta (β) sheets (see Figure 13.1) – that are combined into higher-order motifs, such as the well-known helix–turn–helix, the β hairpin, the β–α–β, and so forth, which are in turn combined to form the higher-level architecture of the folds (Brandon and Tooze, 1999; Lesk, 2001). Each fold is able to assemble itself independently into its native conformation and represents the minimum protein structure capable of folding into a stable molecular conformation (Brandon and Tooze, 1999). Folds are also often referred to as domains (Brandon and Tooze, 1999; Bartlett *et al.*, 2003; Wernisch and Wodak, 2003). Although many proteins, such as myoglobin, β-lactamase, triosephosphate isomerase, ribonuclease, and flavodoxin, are composed of an individual fold, most proteins are composed of several individual folds (or domains) combined into higher-order structures, as in hemoglobin, for example, where four individual globin folds or monomers are combined into a compact quaternary structure.

The elucidation of the atomic structure of so many such complex molecules has been one of the triumphs of twentieth century science. Many different folds have now been identified, and together they form an exotic and beautiful set of organic forms. The strangeness, beauty, and extraordinary diversity – and complexity – of

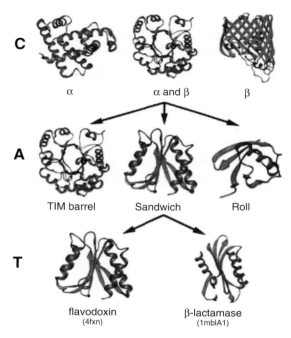

Figure 13.1. Structural classes of protein folds, showing how the folds can be classified into different structural classes. Top row: the three basic fold classes: α, containing only α helices; α and β, containing α helices and β sheets; and β, containing only β sheets. Middle row: three different architectural subclasses of the α and β class: "triosephosphate isomerase (TIM) barrel," "three-layer sandwich," and "roll." Bottom row: two different arrangements of the "three-layer sandwich". The spiral conformations are the α helices, and the broad arrows are the β sheets. (From: Orengo, C. A., Michie, A. D., Jones, S. *et al.* [1997]. CATH – a hierarchic classification of protein domain structures [Figure 2]. *Structure*, **5**, 1093–108. Copyright 1997, Elsevier Science. Reprinted with permission.)

these enigmatic forms (see Figure 13.1) are apparent on even a cursory examination of the various depictions of their atomic structures in major texts (Brandon and Tooze, 1999; Lesk, 2001) and in the various protein databanks, such as the CATH Protein Structure Classification databases (Orengo *et al.*, 1997, 2003), available online at http://cathwww.biochem.ucl.ac.uk/latest/index.html.

The essential molecular nature of the protein folds has not been understood for very long. Indeed, as recently as the early 1950s the fundamental structure of proteins was still largely mysterious. A remarkable testimony to just how little was actually known at that time about even the primary structure of proteins is Sanger's (1952) confession that

... as an initial working hypothesis it will be assumed that the peptide bond theory is valid, in other words, that a protein molecule is built up only of chains of alpha amino ... acids bound together by peptide bonds between their alpha-amino and alpha carboxyl groups. While this peptide theory is almost certainly valid . . . it should be remembered that it is still a hypothesis and has not been definitely proved. Probably the best evidence in support of it is that since its enunciation in 1902 [by Emile Fischer] no facts have been found to contradict it.

Ideas on the three-dimensional structure of proteins were no less preliminary. It was agreed that they were big molecules – macromolecules – probably composed of linear chains of amino acids of up to 100 or more. But just how their primary sequences were determined, how they were synthesized, and how they arrived at their native three-dimensional conformations was a complete mystery (Hunter, 2000, ch. 11). It is a testimony to how little was known about protein function that a majority of biochemists considered that it was proteins, rather then nucleic acids, that played the primary role in heredity and formed the basic aperiodic molecule that made up the material structure of the gene (McCarty, 2003). The question of whether proteins represented a potentially infinite set of Lego®-like assemblages, largely unconstrained by physical law and determined by natural selection, or whether they represented a finite set of natural forms determined mainly by natural law and therefore were "antecedent" to life and evolution (like a set of atoms or crystals), was simply impossible to answer.

The next two decades saw critical advances that were, by the 1970s, to lay the foundations of a rational and ultimately deductive science of protein form and, finally, by the late 1990s, to support the growing realization that the folds represented a new "periodic table" of complex natural molecular forms (Denton and Marshall, 2001; Denton *et al.*, 2002). The term "periodic table" as applied to the protein folds first appeared in *Nature* (Taylor, 2002).

Protein taxonomy

Until the early 1970s, there was no hint that proteins might be a set of lawful structures determined by a set of constructional rules, "laws of protein-fold form," that determine a finite set of natural conformations.

The first three-dimensional protein structures determined by X-ray crystallography provided no obvious support for the concept that the folds were natural forms. The apparent irregularity of the three-dimensional arrangement of the polypeptide chain was one of its most striking features. This was a disappointment for Kendrew and Perutz, who worked out the first structures (Richardson, 1981). Although by the late 1960s the three-dimensional structure of several proteins – including hemoglobin and lysozymes – had been determined, the lack of any apparent regularity in protein structures, and the great dissimilarity between those that had been determined, provided no basis for a rational classification. The picture in those early days was still compatible with the conception that proteins in living organisms on earth might be individual members of a near-infinite set of contingent material assemblages brought together by natural selection over millions of years of evolution, rather than members a finite set of atom-like or crystal-like natural forms determined by physical law.

During the early 1970s, however, as the number of three-dimensional structures began to grow significantly, it first became apparent that the number of protein folds might not be unlimited, that the folds might *not* belong to a potentially infinite set of material configurations. Increasingly, the same folds were found to recur in many different proteins from many distantly related organisms. Moreover, it was also apparent that the three-dimensional structures of individual folds were essentially invariant, some, such as the globin fold and the Rossman fold, for example, having remained essentially unchanged for thousands of millions of years (Richardson, 1981). Fold invariance was a finding that surprised many at the time, as Perutz (1983) comments: "My proposal that the tertiary and quaternary structure of the hemoglobins has been conserved through evolution from fish to mammals has been met with *disbelief among biologists* and others, but it comes as no surprise to protein crystallographers, who have found that homologous proteins in distant species have closely similar structures" (emphasis added).

As well as being essentially invariant, it became apparent during the 1970s, as more structures were determined, that the structure of the folds is also basically hierarchical (see Figure 13.1), consisting of secondary structural elements such as the α helix and β sheet combined into more complex motifs (Rose, 1979) and that the same motifs (helix–turn–helix, β hairpin, etc.) recur in many different proteins (Brandon and Tooze, 1999). The hierarchic nature of fold structure and the recurrence of the same submotifs suggested that physical law is playing a major role in the ordering of global fold structure and further supported the notion that the folds might be a set of natural and lawful structures rather than contingent assemblages of matter.

The fact that the folds can be classified into distinct structural classes (Levitt and Chothia, 1976; Brandon and Tooze, 1999) containing a number of related but variant structural subclasses (Richardson, 1981; Orengo *et al.*, 1997) provided further evidence for their essentially natural status (see Figure 13.1). Levitt and Chothia (1976) derived one of the first structural protein taxonomies, dividing the folds into three main groups: α folds, β folds, and α/β folds. In α structures, the core is built up exclusively of α helices; in the β structures, the core is made up of antiparallel β sheets. The α/β folds are made from combinations of β–α–β motifs that form a predominantly β sheet core surrounded by helices (Brandon and Tooze, 1999).

Another early finding, amply confirmed by subsequent work, is that many different sequences can fold into the same three-dimensional structure (Richardson,

1981; Perutz, 1983); in some cases these sequences give no evidence of sequential homology (Orengo *et al.*, 1994; Holm and Sander, 1996). This suggests multiple different discoveries of the same fold during evolution and also that the same fold may have evolved on many different occasions (Ptitsyn and Finkelstein, 1980). This indicates that the higher-order architecture of the folds may be a real and non-contingent feature of nature (Ptitsyn and Finkelstein, 1980).

A further finding, supporting the notion that the folds are real, natural existents, is that in many instances the same fold has been modified in minor ways to perform many different functions (Orengo *et al.*, 1994; Gerlt and Babbit, 2001). This is most apparent in the case of the so-called superfolds (Orengo *et al.*, 1994). One such superfold (see Figure 13.1) is the triosephosphate isomerase (TIM) barrel fold (an eight-stranded α/β barrel), which has been adapted for more than 61 different enzymic functions (Bartlett *et al.*, 2003), including TIM, enolase, and glycolate oxidase (Brandon and Tooze, 1999). Bartlett *et al.* (2003) comment on the diversity of functions associated with this one fold: "This suggests . . . re-use of the same molecular architecture again and again." Moreover, the fact that in many cases where the same fold is adapted to different functions, no trace of homology – that is, common ancestry – can be detected in their amino acid sequences suggests multiple separate "discoveries" of the same basic structure during the course of evolution (Orengo *et al.*, 1994). This further reinforces the conclusion that the folds are a finite set of ahistorical physical forms.

A picture has emerged of a limited number of ahistorical forms that have been secondarily modified to perform a vast number of adaptive functions. A remarkable feature of these secondary adaptive substitutions is how few seem necessary to cause adaptive shifts in protein function. As Perutz (1983) comments: "In the instances analyzed so far, new chemical functions appear to have evolved by only a few amino acid substitutions in key positions." Thus, changes in only a few amino acid positions in the globin fold have conferred a vast variety of different adaptive functions on the hemoglobins. Other proteins show the same phenomenon: a considerable variety of functions realized by a relatively few minor substitutions. Studies of the evolution of new digestive enzymic functions in monkeys and changes in the absorption spectrum of rhodopsin in deep-sea fish (Shozo, 2002) all show just how very few amino acid changes are necessary in many cases to shift the function of a protein significantly. As Shozo sums up, "avian hemoglobin, coelacanth vision, and langur ribonuclease show that adaptation of proteins can occur by amino acid replacements at a small number of critical sites."

Considered together, their basically invariant forms and fundamental hierarchical structures, the fact that based on purely structural criteria they can be classified into clearly defined groups, the multiple use of the same fold for different functions, and the likelihood that the same folds have occurred independently on several different occasions during evolution point to their being a finite set of "*real* timeless structures" determined primarily by physics.

Toward a rational morphology

The empirical evidence suggesting that the folds are a finite set of natural forms has been largely confirmed over the past three decades by many researchers (Levitt and Chothia, 1976; Ptitsyn and Finkelstein, 1980; Richardson, 1981; Chothia, 1984; Finkelstein and Ptitsyn, 1987; Chothia and Finkelstein, 1990; Przytycka *et al.*, 1999; Srinivasan and Rose, 1999). Their studies have elucidated what amount to a set of rules or laws of protein-fold form that indicate how the physical properties of the amino acid residues in a polypetide chain – and particularly the steric constraints they impose on the conformation of the chain – limit the number of "allowable" folds to a surpisingly small, finite number of about 1000 molecular forms. These constraints or rules that govern the way that the various secondary structural motifs such as α helices and β sheets can be combined and packed into compact three-dimensional structures amount to a *rational and generative morphology of protein form.*

As we now know, proteins exhibit a hierarchical structure consisting of basic motifs, such as the α helix and β sheet, combined into higher-order motifs, which are then combined into the native fold (Brandon and Tooze, 1999; Lesk, 2001). The way in which these ordered motifs are generated and packed together into three-dimensional conformations is limited by steric constraints at all levels of the hierarchy (Ramachandran and Sasisekharan, 1968; Ptitsyn and Finkelstein, 1980; Chothia and Finkelstein, 1990; Srinivasan and Rose, 1999).

These steric constraints arise at the lowest level from the phi (φ) and psi (ψ) torsion angles (Ramachandran and Sasisekharan,1968) and from local interactions between adjacent amino acids in short sections of the chain that limit allowable conformations to a finite number of conformers, including the two key secondary structural elements, the α helix and β sheet (Voet and Voet, 1995). At a higher level, other constraints limit the ways in which helices and sheets may be connected into motifs, and at the highest level different packing constraints determine the ways these submotifs may be folded into higher-order structures (Chothia and Finkelstein, 1990; Srinivasan and Rose, 2002).

Ramachandran and Sasisekharan (1968) established early that most combinations of the φ, ψ torsion angles in a polypeptide are conformationally inaccessible because of steric hindrance between the van der Waals radii of the atoms in the successive amino acids. In a plot of all possible combinations of the torsion angles – the so-called Ramachandran plot – there are only two small regions of sterically allowed φ, ψ combinations that together define what might be considered the first rule or law of protein-fold form. Only one helical conformation simultaneously allows conformation angles and a favorable hydrogen-bonding pattern. This is the α helix, first identified by Linus Pauling by model-building in 1951 (Voet and Voet, 1995). Subsequent studies showed that the α helix is a common secondary structural element of proteins. The other secondary structural element of proteins is the β sheet. This also has repeating ϕ, ψ angles that fall within the allowed regions of the Ramachandran plot and also utilize the full hydrogen-bonding capability of the polypeptide backbone. The β sheet structure was first proposed by Corey and Pauling in 1951 (Voet and Voet, 1995). The two basic motifs, the α helix and the β sheet, are therefore both sterically preferred arrangements for amino acid polymers composed of the twenty proteinaceous amino acids. Because of the constraints imposed by the torsion angles and adjacent side chains, even the supposedly "unfolded chain" is probably more "folded" than commonly assumed (Rose, 2002, 2003).

The higher form of the folds is largely determined by what amounts to another set of rules, which determine the way in which these two key secondary structural elements, helices and sheets, may be combined into higher-order motifs (Voet and Voet, 1995; Brandon and Tooze, 1999). In α helices, several rules explain the different geometric arrangements observed in native protein conformations. As Brandon and Tooze comment:

Rules have been derived that explain the different geometric arrangements of alpha helices observed in alpha domain structures. The helix packing in coiled-coil structures is determined by fitting of the knobs of side chains in the first helix into holes between the side chains of the second helix. For the other α helical structures, the helix packing is determined by fitting ridges of side chains along one α helix into grooves between the side chains of another helix.   *(1999, p. 45)*

Other rules determine the way in which β sheets are arranged into higher-order motifs. As Voet and Voet describe:

Beta sheets invariably exhibit a right-handed twist, which is a consequence of non-bonded interactions between the chiral L-amino acid residues in . . . extended polypeptide chains. These interactions tend to give the polypetide chain a right-handed helical twist.
                                                                                    (1995, p. 151)

This intrinsic right-handed twist of extended sheets leads to further constraints on sheet geometries. For example, it explains why two consecutive parallel β strands almost always have a right-handed helical sense. Finkelstein and Ptitsyn (1987) comment further:

Strong limitations are connected with the handedness of protein structure. . . These limitations often permit distinction between two mirror-image structures. The most well-known limitation is that two parallel beta stands of the same beta layer together with the connection joining them must form the turn of a right superhelix. . . This rule . . . is due to the intrinsic right twist . . . of β sheets from L amino acids residues. The right-handed connection uses the "shortest" way and therefore it is not so "stretched" as the left-handed one.

Altogether, these lower- and higher-level "packing or steric constraints" amount to what is in effect a *rational and generative morphology* consisting of a hierarchy of rules that govern the way in which the various secondary structural motifs such as α helices and β sheets can be combined and packed into compact three-dimensional structures (Chothia and Finkelstein, 1990; Chothia *et al.*, 1997; Przytycka *et al.*, 1999). In the words of Chothia *et al.* (1997):

In most proteins the α helices and β sheets pack together in one of a small number of ways. The connections between secondary structures obey a set of empirical topological rules in almost all cases. . . Subsequently, it was argued that these similarities arise from the intrinsic physical and chemical properties of proteins, and a great deal of work was carried out to demonstrate *that this is the case* [emphasis added].

Understanding the rules that determine protein form may ultimately allow the *ab initio* prediction of three-dimensional form from sequence. Already, hierarchic prediction programs such as LINUS (Srinivasan and Rose, 2002) and ROSETTA (Chivian *et al.*, 2003), based primarily on consideration of the steric constraints that restrict the conformation of short sections of the polypeptide chain, are able to predict higher-order structures with remarkable accuracy.

The various physical constraints that restrict the folded spatial arrangements of linear polymers of amino acids – the laws of protein-fold form – restrict permissible folds is to a very small number of a few thousand (Przytycka *et al.*, 1999). One recent estimate basded on possible arrangements of typical structural elements gave a maximum of 4000 folds (Lindgard and Bohr, 1996). Others have suggested that the maximum is likely to be no more than a few thousand (Chothia *et al.*, 1997). Brenner *et al.* (1997) recently estimated an upper limit in the range of 1000 to 6000. An estimate based on the rate of discovery of new folds, rather than permissible spatial arrangements, suggests that the total number of folds utilized by organisms on earth might not be more than 1000 (Chothia, 1993).

The fact that the total number of theoretically possible protein structures that an individual amino acid chain of 150 residues might adopt – assuming that each peptide group has only three conformations – is 3^{150} or 10^{68} (Brandon and Tooze, 1999) whereas the total number of permissible folds is of the order of 1000 graphically illustrates just how restrictive the laws of protein-fold form are. Whatever the actual figure, the total number of folds is bound to represent a tiny stable fraction of all possible polypeptide conformations, determined by the laws of physics. This further reinforces the notion that the folds, like atoms, represent a finite set of allowable physical structures that would recur throughout the cosmos wherever carbon-based life that utilizes the same twenty amino acids exists.

Recently, Banavar and Maritan (2003a, b; see also Chapter 12, this volume) proposed a novel tube model to account for the character and nature of the protein folds. They argue that polypeptide chains should be considered as tubes of non-zero thickness. When the ratio of tube thickness to the range of attractive interactions uniquely characteristic of polypeptide chains (composed of the twenty proteinaceous amino acids) is taken into account, it turns out that the resultant chain inhabits the vicinity of a unique phase transition between a highly degenerate and disordered compact polymer phase with mutiple inaccessible energy minima and a highly disordered dispersed or fluidic non-condensed phase. The success of this elegant and simple physical model in predicting many of the key motifs seen in native proteins (Banavar and Maritan, 2003a) and the characteristic marginal stability of the folds (see below) reinforces strongly the conception that the folds are indeed natural forms.

A new periodic table

Here it is easy to be reminded of the constructional rules that govern the assembly of subatomic particles into atomic structures, generating the finite set of 92 atoms that make up the periodic table of the elements (Taylor, 2002), or of the rules of grammar, which restrict grammatical letter strings to a tiny finite set of all possible sequences. These laws of protein-fold form represent a set of pre-existing abstract prescriptions that specify a finite set of allowable "material forms." They therefore provide for a rational deductive derivation of *all possible fold morphologies* and thus represent "Laws of Biological Form" of precisely the kind sought after by Goethe, Geoffroy, Carus, and many other biologists of the pre-Darwinian era (Denton *et al.*, 2002; Gould, 2002). And, in perfect conformity with pre-Darwinian thought, protein adaptations would appear to be almost entirely secondary modifications of primary forms, wonderful exemplars of Owen's conception of adaptations as "adaptive masks" (Owen, 1849; Gould, 2002).

Although much remains to be learned about protein structure and evolution, all the available evidence now supports the conclusion that the folds represent a finite natural ensemble of forms, determined by a hierarchic set of physical constructional rules that arise out of the fundamental properties of linear polymers made up of the twenty proteinaceous amino acids, and assemble into their native forms like a set of crystals through a series of phase transitions (Scheraga, 1963; Florey, 1969). And, like any other set of natural forms, such as atoms or crystals, the folds are genuine universals that are *antecedent* to biology and thus to Darwinian selection. In short, the universe of protein forms can be accounted for by physical rules *without reference to biology!* As Przytycka *et al.* (1999) comment: "From the outset chemistry, in promoting structure, *predetermines* the universe of protein folds for polypeptide chains in an aqueous environment" (emphasis added). The folds represent an "abstract changeless" ensemble of natural components of the cosmos in the same sense as Henderson's ensemble of water and carbon dioxide. Just as 92 atoms form the periodic table of chemistry, the 1000 protein folds form another "periodic table" of basic physical, crystal-like structures (Taylor, 2002), which form the underlying fabric of protein-based life.

In itself, this is a landmark discovery. The discovery that the protein universe consists of a finite set of natural forms in a sense completes the molecular biological revolution, revealing finally – five decades after the nature and biological purpose of DNA and RNA were first elucidated – the essential nature of the second great class of biopolymers. It reveals that the purpose of the genetic system is to turn out endless adaptive variants of a set of invariant natural forms. The great complexity of the folds (among the most complex material structures known) indicates, perhaps more clearly than any other previous discovery in the biological sciences, that *very great biological complexity* may be *lawful* and need not necessarily be contingent.

Jacques Monod's bold and well-known claim (1972, p. 49) that the "biosphere does not contain a predicable class of events; that amino acid sequences (pp. 94–7) "show no regularity, special feature or restrictive character . . . but rather appear completely haphazard, each discloses nothing in its structure other than the pure chance of its origin" has proved premature. Indeed, deep restrictions on the amino acid sequences of proteins imposed by the laws of protein-fold form exist, and the biosphere does now contain a class of predictable events. Of course, the proteins may be unique; but their example does raise the distinct possibility of analogous laws of form determining aspects of the supramolecular and higher architecture of life. Note that it has taken nearly two centuries of research to reveal the true nature of proteins and to identify the laws of protein-fold form. At the very least, this suggests that the absence to date of any evidence for analogous laws of form determining cell forms and body plans cannot be taken to support the view that none will come to light as knowledge advances. An intriguing hint that the proteins might be heralds of things to come is the recent evidence that microtubular forms may represent another set of complex organic forms limited by a set of dynamic constructional rules (Surrey *et al.*, 2001). Indeed, we cannot discount the possibility that RNA folds may also turn out to be a finite set of natural forms analogous to the protein folds (Russell *et al.*, 2002).

The finding that the protein folds are lawful natural forms whose inherent biological properties and range of functional propensities are given and delimited by physical law is (as argued earlier) a *necessary* finding if Henderson's argument is to be extended to the protein universe. But it is *certainly not sufficient*. If the folds are to be deemed a uniquely fit, natural ensemble like Henderson's water, carbon dioxide, etc., in addition to their being natural forms *antecedent* to biology, they must also be shown to be uniquely fit. Although their *natural* status is in my view secure, whether they are uniquely fit remains to be resolved.

The fitness of the folds

Because each fold has been subjected to billions of years of selective fine-tuning for specific biochemical functions, efficient folding, and so forth, it is somewhat difficult to judge precisely which properties are universal, generic properties of the folds and which are secondarily evolved features. Nonetheless, as James and Tawfik (2003) point out, "An evolved function can only evolve if it is already present to some extent," and this presumably applies to all characteristics of the folds. Thus, four characteristics that contribute to their fitness are likely to be basic, intrinsic characteristics of the folds themselves: their architectural diversity, marginal stability, robustness, and possession of a hydrophobic core.

The underlying molecular architectures of the folds are, as we have seen, quite amazingly diverse. (This can be seen by turning to the various databanks that depict the folds, such as CATH; see above.) This architectural diversity is a major contributor to their biological fitness, providing the basis for the vast range of structural and functional molecular roles that they play within the cell.

The folds exhibit a combination of robustness and marginal stability, both characteristics that confer important elements of fitness. In terms of marginal stability, the folds are nothing like the rigid conformations conveyed in textbook depictions. In fact, the energy difference between the native conformation of a fold and its denatured state is extraordinarily small – about 5–15 kcal/mol – not much more than the energy level of a single hydrogen bond, which is of the order of 2–5 kcal/mol (Brandon and Tooze, 1999). Studies by various groups, including those of Martin Karpus (Karpus and Petsko, 1990) and Hans Frauenfelder (Frauenfelder *et al.*, 1988) indicate that a protein's native structure consists of a large number of conformational substates. Instead of inhabiting a deep free-energy minimum, a "V-shaped" bowl with steep sides ending in a unique deep pit (as is often depicted in the literature), the folds inhabit a complex energy landscape that is more a "shallow U-shaped bowl" with multiple small depressions on its base (James and Tawfik, 2003). These depressions are the substates, or alternative conformers, available to the fold (Frauenfelder *et al.*, 1988; Karpus and Petsko, 1990), each of near-equivalent stability. Marginal stability is critical during folding, enabling the polypeptide chain to search conformational space for increasingly stable conformations (Dinner *et al.*, 2000). Marginal stability and the characteristic

U-shaped energy landscape arise according to the "tube model" of Banavar and Maritan (2003a, b) from a "novel phase of matter in the vicinity of a phase transition" in which the folds arise (Banavar and Maritan, 2003b; see also Chapter 12, this volume). The tube model also implies that few other polymers may exist that will exhibit discrete, stable, folded conformations associated with their characteristic marginal stability (J. R. Banavar, personal communication).

It is this marginal stability (Brandon and Tooze, 1999; James and Tawfik, 2003) and its consequence, the ability of folds to adopt many slightly different conformations, that have permitted the evolution of allosteric control mechanisms that link logical control circuits with catalysis in the same molecular fabric – a phenomenally sophisticated mechanism that Monod saw as the "second secret of life" (Judson, 1979). In the case of hemoglobin, conformational flexibility allows the globin fold to undergo "breathing movements" that permit oxygen-free access to the heme group by momentarily opening pathways into the tightly packed interior (Karpus and McCammon, 1986). Without this physical ability, the globin fold would be incapable of oxygen uptake or release (Voet and Voet, 1995). More generally, the ability to adopt a variety of related conformations contributes another crucial element of fitness: it greatly facilitates the evolution of new protein functions (James and Tawfik, 2003). Rigid macromolecules not only would be incapable of allosteric regulation or sophisticated catalytic activities, but would also be unfit for the evolutionary exploration of novel functions and their associated variant conformers. Marginal stability, and the consequent flexibility and conformational diversity of the folds, therefore confers a wonderful fitness for adaptive evolution – "evolvability" (James and Tawfik, 2003) – which has been exploited by life to spectacular effect over the past three billion years to produce the huge variety of protein functions exploited by life on earth. It may even, as James and Tawfik (2003) speculate, have allowed "the evolutionary process [to have achieved] great, perhaps unrestricted, functional diversity."

But beneath this marginal stability lies an inherent natural robustness. As self-organizing natural forms, each fold is able to assume its native stable form, as Anfinsen (1973) first showed, through energy minimization without external assistance from any other agency in the cell. This self-forming ability has two consequences that contribute to the fold's fitness. First, it relieves the cell of the considerable burden of having to specify and organize the assembly of 1000 complex three-dimensional atomic architectures, a process that would be costly in energetic and informational terms. Second, it means each fold is able to maintain and regain its native conformation in the turbulent chaos of the cell's interior, even after all manner of momentary conformational disturbances, which may involve anything from the movement of a few atoms to the unfolding of sections of the amino acid chain (Brandon and Tooze, 1999).

Their natural robustness has another consequence. The authors of a recent paper comment: "A protein's function is due to a comparatively small number of residues, suitably interspersed throughout the sequence. This process of imbedding functional resides in a robust framework constitutes a versatile mechanism to confer multiple functions upon a given fold" (Przytycka *et al.*, 1999). The folds are thus able to maintain their core architectures in the face of considerable amino acid sequence variation, and this contributes another important element of fitness: it makes possible adaptive substitutions that do not disrupt the underlying fold architecture, and this facilitates functional molecular evolution. It is the generic robustness of the basic fold frameworks that permits such sequential "tampering" and consequent functional variation.

Another general feature of the protein folds that confers an important element of fitness is that they generally contain a relatively compact hydrophobic core. This provides a "convenient" reaction chamber for organic syntheses that are, for the most part, difficult to carry out unless water is excluded. The dense hydrophobic core may also confer on many folds the ability to stack together into various stable supramolecular assemblies that form the basic elements of the cytoskeleton: nucleosomes, cytoplasmic filaments, microtubules, and so forth.

Finally, the diversity of the bulk properties of proteins is unequaled in any other known polymer class. Proteins form materials as diverse as the hard substance of nails and hair, the transparent substance of the lens, the elastic substance of collagen, and so on. Some of these properties are equaled by polymers in other classes: keratin by the carbohydrate polymer chitin (*N*-acyl-D-glucosamine), the transparency of the lens proteins by the polymer Perspex® (polymethyl methacrylate), the toughness and elasticity of collagen by the polyamide nylon. But no single polymer class has demonstrated such a variety of diverse bulk properties. The compaction of so many diverse bulk properties into one polymer class, polypeptides composed of the twenty proteinaceous amino acids, obviously contributes greatly to their biological fitness.

The fitness of specific folds

Given the great structural diversity exhibited by the 1000 architectures of the folds, one might expect, from first principles, that certain architectures are more suited to particular biochemical functions than others. This is true in the case of certain submotifs, where it is apparent that particular conformations are suited to particular biochemical functions. One example is the helix–turn–helix motif used for DNA binding (Voet and Voet, 1995; Brandon and Tooze, 1999).

What is true of the helix–turn–helix motif may also be true in certain cases of individual folds. Considerable evidence exists for believing that certain folds lend themselves to certain functions. One of the clearest cases is that of the Bin/amphiphysin/Rvs (BAR) domain, a fold found in many diverse proteins involved in membrane remodeling (Lee and Schekman, 2004), whose structure was recently determined (Peter *et al.*, 2004). As Lee and Schekman (2004) comment: "The structure of the domain . . . seems ideally suited to the task of sensing, and perhaps even generating, membrane-bending events during vesicle formation." And later they comment: "If you were to design a protein domain [fold] for detecting or imposing membrane curvature, you would likely come up with something that closely resembles the structure of the *Drosophila* BAR domain now solved by Peter *et al.*" Another instance in which a fold form is clearly suited for a biological function is the membrane-channel-forming proteins – the porins – made up of trimers of β barrels. These folds lend themselves to the construction of solvent-accessible pores through the lipid membrane (Voet and Voet, 1995). Mere cursory observation of the fold structure (essentially a hollow tube) suggests functions related to transport. Similarly, hydrophobic helical folds also lend themselves to the construction of transmembrane proteins (Locher *et al.*, 2003).

In the case of the globin fold, the remarkable way it has been shaped by selection for the carriage and storage of oxygen in vertebrates – involving the "breathing" of the molecule to allow the passage of oxygen into its tightly packed interior (mentioned earlier) – and the many elegant and subtle allosteric control mechanisms (Lesk, 2001) and many functional adaptations in different species (Perutz, 1983) suggest that perhaps few other folds could be so perfectly fine-tuned by selection for these specific functions.

To what extent particular folds are fit for particular functions remains to be established, but already I think it reasonable to suggest that if the laws of protein-fold form were more restricting, if they permitted only a quarter of the existing folds, if they permitted a less rich inventory of architectures (say 250 folds instead of 1000), then despite the adaptability of the folds many key biological functions, indeed protein-based life as we know it, might be impossible. And I think that it is also clear that if the folds were more rigid their fitness would be greatly diminished, whereas if they were less rigid they would be incapable of providing the sufficiently stable scaffolds essential for biochemical function in the cell.

Uniquely fit?

The folds provide the cell with 1000 self-assembling, complex, diverse atomic architectures that are all compacted into a single polymer class. Each exhibits an exquisite balance of robustness and marginal stability, which makes them eminently adaptable and evolvable and allows allosteric control and a suite of intricate properties ideally suited to the vast range of structural and functional biological roles they perform. Moreover, the core of each fold provides a hydrophobic reaction pocket ideally fit for organic synthesis. On top of all this, the bulk properties of proteins are unrivaled by those of any other known polymer class.

Natural selection has exploited this toolkit to dramatic effect during the evolution of life on earth. But are the folds a uniquely fit toolkit for molecular evolution, as Henderson's thesis predicts? Judging their fitness is complicated by the earlier-mentioned challenge of sorting their intrinsic fitness from their evolved fitness. Moreover, in his book Henderson was able to carry out a systematic analysis of the properties of water and carbon dioxide and a relatively exhaustive comparison of their properties with those of other known "alternative" compounds and to conclude that they are indeed uniquely fit. No such comparative method is available in the case of the folds. Thus, although no ensemble of complex molecular architectures of comparable fitness is currently known, the universe of all possible polymers and complex macromolecules that might potentially possess similar characteristics to the folds is mainly unexplored. It is even impossible to judge whether different types of proteins, made up of different sets of amino acids, might be more or less fit than proteins made up of the well-known twenty proteinaceous amino acids! Indeed, experimental assessment of the functional properties of proteins composed of different sets of amino acids provides an obvious means, still largely untested, of assessing the relative fitness of native proteins.

In another direction, we already know that many individual properties of proteins are clearly not unique. Their ability to self-fold is shared by ribozymes (Moore, 1999; Russell *et al.*, 2002) and may be shared by other synthetic polymer types in the future (Oh *et al.*, 2001; Shih *et al.*, 2004). However, RNA polymers lack the biochemical diversity of proteins, being composed of only four similar sub-units against the twenty different amino acids, and RNA molecules also lack the rigid, tightly packed hydrophobic cores of proteins and therefore their hydrophobic chemical-reaction chambers. The lack of a tightly packed hydrophobic core might also render RNA folds less suited to form stable filamentous assemblies, such as microtubules and actin filaments, which play a critical role in the generation of the cell's cytoplasmic architecture. It certainly seems significant that, despite the energetic advantage that would accrue to the cell if RNA polymers were used for its supramolecular assemblies, no cell today utilizes nucleic acid in this way. None the less, research into the biological potential of RNA continues, and many new and unexpected properties of RNA polymers may well be revealed.

In this context, further evidence regarding the uniqueness of the folds will come from future advances in protein chemistry and increased knowledge of alternative polymer types and from research in the fields of artificial life (Langton, 1989; Levy, 1993; Kauffman, 2000; Brooks, 2001; Rasmussen *et al.*, 2004), supramolecular chemistry, nanotechnology, nanorobotics (Lehn, 2002; Seeman, 2003; Shih *et al.*, 2004), and new theoretical approaches to modeling the essential properties of polypeptide chains, such as the "tube model" recently developed by Banavar and Maritan (2003a). Such models may eventually provide the means of estimating what fraction of all polymers might be capable of folding into a set of complex atomic architectures exhibiting the critical mix of robustness and marginal stability of the folds.

Considerating the number and relative stringency of the functional criteria satis-fied by the protein folds – self-organizing robustness in conjunction with marginal stability, diverse architectures, a hydrophobic core fit for organic synthesis, diverse bulk properties, etc. – it seems likely that few other types of polymer will be equally fit. At present, I think that current knowledge is consistent with the possibility that the protein folds represent an ensemble of *natural* forms uniquely fit for the "mech-anism we call life." If correct, this not only would support Henderson's contention that the cosmos is fine-tuned for carbon-based life, but would further restrict this statement to the protein-based variety of life that exists currently on earth.

A final aspect to the fitness of proteins relates to their evolutionary accessibility, especially in the early stages of cellular evolution, that lie close to the origin of life. Henderson makes the point that in the case of water and carbonic acid "In places where life is possible the primary constituents [water and carbonic acid] are necessarily and automatically formed in vast amounts by the cosmic process" (1913, p. 268). Intriguingly, many of the twenty proteinaceous amino acids – out of which the folds are constructed – are among the commonest amino acids found in meteorites and the easiest amino acids to generate in prebiotic syntheses (Denton *et al.*, 2002). Remarkably, the folds – so fit in so many other ways – are not so many steps removed from the synthesis of the elements in the stars, a witness to the fitness of matter for the evolution of protein-based life.

Conclusion

Aside from questions of cosmic fitness or teleology, evidence that the properties of matter may have played a crucial determining role in the generation and evolution of life is an important finding in itself. It echoes ancient animist and pantheist conceptions of nature (Jonas, 2001, pp. 7–9) and the monistic materialism of the pre-Socratic nature philosophers (Gomperz, 1912, pp. 43–79; Copleston, 1946, p. 75), who postulated a living cosmos and matter animated and imbued with the Divine spark, the self-sufficient cause of life. In the *Order of Nature*, Henderson's thought also strays in this animist direction, as is evident in the passage in which he writes: "No idea is older or more common than a suspicion that somehow nature itself is a great, imperfect organism" (1917, p. 205).

Although the question of whether the folds are a uniquely fit ensemble of natural forms remains to be answered, the discovery that this elegant set of atomic architectures represents a set of natural forms determined by physical law (and therefore that many of their generic properties are "antecedent" to biology) not only is profoundly beautiful and intellectually attractive in itself, but has far-reaching consequences. It implies, aside from the question of the actual biological fitness of the folds, that protein-based life is an integral part of nature and may be properly designated an *emergent property of matter*.

Acknowledgments

I am grateful to colleagues in the Biochemistry Department at Otago University, including especially Craig Marshall and John Cutfield, for many illuminating corridor conversations regarding the basic nature of the protein folds and to Craig Marshall for having worked with me specifically on the nature of the folds. Over the past couple of years, I have had many useful discussions with Jayanth Banavar, Amos Maritan, and George Rose at the Abdus Salam Centre for Theoretical Physics in Trieste and with Jayanth Banavar at Penn State. The publications of Cyrus Chothia, Victor Finkelstein, and George Rose have been influential in shaping my views.

References

Anfinsen, C. B. (1973). Principles that govern the folding of protein chains. *Science*, **181**, 223–30.

Banavar, J. R. and Maritan, A. (2003a). Colloquium: geometrical approach to protein folding: a tube picture. *Reviews of Modern Physics*, **75**, 23–34.

Banavar, J. R. and Maritan, A. (2003b). Comment on the protein folds as platonic forms. *Journal of Theoretical Biology*, **223**, 263–5.

Barrow, J. D. and Tipler, F. J. (1986). *The Anthropic Cosmological Principle*. Oxford: Oxford University Press.

Bartlett, G. J., Todd, A. E. and Thornton, J. M. (2003). Inferring protein function from structure. In *Structural Informatics*, ed. P. E. Bourne and H. Weissig. New York, NY: John Wiley, pp. 387–407.

Brandon, C. and Tooze, J. (1999). *Introduction to Protein Structure*, 2nd edn. New York, NY: Garland Publishing.

Brenner, S. E., Chothia, C. and Hubbard, T. J. (1997). Population statistics of protein structures: lessons from structural classifications. *Current Opinion in Structural Biology*, **7**, 369–76.

Brooks, R. A. (2001). The relationship between matter and life. *Nature*, **409**, 409–11.

Car, B. J. and Rees, M. J. (1979). The Anthropic Principle and the structure of the physical world. *Nature*, **278**, 605–12.

Chivian, D., Robertson, T., Bonneau, R. *et al.* (2003). *Ab initio* methods. In *Structural Bioinformatics*, ed. P. E. Bourne and H. Weissig. New York, NY: John Wiley, pp. 547–57.

Chothia, C. (1984). Principles that determine the structure of proteins. *Annual Review of Biochemistry*, **53**, 537–72.

Chothia, C. (1993). One thousand families for the molecular biologist. *Nature*, **357**, 543–4.

Chothia, C. and Finkelstein, A. V. (1990). The classification and origins of protein folding patterns. *Annual Review of Biochemistry*, **59**, 1007–39.

Chothia, C., Hubbard, T., Brenner, S. *et al.* (1997). Protein folds in the all α and all β classes. *Annual Review of Biophysics and Biomolecular Structures*, **26**, 597–627.

Copleston, F. (1946). *The History of Philosophy*, vol. 1. London: Burnes and Oates and Washbourne Ltd.

Davies, P. (1982). *The Accidental Universe*. Cambridge, UK: Cambridge University Press.

Denton, M. J. (1998). *Nature's Destiny*. New York, NY: Free Press.

Denton, M. J. and Marshall C. J. (2001). Laws of form revisited. *Nature*, **410**, 417.

Denton, M. J., Marshall, C. J. and Legge, M. (2002). The protein folds as platonic forms: new support for the pre-Darwinian conception of evolution by natural law. *Journal of Theoretical Biology*, **219**, 325–42.

Dinner, A. R. *et al.* (2000). Understanding protein folding via free energy surfaces from theory and experiment. *Trends in Biochemical Science*, **25**, 331–9.

Eschenmoser, A. (1999). Chemical etiology of nucleic acid structure. *Science*, **284**, 2118–24.

Finkelstein, A. V. and Ptitsyn, O. B. (1987). Why do globular proteins fit the limited set of folding patterns? *Progress in Biophysics and Molecular Biology*, **50**, 171–90.

Florey, P. J. (1969). *Statistical Mechanics of Chain Molecules*. New York, NY: John Wiley.

Frauenfelder, H., Parak, F. and Young, R. D. (1988). Conformational substates in proteins. *Annual Review of Biophysics and Biophysical Chemistry*, **17**, 451–79.

Fruton, J. S. (1999). *Proteins, Enzymes, Genes*. New Haven, CT: Yale University Press.

Gerlt, J. A. and Babbit, P. C. (2001). Divergent evolution of enzymic function. *Annual Review of Biochemistry*, **70**, 209–46.

Gomperz, T. (1912). *The Greek Thinkers*, vol. 1. London: John Murray.

Gould, S. J (2002). *The Structure of Evolutionary Theory*. Cambridge, MA: Harvard University Press.

Henderson, L. J. (1913). *The Fitness of the Environment: An Inquiry into the Biological Significance of the Properties of Matter*. New York: Macmillan. Repr. (1958) Boston, MA: Beacon Press; (1970) Gloucester, MA: Peter Smith.

Henderson, L. J. (1917). *The Order of Nature*. Cambridge, MA: Harvard University Press.

Holm, L. and Sander, C. (1996). Mapping the protein universe. *Science*, **273**, 595–603.

Hoyle, F. (1964). *Galaxies, Nuclei and Quasars*. New York, NY: Harper and Row.

Hoyle, F. (1994). *Home Is Where the Wind Blows*. Mill Valley, CA: University Science Books.

Hunter, G. K. (2000). *Vital Forces*. New York, NY: Academic Press.

James, L. C. and Tawfik, D. S. (2003). Conformational diversity and protein evolution – a 60 year-old hypothesis revisited. *Trends in Biochemical Science*, **28**, 361–8.

Jonas, H. (2001). *The Phenomenon of Life*. Evanston, IL: Northwestern University Press.

Judson. E. (1979). *The Eighth Day of Creation*. New York, NY: Simon and Schuster.

Karpus, M. and McCammon, J. A. (1986). The dynamics of proteins. *Scientific American*, **254**(4), 42–51.

Karpus, M. and Petsko, G. A. (1990). Molecular dynamics simulations in biology. *Nature*, **347**, 631–9.

Kauffman, S. (2000). *Investigations*. New York, NY: Oxford University Press.

Langton, C. G. (1989). Artificial life. In *Artificial Life*, ed. C. G. Langton. (Proceedings of an Interdisciplinary Symposium held in September 1987 in Los Alamos, New Mexico.) New York, NY: Addison-Wesley, pp. 1–48.

Lee, M. C. S. and Schekman, R. (2004). BAR domains go on a bender. *Science*, **303**, 479–80.

Lehn, J. M. (2002). Earth's self-organization and complex matter. *Science*, **295**, 2400–3.

Lesk, A. M. (2001). *Introduction to Protein Architecture*. Oxford: Oxford University Press.

Levitt, M. and Chothia, C. (1976). Structural patterns in protein structures. *Nature*, **261**, 552–8.

Levy, S. (1993). *The Quest for a New Creation*. London: Penguin Books.

Lindgard, P. and Bohr, H. (1996). How many protein fold classes are to be found? In *Protein Folds*, ed. H. Bohr and S. Brunak. New York, NY: CRC Press, pp. 98–102.

Locher, K. P., Bass, R. B. and Ress, D. C. (2003). Breaching the barrier. *Science*, **301**, 603–4.

McCarty, M. (2003). Discovering genes are made of DNA. *Nature*, **421**, 406.

Monod, J. (1972). *Chance and Necessity*. London: Collins.

Moore, P. B. (1999). The RNA folding problem. In *The RNA World*, ed. R. F. Gesteland, T. R. Cech and J. F. Atkins. Cold Spring Harbor, NY: Cold Spring Harbor Laboratory Press, pp. 381–401.

Oh, K., Jeong, K. and Moore, J. S. (2001). Folding-driven synthesis of oligomers. *Nature*, **414**, 889–93.

Orengo, C. A., Jones, D. T. and Thornton, J. M. (1994). Protein superfamilies and domain structures. *Nature*, **372**, 631–4.

Orengo, C. A., Michie, A. D., Jones, S. *et al.* (1997). CATH – a hierarchic classification of protein domain structures. *Structure*, **5**(8), 1093–108. See cathwww.biochem.ucl.ac.uk/latest/index.html.

Orengo, C. A., Pearl, F. M. G. and Thornton, J. M. (2003). The CATH domain structure database. In *Structural Informatics*, ed. P. E. Bourne and H. Weissig. New York, NY: John Wiley, pp. 249–71.

Owen, R. (1849). *On the Nature of Limbs*. London: Jan Van Voorst.

Perutz, M. F. (1983). Species adaptation in a protein molecule. *Molecular Biology and Evolution*, **1**, 1–28.

Peter, B. J., Kent, H. M., Mills, I. G. *et al.* (2004). BAR domains as sensors of membrane curvature: the amphiphysin BAR structure. *Science*, **303**, 495–9.

Popper, K. (1965). *Conjectures and Refutations*. New York, NY: Harper and Row.

Przytycka, T., Aurora, R. and Rose, G. D. (1999). A protein taxonomy based on secondary structure. *Nature Structural Biology*, **6**, 672–82.

Ptitsyn, O. B. and Finkelstein, A. V. (1980). Similarities of protein topologies: evolutionary divergence, functional convergence or principles of folding. *Quarterly Review of Biophysics*, **13**, 339–86.

Ramachandran, G. N. and Sasisekharan, V. (1968). Conformations of polypeptides and proteins. *Advances in Protein Chemistry*, **28**, 283–438.

Rasmussen, S. *et al.* (2004). Transitions from nonliving to living matter. *Science*, **303**, 963–5.

Richardson, J. S. (1981). The anatomy and taxonomy of proteins. *Advances in Protein Chemistry*, **34**, 168–339.

Rose, G. D. (1979). Hierarchic organization of domains in globular proteins. *Journal of Molecular Biology*, **134**, 447–70.

Rose, G. D. (2002). Getting to know U in unfolded proteins. *Advances in Protein Chemistry*, **62**, xv–xxi.

Rose, G. D. (2003). A long day's journey into simplicity. Talk given at the International Workshop on Proteomics: Protein Structure and Interactions (May 5–16) at the Abdus Salam International Centre for Theoretical Physics, Trieste. Published online: http:/agenda.ictp.trieste.it/smr.php?1499*.

Russell, R., Zhuang, X., Babcock, H. *et al.* (2002). Exploring the folding landscape of a structured RNA. *Proceedings of the National Academy of Sciences, USA*, **99**, 155–60.

Sanger, F. (1952). The arrangement of amino acids in proteins. *Advances in Protein Chemistry*, **7**, 1–67.

Scheraga, H. A. (1963). Intramolecular bonds in proteins. II. Noncovalent bonds. In *The Proteins*, ed. H. Neurath, 2nd edn, vol. 1. New York, NY: Academic Press, pp. 477–594.

Schwann, T. (1847). *Microscopical Researches*, trans. H. Smith. London: Sydenham Society. Original published in German (1839).

Seeman, N. C. (2003). DNA in a material world. *Nature*, **421**, 427–31.

Shih, W. M., Quispe, J. D. and Joyce, G. F. (2004). A 1.7-kilobase single stranded DNA that folds into a nanoscale octahedron. *Nature*, **427**, 615–21.

Shozo, Y. (2002). Evaluating adaptive evolution. *Nature Genetics*, **30**, 350–1.

Srinivasan, R. and Rose, G. D. (1999). A physical basis for protein secondary structure. *Proceedings of the National Academy of Sciences, USA*, **96**, 14258–63.

Srinivasan, R and Rose, G. D. (2002). *Ab initio* prediction of protein structure using LINUS. *Proteins: Structure, Function and Genetics*, **47**, 489–95.

Surrey, T., Nedelec, F., Leibler, S. *et al.* (2001). Physical properties determining self-organization of motors and microtubules. *Science*, **292**, 1167–71.

Taylor, W. R. (2002). A periodic table for protein structures. *Nature*, **416**, 657–60.

Trinkaus, J. P. (1984). *Cells into Organs*. Upper Saddle River, NJ: Prentice-Hall.

Voet, D. and Voet, J. G. (1995). *Biochemistry*. New York, NY: John Wiley.

Watson, J. D. (1976). *Molecular Biology of the Gene*. Menlo Park, CA: W. A. Benjamin.

Wernisch, L. and Wodak, S. J. (2003). Identifying structural domains in proteins. In *Structural Informatics*, ed. P. E. Bourne and H. Weissig. New York, NY: John Wiley, pp. 365–85.

Williams, R. J. P. and Fraústo da Silva, J. J. R. (2003). Evolution was chemically constrained. *Journal of Theoretical Biology* **220**, 323–43.

14

Could an intelligent alien predict earth's biochemistry?

Stephen J. Freeland

This is a book about whether our universe is "biocentric." The *Oxford English Dictionary* defines this term as "treating life as a central fact" [1]; thus a biocentric universe is one predisposed towards producing life (life's centrality is implicit if "the fitness of the environment [for life] far precedes the existence of the living organisms" [2]). To date, this unusual idea has been most thoroughly explored (and most widely publicized) under the umbrella term "Anthropic Principle" in physics [3]. In essence, this principle refers to a suite of fundamental physical parameters, dimensionless constants that interact to imbue our universe with such interrelated phenomena as a diverse periodic table of elements, a preponderance of carbon and water, stars that emit energy, and planets that orbit them [4]. It asserts that, without clear explanation at present, the constants responsible for this state of affairs appear finely tuned in our universe to values peculiarly sympathetic with life's emergence.

Even if we accept this view of physics at face value, we remain a long logical leap from establishing truly biocentric credentials for our universe. Understanding "what is" versus "what might have been" for physics must be met by an equivalent understanding in biology. Thus the interface of biochemistry, where physics becomes biology, deserves especially close scrutiny. In this context, the first and perhaps most important point of this chapter is to emphasize that in considering physics and biology, two fundamentally different sets of expectations collide. Explorations of biochemical biocentrism must negotiate the transition from one to the other if they are to form a coherent evaluation.

The reason is that physicists' claims for a finely tuned universe are centered on the assertion of improbable non-randomness. In particular, the handful of physical constants responsible for our universe not only appear finely tuned, but "are, as far as we know, unrelated to one another, that is each can be independently set yet each must fall within a particular range to make a universe in which habitable environments

Fitness of the Cosmos for Life: Biochemistry and Fine-Tuning, ed. J. D. Barrow *et al.*
Published by Cambridge University Press. © Cambridge University Press 2007.

can exist" [5]. Thus, their putative serendipitous values as single entities become all the more remarkable when considered together. But to simply extend this search for improbable non-randomness into biochemistry is fundamentally problematic.

Whereas the abiotic sciences find the confluence of non-random parameters remarkable, biology does not. Indeed, the existence and aggregation of highly non-random attributes within biology is more than unremarkable. It is a primary expectation of a well-characterized process: natural selection. Reduced to a sentence, the theory of natural selection asserts that wherever a population of self-replicating entities inherit characteristics from their parents, traits that positively influence reproductive potential ("adaptations") are likely to increase in frequency. At a straightforward level, such adaptations might include the ability to control the local environment under which self-replication is achieved (e.g. evolution of a cell membrane). Less intuitively, the empirical evidence indicates that, given enough time, adaptation may also include such bizarre strategies as being a gnat or a gnu. All are highly non-random states of matter; but none requires special explanation beyond established evolutionary theory.

Thus, somewhere in fleshing explorations of biocentrism into biochemical detail, we must synthesize the answers to two different sets of questions. Physics may ask: How far forward into biology does the remarkable abiotic non-randomness proceed? (Does some trick of physics explain the mystery of biochemical chirality [6, 7]? Does physics directly define life [8]?) Biology, however, asks: How far back into abiosis does natural selection reach? (Does it explain fundamental biochemical parameters of living systems? Can it contribute to interpreting the physical universe [9]?) A physicist overlooks this fundamental distinction when assuming that improbable non-randomness within living systems is remarkable. A biologist overlooks it when assuming that the universe is just so.

Such misunderstandings are rife. In an otherwise excellent introduction about why evolution is such an important framework for understanding biology, Richard Dawkins writes that "physics is the study of simple things that do not tempt us to invoke design . . . even large physical objects like stars consist of a rather limited array of parts, more or less haphazardly arranged" [10]. In a more infamous faux pas, Sir Fred Hoyle once complained that biologists were willfully ignoring the mathematics of probability when claiming that random mutation could produce higher forms of life [11]. Dawkins found the productions of physics irrelevant to an account of life on earth (or of apparent design in the universe) because, to him, a diverse array of elements and the ready availability of a stable planet, bathed in the sun's radiation and rich in molecules of carbon and water, are background assumptions for a discussion of evolution. Hoyle assumed that biology is lacking in explanatory power because he failed to appreciate that natural selection is a powerful heuristic search algorithm that acts on mutation to directly create non-randomness. Indeed, the misunderstandings persist and have transferred directly to discussions of biocentrism.

Thus, many scientists dismiss any "fit" between the universe and life on earth as a simple *post hoc* fallacy [12], arguing that whatever hand of cards physics had dealt, any life that subsequently emerged under those conditions and acquired the capacity for suitable scientific inquiry would find itself well suited to its universe. If the reader substitutes the word "environment" for the word "universe" to yield the sentence "any life that subsequently emerged . . . would find itself well suited to its *environment*," then it will come as no surprise that such dismissals often stem from evolutionary biologists. But have they correctly understood the claim of a biocentric universe?

One of the most widely touted examples of a biocentric parameter from physics is the so-called cosmological constant: "finely tuned" between a value small enough to have allowed matter to aggregate into stars and planets and large enough to have permitted a leisurely expansion of the universe over billions of years. Could biology really have emerged in a universe without stars, planets, radiation, and elements? Physicists who reject any anthropic principle tend to either opt for a flat denial that any good fit exists between physics and biology [13] or recast the universe itself as the variable, arguing for the existence of an all-possibilities-exist multiverse in which our slice of reality is, quite unsurprisingly, the one peculiarly well adjusted to contain us [14].

What is interesting, then, is that these various denials are more or less mutually exclusive. How can an interesting "fit" between the physical universe and its life be a simple *post hoc* fallacy if some deny such a fit's existence in regard to the one example of life that we know whereas others require a multiverse to render our particular reality (and hence our existence) unexceptional? Indeed, these logically incompatible dismissals combine to outline the central gap in our current knowledge: exactly how are the properties of the universe connected to those of the life that has emerged within it? Put another way: is life necessarily so, and, if it is, then what are the cosmological implications? The answer is of relevance not only to cosmology and general evolutionary theory, but also to exobiology (the young field concerned with the possibility of and search for extraterrestrial life). Even the emerging branch of computer science that takes an evolutionary approach to developing algorithms seeks inspiration here [15].

Having identified the question, where can we look for evidence? Asking questions about what life might look like elsewhere in this universe (or in another universe altogether) is peering into a deep and murky pool. So far, we have exactly one example of the emergence and subsequent evolution of life in precisely one universe (the "$n = 1$ problem" [12]). Yet, as representatives of this evolutionary lineage, we have been studying the story of our descent in great detail. Presumably, our existence, and that of the other living systems that make up earth's natural history, are illustrative of something: the trick is to extract the generalities from this limited data set.

Thus, the significance of complex biological systems lies not in their non-randomness as a collection of abiotic molecules (this is natural theology of an old [16] and discredited [10] flavor), but rather in what they reveal about life in general, about the sort of life that one would expect to emerge elsewhere. It is therefore unsurprising that those most interested in the search for extraterrestrial life (e.g. NASA's exobiology program [17]) are chief patrons of research into the origin of life on earth. Extrapolating from the narrative of terrestrial evolution to general statements about life in the universe is an exercise of lifting current knowledge to a higher level by pulling on our scientific bootstraps. Certainly it requires care, for misleading extrapolations are easy; but it is hardly unscientific by nature. To claim otherwise is to conveniently ignore the extent to which evolutionary theory (and other branches of science) already bootstraps from the actual to the possible, a topic that has been explored at length by philosophers of science for hundreds of years [18]. Thus, rather than ask whether we could predict the fundamental biochemistry of life on another planet, I ask the opposite question: Could intelligent extraterrestrials predict the biochemistry of life on earth? Our detailed knowledge of earth's chemistry and biology at least has us peering in the murky pool at the shallow end.

Other chapters in this book explore the concept of predictable biochemistry both at a general level (for example, once life had emerged, did its subsequent evolution pursue a predictable feedback loop with the inorganic chemistry of the planet [19]?) and at a detailed one (given that life uses proteins for metabolism, did natural selection merely "discover" the shapes that are inherent in this type of chemical polymer [20]?). Here my aim is to ask whether the fundamental biochemical framework for life, namely nucleic acid information (genotype) and protein metabolism (phenotype), was in any sense a predictable outcome. More specifically, I intend to discuss in detail whether and how to broach this topic within the boundaries of standard evolutionary theory. The answer informs the question of whether, given the physical framework of this universe, evolution was bound to produce a certain kind of life.

For those unfamiliar with molecular biology, the next section provides a brief introduction. Readers familiar with DNA, RNA, and proteins may want to skip ahead.

The central dogma of molecular biology

Genetic information is stored in DNA, a chemical chain-like polymer in which each link is a chemical letter (nucleotide) drawn from the nucleic acid alphabet of four

284   *Stephen J. Freeland*

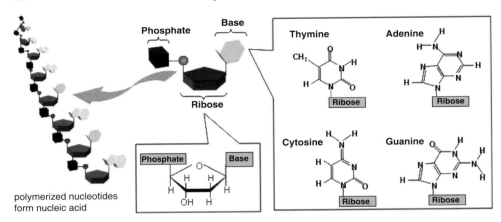

Figure 14.1. The biochemistry of nucleic acid. DNA and RNA are linear polymers in which each building block is a nucleotide. Nucleotides make up a ribose–phosphate scaffold, to which a base is joined. Four bases are used in both DNA and RNA: adenine, cytosine, guanine, and thymine (used in DNA, modified from the chemically similar uracil used in RNA).

letters: **a**denine, **c**ytosine, **g**uanine, and **t**hymine (see Figure 14.1). But DNA is nothing more than information, a "how to" instruction manual for building an organism. In today's organisms, proteins are, to a first approximation, the components that actually interact to make life.

Some proteins are structural (such as collagen in skin and cartilage or the keratin in hair and fingernails), but most are catalytic, steering the chemistry that forms life from energy and raw materials. Thus, the classical definition of a gene is nothing more than a short stretch of DNA that encodes a single protein, and the genome is the total genetic information of an organism.

Each protein is itself a polymer, but here the individual components are amino acids, of which 20 different kinds exist (see Figure 14.2). Thus, both genes and proteins can be represented by strings of symbols (e.g. English letters) that describe the sequence of building blocks of which they are made. Indeed, this is exactly how they are represented in the centralized databases [21] that are currently accumulating the data reported by scientists working around the world whose work entails finding the precise sequences for the genes and proteins of different organisms (see Figure 14.3(A)).

The complete set of rules for decoding (translating) genes into proteins are what is known collectively as the *genetic code*. Given the difference in size between DNA and protein alphabets, the nucleotides of a gene cannot be converted into the amino acids of a protein by a simple 1:1 mapping. Instead, biology uses a system analogous to that of Morse code, in which the letters and punctuation of

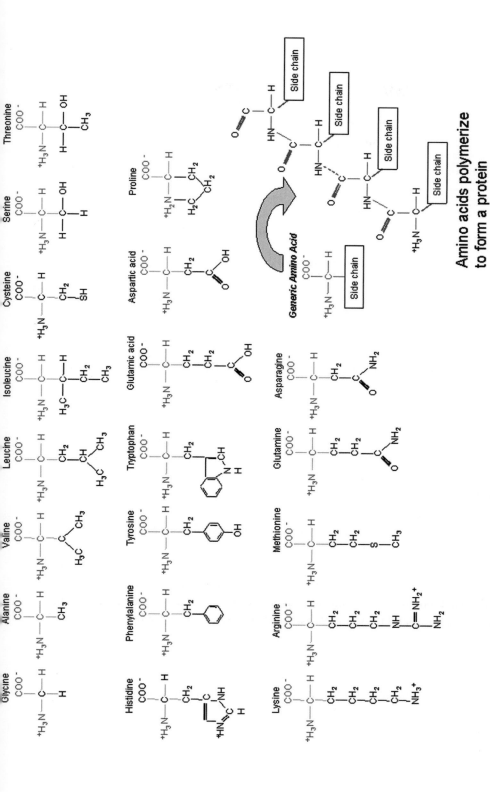

Amino acids polymerize to form a protein

Figure 14.2. The biochemistry of protein. Protein is another linear polymer in which each building block is an amino acid. Amino acids have a central ("alpha") carbon to which an amino group, a carboxyl group, and a variable side chain are joined. Twenty amino acids are encoded within the standard genetic code.

(A)   **(B)**

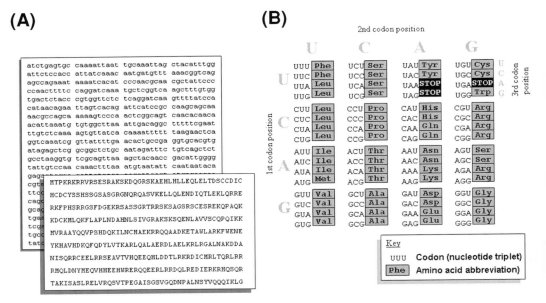

Figure 14.3. (A) Both nucleic acid and protein sequences, as linear polymers, can be represented as strings of English letters. This is, indeed, exactly how they are stored in global, centralized databases of biological data. (B) The genetic code is the system of rules that maps nucleic acid sequences into proteins. Nucleotides are read, three at a time (as "codons"), and converted into a single amino acid by means of tRNAs, specialized adaptor molecules.

the English alphabet correspond to clusters (words) of a binary dot/dash language. The simple, unvarying pattern for gene translation was discovered by an elegant series of experiments in the 1960s [22]: triplets of nucleotides (known as "codons") are read into a single amino acid. Thus, the words "translation" and "decoding" are quite literal. For example, the triplet AAA corresponds in the standard genetic code to the amino acid lysine (Lys or K), whereas AGC corresponds to the amino acid serine (Ser or S), and every gene includes three times as many nucleotides as there are amino acids in the protein that it encodes. Because there are four types of nucleotide, and they are read in triplets, there are a total of $4 \times 4 \times 4 (= 64)$ codons, each assigned to one of 21 translation "meanings": 20 amino acids and a single piece of genetic punctuation – a signal to "stop translating!"

In discovering this translation system, researchers found something quite surprising: the same genetic code is used by organisms as diverse as the bacterium *Escherichia coli* and human beings. In fact, it is now known that this standard genetic code is used by the overwhelming majority of organisms on earth and is thus to be found in every textbook of biology and biochemistry. The usual presentation of the genetic code (see Figure 14.3B) displays codons and their associated amino acid assignments in a $4 \times 4 \times 4$ table, collapsing the third dimension into vertical blocks. In each dimension, the nucleotides are displayed in the same order, U, C, A, then G, and the amino acids are usually represented by their three-letter abbreviations.

However, this overview has so far omitted mention of one other important component: the molecule RNA. RNA is another biopolymer; as its name suggests, it is a sister chemical language to DNA that is also spelled out in four nucleotides. Each nucleotide is a biochemical relative to its equivalent letter in DNA (although for reasons of biochemistry and evolutionary history, the "T" of DNA is equivalent to the "U" [uracil] of RNA [23]). The primary importance of RNA to the system is that DNA is never directly translated into protein. Rather, any gene that is destined for translation is first copied into a temporary RNA form (a process known as *transcription* to reflect the direct copying, nucleotide for nucleotide, from one language into another); the resulting messenger RNA (mRNA) is what is translated by the cell into a protein.

Interestingly, however, this is not the only relevance of RNA to the genetic code. Many metabolically active components of the molecular machinery at the heart of the cell are made of RNA that has folded up into a three-dimensional shape reminiscent of protein. One key example is the ribosome, the subcellular machine of great complexity that supervises the process of translation. The ribosome largely comprises folded, knotted RNA. Although various proteins are embedded within this structure, great interest has surrounded the discovery that the ribosome's primary functional attributes are performed by the RNA [24]. Moreover, the molecular adaptor molecules – known as transfer RNAs (tRNAs) – that effect translation by attaching at one end to a codon and at the other to an amino acid are themselves made of RNA. Indeed, Francis Crick, co-discoverer of the structure of DNA, noted early on that the "tRNA looks like nature's attempt to make RNA do the job of a protein" [25]. A decade later came the suggestion that protein-like molecules at the core of biochemistry could be elegantly explained as "molecular fossils" [26] of a time long before our modern system of DNA and proteins first evolved. In this hypothesized world, all life used just a single biopolymer, RNA, which performed both as an information storage molecule (today deferred to DNA) and as a catalytic molecule (today, largely deferred to proteins). Since then, this observation has been coupled with a growing field of laboratory experiments to form the "RNA world hypothesis" [27]. In fact, it seems likely that the genetic code evolved in the RNA world and that DNA was the last piece of the puzzle to appear [28].

This now completes the brief introduction to the central dogma of molecular biology, the two-step process by which a short stretch of DNA is first transcribed into mRNA and then translated, one triplet codon at a time, into the amino acids of a protein.

A context for the questions of biochemical biocentrism

Imagine a tomorrow in which the world is rocked by NASA's fanfare announcement that, after many false starts and red herrings, its latest robotic probe has unambiguously identified living organisms on Mars. Scientists, philosophers, and theologians scurry to prepare an initial response, thinking of implications for a universe that may be bustling with an unimaginable diversity of life. But even as they mobilize comes the most surprising discovery of all. The sophisticated remote laboratory – equipped not only to detect the telltale signs of self-replication in organisms, but also to study their biochemistry – shows that these life forms are no mysterious aliens: they use the same fundamental biochemical framework as life forms on earth. Each organism has a nucleic acid genome encoding a complex network of proteins that interact to produce metabolism. Even the decoding is eerily familiar, with small adaptor molecules of folded RNA translating successive genetic "codons" into individual amino acids. Mainstream interest quickly subsides as debates refocus on academic issues of how and when a single ancestral life form distributed itself between at least two planets (see, for example, [29]) and whether we are indeed looking at recent contamination by terrestrial bacteria carried to Mars by previous NASA missions [153].

The shift in interpretation reflects current orthodox science: life originating elsewhere would evolve into something very different from what we know as terrestrial biology. In particular, evolutionary biologists deride what we might term the "*Star Trek* Syndrome," biologically naive, unconsciously anthropocentric science fiction in which the aliens have evolved toward the ideal "goal" of humanoid appearance in some sort of synchrony with their evolutionary progress toward self-aware intelligence. Pointed ears, unusual bony skull protuberances, or scaly skin is all that remains to hint at the radically different pathways that each species has traversed en route to achieving our "advanced" state of bipedal curiosity. Indeed, the makers of the recent *Star Trek* series, perhaps embarrassed by criticism of their science, have recently added the plot device in which aliens were guided to this common appearance by a higher intelligence [30].

So, sweeping aside this anthropocentrism leaves an alternative vision: wherever life originates, it will be uniquely and unpredictably fitted to its local environment and will inevitably follow a unique and unpredictable evolutionary trajectory to states that we find difficult to imagine. This sounds altogether more sophisticated and open minded. But what exactly is the scientific evidence for this view? To what extent are the emergence of life and its subsequent evolution contingent on spatial and temporal particularities of the local environment? How does this relate to the trends recorded as historical fact within terrestrial macroevolution? A detailed discussion of the significance of nucleic acid genotype and protein phenotype must be set in the context of a bigger question: does theory really indicate that the "big picture" outcomes of evolution are inherently unpredictable? The occasional optimistic assessment of earth's biochemistry as "common sense" [31] tends to ignore this part of the discussion, so let us start by exorcising a persistent demon of evolutionary orthodoxy.

Evolutionary predictability: a historical context

The simplest and most general arguments against evolutionary predictability assert that the whole point of Darwin's theory was to escape from notions of determinism applied to the development of the biosphere. The corollary here is that all talk of predictability represents the unscientific graft of pre-Darwinian culture onto a legitimate scientific theory. One of the clearest statements to this effect comes from Gould [32], who wrote that the very word *evolution* "has contributed to the most common, current misunderstanding about what is meant" by Darwin's theory – namely, progress. Gould traces this misunderstanding back to the subversion of Darwin by Victorian cultural values, as exemplified in the writings of Herbert Spencer [33]. We read that "Darwin spoke of descent with modification. He shunned evolution as a description for his theory . . ." because of the etymological roots of the word, which lie in the Latin *evolvere*, "to unroll" (especially a book or scroll) [34].

Thus, from the early poetic usage to describe the unfolding of a plan [35], "evolution" originally entered the biological lexicon via "preformationist" theories[1] that interpreted biological development within individual organisms' lifetimes and between successive generations as equally deterministic realizations of a preordained design. For example, Bonnet asserted that an aboriginal female of every species contained the germ (information) for every individual that would subsequently be born to that species [36]. He later expanded this view to incorporate change from one species into another: extinctions were correlated with the developmental migration of species to a higher state such that any and all change in the biosphere was an evolution (unfolding) of this plan [37].

Such thinking had itself grown from the earlier, more general assumption that life's diversity displays an intuitive order, a ladder of progress from lower to higher states of being. This view ultimately traces back to Aristotle's "Great Chain of Being" (*Scala Naturae*), but was subsequently adopted by the Roman Catholic Church and later formalized in Linnaeus' *Systema naturae* [38] as natural theology.

---

[1] Gould [32] ascribes the first usage of "evolution" in this context to Albrecht von Haller. I have been unable to corroborate this source, instead finding a tentative suggestion from his contemporary, Bonnet, that "evolution" describes the process of organic change over time [37]. For our purposes, the discrepancy is inconsequential: both presented preformationist theories of evolution at about the same point in history.

Indeed, the real point of Lamarck's theory of evolution ("transformisme") was not the inheritance of acquired characteristics (which widespread view he shared with most of his contemporaries), but rather that of an innate, preprogrammed behavioral drive of all organisms toward a higher state [39].

From this perspective, the "*Star Trek* Syndrome" has a surprisingly rich cultural history from roots in ancient Greek metaphysics through Roman Catholic theology into early enlightenment science. All told, by the mid nineteenth century, "evolution" had vernacular currency in the English language with the nuanced meaning of an unfurling plan, and it was this that Gould tells us Darwin sought to avoid. And yet, as Gould admits, the *Origin of Species* does contain one use of the word, and it is no casual slip. The final sentence of the final paragraph of the book concludes:

There is grandeur in this view of life, with its several powers, having been originally breathed into a few forms or into one; and that, whilst this planet has gone cycling on according to the fixed law of gravity, from so simple a beginning endless forms most beautiful and most wonderful have been, and are being, **evolved** [40] [emphasis added].

In other words, Darwin chose "evolution" as his masterwork's literal last word on the outcome of "descent with modification." Certainly, Darwin displayed a seemly Victorian correctness in using the passive tense of the verb to reflect its etymological subtext: from our standpoint in time, the biosphere we encounter has indeed evolved (unfolded) according to the laws of natural selection over four billion years. In other words, hindsight reveals how possibility unfurled into fact, whether or not the pathway followed a predictable trajectory.

Likewise, the *a priori* improbability of any particular hand of cards unfolds into certainty as the deck is shuffled and dealing proceeds. But the vernacular and scientific nuance of "evolution" would not have passed unnoticed among either the scientists or the general public at whom Darwin aimed his book. Rather, it would have left them with an obvious question as they finished reading: was this "unfolding" in any way predictable *before* the fact of history? Even when the second and subsequent editions of *Origin* incorporated explicit statements linking the Creator to the emergence and evolution of life on earth [41], Darwin kept evolution in its pointedly ambiguous position, leaving the question as to whether or not the simple algorithm of natural selection produces predictable trends and patterns at a longer timescale. Indeed, Darwin's debates with Asa Gray regarding the roles of purpose and design in evolution clearly reveal his personal agnosticism on this issue of deterministic outcomes [42]. He did not reject the idea that the laws of nature (including natural selection) stemmed from an Ultimate Cause, nor did he deny that natural selection could lead predictably to sentience or humans; he simply denied that the pool of variation on which natural selection worked was directly manipulated by a Higher Hand [43]. Thus, for example, Gould's statement that

"Darwin explicitly rejected the common equation of what we now call evolution with any notion of progress" is hard to reconcile with Darwin's own words in the final chapter of *Origin*: "And as natural selection works solely by and for the good of each being, all corporeal and mental endowments will tend to progress towards perfection" [44].

The issue here is not whether natural selection always pushes every lineage that it influences to states of higher complexity. No one disputes that selection can lead to evolved simplicity (although widespread generalizations are suspect – cave-dwelling animals do not evolve general simplicity relative to their light-dwelling counterparts [45], nor do parasites evolve simplified anatomical design relative to their free-living counterparts [46]). Rather, the issue is what, if any, predictable outcomes would unroll from natural selection? The significance of Darwin's theory was not to demolish notions of evolutionary predictability, but rather to lay the groundwork that (among other things) renders them testable as scientific hypotheses rather than spiritual assumptions.

Distinguishing predictable outcomes from orthogenesis

This is not to deny that since 1859 evolutionary theory has repeatedly had to evaluate suggestions of additional forces that would steer the trajectory of descent with modification toward particular ends. Such attempted introductions form a major class of what Dennett refers to as "skyhooks" [47] – unwarranted and unscientific constructs that seek to lift Darwin's process to a level that can explain the emergence of intelligent human beings, the enormous antlers of Irish elk, or any of a host of other phenomena that defy the imagination as being emergent products of unadulterated natural selection. Evolutionary biologists refer to such skyhooks collectively as "orthogenesis" ("a view of evolution according to which *variations* follow a defined direction and are not merely sporadic and fortuitous" [35]).

Some early examples sought simply to fuse a "hands-on" Creator with Darwin's process, suggesting that God has directly controlled the flow of variation from which natural selection has harvested adaptations (for example, Asa Gray wrote to Darwin that he could not but see the hand of God in steering which mutations led to human beings [43]). Such thinking, exemplified most recently in the cause of Intelligent Design, is a crude attempt to fit an interventionist God into the mechanical universe [48]. It is but a short step away from Lamarck's theory that God imbued the organisms themselves with a desire or drive to "vary" towards ever more perfection. Interestingly, other forms of orthogenesis have taken the other half of Lamarck's equation, arguing that the internal, non-random drive exists, but is not attributable to a divine hand. Rather, they hypothesize unknown mystical properties of the mutation mechanism (R. S. Lull, for example, infamously argued that some genetic factor unleashed a drive toward ever-increasing size in the antlers of the Irish Elk, leading to the eventual extinction of the species [49]).

But although evolutionary research to date has found no supportive evidence for any form of orthogenesis, it is simply wrong to infer that all hints of evolutionary determinism are connected to this unfounded concept. The error lies in a failure to distinguish between the skyhooks of orthogenesis and the potential for a simple algorithm, when applied iteratively over much time or space, to reliably produce patterns that are not immediately obvious. By way of analogy, anyone who has carefully observed the behavior of a colony of social insects (or a wheeling flock of birds or a school of fish) will have been impressed by the high-level patterns of organization that appear somehow programmed into the dynamics of these populations. Indeed, some have gone so far as to interpret the interacting population as a single superorganism [50].

Yet each year that passes brings the scientific community closer to understanding how a relatively few, simple rules followed by each individual but iterated many hundreds of times in a single social grouping could combine to produce such elegant, complex, and unintuitive overall dynamics [51]. The whole is, after all, predictable as the sum of the parts [52]. Why then should we be confident that natural selection, operating in countless lineages over enormous periods of time, could not produce reliable, if unintuitive, outcomes? Why should the whole biochemical framework of life not be a prime example?

Current understanding of unpredictability: from physics to biology

We now know from physics (specifically quantum mechanics) that below a certain resolution deterministic, *a priori* predictions regarding any and all matter (biological or not) do not just lie beyond our current powers of calculation, but are actively incalculable [53]. Thus, whereas the winning sequence of numbers for the Maryland State Lottery unfolds into certainty before many disappointed eyes each week, a deterministic prediction of the draw would require knowledge of the precise position and velocity of all the particles of the Ping-Pong balls in the tombola, and the very nature of the subatomic world precludes such knowledge for even one fundamental particle. Similarly, long-term weather predictions remain informed guesswork as the dynamics of the atmosphere magnify infamous butterfly wing-beats into chaos [54].

Both types of event "evolve" (in the strict etymological sense) from possibility into actuality only with the benefit of hindsight. But there is no reason to believe that quantum uncertainties will automatically penetrate upward into higher-order dynamics – it all depends on which particular higher-order dynamical system one considers. For example, it *is* a highly predictable outcome of human behavior that each week's state lottery will be largely funded by the least financially secure segment of Maryland's population [55]. Indeed, a process as simple as diffusion illustrates how a higher-level dynamical process can tame atomic uncertainties as close to certainty as any statistician would care to calculate. So does evolution amplify or suppress unpredictability? The simple truth is that no one yet knows. Biologists have no reason to expect that lower-level stochasticities of physics are of any relevance to understanding evolution or to suspect that evolution is thereby rendered deterministic.

To understand this ambiguity, start by considering that evolution by natural selection results from two contrary processes. Stochastic mutation acts within a population to create random variation, but natural selection then filters this variation, steering the population towards specific parameter values for specific traits. The continual introduction of new variation matched by a continual selection toward an optimal phenotype will lead, in simple systems, to predictable evolution. Indeed, a computer scientist presented with this description of natural selection as an iteration of twinned processes will immediately recognize the overall process as a simple ("hill-climbing") form of search algorithm – a set of rules that can find an optimal solution to a problem from pretty much any random starting point, given the right conditions. For example, the population might be one of herbivorous insects evolving, under the selection pressure of predation, to the exact shade of green that maximizes camouflage against the leaves on which they sit browsing.

Mutations introduce random variations in color, and predators subsequently filter this gene pool, leaving a disproportionate fraction of the best camouflaged individuals. The important point for our current purposes is that whatever the starting point of the insect phenotype, and whatever the variation provided by mutation, an individual's reproductive success will tend to reflect its physical similarity to the leaf. The stochasticity of mutation is tamed by natural selection into a predictable outcome. However, there is no sense in which the insects are striving toward better camouflage (cf. Bonnet [37] or Lamarck [39]) or that the mutations are leading them toward camouflage (cf. Gray [43] or Lull [49]). Instead, the random variation is being non-randomly filtered by predators, whose appetites unconsciously effect natural selection on the insect population (Darwin [40]). So if natural selection simply accumulates to produce long-term evolutionary change, then isn't it straightforward logic to conclude that all of evolution is deterministic?

This was certainly the caricature of evolutionary biology presented when Gould and Lewontin launched a spirited attack on what they termed the "adaptationist program" of neo-Darwinian evolutionary biology [56]. They argued against what they perceived as a near monomaniacal form of scientific reductionism that was attempting to explain the whole of evolutionary biology as the sum of a series of adaptive parts, each fashioned to perfection by natural selection. The central theme of their criticism was that overzealous "adaptive storytelling" can easily lead to the inappropriate atomization of an organism into a collection of well-adapted traits. When the pieces are reassembled, this in turn would yield a misleading picture of evolutionary determinism to explain the evolved morphology that we encounter as an organism. Gould and Lewontin's analogy for the loyal evolutionary adaptationist was Voltaire's Dr. Pangloss ("It is proved that things cannot be other than they are, for . . . everything is made for the best purpose" [154]). Subsequently, the claim that mid-twentieth-century evolutionary biology was undervaluing the evolutionary importance of biological constraints has proved highly controversial (see, for example, [57]), the logic of the criticism has been questioned (see, for example, [58]), the adaptive program has been vigorously defended (see, for example, [59]), and the general motivation for the critique has itself been questioned (see, for example, [60]). However, in the midst of Gould and Lewontin's work lie two fundamentally non-adaptive features of strict Neo-Darwinism that even their sharpest critics do not dispute.

Noise and drift: the unpredictability in evolution

First, evolution is more than just mutation and natural selection. Evolution is properly defined as "change in genetic material over time." Certainly, natural selection contributes to such change; but so do altogether more stochastic events. These include the sorts of unusual environmental fluctuations, such as an earthquake or meteorite impact, that form "noise" against the "signal" of regular selection pressures to which a population is adapting. No matter how well an insect matches the leaf on which it feeds, it will fail to reproduce (thus contributing to change in the genetic composition of the population, i.e. evolution) if struck by a wandering meteorite. Unlikely? Perhaps. But debates still rage as to the relative importance of extraterrestrial impacts in triggering mass extinctions [61], and as to the role that such events have played in steering the course of long-term evolution [62] (especially near to the time of life's origin[s] [63]).

Moreover, stochastic noise extends (insidiously, although less dramatically) into the very genetic framework by which many organisms transmit their genes to the next generation. August Weismann [64] was among the first to point out that individuals of a sexually reproducing species must necessarily halve their genetic material before reproduction if the fusion of egg and sperm is not to double the genetic material of each successive generation. Gregor Mendel's later experiments demonstrated that genetic material is made up of numerous particulate genes and that each organism of a sexually reproducing species carries two versions (alleles) of every gene. The cellular division process of meiosis, by which such organisms produce gametes with half the "normal" complement of genetic material, is usually a lottery in which each member of a pair of alleles stands a 50 percent chance of making the physical transition into a specific daughter cell.[2] As a result, even in the absence of natural selection, allele frequencies evolve through an extensive binomial sampling process as each generation reproduces. The inherent sampling error ("genetic drift") is as entirely random and without direction as it is absolutely unavoidable. In a large population, the effects are negligible, and population geneticists typically refer to an effective population size above several hundreds as "infinite" in this sense. But stochasticity is built into the system, and who is to say that relatively sudden, relatively drastic reductions in population size have never unleashed genetic drift at an important juncture in the evolution of life on earth (see, for example, [65])?

The second problem with assuming that natural selection accumulates into deterministic macroevolution adds weight to this question mark. It is the observation that the reliability with which natural selection (or any other heuristic search) produces a specific outcome is influenced by key features of the possibility space within which it operates. This can be illustrated by using the notion of a fitness landscape (see Figure 14.4) [67] within which a phenotype composed of n traits is represented as an $(n + 1)$-dimensional surface; the final dimension ("height") is the implied fitness of a phenotype. Returning to the evolving population of herbivorous insects, we might assume that an individual's fitness is defined by two quantifiable traits, say "color" (light absorption wavelength) and "age to sexual maturity."

In such a world, the fitness landscape exists in three dimensions (see Figure 14.4(A)). In this construct, it becomes clear why the simple iteration of mutation and natural selection is called a "hill-climbing" heuristic search algorithm: Mutations form random steps in any direction from the initial genotype, but natural selection favors those that result in an individual of higher fitness (one that resides farther up the hill; see Figure 14.4(B)). The result is that the average genotype of the population progresses toward the optimum. The problem is that hill-climbing algorithms tend to find and become stuck at the nearest optimum to the starting position (i.e. at local optima) even when superior optima exist elsewhere on the landscape. This matters because the adaptive landscape for any real-world organism is likely to be extremely high-dimensional, containing numerous distinct, isolated local optima. Thus, in any "adaptive walk," the precise nature of the first few mutations (each an outcome of a stochastic process) may influence exactly which adaptive peak natural selection subsequently approaches. In other words, were two genetically identical populations to experience exactly the same novel selection pressure(s), then each would follow a different evolutionary pathway to a different optimal solution.

---

[2] In some cases, "ultra-selfish" genes evolve to cheat this process. In particular, alleles known variously as "meiotic drivers" or "segregation distorters" ensure that they end up in the only viable gametes that are produced. Theory and data surrounding the general phenomenon of intragenomic conflict are still relatively new, and the degree to which this potentially important phenomenon influences broad patterns of macroevolution is difficult to assess [66].

Stephen J. Freeland

(A)   (B)

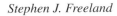
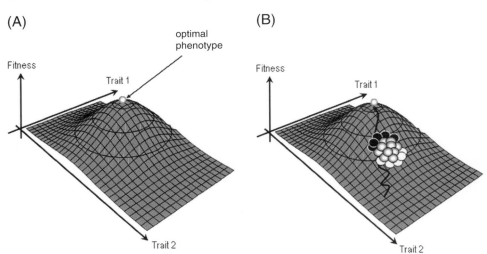

Figure 14.4. (A) The adaptive landscape represents the "possibility space" of phenotypes as surface, with every combination of values for traits $1, \ldots, N$ (co-ordinates in dimensions $1, \ldots, N$ of the landscape) associated with a fitness value (dimension $n + 1$, drawn here as "height"). In this case, two traits define fitness in a simple landscape for which there is only one optimal combination of trait values; thus (B) in an evolving population that exhibits phenotypic variation (shown here as a cluster of circles on the landscape), the fittest individuals are those closest to this optimum (shown as dark circles), and the least fit are those farthest from the optimum (shown as light circles). Thus, through iterated, fitness-dependent reproduction, the population "climbs" to the adaptive peak.

Influential figures of twentieth-century neo-Darwinism have placed central emphasis on this effect for the very process by which new species form. In particular, Mayr's [68] classic model for the formation of new species asserts that a critical first step is the isolation of a small subpopulation during which event stochastic-sampling error (perhaps coupled to a brief period of genetic drift as the population recovers to an appreciable size) triggers adaptive evolution to a new adaptive peak. Thus, extraterrestrial impacts near to life's origin(s) could have reduced primordial populations to a point where genetic drift set the evolutionary course by which fundamental biochemistry emerged.

So, should we not only reject the naïve idea that natural selection inevitably produces long-term evolutionary predictability, but further conclude that the stochasticity of mutation creates the existence of genetic drift and also that the framework of hideously complex adaptive landscapes combine to render evolutionary *un*predictability inevitable? This is perhaps as close to orthodoxy as generalization will allow. For example, in rebutting Gould and Lewontin's [56] critique of the adaptationist program, the arch proponent of natural selection G. C. Williams simultaneously defended the role of the adaptive program in advancing evolutionary theory while emphasizing that adaptation could only ever be to the local conditions [59]. In short, evolution might be producing good design all around, but the global optima that an intelligent engineer might *predict* could be consistently overlooked by the natural selection's myopic tendency for local hill climbing.

Indeed, this viewpoint arguably reached its most famous and focused exposition when, some fifteen years after dismissing evolutionary determinism as a cultural hangover from Victorian philosophy, Gould's book *Wonderful Life* created the catchy sound-bite of "replaying life's tape . . . you press the rewind button . . . then let the [evolution of life on earth] run again and see if the repetition looks anything like the original" [69]. The book's thesis is a protracted argument that the long-term outcomes of evolution are unpredictable and that self-aware humans are an extremely unlikely, happy accident. The claim was based on common sense regarding the unpredictability of the environment in which organisms are adapting. Our world is one in which stochastic noise, from fluctuations in seasonal temperatures and precipitation to the occasional impact of huge meteorites and the intricate maneuverings of plate tectonics, keep throwing the dice for evolving populations. If each population's long-term evolutionary trajectory is contingent on numerous accidents of history, then surely long-term evolution is unpredictable?

Allowing the data to lead

But common sense (more accurately called "intuition") is rarely the friend of science: the data must lead and opinions follow. Indeed, humans historically have displayed an impressive ineptitude for accurate intuition about biological evolution. The primary breakthrough of evolutionary thinking [70] was to expose Aristotle's fundamental "common sense" notion of naturally discrete categorical types as a largely subjective invention of human thought when applied to the natural world of species.[3] Of greater relevance to the issue at hand, the common sense of old-school natural theology's "arguments from design" (that a mechanical universe could not produce organisms so exquisitely suited to their environments) has fallen to data, hypothesis, and experiment. Even Mayr's view of stochastic speciation [68] is coming under increasing pressure from the empirical data [73]. The question for macroevolutionary predictability, then, is whether similarly counterintuitive data are waiting in the wings, ready to redefine common sense.

---

[3] Indeed, field biologists have subsequently uncovered so many examples where two or more "species" blend confusingly into one another [71] that even scientifically literate creationists have found themselves maneuvered into reassigning the point of divine creation somewhat farther up the taxonomic hierarchy: a level somewhere between "family" and "genus" is a typical, if arbitrary, focal point in such circles at present [72].

Interestingly, research from the computer science community has shown that the very process of stochastic drift that forms one half of Neo-Darwinian arguments for evolutionary non-predictability may go some way to solving the problem of multiple adaptive peaks that forms the other half of such arguments. Specifically, if a local hill-climbing algorithm is allowed periodic bursts of random genetic drift, then this can free the evolving population to wander far from local optima. The subsequent re-implementation of natural selection finds the population on the foothills of an alternative adaptive peak [74]. The probability that a short burst of drift will dislodge an evolving population from its current hilltop to the foothills of another peak is inversely proportional to the size of the peak it currently occupies. Because global optima are, by definition, higher than local optima, bursts of drift can actually help populations to find global optima with greater predictability. But simulating evolutionary dynamics on complex landscapes is still a minor field, not least because of difficulties in accurately rendering real-world populations to this theoretical construct. Thus we turn to empirical data.

In this context, Conway Morris' recent book *Life's Solution: Inevitable Humans in a Lonely Universe* [75] brings a broad range of contemporary evidence to challenge common sense. It derives an antithesis to assertions of long-term evolutionary unpredictability by re-evaluating the fundamental notion that no specific pathway through a complex, branching "possibility tree" could be inevitable. To test this assertion, Conway Morris focuses on case studies where the evolution of multiple lineages has seen them converge on similar morphologies, reflecting an optimal solution to the challenges of maximizing fitness (a self-replicating system's self-replication potential). His conclusion is that complex evolutionary convergence is far more widespread than contemporary wisdom would anticipate. The inference is that natural selection seems to navigate through the randomness of this world to "locate" similar, optimal solutions, almost regardless of the starting point. An extreme role for selection pressure at each node in a long series of branching possibilities does not transform blind chance into deterministic certainty, but it does suggest that adaptive evolution can navigate complex fitness landscapes to arrive with surprising predictability at specific outcomes. But how then do we move from a broad consideration of the macroevolutionary landscape to an explicit evaluation of the fundamental biochemistry of life?

Resolving $n = 1$: from comparative analysis to optimality studies

Conway Morris' focus on convergence exemplifies one general approach to studying adaptation: he infers the footprint of natural selection by demonstrating that a particular trait has evolved repeatedly beyond the expectations of chance. The methodology for such analyses has been formalized as the Comparative Method

[76], by which patterns of evolutionary outcomes are mapped onto an associated tree of evolutionary relationships to test their statistical significance. However, the limitation of this approach is that it applies only to evolutionary phenomena for which multiple, independently derived examples exist. Indeed, the very essence of the analysis is to show that similar phenomena have evolved independently with significant regularity.

In considering the question of fine-tuning within biochemistry, some of the most relevant phenomena are of interest because they were universal to the primordial life that subsequently spawned today's biosphere. In particular, during the early evolution of life on earth (defined as the interval between the origin of life and the organism that formed the "Last Universal Ancestor" of all extant life), all organisms followed a similar pattern: all came to use lipid membranes to partition a local environment for metabolism; all came to use nucleic acid to encode genetic information in a linear sequence (Schrödinger's predicted "aperiodic crystal" [8]); all came to use an alphabet of exactly four such chemical letters to spell out their genetic messages; all came to use DNA for genetic information storage and proteins for metabolically active molecules; all came to construct proteins from an alphabet of 20 amino acids; all came to transform genes into living organisms; all came to use an identical genetic code. Although in a few cases these universal constants have subsequently weakened into standards (some viruses have re-evolved an RNA genome [77], some twigs on the tree of life have evolved minor changes to the rules of their genetic code [78], and some lineages have even added one or two amino acids to their protein alphabet [79]), they form the platform on which all subsequent evolution has built. Their significance to understanding the fit between this universe and its life is as fundamental as that of the cosmological constant.

But we cannot interpret the near-ubiquity of such phenomena throughout the many branches of the tree of life as a sign of adaptive value via the comparative method. The problem is that here, widespread distribution derives from shared ancestry rather than from convergence. Although designs of natural selection may be blind with respect to future potential, they are less so to the antecedent(s) on which they build. Specifically, the probability that a mutation is advantageous is inversely proportional to the magnitude of its effect [80]. Consequently, "Evolution is a tinkerer" [81] that tends to construct at the edges of an increasingly constrained foundation. As organisms evolve, characteristics that emerged early on become "locked in" by an increasing number of evolutionary epiphenomena that make adaptive sense only in relation to them. In short, evolutionary convergence from different starting points is far better evidence of adaptation (and hence predictability) than a failure to diverge from a common starting point.

Thus, each biochemical standard may have arisen by chance and then been evolutionarily frozen into place. A striking example of such thinking was Crick's explanation [82] for why the same genetic code is found in organisms as diverse as *E. coli* and *Homo sapiens*. He noted that an organism's entire genome uses one common set of decoding rules to convert genetic information into proteins. Thus, once any genetic code is in place, the fitness of all protein-coding genes in the genome depends implicitly on these decoding rules. Whereas Neo-Darwinism asserts that occasionally a single mutation within a gene might prove advantageous, to mutate any of the decoding rules themselves would amount to simultaneously introducing countless mutations throughout an organism's genetic material. Selection pressure against variation, Crick argued, could explain why the same genetic code was used by all organisms: not because it was an adaptive state in every lineage, but because it was a "frozen accident" of evolutionary history.

However, other interpretations of the code surfaced around the same time. In particular, they focused on the observation that any code mapping 64 codons to just 20 amino acids will be redundant in that at least some amino acids will be specified by more than one codon. The standard genetic code is decidedly non-random in this respect: although redundancy is distributed between most amino acids, synonymous codons (i.e. those that code for the same amino acid) are typically clustered together, sharing two out of three nucleotides (see Figure 14.3(B)). Two researchers suggested that this non-random distribution might buffer genomes against the effects of point mutation (a common type of mutation in which a single nucleotide of DNA is altered) and proposed natural selection as an explanation [83, 84]. Soon after, another researcher noted a more subtle pattern: that amino acids sharing similar biochemical properties are also clustered within the code, such that even where single point mutations *do* lead to a change in amino acid meaning, they tend to substitute an amino acid similar to that specified by the unmutated codon [85]. But then Crick's "wobble hypothesis" proposed that the block pattern of synonymous codons could have resulted from nothing more than biophysical constraints on the specificity with which a tRNA can recognize its cognate codons [86]. The biological community was still reacting to the spectacular collapse of several high-profile hypotheses that had previously predicted how the code would be found to work. Each was extremely elegant, but had been proven wrong by empirical evidence [87]. Thus, the "frozen accident" argument for the impossibility of code evolution, backed by the apparent universality of the standard genetic code, carried the day.

The genetic code: from "frozen accident" to predictable adaptation

More than a decade later, the situation changed when researchers discovered the first example of a naturally occurring, non-standard genetic code [88]. Subsequently, further examples of variations in codon assignments have been uncovered in a disparate group of evolutionary lineages [78]. Each is a minor variation of the standard code, and most occur in simple genomes with few genes. However, these findings joined with plausible mechanisms of codon reassignment [89] to seriously under-mine the intuitively simple logic of the frozen accident hypothesis. This in turn reopened the door to discussions of the possibility that natural selection had steered the evolution of codon assignments within the standard genetic code. The chal-lenge was to go farther than noting features of the standard genetic code that make adaptive sense, to formally test whether the standard genetic code is significantly non-random with respect to plausible alternatives. This approach represents an example of a another well-established, general technique for evaluating evolution-ary outcomes: Optimality Theory [90], a form of what physicists recognize by the description "counterfactual investigation." Optimality theory has been widely used to study evolutionary outcomes that range from animal behavior [91] to genome structure [92]. Thus, Haig and Hurst [93] calculated the average change in amino acid hydrophobicity caused by all possible point mutations within the standard genetic code and compared this value with equivalent measures calculated for a large sample of theoretical alternative codes. Results showed that around one in ten thousand random patterns of codon assignments would buffer genomes against the impact of mutation as well as or better than the standard genetic code. Further work went on to show that more sophisticated calculations of error susceptibility, adjusted to reflect known biases in the patterns within which codons mutate or are mistranslated, reduce this figure by two further orders of magnitude, so that the standard genetic code appears to be literally one in a million [94] (see Figure 14.5).

A full description of how the adaptive theory developed and where it stands today is beyond the scope of this chapter; details are recorded elsewhere [95]. Suffice it to say the evidence that the pattern of codon assignments found within the standard genetic code is a product of natural selection has grown in volume and detail. Indeed, a growing network of analysis is painting a picture of a genetic code that originated through deterministic biochemical interactions between nucleic acid motifs and a few amino acids [96] and grew in complexity by incorporating biochemically modified derivatives of these original amino acids [97] and navigated by natural selection toward a specific, optimal pattern of codon/amino acid assignments [98]. Precise details of one of the fundamental features at the heart of life's biochemistry thus appear to be anything but a frozen accident of evolutionary history. Rather, the discovery of such an "attractor" gives one clear indication of a predictable evolutionary outcome. We might not see exactly the same code a second time around, but perhaps we should expect to see one with strikingly similar features.

Beyond predictable codon assignments

Of course, these findings remain tightly constrained within assumptions (for exam-ple, that we are dealing with nucleic acid genotype mapped to protein phenotype by a code of triplet nucleotides, drawn from an alphabet of 4, each encoding one of

Stephen J. Freeland

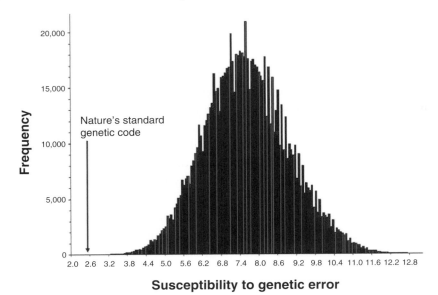

Figure 14.5. The standard genetic code minimizes the effects of point mutation better than the vast majority of alternative plausible genetic codes that could assign 20 amino acids to 64 codons in synonymous codon blocks (adapted from Freeland and Hurst 1998 [94]).

20 amino acids). Every one of these parameters is worthy of exploration for adaptive significance. Together, they pose a network of questions amenable to scientific investigation. Answers here will start to yield a robust picture of the predictability for life's biochemical framework. Many of these biochemical fundamentals have been investigated already to some extent, but progress in formalizing such inquiries and synthesizing the results has been patchy (more thorough expositions of what follows may be found in [99] and [100]).

For example, the amino acid alphabet has received an enormous amount of attention in terms of its origins. The groundbreaking work of Miller and Urey demonstrated that, under plausible prebiotic conditions, close to half of the 20 amino acids found in the standard genetic code can be reliably produced in significant quantities by simple abiotic processes (originally an electric spark applied to a mixture of steam, ammonia, and methane) [101]. Subsequent work has shown that the precise source of energy, and even the precise gaseous mixture, can vary without significantly changing the results [102]; further research suggests that a similar set of amino acids is probably synthesized abiotically throughout the universe [103, 104]. A careful compendium of this evidence indicates a fairly high degree of agreement as to which amino acids were likely available to the earliest life forms and which were "invented" by early metabolism [105]. Although detailed assertions regarding the precise order in which this latter group of amino acids entered the code [106] have proved unpersuasive [107, 108], attempts to collate clues continue [109, 110].

This understanding of how the amino acid alphabet began, how it grew, and by what mechanism it grew only partly addresses the evolution of the code. Several hundred biochemical derivatives of the standard code's 20 amino acids are produced and used in the metabolism of extant life, but have never been incorporated into the code. Moreover, many amino acids are routinely produced in abiotic syntheses, but have no known place in biology. The question of why life "chose" the 20 that we see in the standard genetic code is largely unanswered – indeed, the only direct commentary on nature's choice of amino acids has concluded that only some aspects seem to make sense [111]. Certainly, their ready synthesis under abiotic conditions suggests that they are reasonable components to find at the center of life's biochemistry. Moreover, this analysis offers a simple adaptive explanation as to why analogous organic monomers, also produced abundantly by abiotic syntheses, each with the potential to form linear protein-like polymers, would have been passed over in favor of alpha amino acids by primordial evolution. Specifically, the argument concerns a protein's potential to fold into a stable, three-dimensional shape with specific catalytic or structural properties; amino acid equivalents with longer or shorter "backbones" produce polymers that are too rigid or too flexible to produce useful, reliable structures. Emerging technology is starting to explore the conformational stability of folded biopolymers that contain uncoded amino acids [112]. As computing power and accessibility continue to increase, software that applies this theory to predict the properties of arbitrary folds is growing in power [113]. Weber and Miller's [111] suggestion is ripe for further investigation.

Even less work has tackled the related issue of whether the amino acid alphabet size of 20 is in some way optimal or a mere accident of history. Recent discoveries of some lineages that encode up to 22 amino acids merely re-emphasize that 20 is unlikely to be a frozen accident. Yet to date, the only attempts to test for an optimal amino acid alphabet size have been tucked in as afterthoughts within publications pursuing a fundamentally different theme [114, 115]. The impediment to detailed research here is our generally poor understanding of how amino acid sequence maps into folded structure (and our even weaker general predictive power over how sequence determines function). Previous attempts to produce functional molecules from reduced alphabets of amino acids have met with limited success [116]. However, the enormous scientific [117] and economic [118] motivation to be able to accurately predict protein structure and "rationally design" proteins is leveraging progress [119]. In addition, innovative experimental techniques are starting to broaden possibilities for investigating structure/function maps *in vitro* [120].

Perhaps, in the not-too-distant future, meaningful optimality studies will explore realistically the link between protein structure and the amino acid alphabet. Meanwhile, it would seem pertinent to extend optimality studies of the standard genetic code to explore the properties of codes with fewer, more, or different amino acids with respect to the canonical 20.

For the nucleotides, the situation described for amino acids is almost precisely reversed. We know that many non-biological nucleotides are possible [121]. Indeed, many are used outside of genetics within extant metabolism [122]. Several focused optimality studies have offered detailed explanations for the size and composition of the genetic alphabet used by life [123]. Even the choice of ribose as the sugar component of nucleic acid has been proposed to be optimal for base-pairing strength relative to other prebiotically plausible alternatives [124]. Assuming that nucleic acid is on the table, the details of life's genetic biochemistry appear to be more or less predictable.

Yet this assumption is a problem. The very existence of nucleic acids within biochemistry remains one of the larger mysteries of primordial evolution. It is not at all clear that nucleic acid would have been on the primordial table (at least within the soup appetizer). Certainly, nucleic acid presents some highly attractive features as the genetic [125] and even catalytic [27] start of life. Some regard it as an obvious choice for the core of biochemistry [126]. None the less, prebiotic simulations appear to present serious problems for such a scenario. Specifically, if amino acids are readily produced by abiotic processes, then nucleotides are quite the reverse (a sobering caveat to enthusiastic descriptions of the RNA world [127]). Any single nucleotide comprises three distinct subcomponents: a base (a heterocyclic ring of carbon and nitrogen – that part of the nucleotide that varies and is used to carry genetic information) is linked to a ribose (a cyclic five-membered sugar), which is in turn linked to a phosphate. Each of these components is difficult to understand from the point of view of prebiotic chemistry. To name but a few of the salient problems, not only are the reaction pathways by which the four bases can be synthesized inconsistent with one another, they are downright antagonistic [128]. The most widely accepted abiotic pathway to produce ribose [129] also produces a host of other reaction products, including enantiomers of ribose in a racemic mixture [130] (and all sugars are unstable under what are generally considered to be prebiotic conditions [131]). No plausible source for abiotic phosphate has been found [132]. In fact, it is not only the components that present difficulties. No one has yet produced a convincing, abiotically plausible reaction scheme by which the resulting nucleotide monomers could be assembled from these components [133] (not least because nucleotides themselves are highly unstable, and hence unlikely to have gradually accumulated from ineffective reaction pathways [134]). Moreover, no one has yet proposed how a pool of mixed organic monomers, including amino acids, nucleotides, and sugars, could fail to produce large quantities of heteropolymers (i.e. hybrid chains built from nucleotides, amino acids, and other reactive monomers) on a purely probabilistic basis [135].

Of course, this is exactly the situation where the non-randomness of natural selection might take over from the problems of abiotic improbability: if life got under way using a less complex genetic system [136], then it is easier to understand how the "useful" properties of nucleotides could have been selected for [123, 124, 126, 127]. In fact, there has been no shortage of suggestions as to possible forerunners of RNA. These include peptide–nucleic acid [137], triose nucleic acid [138], polyglyceric acid [139], and even inorganic crystals such as clay [140]. But so far no one has found a meaningful methodology to discriminate these suggestions beyond conjecture. Indeed, even the simple suggestion that perhaps the primitive genetic system comprised fewer than four nucleotides has produced such varied and inconsistent ideas that the different hypotheses combine to undermine one another's credibility (see [99] for references).

Why is biochemistry dichotomous?

Surrounding all of these issues, a subtly distinct and even more fundamental question remains largely untouched. All extant life employs a dichotomy of nucleic acid genetic information and protein metabolism, with an associated suite of complex molecular machinery for translating one into the other. It seems increasingly likely that this situation evolved from an "RNA world" in which nucleic acid performed both roles. Yet our explanation for why a unipolymeric biochemistry evolved into the dichotomy of nucleic acid genetics and protein catalysis currently involves much hand-waving. A typical answer is that the amino acid alphabet permitted greater catalytic specificity than the nucleotide alphabet; for example, DNA could not have evolved (as a more stable form of information storage) until protein catalysts were performing catalytic roles as sophisticated as free-radical biochemistry [141]. But let us rewind this statement a little. Suppose the earliest genetic code consisted of far fewer than 20 amino acids. As ribozymes are being shown to be capable of an unsuspected diversity and sophistication of catalytic properties, all indications are that the nucleic acid alphabet could plausibly have expanded instead [142, 143, 144, 145, 146]. It is unclear why proteins would enter such a situation.

Szathmáry's solution is that the nucleic acid alphabet was constrained in size within an RNA world by the tradeoff demands of carrying information and performing catalysis [123]. A larger nucleic acid alphabet would have permitted more sophisticated metabolism, but would be more prone to copying errors. A smaller nucleic acid alphabet would reverse this situation. In this case, the sort of tradeoff that produced a nucleic acid alphabet of four letters would benefit from the evolution of a dedicated catalytic polymer. This would separate the two competing demands for optimality. However, this cannot account in itself for the initial adaptive

advantage of coding unless it can be clearly shown that the amino acids of the original code opened new catalytic possibilities: evolution cannot plan ahead. In this
context, a very different idea comes from a recent turn in optimality studies of the
standard genetic code. So far, non-randomness in the standard code's assignment
of codons to amino acids has been interpreted as a result of natural selection to minimize the phenotypic impact (on protein function) of genetic errors [95]. Recent
simulations are now offering a challenge to this interpretation. It appears that the precise properties of the standard genetic code (including its error-minimizing effects)
actually speed the general process of adaptive evolution of all proteins [147]. The
tentative suggestion is that the standard genetic code might have evolved by clade
sorting that would favor lineages best able to track changing environmental challenges of the primordial biosphere. Although these findings are preliminary at
present, the potential that coding itself arose as an adaptation for adapting seems
too interesting to ignore.

A speculative prediction for the future biochemistry of earth
(or of visiting aliens)

So far, I have built a cautious argument for the plausibility that, given primordial
earth as a starting point, at least some important aspects of the biochemical basis
for life were predictable. From the beginning, I have argued that in our current
state of scientific knowledge it is easier to explore evolutionary predictability intro-
spectively (could an intelligent alien predict earth's biochemistry?) than through
projection onto the rest of the universe (could we predict the biochemistry of an
alien that was investigating us?). But having pushed this question to its (currently
modest) limits in terms of our current science, I finish with a mix of observation
and speculation about biochemistry such as we might encounter in our intelligent
extraterrestrial observer.

First, the observation. From the vantage point of the twenty-first century, it would
appear that sentience on this planet is moving to incorporate an increasingly important component of disembodied information. The lore of oral culture has given way
to an accumulation of written learning, now merging into a global information technology culture. As Dennett points out, unless we include some supernatural component in our account of how humanity arose, then everything that humans produce
(from organisms that have been "artificially" selected or genetically engineered to
cars and computers) may be considered an outcome of biological evolution, albeit a
kind of meta-evolutionary phenomenon [18]. Put another way, the only distinguishing difference between a molecule that is chemical and one that is biochemical is that
the latter is somehow utilized by living system(s). Viewed in this way, it appears
that the precise details of human information technology culture are becoming
an increasingly important part of the biological information (biochemistry?) of

sentience on this planet. The "representational language" of information associated with this technology is binary (everything is represented as a string of 1s and 0s), whereas that of nucleic acid is based on a quarternary chemical alphabet (U, C, A, G). Certainly, an aperiodic, binary linear sequence is the simplest way in which information can be represented. It might seem that a larger information-carrying alphabet could be advantageous in terms of transporting a denser information load per unit message length, but the increased probability of transmission errors (or cost of increased error-checking processes) renders this unclear. Indeed, binary representation carries several fundamental advantages for storage, transmission, and manipulation of pure information. Thus, if Szathmáry's explanation for the size of the nucleic acid alphabet is correct, then we might interpret this spread of the new, binary "biochemistry" on earth as a further optimization rendered possible once information is freed from the historical constraints of organic chemical structure that initiated it.

Now, the speculation. Because mutation is a necessity for evolution, I would predict that any extraterrestrial life will have evolved by using a form of genetic information prone to errors and decay. As such, the organic phase of alien sentience would be highly unsuitable for the long time periods and harsh conditions implicit in interplanetary space travel. However, I consider it likely that any sentient species capable of space exploration would also have discovered mechanical computation. Like us, such beings would have grown increasingly dependent on this synthetic prosthesis as their science and technology advanced. In other words, wherever sentient life evolves, I think it likely that the spirit of space exploration will find the flesh to be weak and will substitute manufactured technology to do such work. I would therefore consider it likely that if our species ever does interface with extraterrestrial intelligence, it will be by way of the manufactured technological products of an "organic" intelligent alien sentience (its meta-evolutionary products), whether or not the organic aliens themselves still exist. According to such a view, the most predictable feature of any alien that we encounter by dint of its traveling to find us would be its reliance on binary information storage. If so, then we and the aliens would be demonstrating exactly the kind of convergence into optima that Conway Morris has suggested for independently evolving terrestrial lineages.

Conclusion: from evolutionary inevitability to metaphysics

Could an intelligent alien predict earth's fundamental biochemistry? We do not know at present, but some surprisingly encouraging clues are starting to emerge. Aspects of the amino acid alphabet, the size and composition of the nucleic acid alphabet, and the form of the code that maps one to the other show signs that, with enough foreknowledge, these details of earth's biochemistry might well be more or less predictable. Even the existence of the phenotype–genotype split that underpins molecular biology may be an evolutionary "good move" that we would expect to occur wherever life evolves. This hints at a new level to which our universe earns the description "biocentric." Physics as we know it seems to extend meaningfully into the biology that it has produced. But the point of this chapter is not to give a clear, general answer one way or the other. It is to argue that further research seems worthwhile. The questions are scientifically tractable (or at least tractable questions exist that would further our understanding), and there is no reason to dismiss the exercise as contrary to reasonable expectations based on what we know of evolution.

However, it would be remiss to finish without referring to one further factor that colors attitudes toward inquiries into the biocentric credentials of our universe. The issue is that those who perceive a finely tuned aspect to our universe subsequently divide to offer one of two general explanations. Some argue that a biocentric universe suggests the existence of a "designer" God. Others argue that the universe itself must be evolving or be part of a multiverse in which we unsurprisingly find ourselves in the corner best suited to producing us.[4] Either explanation pushes toward questions that have no root in science, and some scientists appear uncomfortable with producing data that stimulates theological or philosophical debate.

Here, my concern was to delve into a straightforwardly scientific question: *given physics*, how can we discern whether this universe was likely to produce life such as what we know on earth? Yet points of philosophical and theological interest lie not far away. One review of Conway Morris book on anthropocentric evolution, quoted on the back cover, concludes that it offers "a welcome antidote to the bleak nihilism of ultra-Darwinists." The description of bleak nihilist hardly fits Darwin himself, and many would be surprised to hear Gould's long-term emphasis on non-adaptive evolutionary luck described as ultra-Darwinian. Perhaps, then, this comment was directed at those evolutionary thinkers, such as Jacques Monod and Richard Dawkins, whose avowed atheism has led them repeatedly to cross from evolutionary theory into statements of a metaphysical nature. However, philosophers have been quick to point out the pitfalls of stretching one belief system to deny the validity of others in this context [149]. Nihilism remains the metaphysical belief it has always been. Just as Kant's "nebula hypothesis" [150] removed the need for an interventionist creation of stars and planets, so modern evolutionary theory removes

---

[4] Alhough proponents of this latter view sometimes portray themselves as liberating cosmology from a temporary abduction by metaphysics, the rescue is not clear-cut. Gingerich has argued with eloquent simplicity that both a multiverse and a Creator are concepts that lie beyond our spacetime reality, such that both are equally inaccessible to science and thus may be considered equally metaphysical. Put more bluntly, once mathematics has shown the plausible existence of alternative universes in which we do not live (because we couldn't), how exactly are we to test this possibility? If their existence is unfalsifiable by experiment, then as an explanation for our finely tuned universe they remain a matter of faith. Indeed, one is tempted to perceive a direct link to Kant's philosophical distinction between noumenal and phenomenal concepts: with only one reality at our disposal, how *could* we bring the existence of multiple alternative realities within the scope of experimental science? Perhaps this will turn out to be a further "God-of-the-gaps" fallacy (for example, see [148]), but this particular debate is for physicists, philosophers, and theologians (with several direct contributions in this volume).

the need to invoke a Cosmic Watchmaker as explanation for the biosphere that we encounter. But, as Kant and many others have pointed out, removing the necessity for an interventionist creator hardly disproves the existence of God. Likewise, Darwinism (ultra- or otherwise) is currently linked to metaphysical opinion only by the faith of its proponents, and some of its most vociferous proponents have placed their faith in atheism. This is not to imply that the equally simplistic conclusion of *de facto* mutual independence for faith and science [151] is any more defensible, but I happily relinquish further discussion of this territory to those whose thinking begins here [152]. My point is simply that ultra-Darwinism, if this phrase is taken to mean methodical application of neo-Darwinist theory, by way of optimality theory, is well suited to objective investigations of the significance of earth's biochemistry. Proper metaphysical interpretation lies beyond science, as the task of scholars with appropriate expertise in such matters.

Dawkins accurately describes natural selection's "utility function" in modern biology as the preservation of genes. Yet nothing in this description confirms or denies whether selection played a role in shaping life's biochemical foundation, with consequences that profoundly influenced the ultimate outcome of evolution. It is simply unscientific to regard a lack of evidence as the final word. I therefore advocate neo-Darwinian optimality investigations as appropriate lines of inquiry for evolutionary biologists to study biochemical fine-tuning. These will include investigations into the significance of the size and composition of the nucleic acid and amino acid alphabets, the "choice" of nucleic acid and protein biopolymers, the organization of the standard genetic code, and the evolution of a dichotomy of nucleic acid genotype and protein phenotype. If evolutionary biologists dismiss such scientific inquiry on the grounds of metaphysical distaste, then they follow in the footsteps of creationists who ignore related areas of scientific inquiry for essentially the same reason.

References

[1]   *Oxford English Dictionary*, online edn. 2E: http://dictionary.oed.com/.
[2]   L. J. Henderson. *The Fitness of the Environment: An Inquiry into the Biological Significance of the Properties of Matter* (New York, NY: Macmillan (1913); repr. Boston, MA: Beacon Press (1958); Gloucester, MA: Peter Smith (1970), p. 278.
[3]   J. D. Barrow and F. J. Tipler. *The Anthropic Cosmological Principle*. Oxford: Oxford University Press (1986).
[4]   M. J. Rees. *Just Six: The Deep Forces That Shape the Universe*. New York, NY: Basic Books (2000).
[5]   O. Gingerich. "God's goof," and the universe that knew we were coming. In *Science and Religion: Are They Compatible?* ed. P. Kurtz., B. Karr and R. Sandhu. Amherst, MA: Prometheus Books (2003), p. 53.
[6]   G. E. Tranter. Parity-violating energy differences of chiral minerals and the origin of biomolecular homochirality. *Nature*, **318** (1985), 172.

310   *Stephen J. Freeland*

[7]   A. J. MacDermott. Electroweak enantioselection and the origin of life. *Origins of Life and Evolution of Biospheres*, **25** (1995), 191–9.

[8]   E. Schrödinger. *What Is Life? The Physical Aspect of the Living Cell.* Cambridge, UK: Cambridge University Press (1944).

[9]   L. Smolin. *The Life of the Cosmos.* Oxford: Oxford University Press (1997).

[10] R. Dawkins. *The Blind Watchmaker: Why the Evidence of Evolution Reveals a Universe without Design.* New York, NY: W. W. Norton (1986). (Quote here is to be found on pp. 1 and 2 of Norton's [1996] 2nd paperback edn.)

[11] Reported in *Nature*, **294** (1981), 10: " . . . Sir Fred Hoyle [offered] a statement for disbelieving conventional views about the evolution of the universe . . . the essence of his argument was that the information content of the higher forms of life is represented by the number 1040,000 . . . Evolutionary processes would, Hoyle said, require several Hubble times to yield such a result. The chance that higher life forms might have emerged in this way is comparable with the chance that 'a tornado sweeping through a junkyard might assemble a Boeing 747 from the materials therein' . . . Of adherents of biological evolution, Hoyle said he was at a loss to understand 'biologists' widespread compulsion to deny what to me seems obvious.'"

[12] For example, see D. E. Erwin. The Goldilocks Hypothesis [review of S. Conway Morris's *Life's Solution: Inevitable Humans in a Lonely Universe*], *Science*, **302** (2003), 1682–3.

[13] See, for example, S. Weinberg. A designer universe? *New York Review of Books*, **46** (1999), 46–8.

[14] For example, see S. Coleman. Black holes as red herrings: topological fluctuations and the loss of quantum coherence. *Nuclear Physics*, **B307** (1988), 867.

[15] For example, see J. E. Smith, A. E. Eiben and J. D Smith. *Introduction to Evolutionary Computing.* New York: Springer-Verlag (2003).

[16] W. Paley. *Natural Theology; or, Evidences of the Existence and Attributes of the Deity.* London: Printed for J. Faulder (1809); 12th edn. Now available in the public domain: www.hti.umich. edu/cgi/p/pd-modeng/pd-modeng-idx?type=header& byte=53049351.

[17] See http://exobiology.arc.nasa.gov/.

[18] D. Dennett. Possibility naturalized. In *Darwin's Dangerous Idea.* New York: Simon and Schuster (1995), pp. 118–23.

[19] R. J. P. Willams, and J. J. R. Fraústo da Silva. Evolution revisited by inorganic chemists: Chapter 21, this volume.

[20] M. J. Denton. Protein-based life as an emergent property of matter: the nature and biological fitness of the protein folds. Chapter 13, this volume.

[21] The American portal to this information is provided by the National Center for Biotechnology Information at www3.ncbi.nlm.nih.gov/.

[22] See L. Frisch, ed. The genetic code. *Cold Spring Harbor Symposia on Quantitative Biology*, **1** (1966), 747.

[23] See A. Lazcano *et al.* The evolutionary transition from RNA to DNA in early cells. *Journal of Molecular Evolution*, **27** (1988), 283–90.

[24] See T. R Cech. The ribosome is a ribozyme. *Science*, **289** (2000), 878–9.

[25] F. H. C. Crick. Protein synthesis directed by DNA phage messengers. *Cold Spring Harbor Symposia on Quantitative Biology*, **31** (1966), 157–71. See also comments regarding an RNA world in F. H. C. Crick. The origin of the genetic code. *Journal of Molecular Biology*, **38** (1968), 367–79.

[26] H. B. White. Coenzymes as fossils of an earlier metabolic state. *Journal of Molecular Evolution*, **7** (1976), 101–4.

[27] R. F. Gesteland and J. F. Atkins. *The RNA World*. Cold Spring Harbor, NY: Cold
       Spring Harbor Laboratory Press (1993). For updated progress, see also F. R.
       Gesteland, T. R. Cech and J. F. Atkins. *The RNA World*, 2nd edn., Monograph 37.
       Cold Spring Harbor, NY: Cold Spring Harbor Laboratory Press (2000).
[28] S. J Freeland, R. D. Knight and L. F. Landweber. Do proteins predate DNA?
       Science, **286** (1999), 690–2.
[29] G. Cooper *et al*. Carbonaceous meteorites as a source of sugar-related organic
       compounds for the early Earth. *Nature*, **414** (2001), 879–83.
[30] *Star Trek: The Next Generation*, Episode No. 246: "The Chase" (Stardate:
       46731.5); aired April 26, 1993.
[31] N. R. Pace. The universal nature of biochemistry. *Proceedings of the National
       Academy of Sciences, USA*, **98** (2001), 805–8.
[32] S. J. Gould. Darwin's dilemma: the odyssey of evolution. In *Ever since Darwin*
       (1977). Repr. London: Penguin (1991).
[33] H. Spencer. *A System of Synthetic Philosophy*, 2 vols. London: Williams and
       Norgate (1864, 1867), 2nd edn., pp. 1898–9.
[34] See, for example, http://catholic.archives.nd.edu/cgi-bin/lookup. pl?stem=evolv&
       ending=ere.
[35] *Oxford English Dictionary*, online edn. 2E: http://dictionary. oed.com/.
[36] C. Bonnet. *Considérations sur les corps organisés*. 2 vols. (Amsterdam, M. M.
       Rey: 1762): http://visualiseur. bnf.fr/Visualiseur?Destination=Gallica&O=
       NUMM-87656.
[37] C. Bonnet. *La Palingénésie philosophique*. Geneva: Philibert et Chirol (1769):
       http://home.tiscalinet.ch/biografien/sources/bonnet palingenesie.htm. The word
       "evolution" makes its debut in Partie 6: "Abuserois-je de la liberté de conjecturer,
       si je disois, que les plantes et les animaux qui éxistent aujourd'hui, sont provenus
       par une sorte d'évolution naturelle des êtres organisés qui peuploient ce premier
       monde sorti immédiatement des mains du créateur?" ("Would I abuse my liberty to
       conjecture if I said that the plants and animals that exist today were created by a
       sort of natural evolution from the organized beings who populated the initial world
       that came out immediately from the hands of the creator?")
[38] P. J. Bowler. *Evolution, the History of an Idea*. (1983). Repr. Berkeley and Los
       Angeles, CA: University of California Press (1989), pp. 59–63.
[39] Id, pp. 82–9.
[40] C. Darwin. *On the Origin of Species by Means of Natural Selection*. London: John
       Murray (1859). An online version of this text is available in the public domain at
       www.literature.org /authors/darwin-charles/the-origin-of-species/.
[41] For an excellent summary of the changes Darwin introduced over the course of six
       editions, see the introduction by G. Beer in World's Classics Paperbacks *Origin of
       Species*. Amherst, MA: Prometheus (1996), pp. vii–xxix.
[42] S. J. Miles. Charles Darwin and Asa Gray discuss teleology and design.
       Perspectives on Science and Christian Faith, **53** (3) (2001), 196–201.
[43] For example, see C. Darwin (1860). *The Correspondence of Charles Darwin*,
       vol. 8. Cambridge, UK: Cambridge University Press (1993), p. 496; from a letter to
       Asa Gray, "But I grieve to say that I cannot honestly go as far as you do about
       Design . . . [Y]ou lead me to infer that you believe 'that variation has been led
       along certain beneficial lines.'"
[44] This statement appears in every edition of *Origin* (full reference found in [40] as
       the final sentence of the penultimate paragraph of the book).
[45] T. L. Poulson. Adaptations of cave fishes with some comparisons to deep sea
       fishes. *Environmental Biology of Fishes*, **62** (2001), 345–64.

[46] D. R. Brooks and D. A. McLennan. *Parascript*. Washington, DC: Smithsonian Institution Press (1993).

[47] D. Dennett. *Darwin's Dangerous Idea*. New York, NY: Simon and Schuster (1995). The term "skyhook" is defined on p. 74, but is integral to the whole thesis of the book and is discussed intermittently throughout.

[48] For a context to this statement, see D. Cupitt. Chapter 2, The mechanical universe. In *The Sea of Faith*. Cambridge, UK: Cambridge University Press (1988), pp. 36–55.

[49] R. S. Lull. *Organic Evolution*. New York, NY: Macmillan (1925).

[50] D. S. Wilson and E. Sober. Reviving the superorganism. *Journal of Theoretical Biology*, **36** (1989), 337–56.

[51] See, for example, D. C. Queller and J. E. Strassmann. The many selves of social insects. *Science*, **296** (2002), 311–13.

[52] S. Camazine, J.-L. Deneubourg, N. R. Franks *et al. Unveiling Mechanisms of Collective Behavior*. Princeton, NJ: Princeton University Press (2001).

[53] W. Heisenberg. Über den anschaulichen Inhalt der quantentheoretischen Kinematik und Mechanik. *Zeitschrift für Physik*, **43** (1927), 172–98.

[54] E. Lorenz. Predictability: does the flap of a butterfly's wings in Brazil set off a tornado in Texas? Presented at the Meeting of the American Association for the Advancement of Science, Washington, DC (1972). (Full text in Appendix in E. Lorenz. *The Essence of Chaos*. Seattle, WA: University of Washington Press [1993].)

[55] J. W. Welte *et al.* Gambling participation in the U. S. – results from a national survey. *Journal of Gambling Studies*, **18** (2002), 313–37.

[56] S. J. Gould and R. C. Lewontin. The spandrels of San Marco and the Panglossian paradigm: a critique of the adaptationist programme. *Proceedings of the Royal Society*, B**205** (1979), 581–98.

[57] J. Maynard Smith *et al.* Developmental constraints and evolution. *Quarterly Review of Biology*, **60** (1985), 265–87.

[58] See D. Dennett. The spandrel's thumb. In *Darwin's Dangerous Idea*. New York, NY: Simon and Schuster (1995), pp. 267–82.

[59] G. C. Williams. A defense of reductionism in evolutionary biology. *Oxford Surveys in Evolutionary Biology*, vol. 2, ed. R. Dawkins and M. Ridley. Oxford: Oxford University Press (1985).

[60] D. C. Queller. The spaniels of St. Marx and the Panglossian paradox: a critique of a rhetorical programme. *Quarterly Review of Biology*, **70** (1995), 485–9.

[61] See, for example, R. A. Kerr. Mass extinction. Extinction by a whoosh, not a bang? *Science*, **302** (2003), 1315, and references therein.

[62] For example, contrast S. J. Gould's *Wonderful Life*. New York, NY: Norton (1989), ch. 5, section on mass extinction, with S. Conway Morris's *Life's Solution: Inevitable Humans in a Lonely Universe*. Cambridge, UK: Cambridge University Press (2003), pp. 94–5.

[63] See, for example, G. R. Byerly *et al.* An archean impact layer from the Pilbara and Kaapvaal cratons. *Science*, **297** (2002), 1325–7, and references therein.

[64] For example, see A. Weismann (1883). In E. B. Poulton, S. Sholand and A. E. Shipley, eds. *Essays upon Heredity and Kindred Biological Problems*. Oxford: Oxford University Press (1993), pp. 1–66.

[65] M. Lynch and J. S. Conery. The origins of genome complexity. *Science*, **302** (2003), 1401–4.

[66] L. Partridge and L. D. Hurst. Sex and conflict. *Science*, **281** (1998), 2003–8.

[67] S. E. Wright. Roles of mutation, inbreeding, crossbreeding and selection in evolution. *Proceedings of the sixth Annual Congress of Genetics*, **1** (1932), 356–66. The construct has been criticized in W. B. Provine. *Sewall Wright and Evolutionary Biology*. Chicago, IL: University of Chicago Press (1986). It has been defended in M. Ruse. Are pictures really necessary? The case of Sewall Wright's "adaptive landscapes." *PSA: Proceedings of the Biennial Meeting of The Philosophy of Science Association*, Vol. 2: *Symposia and Invited Papers* (1990), pp. 63–77. Suffice it to say that the simple rendition described here is the adaptive landscape at its simplest and most defensible.

[68] E. Mayr. *Systematics and the Origin of Species*. New York, NY: Columbia University Press (1942).

[69] S. J. Gould (1989). *Wonderful Life*. New York, NY: Norton (1989). The specific quote may be found in ch. 1, p. 48 of the Penguin edn. (repr.) London: Penguin (1991).

[70] Usually attributed to the Comte de Buffon. For example, see A. O. Lovejoy. Buffon and the problem of species. In *Forerunners of Darwin*, ed. B. Glass *et al.* Baltimore, MD: Johns Hopkins University Press (1968), pp. 84–113 (esp. p. 111).

[71] For example, see the recent review by K. Omland and D. Funk. Species level paraphyly and polyphyly. *Annual Reviews in Ecology, Evolution and Systematics*, **34** (2003), 397–423.

[72] For example, see the *Answers in Genesis* web resource: www.answersingenesis.org.

[73] For example, see H. D. Rundle *et al.* Natural selection and parallel speciation in sticklebacks. *Science*, **287** (2000), 306–8. Also see R. E. Glor *et al.* Phylogenetic analysis of ecological and morphological diversification in Hispaniolan trunk-ground anoles (*Anolis cybotes* group). *International Journal of Organic Evolution*, **57** (2003), 2383–97.

[74] See, for example, the lucid description given in ch. 2 of J. Felsenstein. *Inferring Phylogenies*. Sunderland, MA: Sinauer Associates (2004). It is purely coincidental that the programming algorithms being described are inferring biological evolutionary relationships; the point is that the quasi-natural-selection approach used by the computer software to find the best possible tree of relationships is one that benefits from stochastic noise.

[75] S. Conway Morris. *Life's Solution: Inevitable Humans in a Lonely Universe*. Cambridge, UK: Cambridge University Press (2003).

[76] P. Harvey and M. Pagel. *The Comparative Method of Evolutionary Biology*. Oxford: Oxford University Press (1991).

[77] Including, of course, many of contemporary concern, such as HIV and SARS. A good overview is given in A. Moya, E. C. Holmes and F. Gonzalez-Candelas. The population genetics and evolutionary epidemiology of RNA viruses. *National Review of Microbiology*, **2** (2004), 279–88.

[78] R. D. Knight, S. J. Freeland and L. F. Landweber. Rewiring the keyboard: evolvability of the genetic code. *Nature Reviews Genetics*, **2** (2001), 49–58.

[79] See J. Atkins and R. Gesteland. The 22nd amino acid. *Science*, **296** (2002), 1409–10 and references therein.

[80] First predicted by R. A Fisher in *The Genetical Theory of Natural Selection*. Oxford: Oxford University Press (1930). For recent empirical corroboration, see, for example, C. Burch and L. Chao. Evolution by small steps and rugged landscapes in the RNA virus phi6. *Genetics*, **151** (1999), 921–7.

[81] F. Jacob. Evolution and tinkering. *Science*, **196** (1997), 1161–6.

[82]   F. H. C. Crick. The origin of the genetic code. *Journal of Molecular Biology*, **38** (1968), 367–79 (see esp. pp. 369–70).

[83]   T. M. Sonneborn. Degeneracy in the genetic code: extent, nature and genetic implications. In *Evolving Genes and Proteins*, ed. V. Bryson and H. J. Vogel. New York, NY, and London: Academic Press (1965).

[84]   E. Zuckerkandland and L. Pauling. Evolutionary divergence and convergence in proteins. In *Evolving Genes and Proteins*, ed. V. Bryson and H. J. Vogel. New York, NY, and London: Academic Press (1965).

[85]   C. R. Woese. On the evolution of the genetic code. *Proceedings of the National Academy of Sciences, USA*, **54** (1965), 1546–52.

[86]   F. H. C. Crick. Codon–anticodon pairing: the wobble hypothesis. *Journal of Molecular Biology*, **19** (1966), 548–55.

[87]   For an excellent history of these strange theories and the experimental evidence that turned them on their head, see B. Hayes. The invention of the genetic code. *American Scientist*, **86** (1998), 8–14.

[88]   B. G. Barrell, A. T. Bankier and J. Drouin. A different genetic code in human mitochondria. *Nature*, **282** (1979), 189–94.

[89]   For example, see S. Osawa and T. H. Jukes. Codon reassignment (codon capture) in evolution. *Journal of Molecular Evolution*, **21** (1989), 271–78. For an alternative view, see M. A. Santos *et al*. Driving change: the evolution of alternative genetic codes. *Trends in Genetics*, **20** (2004), 95–102.

[90]   J. Maynard Smith. Optimization theory in evolution. *Annual Review of Ecology and Systematics*, **9** (1978), 31–56.

[91]   For example, see E. L. Charnov. Optimal foraging: attack strategy of a mantid. *American Naturalist*, **110** (1976), 141–51, for a simple, elegant, and pioneering contribution in this field.

[92]   For example., X. Xia. Body temperature, rate of biosynthesis, and evolution of genome size. *Molecular Biology and Evolution*, **12** (1995), 834–42.

[93]   For example, D. Haig and L. D. Hurst. A quantitative measure of error minimisation within the genetic code. *Journal of Molecular Evolution*, **33** (1991), 412–17.

[94]   S. J. Freeland and L. D. Hurst. The genetic code is one in a million. *Journal of Molecular Evolution*, **47** (1998), 238–48.

[95]   A comprehensive review is given in S. J. Freeland, T. Wu and N. Keulmann. The case for an error minimizing standard genetic code. *Origins of Life and Evolution of Biospheres*, **33** (2003), 457–77.

[96]   R. D. Knight and L. F. Landweber. The early evolution of the genetic code. *Cell*, **101** (2000), 569–72.

[97]   J. T. Wong. Evolution of the genetic code. *Microbiological Science*, **5** (1988), 174–81.

[98]   S. J. Freeland, R. D. Knight, L. F. Landweber *et al*. Early fixation of an optimal genetic code. *Molecular Biology and Evolution*, **17** (2000), 511–18.

[99]   R. D. Knight, S. J. Freeland and L. F Landweber. Adaptive evolution of the genetic code. In *The Genetic Code and the Origin of Life*, ed. L. R. Pouplana. Georgetown/New York, NY: Landes Bioscience and Kluwer Academic/Plenum (2004), pp. 204–23.

[100] Chapters 3 and 4 of S. Conway Morris's *Life's Solution: Inevitable Humans in a Lonely Universe*. Cambridge, UK: Cambridge University Press (2003), pp. 32–63, give a somewhat more skeptical overview of these topics; in particular, the author's views are pessimistic regarding the ready availability of amino acids in a prebiotic world.

[101] S. L. Miller and H. C. Urey. Production of amino acids under possible primitive earth conditions. *Science*, **117** (1953), 528–9; S. L. Miller and H. C. Urey. Production of some organic compounds under possible primitive earth conditions. *Journal of the American Chemical Society*, **77** (1955), 2351–61.

[102] S. L. Miller, H. C. Urey and J. Oro. Origin of organic compounds on the primitive earth and in meteorites. *Journal of Molecular Evolution*, **9** (1976), 59–72.

[103] K. Kvenvolden, J. Lawless, K. Pering *et al.* Evidence for extraterrestrial amino-acids and hydrocarbons in the Murchison meteorite. *Nature*, **5** (1970), 923–6.

[104] See, for example, M. P. Bernstein, J. P. Dworking, S. A. Sandford *et al.* Racemic amino acids from the ultraviolet photolysis of interstellar ice analogues. *Nature*, **416** (2002), 401–3.

[105] See comparison tables offered in J. T.-F. Wong and P. M. Bronskill. Inadequacy of pre-biotic synthesis as the origin of proteinaceous amino acids. *Journal of Molecular Evolution*, **13** (1979), 115–25.

[106] J. T.-F. Wong. A co-evolution theory of the genetic code. *Proceedings of the National Academy of Sciences, USA*, **72** (1975), 1909–12.

[107] R. Amirnovin. An analysis of the metabolic theory of the origin of the genetic code. *Journal of Molecular Evolution*, **44** (1997), 473–6.

[108] T. A. Ronneberg, L. F. Landweber and S. J. Freeland. Testing a biosynthetic theory of the genetic code: fact or artifact? *Proceedings of the National Academy of Sciences, USA*, **97** (2000), 13690–5.

[109] E. N. Trifonov. Consensus temporal order of amino acids and evolution of the triplet code, *Gene*, **261** (2000), 139–51.

[110] M. Di Giulio. Genetic code origin: are the pathways of type Glu-tRNA(Gln) → Gln-tRNA(Gln) molecular fossils or not? *Journal of Molecular Evolution*, **55** (2002), 616–22.

[111] A. L. Weber and S. L. Miller. Reasons for the occurrence of the twenty coded protein amino acids. *Journal of Molecular Evolution*, **17** (1981), 273–84.

[112] For example, P. Mathur, S. Ramakumar and V. S. Chauhan. Peptide design using alpha, beta-dehydro amino acids: from beta-turns to helical hairpins. *Biopolymers*, **76** (2004), 150–61.

[113] Although mainstream protein structural prediction methods are turning ever more to "learning" patterns from proteins that have evolved in nature, first principles prediction methods are steadily improving; for example, see C. A. Rohl *et al.* Modeling structurally variable regions in homologous proteins with rosetta. *Proteins*, **55** (3) (2004), 656–77.

[114] J. T.-F. Wong. The evolution of a universal genetic code. *Proceedings of the National Academy of Sciences, USA*, **73** (1976), 2336–40.

[115] Suggested by E. Szathmáry. Four letters in the genetic alphabet: a frozen evolutionary optimum? *Proceedings of the Royal Society*, B**245** (1991), 91–9.

[116] For example, see Y. Wei and M. H. Hecht. Enzyme-like proteins from an unselected library of designed amino acid sequences. *Protein Engineering Design and Selection*, **17** (2004), 67–75.

[117] For example, see E. N. Baker, V. L. Arcus and J. S. Lott. Protein structure prediction and analysis as a tool for functional genomics. *Applied Bioinformatics*, **2** (suppl. 3) (2003), S3–10.

[118] For example, see P. L. Elkin. Primer on medical genomics part V: bioinformatics. *Mayo Clinic Proceedings*, **78** (2003), 57–64.

[119] P. E. Bourne. CASP and CAFASP experiments and their findings. *Methods of Biochemical Analysis*, **44** (2003), 501–7.

[120] A. D. Keefe *et al.* One-step purification of recombinant proteins using a
       nanomolar-affinity streptavidin-binding peptide, the SBP-Tag. *Protein Expression
       and Purification*, **23** (2001), 440–6.
[121] For example, see J. Piccirilli, T. Krauch, S. Moroney *et al.* Enzymatic
       incorporation of a new base pair into DNA and RNA extends the genetic alphabet.
       Nature, **343** (1990), 33–7; D. E. Bergstrom, P. Zhang and W. T. Johnson.
       Comparison of the base pairing properties of a series of nitrozole nucleobase
       analogs in the oligodeoxyribonucleotide sequence 5′-d(CGCXAATTYGCG)-3′.
       Nucleic Acids Research, **25** (1997), 1935–42, and references therein; J. C. Delaney
       et al. High-fidelity *in vivo* replication of DNA base shape mimics without
       Watson–Crick hydrogen bonds. *Proceedings of the National Academy of Sciences,
       USA*, **100** (8) (2003), 4469–73.
[122] H. Grosjean and R. Benne. *Modification and Editing of RNA*. Washington, DC:
       American Society for Microbiology Press (1998).
[123] E. Szathmáry. Why are there four letters in the genetic alphabet? *Nature Reviews
       Genetics*, **4** (2003), 995–1001. See also P. P. Gardner, B. R. Holland, V. Moulton,
       D. Hendy and D. Penny. Optimal alphabets for an RNA world. *Proceedings of the
       Royal Society*, B**270** (2003), 1177–82.
[124] A. Eschenmoser. Chemical etiology of nucleic acid structure. *Science*, **28** (1999),
       2118–24.
[125] E. Szathmáry and J. Maynard Smith. *The Major Transitions in Evolution* (Oxford:
       Oxford University Press, 1995); see chapters 3–5 for discussions of the inherent
       advantages of template-based genetics.
[126] N. R. Pace. The universal nature of biochemistry. *Proceedings of the National
       Academy of Sciences, USA*, **98** (2001), 805–8.
[127] D. Bartel and P. Unrau. Constructing an RNA world. *Trends in the Biochemical
       Sciences*, **24** (Millennium Issue) (1999), M9–M13.
[128] For example, R. Shapiro. The prebiotic role of adenine: a critical analysis. *Origins
       of Life and Evolution of Biospheres*, **25** (1995), 83–98.
[129] R. Shapiro. Prebiotic ribose synthesis, a critical analysis. *Origins of Life and
       Evolution of Biospheres*, **18** (1988), 71–85.
[130] A graphic account is given in P. Decker, H. Schweer and R. Pohlmann.
       Identification of formose sugars, presumable prebiotic metabolites, using
       capillary gas chromatography/gas chromatography–mass spectrometry of
       n-butoximine trifluoroacetates on OV-225. *Journal of Chromatography*, **244**
       (1982), 281–91.
[131] R. Larralde, M. P. Robertson and S. L. Miller. Rates of decomposition of ribose
       and other sugars: implications for chemical evolution. *Proceedings of the National
       Academy of Sciences, USA*, **84** (1995), 4398–402.
[132] A. Keefe and S. L. Miller. Potentially pre-biotic syntheses of condensed
       phosphates. *Origins of Life and Evolution of Biospheres*, **26** (1996), 15–25; but see
       also D. Glindemann, R. M. de Graaf and A. W. Schwartz. Chemical reduction of
       phosphate on the primitive earth. *Origins of Life and Evolution of Biospheres*, **29**
       (1999), 555–61.
[133] An excellent, if ultra-skeptical, overview is given in R. Shapiro. *Origins: A
       Skeptic's Guide to the Creation of Life on Earth*. New York, NY: Bantam (1987),
       pp. 182–4.
[134] M. Levy and S. L. Miller. The stability of the RNA bases: implications for the
       origin of life. *Proceedings of the National Academy of Sciences, USA*, **95** (1998),
       7933–8.

[135] R. Shapiro. A replicator was not involved in the origin of life. *International Union of Biochemistry and Molecular Biology Life*, **49** (2000), 173–6.

[136] G. F. Joyce, A. W. Schwartz, S. L. Miller *et al.* The case for an ancestral genetic system involving simple analogs of the nucleotides. *Proceedings of the National Academy of Sciences, USA*, **84** (1987), 107–19.

[137] K. E. Nelson, M. Levy and S. L. Miller. Peptide nucleic acids rather than RNA may have been the first genetic molecule. *Proceedings of the National Academy of Sciences, USA*, **97** (2000), 3868–71.

[138] K.-U. Schöning, P. Scholz, S. Guntha *et al.* Chemical etiology of nucleic acid structure: the alpha-threofuranosyl-(3′→ 2′) oligonucleotide system. *Science*, **290** (2000), 1347–51.

[139] A. L. Weber. Thermal synthesis and hydrolysis of polyglyceric acid. *Origins of Life and Evolution of Biospheres*, **19** (1989), 7–19.

[140] A. G. Cairns-Smith. *Genetic Takeover and the Mineral Origins of Life.* Cambridge, UK: Cambridge University Press (1982).

[141] S. J. Freeland, R. D. Knight and L. F. Landweber. Do proteins pre-date DNA? *Science*, **286** (1999), 690–2.

[142] J. A. Picirilli, T. Krauch, S. E. Moroney *et al.* Enzymatic incorporation of a new base into DNA and RNA extends the genetic alphabet. *Nature*, **343** (1990), 33–7.

[143] Y. Wu *et al.* Efforts toward expansion of the genetic alphabet: optimization of interbase hydrophobic interactions. *Journal of the American Chemical Society*, **122** (2000), 7621–32.

[144] S. A. Benner, T. R. Battersby, B. Eschgfaller *et al.* Redesigning nucleic acids. *Pure and Applied Chemistry*, **70** (2) (1998), 263–6.

[145] M. Berger, Y. Wu, A. K. Ogawa *et al.* Universal bases for hybridization, replication and chain termination. *Nucleic Acids Research*, **28** (15) (2000), 2911–14.

[146] M. Levy and S. L. Miller. The prebiotic synthesis of modified purines and their potential role in the RNA world. *Journal of Molecular Evolution*, **48** (1999), 631–7.

[147] W. Zhu and S. J. Freeland. The standard genetic code enhances adaptive evolution of proteins. *Journal of Theoretical Biology*, **239** (2006), 63–70.

[148] D. F. Siemens, Jr. On Moreland: spurious freedom, mangled science, muddled philosophy. *Perspectives on Science and Christian Faith*, **49** (1997), 196–9.

[149] M. Midgley. Evolution as religion, a comparison of prophecies. *Zygon*, **22** (1987), 179–94.

[150] I. Kant. *Sammtliche Werke.* In chronologischer Reihenfolge herausgegeben von G. Hartenstein. Leipzig: Leopold Voss (1867) [–68]; vol. i, pp. 207–345.

[151] For example, the non-overlapping magisteria argument presented by Gould. *Rocks of Ages: Science and Religion in the Fullness of Life.* New York, NY: Ballantine Books (1999).

[152] For example, the lucid discussion from J. F. Haught, Is fine-tuning remarkable? Chapter 3, this volume.

[153] For example, see discussion in D. P. Glavin and J. L. Bada. Survival of amino acids in micrometeorites during atmospheric entry. *Astrobiology*, **1** (2001), 259–69.

[154] Voltaire. *Candide* (1759), ed. Stanley Appelbaum. Mineola, NY: Dover Publications (1991).

15

Would Venus evolve on Mars? Bioenergetic constraints, allometric trends, and the evolution of life-history invariants

Jeffrey P. Schloss

If there is ever a time in which we must make profession of two opposite truths, it is when we are reproached for omitting one.

– Pascal, *Pensées*[1]

Introduction

A famous metaphor in integrative biology refers to the relationship between environmental constraint and evolutionary change as "the ecological theater and the evolutionary play" (Hutchinson, 1965). In the forty years since the penning of that phrase, the relationship between play and stage has been a matter of vigorous and fascinating debate. The issues have profound implications, not only for our scientific account of the evolutionary process, but also for our expectation of what life might look like on other planets and, indeed, for our philosophical and theological understandings of what it might mean on this one.

The controversy involves differing conclusions about the roles of contingency and constraint in evolutionary history, including, among other things, the way in which fundamental regularities of the physico-chemical environment, or "stage," influence the unfolding of the evolutionary drama. On the one hand, many prevalent expositions of the evolutionary play suggest that the fundamental or ultimate actors are genes, not organismic (much less mental) agents (Dawkins, 1976, 1998; Dennett, 1995). The drama itself is a theater of the absurd, a plotless improvisation using whatever props are contingently provided by the environment (Gould, 1989, 1996). Contingency is held to exert determinative influence on history, and, according to Stephen Gould's widely cited metaphor, "we would probably never arise again even if life's tape could be replayed a thousand times" (1989, p. 234). Thus, our existence

---

[1] Pascal (1958). Pensée no. 865.

Fitness of the Cosmos for Life: Biochemistry and Fine-Tuning, ed. J. D. Barrow *et al.*
Published by Cambridge University Press. © Cambridge University Press 2007.

is "utterly unpredictable" and "entirely contingent." Come back to the theater on another night, and one is likely to see an entirely different play. Or none.

On the other hand, some accounts of evolutionary change emphasize a scientifically explicable thematic continuity and even a plausible, perhaps inevitable, directionality (Carroll, 2001; Conway Morris, 1998, 2003; de Duve, 2002; Denton, 1998; Salthe, 1993; Szathmáry and Maynard Smith, 1995; Williams and Fraústo da Silva, 2003; but see Szathmáry, 2002). Like a good melodrama, revisit the theater and one will see variations on a recurrent theme.

There are two classes of explanation for this, both of which emphasize evolutionary constraint. One approach to this issue focuses on the preconditional fitness of the *abiotic* environment – the cosmos, earth, and/or physico-chemical variables – for life. Structural design of the theater predictably constrains the plot of the play, as environmental ordering trumps or at least mitigates contingent influences on evolution. This could occur through intrinsic, prebiotic properties of the stage. Or it could occur interactively though modification of the environment by organismic agents, which directionally transform stage construction by their chemical metabolism (Williams and Fraústo da Silva, 2003). Another approach involves fundamental *biotic* constraints, emerging, of course, from underlying chemistry. These constraints operate at the internal level of bioenergetic (West *et al.*, 1999a, 2004) or developmental (Maynard Smith *et al.*, 1985) limits, or at the interactive level of selectionally optimized or invariant strategies for allocating resources across life history (Kozlowski and Weiner, 1997).

Of course, metaphor may help frame (or serve to obscure!) a question, but it will not answer it. In this chapter I hope to clarify the contingency/constraint debate and its relationship to environmental fitness and directionality by focusing on several conceptually related examples of convergent functional trends in physiological and evolutionary ecology. Although all cases of convergence or directionality entail questions of physico-chemical constraint, the effects may be manifest at various levels of biotic organization, including the molecular mechanisms of replication and metabolism, structures of cellular integration, multicellular functional complexity, and even intelligence. In fact, Williams and Fraústo da Silva (2003, p. 323) recognize this organizational hierarchy itself as an "inevitable progression" driven by the sequential oxidation of the environment through living cells' reductive organic chemistry.

Various contributions to this volume address each of these levels. I wish to emphasize the "upper" end of the continuum by focusing not on the traditional apex of intelligence, but on the evolved capacity for interorganismal investment or "inter-subjective commitment" (Nesse, 2001). I will argue that, within bioenergetic constraints of metabolism and selective constraints of fitness tradeoffs in energy budgets, evolution has predictably, perhaps inevitably (although not eliminatively) converged on life history strategies with particular relational significance. These strategies involve the capacity to recognize other individuals and make significant, protracted investments in their welfare. In life history theory, this has culminated in what has somewhat euphemistically, although not altogether facetiously, been called the transfer of benefits or "live to give" hypothesis (Lee, 2003; Wade, 2003).

Vigorous controversies notwithstanding, I must begin by acknowledging that the roles of constraint and contingency in evolutionary processes are, of course, not mutually exclusive. In fact they are mutually necessary (Carroll, 2001). Nevertheless, a significant amount of both the scientific and interdisciplinary literature on these issues has been characterized by an unnecessary dichotomization of causal explanations and a regrettable polarization of rhetoric. There are two reasons for this, both of which I hope at least to avoid, if not to redress.

First, these questions have obvious entanglements with ideological and metaphysical issues. On the one hand, philosophical precommitments influence the plausibility criteria by which theories are assessed. On the other hand, scientific conclusions have implications for the justification of theological and moral belief. We will clearly see these dynamics at work in the controversies that follow.

Understandably, extreme assertions have begotten extreme responses. The cliché exhortation here would be to disentangle the science from metaphysics; but given the nature of the issues, it is not clear that this is entirely possible, or even desirable. A more modest and potentially more constructive response would be to explore the issues in light of an explicit and nuanced recognition of the logical entailments. As I will argue below, and contrary to prominent assertions (Gould, 1989, 1996; Ruse, 1996, 2003), the dichotomized scientific accounts are not diametrically opposed in their theological implications. Either can be, and each has been, employed on behalf of both theologically hostile and hospitable arguments. This does not mean the issues are irrelevant, and in fact I will argue that the very ambiguity is theologically significant.

Second, and more specifically, two logically separate but related scientific aspects of these questions have frequently been intertwined, if not conflated. However, they have quite different evolutionary foci and interdisciplinary ramifications. These involve questions of what might be considered the *necessary* versus *sufficient* conditions for the evolution of life. The first question involves the issue of which features of the abiotic environment are requisite to the origin and manifest diversification of life. This has been the traditional domain of "fine-tuning" arguments. A less metaphysically laden and more historically consistent way to speak of it would be in terms of "the fitness of the environment," as emphasized nearly a century ago by Lawrence J. Henderson (1913). Darwinian theory typically addresses the fitness of organisms to the environment. A complementary question – which at face value is *a priori* to the question of natural selection – is the fitness of the environment to life, or the preconditional requirements of the "theater," to support an evolutionary drama.

This question may be asked at the level of the physical constants of the cosmos, the life-sustaining features of our planet, or the unique aspects of chemistry that are fundamental to life. If, on empirical grounds, those preconditions appear to be highly specific or biophilically constant, it raises the teleological question of the environment's fitness not just *to* life, but *for* life.[2] Furthermore, if – on non-empirical but some would argue rationally justifiable grounds – the specificity seems to be highly unlikely (a complicated and perhaps intractable question; see Barrow, 2002; Manson, 2000), it raises the metaphysical question of fine-tuning or divine design. It is possible to affirm the former without affirming the latter (Henderson, 1913; Denton, 1998).

In contrast to the question of necessity, which entails the characteristics of the environment (E), required for life (L) to evolve, given what we know of life (L → E), the question of sufficiency entails the characteristics of life that must have come to pass, given what we know of the environment (E → L). This involves precisely the question of inevitability in playing the tape over and getting the same results. However the environment came about, and however finely tuned its properties are or are not, does the environment that exists so constrain biochemistry that the origin of life and the direction of its evolution were highly probable, if not certain? Is evolutionary "'unfolding' in any way predictable *before* the fact of history?" (Freeland, Chapter 14, this volume). The idea that the world is so constructed as to guide the historical unfolding of life in a way that is open to, if not suggestive of, developmental or teleological interpretation has attracted vigorous theological interest. However, although these issues are theologically significant, there is no theistically "right" answer here. One can answer this fascinating question affirmatively, without advocating theistic precommitments or conclusions (de Duve, 2002; Denton, 1998; Wright, 2000), or negatively, while accepting them (Behe, 1996; Behe *et al.*, 2000; Dembski, 2001, 2002).

The two different issues of sufficiency and necessity – if, and only if – are alluded to in the subtitle of Simon Conway Morris's (2003) book, *Life's Solution: Inevitable Humans in a Lonely Universe*, and involve, respectively, determinative and requisite constraints in evolution. This chapter will focus on the question of sufficient or determinative constraint. The question of whether the environment is suited or fit to ensure the arising of complex, socially affiliative life is both scientifically and theologically significant. I will argue that directional and convergent trends in the evolution of life history strategies reflect an inherent tilt toward increased

---

[2] Lest the "to" versus "for" distinction seem obscure, let me illustrate by pointing to the differing implications of saying the heart is fit *to* pump blood and the heart is *for* pumping blood. Or a Stradivarius may be fit *to* burn but is not *for* kindling. Although teleology in terms of Final Causes has been extruded quite legitimately and fruitfully from science, there is significant debate about whether teleological explanations remain helpful, perhaps essential, in organismic or evolutionary biology (Fodor, 1998; Bekoff and Allen, 1998). Henderson posited such a teleology without supernatural metaphysical entailments.

interorganismal investment. In so doing, I will make some general comments about inferences of directional trends and constraint, survey three major and interacting domains of evolutionary trend, assess current debates over their possible relationship to intrinsic constraints, and conclude with some theological comments on how these scientific issues relate to questions of cosmic purpose, or "plot" in the evolutionary play.

Of trends and *telos*

The issue of evolutionary directionality as it reflects finely tuned constraints involves three levels of disagreement.

Trends or no trends?

The first issue is whether there even *are* meaningful evolutionary trends suggestive of constrained directionality rather than contingent variability. Of course, evolutionary history exhibits directional change: we have life now and did not have it four billion years ago. And since the origin of life, organisms have become, among other things, larger, multicellularly complex, taxonomically diverse, and energetically intensive (Bonner, 1988; Maynard Smith and Szathmáry, 1995; McShea, 2001b; McShea and Changizi, 2003; Szathmáry and Maynard Smith, 1995; Williams and Fraústo da Silva, 2003). The question here is whether this represents the merely contingent directionality of a few "major evolutionary transitions" (Janis, 1993) precipitated by unpredictable geo- or climatological cataclysms or whether directional change representative of consistent evolutionary trends, involves, in Eldredge and Gould's terms, "biostratigraphic character gradients" (1988, p. 211)? If so, and most important, does this reflect the fact that "life was in a physical chemical tunnel and there was only one way to go" (Williams and Fraústo da Silva, 2003, p. 335)?

Although I will assess evidence for the latter, the question has been complicated by theological entanglements. The twin issues of evolutionary progress and directional inevitability have been invested with substantial significance in the wake of natural theology's displacement by Darwinian naturalism. Once organismic *products* could no longer be construed as evidence of intervention by a designing deity, the evolutionary *process* was argued to represent the historical drama of a divine playwright (Ruse, 1996). Thus, debates over progress and inevitability are frequently represented as disputes between theism-friendly and theism-hostile versions of evolution (Gould, 1989, 1996, 2002; Ruse, 1996, 2005). But this is not necessarily the case on either count. With respect to progress, although there has been a flourishing of evolutionary eschatologies since Teilhard de Chardin (Schloss, 2002), no intrinsic theological warrant exists for believing that historical change must naturally incline toward improvement. In fact, some thoughtful traditions reject a ruddy optimism gleaned from extrapolation of progressive trends in favor of theological hope for redemption of a fundamentally ambiguous creation (Barth, 2000; Kierkegaard, 1995; Pascal, 1958). Perhaps it is even tilted toward chaos, or at least vulnerable to demise (Polkinghorne and Welker, 2000; Russell, 2002; Schloss, 2002; Watts, 2000). With respect to inevitability, it has almost become a shibboleth to say that contingency is irreconcilable with providence. In his seminal and oft-quoted *The Meaning of Evolution*, George Gaylord Simpson maintained that "a purposeless and materialistic process that did not have man in mind . . . [means] he was not planned" (Simpson, 1967, p. 344). Similarly, Stephen Gould concludes that seeing another drama on replaying the tapes is incommensurate with notions of Providence (Gould, 1996). But this is uncritically to infer metaphysical from historical contingency. It also conflates the lack of intentionality in a mechanistic cause (or process) with the lack of a final cause or divine purpose. Ironically, these are precisely the errors that are justifiably criticized in creationists' rejection of evolution.

Moreover, if contingency is not necessarily the foe of providence, inevitability is surely not the friend (Gingerich, 2005). With the rise of seventeenth century mechanism, this very issue was the focus of debate between Descartes and Gassendi, whose views were rooted in the even more ancient differences of Thomists and nominalists (Osler, 1994). Indeed, with no contingency at all, there is no room for final cause or purposeful agency. Einstein's aphorism reflects longstanding theological questions of causal necessity: "I want to know whether God had any choice in creating the universe." These same issues are still unresolved between two very different (and equivalently over-simplistic) contemporary approaches to natural theology. One argues that the inevitability of life's origin and evolutionary history is testimony of divine purpose, and another infers divine design precisely from the wildly contingent improbability of obtaining what we have (Behe, 1996; Dembski, 1998, 2001; Ross, 2001). My point is that the contingency/constraint debate is not a dispute between science and religion, but is a fundamental issue of ambiguity existing within both theological and scientific interpretations of nature.

Biased vs. driven trends

Even if one tentatively grants the existence of significant and constrained evolutionary trends (Bonner, 1988; Gould, 1988a,b; Knoll and Bambach, 2000; Maynard Smith, 1970; McShea, 1998, 2001a; McShea and Changizi, 2003; Stanley, 1973; Szathmáry and Maynard Smith, 1995; Wagner, 1996), differing understandings remain about (a) the nature of this directionality and (b) the factors that constrain it. One involves debate over whether the trends involve passive or driven evolutionary change (McShea, 1994, 1998; Wang, 2001), the other over whether factors constraining major trends are primarily internal or external (Carroll, 2001).

Driven change would involve biased replacement or movement away from one pole and toward another pole of a character gradient, mediated by consistent selective advantages and/or fundamental biomechanical–developmental constraints. Conversely, a passive trend would involve no evolutionary bias at all, but rather random diffusion away from a minimum functional threshold of size, complexity, or other quality (Bonner, 1988; Gould, 1988b; Knoll and Bambach, 2000; McShea, 1994, 1998; Stanley, 1973). In fact, in the absence of such a minimum, there would be no directional trend at all, but simply an adirectional increase in phenotypic variance (Gould, 1988b).

As above, this debate has involved both ideological and theological issues. Ideological implications include the concern that claims of a strong evolutionary or "natural" bias toward particular biological characteristics have been used to justify, and have also often reflected, an ideologically based social vision (Gould, 1981, 1988a; Ruse, 1996). This is regrettably true in historic cases such as Social Darwinism, eugenics, Marxist genetics, and some theories of gender and social stratification (Gould, 1981; Kaye, 1986; Larson, 1995; Lewontin *et al.*, 1984; Sahlins, 1976; Sayers, 1982). For this reason, Stephen Gould has been a strong proponent both of largely contingent evolutionary history and, in clear cases of constrained trends, of passive rather than driven change. His powerful metaphor for this is a drunk's random stumbling down a street with a wall on one side. This strong rejection of progress or any kind of directional bias in evolution contrasts with his earlier views, congenial to the argument presented here, that posited a life history trend toward higher and more selective investment in other organisms (Gould, 1977). His position appears to have emerged while writing his 1981 treatise on biological racism, *The Mismeasure of Man* (Ruse, 1996).

It is indeed important to avoid using a theory of how things are as a justification for how things should be – the naturalistic fallacy (Moore, 1903). But it is equally important not to reject an idea on the basis of its implications if misused – the consequentialist fallacy. In fact, there is a long (though controversial) history, from Thomas Huxley (1894) to Richard Dawkins (1976) and George Williams (1993), of recognizing the morally ambiguous, even objectionable, nature of the evolutionary process and many of its products. Far from rejecting the underlying science on this basis, these writers have argued that it can be used to inform and exhort a human morality founded on selectively resisting, rather than uncritically endorsing, all aspects of nature.

Theologically, the image of a drunk stumbling down the road is certainly provocative, and the metaphor of passive diffusion rather than driven change is less connotive of evolutionary *telos* or inevitability. Connotations notwithstanding, I should point out that although differences between passive and driven trends have important implications for understanding the specific causes of directional change in the evolution of various characters, each view harbors completely equivalent implications for the issue of inevitability. A diffusional process is entirely non-contingent, and although I will argue that the following trends reflect strong biases, they would be no less certain if they entailed passive, diffusional increases in maxima. "Passive trends towards increases in organismal size, complexity, and diversity from some initial minima are certain to prevail in any system" (Carroll, 2001, p. 1108). Moreover, the metaphor of diffusion (and metaphor it is) functions merely as a description of pattern, but not an explanation of cause (McShea, 1994). Even if trends are passive, we still need to examine the reasons for constrained minima and inevitable movement away from this boundary (McShea, 1998).

Internal vs. external constraints

There is increasing discussion of whether major evolutionary trends, particularly patterns in life history evolution, reflect constraints that are primarily internal or external in nature. "Internal" constraints are taken to be inherent in the physico-chemical, metabolic, or developmental limits of the organism and in some cases are posited to entail universal first principles underlying all organisms. "External" constraints are understood in terms of what will reproductively flourish or "work" in any given environment. They involve selectional tradeoffs between solutions to a variety of challenges imposed by different abiotic and biotic environmental factors. Many of these factors, such as dispersals of predators and competitors, or geological and climatological cataclysms, may be contingent.

Like the above two controversies, discussions of this issue have become somewhat intellectually dichotomized and rhetorically polarized.[3] Although advocates of the positions seem to represent differences not so much in ideological as in disciplinary commitments, the arguments reflect profoundly contrasting ways of understanding the history and, indeed, perhaps the nature of life. On the one hand, the impressive and rapidly growing literature on internal constraint, involving significant contributions by physicists and chemists, strongly maintains that physico-chemical principles are sufficient to determine major evolutionary patterns. They entail provocative proposals for a universal or master equation of metabolism (Gillooly *et al.*, 2001; Savage *et al.*, 2004) or an oxidatively driven "inevitable progression" from prokaryotes through metazoan nervous systems to human beings (Williams and Fraústo da Silva, 2003). On the other hand, the notion of external

---

[3] For an excellent representation of the arguments in one recent and particularly fascinating instantiation of this debate, see the theme issue of *Functional Ecology* (**18** 2, 2004) dedicated to internalist and externalist understandings of allometric trends in metabolic rate.

constraints, emphasized by many comparative physiologists and evolutionary ecologists, acknowledges far less clarity and inevitability in evolutionary history. It views directional patterns as being underdetermined by physical necessity and reflecting the stochastic outcomes negotiated by natural selection (Bokma, 2004; Charnov, 1993; Hochachka and Somero, 2002; Kozlowski and Konarzewski, 2004; Kozlowski and Weiner, 1997; Stearns, 1992).

This issue is significant for two reasons. First, providing an account for any set of generalized patterns that goes beyond statistical description to causal explanation in terms of first principles would represent a revolutionary contribution to the biological sciences. This is especially true for ecology and evolutionary biology, which have been markedly recalcitrant to such approaches (Whitfield, 2001). This question is germane to the themes of this volume, because it is central to the concept of biochemical fine-tuning or the Hendersonian notion of environmental fitness: it is physico-chemical first principles that would constitute sufficient constraints, i.e. the suitability of the prebiotic cosmos to predictably generate the kind of life we observe.

Second, because internal constraint would presumably influence evolutionary unfolding in a way less vulnerable to contingent variation than selection-mediated tradeoffs in solutions to independent environmental challenges (Clarke, 2004; Clarke and Fraser, 2004; West and Brown, 2004; West *et al.*, 2004), it has implications for the theologically charged question of inevitable evolutionary directionality or divine *telos*. That is undoubtedly one reason for the interdisciplinary nature of this volume. However, I have already argued that inevitability is theologically ambiguous. In what follows, I will also argue that the vigorous and fascinating debate over internal and external constraint – prominent in each of the following issues – entails but does not reduce to necessitarian versus contingent views of evolutionary history.

Bioenergetic trends

Two of the most significant trends in evolutionary history – and perhaps two of the most notable generalizations in all the biological sciences – entail evolutionary increases in body size (Cope's Rule) and energetic intensiveness across and within the major taxa. Although the former relates to the latter in several ways, I want to discuss two manifestations of energetic trends across evolutionary history that have significant implications for life history strategy. First, energy flow through biota and individual organisms has generally escalated. Biomass density, gross primary productivity, production efficiency, and secondary productivity have all increased. More significantly, energy utilization per organism has increased with increasing body mass, and mass-specific metabolic rate has increased from unicellular organisms, through ectothermic metazoans, to endothermic metazoans.

These trends represent a fascinating and well established instance of necessary conditions having been progressively generated by the action of life itself upon the environment (see, for example, Williams and Fraústo da Silva, 2003). Life-generated prerequisites to these trends include increases in atmospheric free oxygen, biomass density, and biomineralization over evolutionary time. But what drives utilization of these substrates in the direction of increased energetic intensity? Approaches to this involve internal and external constraints. Vermeij proposes a co-evolutionary arms race, in which defensive armaments and attendant counter-measures result in a continuing "evolutionary escalation" (1987, 1994). Selection favors energetically intensive adaptations such as locomotor performance, toxicity, armor, high growth rates, and metabolically demanding increases in information gathering and processing. This involves intrinsic directionality, perhaps spiked by episodic and contingent environmental increases in nutrient availability (Vermeij, 1987). Alternatively, Stanley Salthe (1993) has an elegant proposal for an intrinsic increase of energy utilization in developing systems, followed by a decrease during senescence. These developmental changes occur with concomitant changes in information and organizational complexity and apply to developmental processes at organismal, ecological, and evolutionary scales (Salthe, 1993). The above approaches and others involving external and internal constraints need not be mutually exclusive, and may even be nested. Importantly, all posit intrinsic and sufficient conditions for evolutionary increase in energetic intensiveness.

The second and more specific bioenergetic trend involves thermoregulation. Beyond the fact that its very emergence represents a historic vector (primitive organisms are thermal conformers), there are two fascinating and sometimes conflated aspects of a trend toward thermoregulatory escalation. First is the increase in body temperature, or differentials between organismal and ambient temperature (Hamilton, 1973). This consistent progression is evident along an extensive phyletic continuum that includes invertebrates, amphibians, reptiles, monotremes, marsupials, eutherian mammals, passerines, and non-passerine birds. It is a dramatic example of a consistent directional trend across major taxa, but even within taxa it is tempting to describe temperature relations as reflecting a drive toward maximization or "maxithermy" (Hamilton, 1973). The preferred temperatures and thermal performance curves of most organisms are not normal but skewed left, with temperature optima that approach the highest sustainable in a given environment (Hamilton, 1973) and that are often quite close to upper-critical and even lethal temperatures (Huey and Bennett, 1987).

Why is this the case? Numerous studies provide evidence for the selective advantage of increased body temperature for food capture, prey avoidance, assimilation efficiency, and growth rates (Avery, 1984; Avery *et al.*, 1982; Christian and Tracy, 1981; Greenwald, 1974). Locomotor performance in a wide variety of species is optimized at temperatures higher than ambient, and compensation for the depressing effects of decreased temperature does not appear to be effective: "in regards to locomotion, warmer appears to be better" (Bennett, 1987, p. 422). This appears to entail a driven directional and convergent trend across major evolutionary transitions, that involves the direct physico-chemical effects of temperature.

A question that is related to but separate from the issue of temperature increase is that of convergence: why does body temperature in independent endotherms approximate 36–38 °C? Paul (1986) argued that the body temperatures of most endotherms approximate the 36 °C temperature at which the specific heat of water is lowest, and hence the temperature at which the least heat would be lost to the environment. However, the problem with this proposal is that the actual rate of heat loss, and hence the behavioral or metabolic costs of heat replacement, is entirely independent of specific heat. Thermal flux is determined by the driving force or the temperature differential between organism and environment, and elevating the temperature of endotherms to 36 °C actually results in greater heat loss. An alternative physico-chemical proposal, based on this observation, is that raising the organism–environment temperature differential, and hence the driving force, facilitates the dissipation of excess metabolic heat in birds and mammals, by conduction, convection, and radiation, rather than by costly evaporative loss of water (Calder, 1986). A complementary proposal posits the optimization of water's viscosity, which decreases with temperature, and the solubility of hydrophobic molecules, which increases with temperature. The intersection of these curves occurs at approximately 36 °C (Duntee and Benner, 1986).

The significant bottom line is that fundamental physical and chemical properties of the abiotic environment appear related to both an increase and a convergence of body temperatures. But how do organisms maintain these temperatures? The second aspect of thermoregulatory trends involves escalation, not of the set-point, but of the modes of temperature regulation, which have become increasingly metabolically and behaviorally costly. One aspect of this involves the maintenance of body temperature by internal (metabolic) rather than external means, the progressive employment of endo- versus ectothermy, which convergently arises across vertebrate, invertebrate, and even plant taxa. Another aspect involves increasing precision of temperature regulation, the transition from poikilo- to homeothermy, entailing higher peak performance and narrower thermal performance breadth. These strategies represent not dichotomies but continua, which have increased and been coupled over evolutionary history.[4]

---

[4] Although some primitive organisms inhabiting thermally stable environments have been referred to as nominal homeotherms, this is a misnomer as such organisms are not thermal regulators at all, much less precise ones. Some dinosaurs may have been inertial homeotherms, relying on their large thermal mass (McNab, 2002).

Three classic explanations for the evolution of endothermic homeothermy all entail bioenergetic constraint: increased stability of body temperature enabling refinement of optimal performance, increased spatial and temporal independence of body temperature enabling expanded domains of activity, and increased metabolic scope enabling higher and more sustained peak performance (McNab, 2002). Because these advantages obviously entail tradeoffs between costs and benefits, the directional trend they are proposed to explain is not eliminative : although there is an increase in maximal thermoregulatory investment, there are still plenty of ectotherms and thermal conformers. Crucial to our focus, though, the trend is very strongly driven to converge with other life history traits, particularly parental care. The higher energetic requirements of this thermal strategy demand substantially greater parental input into young, especially because the surface area : volume ratio of juveniles involves high heat loss. Thus, while natural selection has produced and sustained a variety of thermoregulatory strategies, there have been a directional increase and convergent arising of maxithermy, endothermy, and homeothermy – which are associated with one another and have driven concomitant increases in parental care.[5] If this account is correct, thermodynamics entails parental investment.

Three fundamental affirmations are relevant at this point. First, there are clear directional and convergent trends of increase in energetic and thermoregulatory intensity within and across lineages. These trends appear to have both passive and driven components, but in both cases increased maxima reflect evolutionary inevitabilities. Second, debated explanations of these trends employ internal and external constraints, but little theoretical or empirical warrant exists for considering them mutually exclusive, nor is there any credible account that is wholly contingent. Third, both energetic and size increases harbor significant implications – separately and in their interaction – for parental care and, as we shall see, other major life history trends.

Allometry: of mice and men

Although trends in body size and energy utilization are significant in themselves, the scaling relationship between mass and metabolism represents a crucial, some argue universal (Savage *et al.*, 2004; West and Brown, 2004), convergent evolutionary pattern. Not only is it one of the most fundamental and widely discussed regularities

---

[5] A recent revisionist theory has suggested that instead of the selective advantages of endothermy driving the evolution of parental care, the reverse was the case. High parental attentiveness drove the evolution of endothermy by the requirements of a stable temperature during development and sustained performance levels for postnatal parental provisioning (Farmer, 2000; Watanabe, 2005). These approaches may not be mutually exclusive, as it is possible to have autocatalytic effects. But while the fascinating question of the evolution of endothermy remains unresolved, the undisputed and for our purposes crucial point is the tight linkage with parental care.

in the biological sciences, but also it relates – empirically if not causally – to the important life history characteristics under discussion here.

Bergmann (1847) first noted that metabolic rate in mammals appeared to scale with mass to a 2/3 power. Rubner (1883) observed that heat production appeared to be more closely correlated with surface area than with mass, positing the widely cited Rubner's rule or "surface law." Huxley (1932) advocated fitting metabolic rate to a power function of mass, and based on the most extensive empirical analysis to date, Kleiber (1932) concluded that the exponent was 3/4. By mid-century we had the "3/4 rule." Surface area explanations of metabolic scaling were abandoned. For one thing, they required a 2/3 exponent (representing the linear dimension squared) that did not fit the data. In addition, metabolic rate scaled to body mass by a similar power function in ectotherms, for which the metabolic replacement of surface-mediated heat loss was not relevant.

In spite of the general acceptance of the 3/4 rule, which has become a prominent "textbook example" of scaling, discussion has continued about whether this value for the exponent is reliable. The debate has been renewed over whether the exponent is 2/3 (Heusner, 1982a,b) or 3/4 (MacMahon, 1973, 1975; Feldman and MacMahon,1983). There has also been concern about whether the interspecific slope reflects experimental or statistical artifact (Elgar and Harvey, 1987; Heusner, 1991; McNab, 1986, 1988). An adequate mathematical understanding of these relations has been described as "the central question in comparative physiology" (Heusner, 1991, p. 34). Indeed, until recently we have lacked not only a consensus description of how metabolism and mass scale, but also, and more importantly, a coherent mechanistic proposal for why they do so (Bennett, 1988).

That has changed dramatically over the past several years, with the groundbreaking work of West and co-workers (West *et al.*, 1997, 1999a,b, 2003; West and Brown, 2004; see also Brown *et al.*, 1993; Enquist *et al.*, 1998, 1999; Savage *et al.*, 2004). In two seminal papers, West *et al.* (1997, 1999a) accept that the 3/4 law is empirically warranted, and assume that natural selection maximizes resource exchange and minimizes time and energy for resource transport. They develop a model that describes resource transport through a branched, fractal-like network of circulatory tubes (1997) or, more generally, a space-filling distributional surface area (1999a), corresponding to every level of biotic exchange from digestive cavities or plant leaves to transport vessels to cell surfaces, mitochondria, and even molecules. Unlike external surface area, which scales through the square of the linear dimension, an internal, space-filling surface area scales as the cube. Energy utilization will scale as a function of $D/(D+1)$, where D = the number of dimensions. The adaptations to provisioning an internal, "three-dimensional" surface area effectively endow living systems with an additional, "fourth" dimension.

The authors conclude that these geometric and physical constraints so limit the possibility space on which selection can act, that all organisms reflect quarter-power scaling laws – 3/4 for metabolic rate, 1/4 for internal time and distance – that are as universal as biochemical pathways or the genetic code. However, unlike the last two cases, which may reflect the retention of an underdetermined initial character that arose but once, adaptations that effectively contribute a fourth dimension – lungs, gills, guts, kidneys, organelles, as well as branching whole-organism morphologies – have convergently arisen at numerous times in evolutionary history.

This proposal both points out and explains the convergent scaling of numerous physiological adaptations, and it also has been extended in an impressively ambitious and promising attempt to explain a wide range of previously disparate relations in population and community ecology. This includes: proposals for universal temperature dependence of life; scaling relations in plant morphology, productivity, and life history; energy flux, carbon turnover, and biomass density in plant populations; a general allometric model for growth and development; energy flux in ecosystems and across food webs; rates of mutation and carcinogenesis; and patterns in global biodiversity (Allen *et al.*, 2002; Brown and Gillooly, 2003; Enquist, 2002; Enquist *et al.*, 1998, 1999, 2003; Gillooly *et al.*, 2001, 2002; Niklas and Enquist, 2001; West *et al.*, 1999b, 2001, 2004). It is the first attempt to provide an account of a wide array of biological and ecological generalizations on the basis of first principles.

The formulation of inherent biophysical constrains represents an emphatic alternative to radical contingency: "Does some fixed point or deep basin of attraction in the dynamics of natural selection ensure that all life is organized by a few fundamental principles and that energy is a prime determinant of biological structure and dynamics among all possible variables?" (West and Brown, 2004, p. 42). The proposal for fundamental internal constraints is one of the most broadly unifying and mechanistically based explanations advanced in evolutionary ecology. It has been widely lauded – "if it holds up, it's going to rewrite our evolutionary biology" (Klarreich, 2005) – and also severely criticized.

The disagreement falls into three main classes. First, some researchers still dispute the 3/4 characterization of metabolic scaling (Bokma, 2004; Dodds *et al.*, 2001; White and Seymour, 2003). However, in an analysis of what is arguably the most extensive and rigorously selected data set to date, the West team confirmed 3/4 scaling across mammals and a variety of other taxa (Savage *et al.*, 2004). Moreover, they point out that deviations from 3/4 scaling, found in smaller mammals by other studies, are actually predicted by their initial model on the basis of dominance of Poiseuille rather than pulsatile flow in smaller-diameter, originating vessels.

Second, some workers affirm the 3/4 rule and posit an explanation for it that involves intrinsic constraints, but disagree with West and co-workers' proposal. Banavar *et al.* (1999, 2002) advocate an account that is amenable to but more general than West's, rejecting the notion of an effective fourth dimension to life. Instead, allometric scaling is intrinsic to all systems with directed flow and circulation times that are proportional to length but not size. However, a problem with this is that it does not provide a conciliatory explanation for all the other organismal characteristics that scale allometrically.

Darveau *et al.* (2002, 2003) propose that the relationship between body mass and metabolic rate reflects the contribution of multiple factors – ATP-utilization processes in parallel, supply processes in series – that each have different power functions. This hierarchical layering results in an "allometric cascade" that has different scaling implications for different measures of metabolism. They contend that only a multiple-factor account, and not West's (or any) single-cause account, can explain the scaling difference between basal and maximum metabolic rate (Bishop, 1999). However, the mathematical formulation of their model has been severely criticized (Banavar *et al.*, 2003; West *et al.*, 2003), and in any case it does not provide an account of why individual processes scale as power functions of mass or why the causal cascade results in a whole-organism metabolism that approximates the 3/4 rule (Bokma, 2004; West *et al.*, 2003; West and Brown, 2004).

Third, dispute exists over whether there even is a universal scaling relation and whether ostensible patterns reflect internal biophysical constraints of any kind (Clarke, 2004; Clarke and Fraser, 2004; Kozlowski, 1996; Kozlowski and Weiner, 1997; Kozlowski and Konarzewski, 2004). This is asserted for two reasons. For one thing, it is claimed that the data do not support the model. This is because the 3/4 rule itself is an artifact of an interspecific regression that traverses heterogeneous intraspecific regressions with independent slopes and intercepts and also because functionalist explanations do not adequately account for the extremely large and highly correlated residual variation (Kozlowski and Weiner, 1997).[6]

For another thing, it is maintained on theoretical grounds that metabolic and life history variables interact with one another in the ecological context of each species, rather than being deterministically related to body mass for all species. The balance between these factors, and their relationship to body mass, reflects the ecological conditions unique to each species and, in principle, no meaningful interspecific generalization can be formulated. Several models of evolutionary tradeoffs have been proposed to explain the same data used by West and co-workers (Kozlowski and Weiner, 1997; Clarke, 2004). However, like the above multiple-factor internalist accounts, what the approach of viewing allometric trends as by-products does not do is to explain why metabolic rate scales to body mass – even within species – as a power function to begin with. Nor does it explain the generalized patterns

---

[6] For any given mass, variation in metabolic rate may span an order of magnitude, and the same metabolic rate may be found in organisms with 20–30-fold differences in mass. More importantly, the residual variation itself is not random, and many life history parameters are still highly correlated even after being adjusted for mass (Kozlowski and Weiner, 1997; Kozlowski and Konarzewski, 2004).

across taxa, i.e. why certain proportions are more likely than others (Kozlowski and Weiner, 1997) and what constrains the distribution of possibilities on which selection can act.

So what can we conclude? The proposal for internal constraints represents a path-breaking attempt to explain long-observed patterns in biology, on the basis of physical first principles. And it not only fits the basic, coarse-grained allometric data, but also has fruitfully generated new hypotheses and has been used to explain a wide array of other observations as well. None of the alternative proposals accomplishes this. However, it does not explain fine-grained patterns and significant deviations from predictions, which appear to represent not mere noise but residual variation explainable by other – most likely external – factors.

To some extent, a disciplinary divide is at work here, as "probabilistic models derived from population biology and selection theory differ fundamentally from engineering models, which depend on . . . the surface area of isometric bodies, or the structure of branching networks" (McNab, 2002, p. 35). This divide entails differences not only in analytic approach, but also in evaluative criteria that have both polarized the dispute and made it difficult to resolve empirically. However, my point is that these tensions do not require a forced choice between explanatory accounts, which are not intrinsically irreconcilable. Internal constraints may fix the allometric baseline, which selection may modify under certain circumstances. One of the postulates of West and co-workers' model is that "organisms evolve toward an optimal state in which the energy required for resource distribution is minimized" (West and Brown, 2004, p. 38). "Toward" is the key word here, and the extent to which evolution attains any particular optimality target often reflects compromise with other selective demands: physical first principles may constrain what is optimal, but do not always determine what is actual.

Therefore, we can summarize with three conclusions that are analogous to what was affirmed in the previous section. First, there are clear allometric trends that represent convergence of scaling relations across a wide range of both taxa (bacteria, plants, invertebrates, and vertebrates) and organizational levels (organelles, cells, bodies, and ecosystems). Second, although the relative contribution of "internal" (biophysical) and "external" (selection) constraints is debated, no account of these trends – which include invariant relationships – interprets them as products of primarily contingent causes or passive radiation. And third, allometric trends are strongly related to, and to some extent constitutive of, important life history trends to which we now turn our attention.

Life history across life's history

"Life history" refers to the timing and allocation of resources for growth, maintenance, and reproduction over an organism's lifetime. The size and number of offspring; investment in parental care; ovi- and viviparity; growth rate and adult body size; and ages of reproductive maturity, reproductive cessation, and death are presumably all under control of selection to optimize reproduction. These variables do not evolve independently of one another. Some are related by invariant ratios, and most occur in strategic constellations that represent significant directional and convergent evolutionary trends.

What I want to focus on is the capacity for selective interorganismal investment. This is represented by high degrees of parental care and social reciprocity, which are positively associated with body size and lifespan and negatively associated with specific metabolic rate and lifetime fecundity. A clear trend of investment increase over evolutionary time has been observed, although it involves an increase in maximum and not mean investment. It represents a classic example of non-eliminative, passive – although apparently inevitable – diffusion away from a minimum. On the other hand, its coupling with other parameters is a driven or strongly constrained trend.

As with the trends previously mentioned, proposals have been promulgated for internal and external constraints. At first pass, it is tempting to account for relations between life history variables almost purely on the basis of fundamental allometric constraints. Metabolic rate, lifespan, fecundity, age at maturity, and maternal investment all vary with body mass as power functions. In fact, relations are invariant between some of these variables. For example, lifespan scales with body mass by a $1/4$ power, and heart rate (or the rate of ATP synthesis) scales with body mass by a $-1/4$ power. The product yields an approximately constant number of "metabolic events" in mammal species, independent of body mass or lifespan. Age at maturity / lifespan, and annual maternal investment / lifespan (for indeterminate growers), are also invariant ratios (Charnov, 1993; Charnov *et al.*, 2001; Stearns, 1992). West and Brown (2004) point out that invariant ratios, and universal quarter-power allometric trends in general, suggest underlying physical first principles. They employ their model to explain these life history relations (Enquist *et al.*, 1999; Niklas and Enquist, 2001; West *et al.*, 2001).

However, the two types of criticism leveled (as I have argued, unconvincingly) at their application of physical principles to scaling of metabolic rate are more telling against this being an adequate *single-cause* account of all these variables' relations to body size and one another. First, selection theory argues against an inflexible coupling of life history variables to mass (Calder, 1984; Charnov, 1993; Charnov et al. 2001; Kozlowski, 1996; Kozlowski and Weiner, 1997; Kozlowski and Konarzewski, 2004; Stearns, 1992). This has recently been empirically confirmed by an elegant analysis that related life history traits in mammalian carnivores to body sizes of extant species and their direct ancestors (Webster *et al.*, 2004). In those species that have recently undergone size change, life history variables were more closely correlated with ancestral than present size; for example, phyletic giants have smaller offspring, in larger, earlier, and more frequent litters, than similarly sized species that have not recently become large. Therefore, life history variables, which do scale to body mass, may be coupled loosely by selection rather than rigidly by biophysics.

Second, and perhaps more importantly, both phyletic differences and residual analysis of general relations have important stories to tell. For example, although age at maturity scales directly with average adult lifespan, the ratio decreases across vertebrate classes (fish, reptiles, mammals, birds), reflecting a gross trend toward iteroparity, or increased proportion of lifetime over which reproduction occurs. Regarding residuals, although length of maternal investment and annual fecundity scale with body mass (and hence each other), the residuals also scale with one another. That is, a strong negative relation exists between size-adjusted or relative fecundity and relative maternal investment (Stearns, 1992). This is a crucial life history trend in its own right, having an allometric but also an entirely size-independent component. Bats and primates are at the upper end of the regression and also occupy similar positions in residual analysis of lifespan and body mass. There is an evolutionary trend toward protracted lifespans, reduced fertility, and increased parental investment beyond that predicted by first principles, and a linking of these features in a fashion that is not allometrically mediated.

Here again the literature has become highly polarized, though it is not only possible but also necessary to avoid a false dichotomy. It is precisely the contribution of both internal and external constraints that is most noteworthy in these trends. The important conclusion is that natural selection has found a way of both employing and "amplifying" allometrically mediated relations between life history variables to generate organisms that invest more resources, with more selectivity, over greater timescales. Physical first principles underlie but do not wholly determine this.

A salient example of interaction between internal and external constraints in life history trends involves the issue of lifespan or the timing of senescence and death, which also turns out to have newly posited implications for investment. On the one hand, lifespan has been interpreted as reflecting fundamental allometric constraints (West *et al.*, 2001; West and Brown, 2004). Larger organisms have lower specific metabolic rates and tend to live longer. This is consistent with the quarter power account of biological time (West and Brown, 2004) and concords with metabolic damage (e.g. telomere damage, free-radical accumulation) or "rate of living" theories of aging (Finch, 1990; Rose, 1991).

On the other hand, evolutionary theories of aging posit that the timing of senescence is mechanistically determined not by metabolically induced damage, but rather, by selection's response to *external* mortality factors. In fact, there should be no senescence at all in the absence of external causes of death. The extent to which an organism invests in repairing damage will reflect the externally mediated likelihood that an investment in maintenance at any given point in the lifespan will result in sufficient additional reproductive opportunities to pay the investment back in future progeny. More generally, selection will allow the accumulation of deleterious mutations that are expressed late in life, when an organism's reproductive potential is low (Medawar, 1952; Hamilton, 1966). And it will actually promote, through antagonistic pleiotropy, traits that increase reproduction early in life at the expense of dysfunction later in life (Kirkwood, 1977; Williams, 1957). Various experimental (Rose, 1984, 1991; Tyner *et al.*, 2002) and observational (Keller and Genoud, 1997; Sherman and Jarvis, 2002) studies support this, which are not made sense of by first-principle, rate of living accounts.

As before, both internal and external constraints – entailed in this case by rate of living and evolutionary theories of aging – can operate simultaneously, indeed interactively. But neither has done a good job of explaining certain unexpected mortality patterns that some highly social species exhibit at the beginning and end of life. Although both approaches predict that mortality rates will increase with age across the lifespan, in some species with high parental care infant mortality actually decreases as juveniles mature, and these species also exhibit protracted post-reproductive longevity. Enter the first theory to incorporate parental care – or "transfer of benefits" as opposed to brute fertility – as the major determinant of mortality, and of life history under some circumstances (Lee, 2003). In demographic conditions where intensive parental care is essential to survival of progeny, selection may more effectively eliminate deleterious traits in offspring that have already incurred high parental investment (hence the declining juvenile mortality with age). This approach also explains how selection can operate against antagonistic pleiotropy or accumulation of mutations that are expressed later in life. Anticipated by the notion of a "grandmother effect," in which post-reproductive individuals can still very significantly enhance fitness by caring for progeny or their kin (Williams, 1957), Lee formalizes this in a model that integrates economic exchange, demography, and population genetics. Because it provides a unified account for mortality patterns across the entire lifespan, it has been referred to as "the most comprehensive evolutionary theory of aging that we have seen to date" (Rogers, 2003, p. 9115).

In fact, under certain circumstances, the ability to "transfer benefits" to progeny or kin, rather than the maximization of lifetime fertility, becomes the primary controlling parameter for other life history characters. An organism's entire life history strategy may be optimized for this. In addition to providing a unified explanation for patterns that previously appeared anomalous, this life history theory is significant in that it explicitly posits a driven, directional trend. It entails "a positive feedback loop that selects for reduced fertility, higher consumption, greater investments in juveniles, and longer life. This describes the evolution of primates and other kinds of species with low fertility, heavy investment in offspring, and long adult life" (Lee, 2003, p. 9640). In contrast to some representations of the debate over internal/external constraints, this feedback loop involves an inevitable convergence that is driven not just by internal biophysical, but by external selective, constraints.

A distinctive, and I will suggest a beautiful, implication of the theory is its emphasis not only on the phylogenetic, but also on the ontogenetic intensification of investment or resource transfer. The progressive transition from receiving to giving benefits is itself a life history parameter that changes developmentally over the lifespan, a notion Lee considered calling the "live to give" theory (Wade, 2003). Is there, in this theory, a ground for rehabilitating Alfred Tennyson's dashed hope in "love Creation's final law"? Although that affirmation of Romanticism may be a bit too saccharine in light of evolutionary ambiguity, a profound and escalating other-orientation does appear to constitute a fundamental biotic strategy.

Conclusion: trends, progress, and purpose in evolution

I began this chapter by observing, and lamenting, the fact that discussions of evolutionary trends have frequently been polarized by divergent theological or philosophical precommitments. However, seeking to avoid metaphysical biases need not mean refusing to reflect on theological implications. I want to conclude with some brief comments on the significance of these issues. To do so, I distinguish between evolutionary trends (change in a given direction), progress (change in a valued direction; Ayala, 1988), and purpose (regulated change in a targeted direction).

First, fashionable pronouncements of disteleology notwithstanding, there is little room to doubt the existence of significant life history trends in evolutionary history. This entails the convergent arising, biased retention, and directional elaboration of fundamental organic structures and functional strategies that increase selective investment. Such increases appear to be the inevitable result of both fundamental biophysical and selective constraints. Moreover, these trends involve a genuinely interesting story, or to return to our initial metaphor, an evolutionary play whose plot involves increasing size, energy, complexity, diversity, longevity, and affiliative investment. This is, in a real sense, increasingly abundant *life* (Jonas, 1966).

As I mentioned at the beginning of the chapter, it is possible, even tempting, to reflect on this in a couple of ways. First, one may marvel at the stage and calculate the odds of its having been designed without input from the playwright. This is the domain of environmental fitness or fine-tuning arguments, emphasizing what I have called "necessary preconditions." Second, one may marvel at the way the stage supports this *particular* drama and wonder whether it could have turned out differently in this theater. This is the domain of inevitability arguments entailing sufficient constraints or preconditions, which I have been emphasizing. Both of these enterprises are legitimate, and each is interesting. But both may fail to focus on the message of the play itself. Like some forms of biblical fundamentalism that seem to place more emphasis on the method of textual inspiration than on the larger meaning of scriptural revelation, one may be so caught up in the architecture of the physico-chemical stage that one does not adequately engage the drama of the evolutionary play.

Second, what then can be said about the "message" of the script itself? Here again I want to question the *Zeitgeist* and maintain that, by any reasonable sense of the word "progress," evolutionary drama is filled with it, by virtue of an ongoing, serial intensification of the wondrous qualities of life itself. Theologian John Haught, who contributed a chapter to this volume, maintains that "Surely, by any objective standard of measurement, something momentous has been going on here . . . an over-arching inclination . . . toward the heightening of beauty" (Haught, 2005, p. 520). Philosopher Hans Jonas (1966) refers to the evolutionary epic as the "the ascent of soul." In response to qualms that such a judgment is parochially anthropomorphic, he contends it is appropriately *zoo*morphic. I would suggest that all of the major trends discussed in this paper build on one another to constitute a suite of biotic capacities that are jointly necessary for what we would call "love" or "relational commitment." This surely constitutes progress, change in a valued direction.

However, I want to point out two profound ambiguities in this "progress." First, as indicated earlier, none of these trends is eliminative. Even if they constitute progress, the "world" is not getting better, in terms of either a modal increase in beneficence or an attenuation of natural evil. Second, these capacities for affilia-tive investment that underlie love and commitment are necessary, but by no means sufficient. For one reason, other capacities are required that lie beyond the scope of this paper (and may or may not be outside the domains of existing biological theory; [Schloss, 2004]). For another and more important reason, these capaci-ties are just that – capacities, and not deterministic inclinations – to act in certain ways. In fact, behavioral plasticity increases along with the strategic assemblage of these capacities (Changizi, 2003). Thus, the cognitive and affective capacities, and life history and social contexts, that give rise to attachment, altruism, or moral commitment, may also facilitate manipulation, spite, or deceptive moral posturing. Ultimately, evolutionary "progress" is morally ambiguous. Like the biblical story of creation, it involves the origin of potentiality for, but not inevitability of, creaturely goodness, and the increase of this biological capacity is also coupled with an increased capacity for suffering (Schloss, 2002).

Finally, does it make sense to speak of evolutionary purpose? At one level, the answer to this question is easy. Like all mechanistic processes, evolution is non-purposive. In fact, natural selection is explicitly formulated as a non-teleological alternative to explanatory accounts invoking purposeful design. But supernatural metaphysics aside, there is a naturalistic context in which it could make scientific sense to speak of evolutionary purpose in a way analogous to the teleonomic, goal-oriented, or purposive behavior of living organisms. Living systems exhibit "regulated change in a targeted direction." Evolutionary change could be construed as purposive if: (a) its directionality were targeted or specified (perhaps by pre-existing or emergent environmental information) and (b) the process were adjusted by regulatory feedback loops (perhaps involving ecosystem- or biospheric-level interactions, e.g. Gaia). Although notions of environmental fitness or biochemical fine-tuning could and probably do contribute to the first requirement, the second, although it is fully naturalistic in principle, lies outside our current ability to assess empirically. Lee (2003) and Demetrius (1997, 2000) do provide conceptual proposals for how such directional feedback might occur in selection.

But the question of evolutionary purpose has another meaning, which involves not so much the purposes *of* evolution, but the possible purposes *behind* evolution. Can evolution, particularly the trends discussed here, be reasonably viewed as reflecting, even instantiating, purposes that transcend the process itself? Here again the spirit of the age is far from shy. A virulent natural atheology (Lustig, 2004) asserts that ascribing divine purpose to evolution is not only methodologically illegitimate, but also rationally incoherent (Dawkins, 2003; Gould, 1989, 1996; Rachels, 1990; Williams, 1993). Evolutionary history is represented as involving such great waste and suffering, resulting in such little if any directional progress, it is asserted that one who wishes to believe in a God behind evolution should be prepared to believe in "a monster" (Williams, 1993). Sadly, many traditional theists have either uncritically accepted or actively promoted the same view. I want to close by maintaining that the emerging understandings of evolutionary constraint described in this chapter, and in other chapters in this volume, are consistent with, even suggestive of, divine purpose. But they are by no means demonstrative of or simplistically concordant with it. This very ambiguity is itself quite important for science and theology, both of which suffer when science is distorted to promote a false nihilism or co-opted to bolster a mechanistic optimism about the fate of the cosmos. As Pascal maintained centuries before evolutionary theory, the natural world affords enough light to allow, but enough shadow to require, theological hope rather than religious or nihilistic certainty.

References

Allen, A. P., Brown, J. and Gillooly, J. (2002). Global biodiversity, biochemical kinetics, and the energetic-equivalence rule. *Science*, **297**, 1545–8.

Avery, R. (1984). Physiological aspects of lizard growth: the role of thermoregulation. *Symposia of the Zoological Society of London*, **52**, 407–24.

Avery, R., Bedford, J. and Newcombe, C. (1982). The role of thermoregulation in lizard biology: predatory efficiency in a temperate diurnal basker. *Behavioral Ecology and Sociobiology*, **11**, 261–7.

Ayala, F. J. (1988). Can "progress" be defined as a biological concept? In *Evolutionary Progress*, ed. M. Nitecki. Chicago, IL: University of Chicago Press, pp. 75–96.

Banavar, J., Maritan, A. and Rinaldo, A. (1999). Size and form in efficient transportation networks. *Nature*, **399**, 130–2.

Banavar, J. R., Damuth, J., Maritan, A. *et al.* (2002). Modelling universality and scaling. *Nature*, **420**, 626–7.

Banavar, J. R., Damuth, J., Maritan, A. *et al.* (2003). Allometric cascades. *Nature*, **421**, 713–14.

Barrow, J. (2002). *The Constants of Nature.* London: Jonathan Cape.

Barth, K. (2000). *Doctrine of Creation, the Creator and His Creature,* ed. Thomas Torrance. London: T. and T. Clark.

Behe, M. (1996). *Darwin's Black Box: The Biochemical Challenge to Evolution.* New York, NY: Simon and Schuster.

Behe, M., Dembski, W. and Meyer, S. (2000). *Science and Evidence for Design in the Universe.* San Francisco, CA: Ignatius Press.

Bekoff, M and Allen, C. (1998). *Nature's Purposes: Analysis of Function and Design in Biology.* Cambridge, MA: Bradford Books/Massachusetts Institute of Technology Press.

Bennett, A. (1987). Evolution of the control of body temperature: is warmer better? In *Comparative Physiology: Life in Water and on Land*, ed. P. Dejours, L. Bolis, C. R. Taylor *et al.* Padova: Liviana Press, pp. 421–31.

Bennett, A. (1988). Structural and functional determinants of metabolic rate. *American Zoologist*, **28**, 699–708.

Bergmann, C. (1847). Über die Verhältnisse der Wärmeökonomie der Tiere zu ihrer Grösse. (On the relationship of the heat budget of animals to their size.) *Göttinger Studienanfänger*, **3**, 595–708.

Bishop, C. M. (1999). The maximum oxygen consumption and aerobic scope of birds and mammals: getting to the heart of the matter. *Proceedings of the Royal Society of London*, B**266**, 2275–81.

Bokma. F. (2004). Evidence against universal metabolic allometry. *Functional Ecology*, **18**(2), 184–7.

Bonner, J. T. (1988). *The Evolution of Complexity by Means of Natural Selection.* Princeton, NJ: Princeton University Press.

Brown, J. and Gillooly, J. (2003). Ecological food webs: high quality data facilitate theoretical unification. *Proceedings of the National Academy of Sciences, USA*, **100**, 1467–8.

Brown, J., Marquet, P. and Taper, M. (1993). Evolution of body size: consequences of an energetic definition of fitness. *American Naturalist*, **142**, 573–84.

Calder, W. (1984). *Size, Function and Life History.* Cambridge, MA: Harvard University Press.

Calder, W. (1986). Body temperature and the specific heat of water. *Science*, **324**, 418.

Carroll, S. B. (2001). Chance and necessity: the evolution of morphological complexity and diversity. *Nature*, **409**, 1102–9.

Changizi, M. A. (2003). Relationship between number of muscles, behavioral repertoire, size, and encephalization in mammals. *Journal of Theoretical Biology*, **220**, 157–68.

Charnov, E. L. (1993). *Life History Invariants: Some Explorations of Symmetry in Evolutionary Ecology.* New York, NY: Oxford University Press.

Charnov, E. L., Turner, T. F. and Winemiller, K. O. (2001). Reproductive constraints and the evolution of life histories with indeterminate growth. *Proceedings of the National Academy of Sciences, USA*, **98**, 9460–4.

Christian, K. and Tracy, C. R. (1981). The effect of the thermal environment on the ability of hatchling Galapagos land iguanas to avoid predation during dispersal. *Oecologia*, **49**, 218–23.

Clarke, A. (2004). Is there a universal temperature dependence of metabolism? *Functional Ecology*, **18**, 252–6.

Clarke, A. and Fraser, K. P. (2004). Why does metabolism scale with temperature? *Functional Ecology*, **18**, 243–51.

Conway Morris, S. (1998). *The Crucible of Creation: The Burgess Shale and the Rise of Animals*. Oxford: Oxford University Press.

Conway Morris, S. (2003). *Life's Solution: Inevitable Humans in a Lonely Universe*. Cambridge, UK: Cambridge University Press.

Darveau, C. A., Suarez, R. K., Andrews, R. D. *et al.* (2002). Allometric cascade as a unifying principle of body mass effects on metabolism. *Nature*, **417**, 166–70.

Darveau, C. A., Suarez, R. K., Andrews, R. D. *et al.* (2003). Response to critics. *Nature*, **421**, 714.

Dawkins, R. (1976). *The Selfish Gene*. Oxford: Oxford University Press.

Dawkins, R. (1998). *The Blind Watchmaker: Why the Evidence of Evolution Reveals a Universe without Design*. New York, NY: W. W. Norton.

Dawkins, R. (2003). *A Devil's Chaplain: Reflections on Hope, Lies, Science, and Love*. New York, NY: Houghton Miflin.

de Duve, C. (2002). *Life Evolving: Molecules, Mind, and Meaning*. Oxford: Oxford University Press.

Dembski, W. (1998). *The Design Inference: Eliminating Chance through Small Probabilities*. Cambridge, UK: Cambridge University Press.

Dembski, W. (2001). *No Free Lunch: Why Specified Complexity Cannot Be Purchased without Intelligence*. Lanham, MD: Rowman and Littlefield.

Dembski, W. (2002). *Intelligent Design: The Bridge between Science and Theology*. Downers Grove, IL: InterVarsity Press.

Demetrius, L. (1997). Drectionality principles in thermodynamics and evolution. *Proceedings of the National Academy of Sciences, USA*, **94**, 3491–8.

Demetrius, L. (2000). Directionality theory and the evolution of body size. *Proceedings of the Royal Society of London*, B**267**, 2385–91.

Dennett, D. (1995). *Darwin's Dangerous Idea: Evolution and the Meanings of Life*. New York, NY: Simon and Schuster.

Denton, M. (1998). *Nature's Destiny: How the Laws of Biology Reveal Purpose in the Universe*. New York, NY: Free Press.

Dodds, P. S., Rothman, D. H. and Weitz, J. S. (2001). Re-examination of the "3/4-law" of metabolism. *Journal of Theoretical Biology*, **209**, 9–27.

Duntee, J. and Benner, S. (1986). Body temperature and the specific heat of water. *Science*, **324**, 418.

Eldredge, N. and Gould, S. J. (1988). Punctuated equilibrium prevails. *Nature*, **332**, 211–12.

Elgar, M. A. and Harvey, P. H. (1987). Basal metabolic rates in mammals: allometry, phylogeny and ecology. *Functional Ecology*, **1**, 25–36.

Enquist, B. (2002). Universal scaling in tree and vascular plant allometry: toward a general quantitative theory linking plant form and function from cells to ecosystems. *Tree Physiology*, **22**, 1045–64.

Enquist, B., Brown, J. H. and West, G. B. (1998). Allometric scaling of plant energetics and population density. *Nature*, **395**, 163–5.

Enquist, B. J., West, G. B., Charnov, E. L. *et al.* (1999). Allometric scaling of production and life-history variation in vascular plants. *Nature*, **401**, 907–11.

Enquist, B., Economo, E. P., Huxman, T. E. *et al.* (2003). Scaling metabolism from organisms to ecosystems. *Nature*, **423**, 639–42.

Farmer, C. G. (2000). Parental care: the key to understanding endothermy and other convergent features in birds and mammals. *The American Naturalist*, **155**(3), 326–34.

Feldman, H. A. and MacMahon, T. A. (1983). The 3/4 mass exponent for energy metabolism is not an artifact. *Respiratory Physiology and Neurobiology*, **52**, 149–63.

Finch, C. E. (1990). *Longevity, Senescence and the Genome.* Chicago, IL: University of Chicago Press.

Fodor, J. (1998). *In Critical Condition: Polemical Essays on Cognitive Science and the Philosophy of Mind.* Cambridge, MA: Massachusetts Institute of Technology Press.

Gillooly, J. F., Brown, J, West, G. *et al.* (2001). Effects of size and temperature on metabolic rate. *Science*, **293**, 2248–51.

Gillooly, J. F., Charnov, E. L., West, G. B. *et al.* (2002). Effects of size and temperature on developmental time. *Nature*, **417**, 70–3.

Gingerich, O. (2005). In praise of contingency: chance versus inevitability in the universe we know. In *Spiritual Information: 100 Perspectives on Science and Religion,* ed. C. L. Harper, Jr. Conshohocken, PA: Templeton Foundation Press, pp. 59–62.

Gould, S. J. (1977). *Ontogeny and Phylogeny.* Cambridge, MA: Belknap Press.

Gould, S. J. (1981). *The Mismeasure of Man.* New York, NY: W.W. Norton.

Gould S. J. (1988a). On replacing the idea of progress with an operational notion of directionality. In *Evolutionary Progress*, ed. M. Nitecki. Chicago, IL: University of Chicago Press, pp. 319–38.

Gould S. J. (1988b). Trends as changes in variance: a new slant on progress and directionality in evolution. *Journal of Paleontology*, **62**, 319–29.

Gould, S. J. (1989). *Wonderful Life: The Burgess Shale and Natural History.* New York, NY: W.W. Norton.

Gould, S. J. (1996). *Full House: The Spread of Excellence from Plato to Darwin.* New York, NY: Harmony.

Gould, S. J. (2002). *Rocks of Ages: Science and Religion in the Fullness of Life.* New York, NY: Random House.

Greenwald, O. (1974). Thermal dependence of striking and prey capture by gopher snakes. *Copeia*, 1974, 141–8.

Hamilton, W. D. (1966). The moulding of senescence by natural selection. *Journal of Theoretical Biology*, **12**, 12–45.

Hamilton, W. J. (1973). *Life's Color Code.* New York, NY: McGraw-Hill.

Haught, J. (2005). Reading an unfinished universe: science and the question of cosmic purpose. In *Spiritual Information: 100 Perspectives on Science and Religion,*ed. C. L. Harper, Jr. Conshohocken, PA: Templeton Foundation Press, pp. 519–23.

Henderson, L. J. (1913). *The Fitness of the Environment: An Inquiry into the Biological Significance of the Properties of Matter.* New York, NY: Macmillan. Repr. (1958) Boston, MA: Beacon Press; (1970) Gloucester, MA: Peter Smith.

Heusner, A. (1982a). Energy metabolism and body size. I: is the 0.75 mass exponent of Kleiber's equation a statistical artifact? *Respiratory Physiology and Neurobiology*, **48**, 1–12.

Heusner, A. (1982b). Energy metabolism and body size. II: dimensional analysis and energetic non-similarity. *Respiratory Physiology and Neurobiology*, **48**, 13–25.

Heusner, A. (1991). Size and power in mammals. *Journal of Experimental Biology*, **160**, 25–54.

Hochachka, P. and Somero, G. (2002). *Biochemical Adaptation: Mechanism and Process in Physiological Evolution*. Oxford: Oxford University Press.

Huey, R. and Bennett, A. (1987). Phylogenetic studies of coadaptation: preferred temperatures versus optimal performance temperatures of lizards. *Evolution*, **41**(5), 1098–1115.

Hutchinson, G. E. (1965). *The Ecological Theater and the Evolutionary Play*. New Haven, CT: Yale University Press.

Huxley, J. (1932). *On Relative Growth*. London: Methuen.

Huxley, T. H. (1894). *Evolution and Ethics*. Princeton, NJ: Princeton University Press. Repr. (1989).

Janis, C. M. (1993). Tertiary mammal evolution in the context of changing climates, vegetation, and tectonic events. *Annual Review of Ecology and Systematics*, **24**, 467–500.

Jonas, H. (1966). *The Phenomenon of Life: Toward a Philosophical Biology*. New York, NY: Harper and Row.

Kaye, H. L. (1986). *The Social Meaning of Modern Biology: From Social Darwinism to Sociobiology*. New Haven, CT: Yale University Press.

Keller, L. and Genoud, M. (1997). Extraordinary lifespans in ants: a test of evolutionary theories of ageing. *Nature*, **389**, 958–60.

Kierkegaard, S. (1995). *Works of Love,* ed. and transl. H. V. Hong and E. H. Hong. Princeton, NJ: Princeton University Press.

Kirkwood, T. B. (1977). Evolution of aging. *Nature*, **270**, 301–4.

Klarreich, E. (2005). Life on the scales: simple mathematical relationships underpin much of biology and ecology. *Science News*, **167**, 106–8.

Kleiber, M. (1932). Body size and metabolism. *Hilgardia*, **6**, 315–53.

Knoll, A. H. and Bambach, R. K. (2000). Directionality in the history of life: diffusion from the left wall or repeated scaling of the right? *Paleobiology*, **26** (suppl.), 1–14.

Kozlowski, J. (1996). Energetic definition of fitness? Yes, but not that one. *American Naturalist*, **147**, 1087–91.

Kozlowski, J. and Konarzewski, M. (2004). Is West, Brown, and Enquist's model of allometric scaling mathematically correct and biologically relevant? *Functional Ecology*, **18**(2), 283–9.

Kozlowski, J. and Weiner, J. (1997). Interspecific allometries are by-products of body size optimization. *American Naturalist*, **149**, 352–80.

Larson, E. J. (1995). *Sex, Race and Science: Eugenics in the Deep South*. Baltimore, MD: Johns Hopkins University Press.

Lee, R. (2003). Rethinking the evolutionary theory of ageing: transfers, not births, shape senescence in social species. *Proceedings of the National Academy of Sciences, USA* **100**(16), 9637–42.

Lewontin, R., Rose, S. and Kamin, L. (1984). *Not in Our Genes*. New York, NY: Knopf Publishing.

Lustig, A. J. (2004). Natural atheology. In *Darwinian Heresies*, ed. A. Lustig, R. Richards and M. Ruse. Cambridge, UK: Cambridge University Press, pp. 69–83.

MacMahon, T. (1973). Size and shape in biology. *Science*, **179**, 1201–4.

MacMahon, T. (1975). Allometry and biomechanics: limb bones in adult ungulates. *American Naturalist*, **109**, 547–63.

Manson, N. A. (2000). There is no adequate definition of "fine-tuned for life." *Inquiry*, **43**, 341–52.

Margulis, L., Matthews C. and Haselton, A., eds. (2000). *Environmental Evolution: Effects of the Origin and Evolution of Life on Planet Earth*. Cambridge, MA: Massachusetts Institute of Technology Press.

Maynard Smith, J. (1970). Time in evolutionary process. *Studium Generale*, **23**, 266–72.

Maynard Smith, J. and Szathmáry, E. (1995). *The Major Transitions in Evolution*. Oxford: Freeman.

Maynard Smith, J., Burian, R., Kauffman, S. *et al.* (1985). Developmental constraints and evolution: a perspective from the Mountain Lake Conference on Development and Evolution. *Quarterly Review of Biology*, **60**, 265–87.

McNab, B. (1986). The influence of food habits on the energetics of eutherian mammals. *Ecological Monographs*, **56**, 1–19.

McNab, B. (1988). Complications inherent in scaling basal rate of metabolism in mammals. *Quarterly Review of Biology*, **63**, 25–54.

McNab, B. (2002). *The Physiological Ecology of Vertebrates: A View from Energetics*. Ithaca, NY: Cornell University Press.

McShea, D. W. (1994). Mechanisms of large-scale evolutionary trends. *Evolution*, **48**(6), 1747–63.

McShea, D. W. (1998). Possible largest scale trends in organismal evolution: eight "live hypotheses." *Annual Review of Ecology and Systematics*, **29**, 293–318.

McShea, D. W. (2001a). The minor transitions in hierarchical evolution and the question of directional bias. *Journal of Evolutionary Biology*, **14**(3), 502–18.

McShea, D. W. (2001b). The hierarchical structure of organisms: a scale and documentation of a trend in the maximum. *Paleobiology*, **27**, 392–410.

McShea, D. W. and Changizi, M. A. (2003). Three puzzles in hierarchical evolution. *Integrative and Comparative Biology*, **43**, 74–81.

Medawar, P. B. (1952). *An Unsolved Problem of Biology*. London: H. K. Lewis.

Moore, G. E. (1903). *Principia ethica*. Cambridge, UK: Cambridge University Press.

Nesse, R., ed. (2001). *Evolution and the Capacity for Commitment*. New York, NY: Russell Sage Foundation Publications.

Niklas, K. J. and Enquist, B. J. (2001). Invariant scaling relationships for interspecific plant biomass production rates and body size. *Proceedings of the National Academy of Sciences, USA* **98**, 2922–7.

Norris, R. D. (1991). Biased extinction and evolutionary trends. *Paleobiology*, **17**, 388–99.

Osler, M. (1994). *Divine Will and the Mechanical Philosophy: Gassendi and Descartes on Contingency and Necessity in the Created World*. Cambridge, UK: Cambridge University Press.

Pascal, B. *Pensées* (1958). T. S. Eliot, commentator. New York: E. P. Dutton.

Paul, J. (1986). Body temperature and the specific heat of water. *Science*, **323**, 300.

Polkinghorne, J. and Welker, M. (2000). Science and theology on the end of the world and the ends of God. In *The End of the World and the Ends of God: Science and Theology on Eschatology*, ed. J. Polkinghorne and M. Welker. Harrisburg, PA: Trinity Press International, pp. 29–41.

Rachels, J. (1990). *Created from Animals: The Moral Implications of Darwinism*. Oxford: Oxford University Press.

Rogers, A. R. (2003). Economics and the evolution of life histories. *Proceedings of the National Academy of Sciences, USA* **100**, 9114–15.

Rose, M. R. (1984). Laboratory evolution of postponed senescence in *Drosophila melanogaster*. *Evolution*, **38**, 1004–10.

Rose, M. R. (1991). *Evolutionary Biology of Aging*. Oxford: Oxford University Press.

Ross, H. (2001). *Creator and Cosmos: How the Greatest Scientific Discoveries of the Century Reveal God*. Colorado Springs, CO: NavPress.

Rubner, M. (1883). Über den Einfluss der Körpergrösse auf Stoff- und Kraftwechsel. (On the influence of body size on metabolism and force change.) *Zeitung für Biologie*, **19**, 535–65.

Ruse, M. (1996). *Monad to Man: The Concept of Progress in Evolutionary Biology*. Cambridge, MA: Harvard University Press.

Ruse, M. (2003). *Darwin and Design: Does Evolution Have a Purpose?* Cambridge, MA: Harvard University Press.

Ruse, M. (2005). *The Evolution–Creation Struggle*. Cambridge, MA: Harvard University Press.

Russell, R. (2002). Bodily resurrection, eschatology, and scientific cosmology. In *Resurrection: Theological and Scientific Assessments*, ed. T. Peters, R. J. Russell and M. Welker. Grand Rapids, MI: William Eerdmans, pp. 3–30.

Sahlins, M. (1976). *The Use and Abuse of Biology*. Ann Arbor, MI: University of Michigan Press.

Salthe, S. (1993). *Development and Evolution: Complexity and Change in Biology*. Cambridge, MA: Massachusetts Institute of Technology Press.

Savage, V. M., Gillooly, J. F., Woodruff, W. H. *et al.* (2004). The predominance of quarter-power scaling in biology. *Functional Ecology*, **18**(2), 257–82.

Sayers, J. (1982). *Biological Politics: Feminist and Anti-feminist Perspectives*. New York, NY: Routledge.

Schloss, J. P. (2002). From evolution to eschatology. In *Resurrection: Theological and Scientific Assessments*, ed. T. Peters, R. J. Russell and M. Welker. Grand Rapids, MI: William Eerdmans, pp. 56–85.

Schloss, J. P. (2004). Evolutionary ethics and Christian morality: surveying the issues. In *Evolution and Ethics: Human Morality in Biological and Religious Perspective*, ed. P. Clayton and J. P. Schloss. Grand Rapids, MI: William Eerdmans, pp. 1–24.

Sherman, P. and Jarvis, J. (2002). Extraordinary life spans of naked mole-rats. *Journal of Zoology*, **258**(3), 307–11.

Simpson, G. G. (1967). *The Meaning of Evolution*, rev. edn. New Haven, CT: Yale University Press.

Stanley, S. M. (1973). An explanation for Cope's rule. *Evolution*, **27**, 1–26.

Stearns, S. (1992). *The Evolution of Life Histories*. Oxford: Oxford University Press.

Szathmáry, E. (2002). The gospel of inevitability: was the universe destined to lead to the evolution of humans? *Nature*, **419**, 779–80.

Szathmáry, E. and Maynard Smith, J. (1995). The major evolutionary transitions. *Nature*, **374**, 227–32.

Tyner, S., Venkatachalam, S., Chol, J. *et al.* (2002). Mutant mice that display early ageing-associated phenotypes. *Nature*, **415**, 45–53.

Vermeij, G. (1987). *Evolution and Escalation: An Ecological History of Life*. Princeton, NJ: Princeton University Press.

Vermeij. G. (1994). The evolutionary interaction among species: selection, escalation, and coevolution. *Annual Review of Ecological Systems*, **25**, 125–52.

Wade, N. (2003). Why we die, why we live: a new theory of aging. *New York Times*, July 15, Section F, p. 3.

Wagner, P. J. (1996). Contrasting the underlying patterns of active trends in morphological evolution. *Evolution*, **50**, 990–1007.

Wang, S. C. (2001). Quantifying passive and driven large-scale evolutionary trends. *Evolution*, **55**, 849–58.

Watanabe, M. (2005). Generating heat: new twists in the evolution of endothermy. *BioScience*, **55**, 470–5.

Watts, F. (2000). Subjective and objective hope: propositional and attitudinal aspects of eschatology. In *The End of the World and the Ends of God: Science and Theology on Eschatology*, ed. J. Polkinghorne and M. Welker. Harrisburg, PA: Trinity Press International, pp. 47– 60.

Webster, A. J., Gittleman, J. L. and Purvis, A. (2004). The life history legacy of evolutionary body size change in carnivores. *Journal of Evolutionary Biology*, **17**(2), 396–407.

West, G. B. and Brown, J. H. (2004). Life's universal scaling laws. *Physics Today*, **57**, 36–42.

West, G. B., Brown, J. H. and Enquist, B. J. (1997). A general model for the origin of allometric scaling laws in biology. *Science*, **276**, 122–6.

West, G. B., Brown, J. H. and Enquist, B. J. (1999a). The fourth dimension of life: fractal geometry and allometric scaling of organisms. *Science*, **284**, 1677–9.

West, G. B., Brown, J. H. and Enquist, B. J. (1999b). A general model for the structure and allometry of plant vascular system. *Nature*, **400**, 664–7.

West, G. B., Brown, J. H. and Enquist, B. J. (2001). A general model for ontogenetic growth. *Nature*, **413**, 628–31.

West, G. B., Savage, V. M., Gillooly, J. *et al.* (2003). Why does metabolic rate scale with body size? *Nature*, **421**, 713.

West, G. B., Brown, J. H. and Enquist, B. J. (2004). Growth models based on first principles or phenomenology? *Functional Ecology*, **18**(2), 188–96.

White, C. R. and Seymour, R. (2003). Mammalian basal metabolic rate is proportional to body mass$^{2/3}$. *Proceedings of the National Academy of Sciences, USA* **100**, 4046–9.

Whitfield, J. (2001). All creatures great and small. *Nature*, **413**, 342–4.

Williams, G. (1957). Pleiotropy, natural selection, and the evolution of senescence. *Evolution*, **11**, 398–411.

Williams, G. C. (1993). Mother nature is a wicked old witch. In *Evolutionary Ethics*, ed. M. H. Nitecki and D. V. Nitecki. Albany, NY: State University of New York Press.

Williams, R. J. P. and Fraústo da Silva, J. J. R. (2003). Evolution was chemically constrained. *Journal of Theoretical Biology*, **220**, 323–43.

Wright, R. (2000). *Nonzero: The Logic of Human Destiny*. New York, NY: Pantheon Books.

Part IV

The fitness of the chemical environment

16

Creating a perspective for comparing

Albert Eschenmoser

Introduction

Any chemist looking at the molecular workings of a living cell from the vantage point of organic chemistry may have moments in which he desists from scientific business-as-usual and finds himself standing in awe before so much "molecular ingenuity" and sheer chemical beauty. If anyone, besides the biochemist, may be fit to recognize such marvels on the molecular level and put them into a proper perspective, it is the synthetic organic chemist, who tends to judge any new discovered molecular structure or process by the criterion of whether he could do such a thing himself: "If I had to, could I make this?" The question reflects a dichotomy, epistemological in nature, that has been with the science of organic chemistry from the very beginning: the two-fold task of studying molecules occurring in nature and creating by chemical synthesis molecules that have never existed before. Chemical synthesis has traditionally been the organic chemist's major tool for exploring the molecular world: the ability to synthesize molecules of ever-increasing complexity that mirror those produced by living nature has been a significant measure of progress in organic chemistry as a whole. Yet, the gap between what chemists are able to create by chemical synthesis and what nature achieves in biosynthesis remains immense.

Fortunately, it is inspiration rather than resignation that chemists are drawing from this gap, and they are encouraged to do so by considering how chemical thought and the chemist's ability to make molecules have changed in the course of the past two centuries. Today, many organic chemists are shifting their research interest towards the chemistry–biology interface. There, besides encountering a multitude of new scientific and technological challenges, they come in contact with the principles and problems of evolutionary thinking, a different way of looking at

Fitness of the Cosmos for Life: Biochemistry and Fine-Tuning, ed. J. D. Barrow *et al.*
Published by Cambridge University Press. © Cambridge University Press 2007.

the world and one that has been historically completely foreign to their core field. One of the purely scientific challenges at this interface is etiological in nature, namely, the quest for chemists to join biologists in addressing problems of origin. Their challenge here is to pursue such problems on the territory of chemistry, to define them at the chemist's level of looking at molecules, and to tackle them by using the chemist's tools – not the least of which is chemical synthesis. The ultimate focus of such etiological research on the territory of chemistry will be the search for the chemistry of life's origin, the problem at which chemistry and biology may be expected to mesh in the most fundamental way.

Modern molecular biology's mechanistic demystification of the material aspects of life does not mean that we have lost our ability to stand in awe before the molecular "miracles" happening within living cells; such emotions differentiate us from "research robots." Yet in no way are these emotions supposed to imply or express doubts about the basic paradigm of contemporary science. Surely, it must have been such "standing in awe" before thought-provoking coincidences that induced Lawrence J. Henderson, almost one hundred years ago, to write *The Fitness of the Environment* (1913). However, his central message, although it remains thought-provoking today, failed to become an explicit part of mainstream science because it seemed to transcend limits that the chemistry and biology of the twentieth century had set for themselves as fundamental natural sciences. Specifically, these disciplines showed a tendency to deny, ignore, postpone, or "outsource" interest in phenomena, questions, and views that paradigmatic reasoning concluded were not, or perhaps not yet, accessible to scientific inquiry. More recently, however, Henderson's bold vision of an innate biocentricity implicit within our universe has been forcefully echoed in debates within contemporary cosmology. Such debates concern the epistemological status of provocative pronouncements such as the "anthropic principle(s)" or, more mildly, the role in science of what is referred to as "anthropic reasoning," a topic amply discussed in other chapters of this volume. Almost needless to say, any debate such as this is bound to emanate beyond the limits of natural science into other realms of human thought. Therefore, it seems consistent and timely that the John Templeton Foundation has responded to an initiative of interested cosmologists by organizing an interdisciplinary workshop commemorating the 1913 publication of Henderson's book. The workshop juxtaposed the thoughts and expertise of cosmologists, biochemists, and evolutionary biologists with regard to the question of whether the cosmologists' perception of "fine-tuning of the fundamental parameters of physics toward life" has any counterpart on the level of contemporary biochemistry.

In attempting to retrospectively analyze the goal and outcome of the workshop from the point of view of the chemist, it is helpful to begin with the original wording of three (of the five) core questions that this workshop was asked to address:

1. Do analogs to the fine-tuned biocentric parameters in physics and cosmology exist in biochemistry?
2. How can this question be addressed adequately within biochemistry?
3. Might fine-tuning in biochemistry be fundamentally different from fine-tuning in physics and cosmology?

With hindsight, I think that the discussions seeded by these questions at the workshop turned out to be handicapped by an underlying misunderstanding between the cosmologists (interrogators) and the biochemists (those being interrogated). The biochemist tends to use the term "fine-tuned" intuitively and qualitatively, thinking of the physical and chemical properties of matter, very much as Henderson did. To the cosmologist, a statement such as "the fundamental parameters of physics seem fine-tuned toward life" is based on data obtained by counterfactual variation of fundamental physical parameters and theoretical calculations of the effects such variation would have on the structure and the properties of the universe. In essence, the initiators of the workshop expected their biochemical colleagues to comment on two questions: first, whether the effects of counterfactual variation in fundamental physical parameters could be delineated on the biochemical level and whether signs of "fine-tuning" might be discernible on this level; and, second, whether the tool of counterfactual variation itself could be applicable at the level of biochemistry (in the sense of varying counterfactually physico-chemical parameters and evaluating by theory the effect of such variation on vital biochemical processes).

In this context, it is illuminating to remember that chemistry in the eighteenth and nineteenth centuries became an independent branch of natural science because physicists had "outsourced" the study of the transformations of "chemical matter" to chemists; such phenomena seemed too complex and too foreign to common experience at the time to be rendered in the physicist's language of mathematics. Analogously, in the late nineteenth and in the twentieth century, biochemistry and molecular biology became independent branches of natural science as chemists had "outsourced" investigations of biopolymers and the living cell as structures with properties too complex to be analyzed by chemical methods and expressed in the language of chemistry. Today, of course, we know better. Chemistry, with the help of physics, has become a widely quantifying science; biochemistry and molecular biology, largely with the help of physical methods, have become the sciences that deal with the chemistry of the living cell. This nothwithstanding, no contemporary biochemist (at least of the ilk of this author) would think that counterfactual variations of fundamental physical parameters could be translated

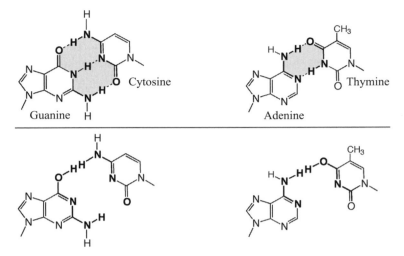

Figure 16.1. Dependence of the existence of Watson–crick base-pairing on the position of the equilibria between tautomers of the canonical nucleobases. Watson–Crick base pairing would not exist if the bond energy of the C=O double bond were lower by only a few kcal/mol relative to that of the C=C and C=N bonds.

by calculation into predictions regarding biochemical processes at the molecular level; that is, consequential variations of chemical, let alone biological, properties of complex organic molecules. With respect to this degree of challenge, chemistry remains an empirical science, biology even more clearly so. In this sense, there would have been no way that the biochemists present at the workshop could have lived up to the optimistic expectations of the cosmologists.

Thus remains the question of whether the tool of counterfactual variation, applied to parameters at the level of biochemistry (implying physico-chemical parameters) could provide specifically biochemical evidence for matter being fine-tuned toward life. In principle, Henderson had done just that in an intuitive and qualitative way a century ago. But is there not more to be said today? My view is that the underlying problem remains unchanged. I want to illustrate the point in more detail by means of a recently considered example (see Figure 16.1): Watson–Crick base-pairing of oligonucleotides (Eschenmoser, 2001; Groebke *et al.*, 1998). Discussion of this example at the workshop highlighted the basic difficulty of attempting to predict variation in biochemical parameters that would result from counterfactual variation in physico-chemical parameters.

Amazing Watson–Crick base-pairing

The existence of Watson–Crick base-pairing in DNA and RNA is crucially dependent on the position of the chemical equilibria between tautomeric forms of the nucleobases.[1] These equilibria in both purines and pyrimidines lie sharply on the side of amide- and imide-forms containing the (exocyclic) oxygen atoms in the form of carbonyl groups ($C=O$) and (exocyclic) nitrogen in the form of amino groups (NH_2). The positions of these equilibria in a given environment are an intrinsic property of these molecules, determined by their physico-chemical parameters (and thus, ultimately, by the fundamental physical constants of this universe). The chemist masters the Herculean task of grasping and classifying the boundless diversity of the constitution of organic molecules by using the concept of the "chemical bond." He pragmatically deals with the differences in the thermodynamic stability of molecules by using individual energy parameters, which he empirically assigns to the various types of bonds in such a way that he can simply add up the number and kind of bonds present in the chemical formula of a molecule and use their associated average bond energies to estimate the relative energy content of essentially any given organic molecule.[2] As it happens, the average bond energy of a carbon–oxygen double bond is about 30 kcal per mol higher than that of a carbon–carbon or carbon–nitrogen double bond, a difference that reflects the fact that ketones normally exist as ketones and not as their enol-tautomers.[3] If (in the sense of a "counterfactual variation") the difference between the average bond energy of a carbon–oxygen double bond and that of a carbon–carbon and carbon–nitrogen double bond were smaller by a few kcal per mol, then the nucleobases guanine, cytosine, and thymine would exist as "enols" and not as "ketones," and Watson–Crick base-pairing would not exist – nor would the kind of life we know.

It looks as though this is providing a glimpse of what might appear (to those inclined) as biochemical fine-tuning of life. However, I agree with Paul Davies' comment at the workshop: in order for the proposed change of the bond energy of a carbon–oxygen double bond to be a proper counterfactual variation of a physico-chemical parameter, we concomitantly would have to change the bond energies of all other bonds occurring in the chemical formulae of the nucleobases in such a way that we would remain internally consistent within the frame of molecular physics. To do this in a theory-based way is not feasible because the average energies assigned to (isolated) chemical bonds are empirical parameters that have no direct equivalents in quantum-mechanical models of organic molecules. Without the

---

[1] This statement has a remarkable precedent in the history of the discovery of the DNA double helix: Jim Watson and Francis Crick's efforts to build a model of DNA remained futile as long as Jerry Donohue had not told them that they were using the wrong tautomers for the nucleobases (Watson, 1968).

[2] Such a sum of the average bond energies for a given organic molecule approximates the energy change involved in the formation of one mole of the molecule from the corresponding free atoms (gaseous). Average bond energies of chemical bonds symbolized in the chemical formulae of organic molecules are in the range of 35–120 kcal/mol (single bonds) and 100–180 kcal/mol (double bonds). The bond-energy values for carbon–carbon double bonds are around 145 kcal/mol and those of carbon–oxygen double bonds (carbonyl bonds) around 175 kcal/mol (Roberts and Caserio, 1977).

[3] Tautomers are isomers that differ in the position of hydrogen atoms and can rapidly transform themselves into each other.

possibility of calculating bond energies from first principles, average bond energies cannot be meaningfully used as a parameter for counterfactual variation.

On the other hand, calculating the position of tautomeric equilibria in nucleo-bases is certainly within the grasp of contemporary quantum chemistry, and semi-empirical physico-chemical parameters on which the positions of these equilibria might most sensitively depend could presumably be identified. Whether in this special case it would be feasible and conceptually proper to attempt an internally consistent variation of physico-chemical parameters followed by calculation of associated properties for resulting virtual nucleobases is a question to be answered by a quantum chemist rather than an experimentalist. It nevertheless would seem that Watson–Crick pairing is a promising target (for those so inclined) in a theory-consistent search for a biochemical example of fine-tuning of chemical matter toward life. It represents an example of a question referring to existence that might be reduced to a question of the position of chemical equilibrium between tautomers.

Irrespective of the outcome of such a search, the cascade of coincidences embod-ied in nature's canonical nucleobases will remain, from a chemical point of view, an extraordinary case of evolutionary contingency on the molecular level (even to those unconcerned about the question of a biocentric universe). The generational simplicity of these bases when compared with their relative constitutional com-plexity,[4] their capacity to communicate with one another in specific pairs through hydrogen bonding within oligonucleotides, and, finally, the role they were to take over at the dawn of life and to play at the heart of biology ever since is extraordinary. I have little doubt that Henderson – could he have known it – would have added these coincidences to his list of facts that were, to him, convincing evidence for the environment's fitness to life.

Let us then assume, for the sake of argument, that the equilibria between the tautomers of the nucleobases prevented Watson–Crick base-pairing of the kind we know. Would there be an alternative higher form of life? If we were to answer in the affirmative – aware of the immense diversity of the structures and properties of organic molecules and conscious of the creative powers of evolution – could we have any idea of what such a life form might look like, chemically? The helplessness that overwhelms us as chemists in being confronted with such a question can give rise to two different reactions. Some of us would seek comfort in declaring that such questions do not belong to science, and others would simply be painfully reminded of how little we really know and comprehend of the potential of chemical matter to become and to be alive. Our insight into the creativity of biological evolution

[4] The constitution of the canonical purines and pyrimidines is relatively complex with regard to content and positioning of carbon, hydrogen, nitrogen, and oxygen atoms in their molecules, yet simple from the point of view of their (experimentally documented) formation from elementary (potentially geochemical) starting materials such as hydrocyanic acid, cyanamide, cyanoacetylene, and water (see, for example, Ferris and Hagan, 1984). The most remarkable member is adenine, a "simple" pentamer of HCN (Oro and Kimball, 1960).

on the molecular level is far too narrow for us to judge by biochemical reasoning what would have happened to the origin and the evolution of life if they had had to occur and operate in a world of (slightly) different physico-chemical parameters. I shall return to this point below.

Statements about fine-tuning toward life in cosmology referring to criteria such as the potential of a universe to form heavy elements and planets are in a category fundamentally different from statements about fine-tuning of physico-chemical parameters toward life at the level of biochemistry. Whatever biological phenomena appear fine-tuned can be interpreted in principle as the result of life having fine-tuned itself to the properties of matter through natural selection. Indeed, to interpret in this way what we observe in the living world is mainstream thinking within contemporary biology and biological chemistry. Thus, life science and cosmology are in very different positions when it comes to the question of how to interpret, or even identify, data that point to fine-tuning. To return to our example: in biology, the existence of a central feature such as Watson–Crick base-pairing may be seen as an achievement of life's evolutionary exploration of, and adaption to, the chemical potential of matter on planet earth. In cosmology, there is no corresponding way to interpret the formation of, let us say, a planet, and proposals of "evolutionary" universe-selection imposed on multiverse models would fall short of creating a correspondence.

Mainstream thinking in biology that unconditionally traces everything observed in the living world back to the known and (still) unknown powers of Darwinian evolution is, at its roots, the expression of a "dogmatized Occamism" that fundamentally dominates natural science. Needless to say, at the edges of scientific inquiry, such as the potential biocentricity of our universe, cultural imprinting, preconceptions, and personal inclinations are bound to participate, admittedly or not, to an unusually large degree in our scientific reasoning. Prejudices at a dynamic borderline of science are legitimate and can be important (provided its supporters accept them to be potentially transient) because research, stimulated by prejudice and taken up and forcefully pursued by the prejudiced, may eventually lead to insights that become a generally accepted part of science. Such prejudicial inquiry becomes harmful to scientific progress only when it fails to cede in the light of continuing negative evidence.

On the question of life's biochemical uniqueness

If fundamental parameters of physics, acting through the properties of existing chemical matter, were fine-tuned toward life, then how sharply determined are the parameters of the life to which these physical parameters are "fine-tuned"? Put another way, how sharply has the life we know "fine-tuned" itself to the properties of chemical matter? How biochemically diverse *could* a library of different life chemistries be while maintaining compatibility with the physics that we know?

At least in principle, chemistry could offer a strategy to find out, such as by creating, through chemical synthesis, alternative biologies and systematically comparing their biochemistries with that of "natural" biology. Needless to say, such a thing is (at present) out of reach. But within reach here and now is the ability to adhere to the general strategy of *creating alternatives to apparent uniqueness for the sake of gaining a perspective for comparing* by applying this strategy to modest, yet nevertheless significant, biomolecular targets such as the structure type of life's basic informational polymers, the nucleic acids.

For such a project, the strategy may read as follows. Conceive (through chemical reasoning) potentially natural alternatives to the structure of RNA; synthesize such alternatives by chemical methods; compare them with RNA with respect to those chemical properties that are fundamental to its biological function (Eschenmoser, 1999). Fortunately for this special case of the nucleic acids, it is not at all problematic to decide what the most important of these properties has to be: it must be the capability to undergo informational Watson–Crick base-pairing. The relevance of the perspective created in such a project will strongly depend on the specific choice of the alternatives' chemical structures. The quest is to focus on systems deemed to be potentially natural in the sense that they could have formed, according to chemical reasoning, by the very same type of chemistry that (under unknown circumstances) must have been operating on earth (or elsewhere) at the time when and at the place where the structure type of RNA was born. Candidates that lend themselves to this choice are oligonucleotide systems, the structures of which are derivable from $(CH_2O)_n$ sugars ($n = 4, 5, 6$) by the type of chemistry that allows the structure of natural RNA to be derived from the C_5-sugar ribose (see Figure 16.2). This approach is based on the supposition that RNA structure originated through a process that was combinatorial in nature with respect to the assembly and functional selection of an informational system within the domain of sugar-based oligonucleotides. In a way, the investigation is an attempt to mimic the selection filter of such a natural process by chemical means, irrespective of whether RNA first appeared in an abiotic or a biotic environment.

In retrospect, the results of systematic experimental investigations carried out along these lines justify the effort (see Figure 16.3). It is found that hexopyranosyl analogs of RNA (with backbones containing six carbons per sugar unit instead of five carbons and six-membered pyranose rings instead of five-membered furanose rings) do not possess the capability of efficient informational Watson–Crick base-pairing. Therefore, these systems could not have acted as functional competitors of RNA in nature's choice of a genetic system, even though these six-carbon alternatives of RNA should have had a comparable chance of being formed

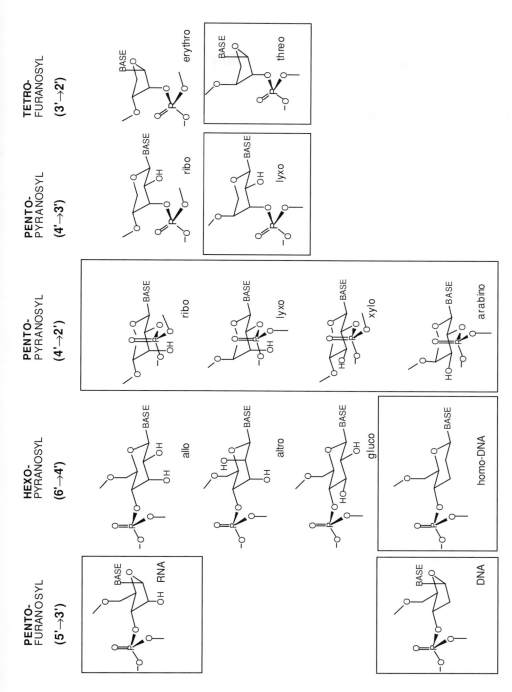

Figure 16.2. Chemical formulas of the monomer units (in their pairing conformation) of the natural nucleic acids DNA and RNA, as well as of the nucleic-acid alternatives that have been investigated experimentally.

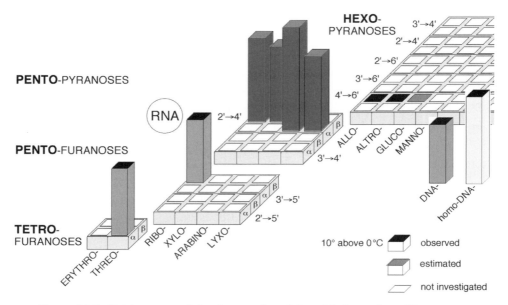

Figure 16.3. Pairing-strength landscape of nucleic-acid alternatives. The squares that form the four platforms represent the diversity of (formally) possible alternatives of RNA containing four (tetrafuranoses), five (pentafuranoses and pentapyranoses), and six (hexapyranoses) carbon atoms per sugar unit. Squares under columns indicate which of the possible alternatives have been experimentally investigated; relative heights of columns stand for relative strength of base-pairing; different shading of the columns corresponds to different base-pairing languages.

under the conditions that formed RNA. The reason for their failure revealed itself in chemical model studies: six-carbon-six-membered-ring sugars are found to be too bulky to adapt to the steric requirements of Watson–Crick base-pairing within oligonucleotide duplexes. In sharp contrast, an entire family of nucleic acid alternatives in which each member comprises repeating units of one of the four possible five-carbon sugars (ribose being one of them) turned out to be highly efficient informational base-pairing systems. All of them contain their five-carbon sugar in the six-membered pyranose form, all of them show a base-pairing strength that is higher than that of natural RNA, and all of them possess duplex structures quite distinct from that of RNA and DNA (quasi-linear ladder structures; see Figure 16.4). They all are able to communicate by Watson–Crick cross-pairing among each other, but are unable to do so with RNA or DNA. They "speak another language."

More recently, the systematic exploration of alternative nucleic-acid systems from RNA's structural neighborhood has presented us with a further challenging observation. TNA, the threose-analog of RNA (a system containing only four instead of five carbon atoms in its sugar building block), turns out to be a marvelous Watson–Crick base-pairing system. Indeed, not only is TNA very similar to

RNA in its base-pairing properties, but it is also much more resistant to hydrolytic decay, and, above all (in sharp contrast to the pyranose form of RNA), it speaks the same "base-pairing language" as RNA and DNA (Schöning *et al.*, 2000). TNA is a provocative alternative nucleic-acid system, not only because of its RNA-like properties, but also because of its structural simplicity. With its four-carbons-only sugar building block, it is a simpler type of molecule than RNA (the simplest synthetic pathway to the four-carbon sugar threose requires one single C_2-starting material, that to a five-carbon sugar requires in addition a C_1-component). Not surprisingly, therefore, the question has been raised as to whether this system could have fulfilled the role of the, or an, ancestor of RNA (Orgel, 2000). The interest in such a conjecture may be seen to lie primarily in the question of what kind of experiments it induces. Rather remarkable results have already come forward in other laboratories from such experiments. Certain DNA polymerases were found to be able to faithfully transcribe TNA sequences into complementary DNA sequences and vice versa (Chaput and Szostak, 2003; Kempeneers *et al.*, 2003). While of interest from an etiological point of view, such observations also throw new light on the amazing structural versatility of modern DNA polymerases and may open the door to applying the methods of *in vitro* evolution to the nucleic-acid alternative TNA (Chaput and Szostak, 2003).

What lessons do the observations made so far on the base-pairing properties of potential nucleic-acid alternatives teach us? First, *the capability of Watson–Crick base-pairing is by no means unique to the natural nucleic acids* and is not bound to the type of double-helix structure occurring in DNA, as had been quite generally assumed before.[5] Furthermore, *nature did not choose RNA among possible alternatives by pursuing the strategy of maximizing base-pairing strength.* For instance, starting from the very same building blocks and using the same type of chemistry that led to RNA, nature could have easily arrived at the pyranose form of RNA (see Figure 16.4). Watson–Crick pairing would have been not only stronger, but even more selective than in the natural isomer. Pyranosyl-RNA is a system that would also have been more resistant to hydrolytic decay than RNA. Hence, this six-membered ring isomer of RNA is perhaps the most relevant reference point that we have today for identifying assets of natural RNA structure that made it the (putatively) superior system among the alternatives from RNA's structural neighborhood and, therefore, the one that became part of the biology we know today. Can we see hints of why nature chose one and not the other? Properties such as relative backbone flexibility (greater in RNA) and base-pairing strength (greater in pyranosyl-RNA) come immediately to mind

---

[5] This insight has been extensively supported and extended by the results of the contemporary search in medicinal chemistry for structural nucleic-acid variants that may have the potential of acting as antisense agents (see, for example, Hyrup and Nielsen, 1996; Herdewijn, 1996; Leumann, 2002).

(see Figure 16.2). High base-pairing strength can facilitate template-assisted growth of oligomer strands in duplexes, but can hamper replication because of stronger product inhibition. However, when we consider in detail the task of comparing and evaluating assets, it becomes clear that really nothing is straightforward. This is because chemical fitness does not necessarily parallel biological fitness, and we do not know how far life's evolution had advanced by the time RNA became selected. Take the difference in backbone flexibility between the pyranose form of RNA and natural RNA: in the former, two out of the six chemical bonds per backbone unit are rotationally constrained (part of a ring), whereas RNA has only one such constrained bond (see Figure 16.2). How can we know what this "small" constitutional difference would imply for the diversity of molecular shapes in long base sequences and, in turn, for the diversity of catalytic capabilities of these shapes and, ultimately, for the evolvability of an emerging biological system that had to operate with either of these alternative nucleic-acid systems? Such questions forcibly remind us of limitations associated with the task of looking at the molecular level for the existence of cosmophysical and physico-chemical fine-tuning toward the uniqueness of the life that we know. Although our experimental observations indicate that nature, in selecting the molecular structure of her genetic system, had other options available besides RNA, the notion we naturally would be inclined to consider – namely, that RNA might be the biologically fittest of them all – remains but a conjecture.

The work done thus far has been deliberately restricted to nucleic acid backbone structures that belong to the close structural neighborhood of RNA. There is an important reason for this. The closer the structural relationship between a chosen alternative and RNAs, the more reliably we can reason that the alternative represents a potentially natural molecular structure and thus could have been an evolutionary competitor of RNA (irrespective of whether the RNA structure is of prebiotic or biotic origin). The etiological relevance of nucleic-acid alternatives that are structurally *unrelated* to RNA is a much greater problem because of our increased uncertainty regarding their generation under environmental conditions that prevailed at RNA's origin. It nevertheless behoves us to ask whether nature could have had at its disposal genetic oligomers that are built not of sugar (phosphate) units, but of other types of building blocks. Inescapably, such inquiry will eventually touch on one of contemporary life's most fundamental features: the structural and functional separation of genotype and phenotype at the molecular level. From a chemical point of view, its realization in the course of evolution demanded so extraordinary an innovation – the entanglement of genotype and phenotype by means of the former codifying the synthesis of the latter – that its emergence from a non-living environment seems inconceivable, at least in the form that we know today. A conceptually attractive escape from the dilemma is the idea of an RNA

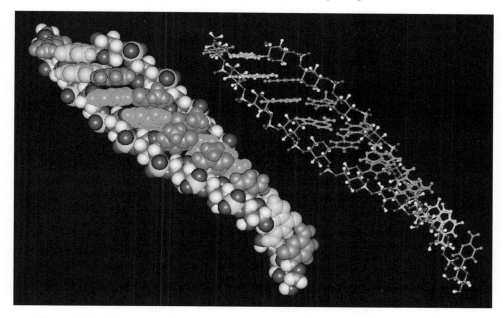

Figure 16.4. Structure of a pyranosyl-RNA duplex containing an oligonucleotide backbone built of ribose monomer units in their pyranose form. Note the drastic difference from the structure of a conventional RNA or DNA duplex.

world that preceded the present DNA/RNA/protein world (Woese, 1967; Crick, 1968; Orgel, 1968). This idea, since the time of its resurrection (Gilbert, 1986) in the wake of the discovery of ribozymes (Kruger *et al.*, 1982; Guerrier-Takada *et al.*, 1983), has received an impressive amount of empirical support (Gesteland *et al.*, 1999). To the chemist, its conceptual attractiveness resides in its replacing the need for a complex *chemical* information-transfer *between* molecules (taking place in today's ribosomes) with that for a ubiquitous type of *physical* process *within* molecules. In other words, in any RNA molecule that folds itself into a catalytically active structure, the constitution (base sequence) automatically "codes" for the molecule's conformation by reasons having their roots in molecular physics. This, in turn, co-determines the molecule's shape and, therewith, its functional (catalytic) capabilities.

The concept of the RNA world forcibly raises the question of RNA's origin. Experience over the years with the demanding requisites of the generational chemistry of the different types of oligonucleotides and their building blocks makes me share the belief of those who think that RNA could hardly have assembled itself and initiated an evolutionary process in a prebiotic world (Joyce and Orgel, 1999). This means that we must consider RNA as originating from a biotic environment, the chemobiological nature of which is completely unknown to us. To throw light on this unknown territory is a major challenge to chemistry. It may be nothing less than the quest for the chemical roots of life. Thus, the systematic chemical study of nucleic-acid alternatives, originally directed at the question of life's chemical uniqueness, will eventually have to merge with the search for the chemistry of life's origin.

The search for the chemistry of life's origin

The search for the chemistry of life's origin has been going on under the heading "prebiotic chemistry" since the advent of the Miller–Urey experiment half a century ago (Miller, 1953). Today, the elementary and potentially prebiotic nature of the basic building blocks of contemporary life's fundamental biopolymers, the alpha-amino acids, the sugars, and the nucleobases, is considered to be established (Oro and Kimball, 1960; Miller and Orgel, 1974). On a purely chemical level of experimentation, it has also been shown that oligonucleotide base sequences can be transcribed into complementary sequences without enzymes (Sulston *et al.*, 1968; Orgel, 1992) and that some short sequences are even able to replicate autocatalytically (von Kiedrowski, 1986; Orgel, 1992). However, these experiments were carried out under constraints and conditions that are so specific that they do not necessarily demonstrate the capability of nucleic acids to self-replicate autocatalytically under potentially prebiotic conditions. Remarkably, certain oligopeptides have also been shown to possess the capability of acting as templates that autocatalyze their own formation by ligating hemi-complementary oligopeptides, directed by sequence-dependent recognition factors such as lipophilicity and hydrophilicity of amino-acid side chains (Lee *et al.*, 1996). The exploratory chemical studies on self-replication in both the oligonucleotide and oligopeptide series reflect a classical dichotomy concerning the origin of the kind of life we know – namely, the old question of what might have been first: nucleic acids or proteins? Although this dichotomy is narrowly confined to structures that are known from contemporary biology, it reflects a broader (although not necessarily closely related) dichotomy that refers to function: should the search for the chemistry of life's origin focus on primordial genetic or primordial metabolic function? The "geneticists'" school of thought (Eigen, 1971) maintains that life began with the first successful informational replicator, whereas the "metabolists" (e.g. Wächtershäuser, 1990; Morowitz *et al.*, 2000) consider genetic function as a later evolutionary achievement of a life that began as autocatalytic-metabolic cycles. A third prejudice is propounded by the "compartmentalists," (e.g. Morowitz *et al.*, 1988) who emphasize the fact that all life on earth is cellular, and therefore autocatalytic systems were not alive until they operated in cellular compartments.

The literature concerning the problem of life's origin is littered with speculations. What is badly needed, if our scientific approach to this problem is to evolve, is experimental research that will create a platform of relevant new facts, facts that reflect the whole range of chemical possibilities. Not only is there no harm in letting such research proceed in the different directions that reflect the whole spectrum of existing (hopefully potentially transient) preconceptions, but this plurality is absolutely required at this stage of our ignorance. The overall strategy for continuing the search for the chemistry of life's origin is to conceive or discover, and above all experimentally explore, any structural and functional type of system that expresses elements of chemical matter's potential for self-organization. The harvest expected to result from concerted efforts of devoted geneticists, metabolists, and compartmentalists, each exploring their own molecular space of structural and functional diversity, may reveal to us which of these approaches (or perhaps alternatives, thus far unrecognized) deserves our ultimate focus.

In this concert of efforts, one of the geneticist's tasks is to extend comparative studies of potentially natural nucleic-acid alternatives into explorations of informational oligomers that are of a broader structural variety than merely alternative oligonucleotide systems taken from RNA's structural neighborhood. Informational oligomers with polyamide backbones are already known (Nielsen, 1993). From the chemical point of view, systems of lesser generational complexity can be envisaged and, among them, systems that contain nucleobases different from the canonical ones. In contrast to the strategy followed so far, a structurally unconstrained search for informational oligomers opens up such broad possibilities that it cannot remain systematic; yet the major criterion for selecting a given system for study, the potential for constitutional self-assembly, will remain the same. What we need is a map, based on experimental facts, of all structure types of molecular or supramolecular systems that could be relevant to the geneticist's "dream of a primordial replicator."

A comprehensive search pursued at the chemical level not only geneticists, but also by metabolists and compartmentalists, may eventually open up the central bottleneck to making progress in the origin-of-life field – namely, our uncertainty about the contingent geochemical boundary conditions of biogenesis. Assessing the requisites of an entire *library* of chemically validated candidate systems and juxtaposing it with a *library* of geologically and geochemically "allowed" boundary conditions might uncover coincidences that, in the best case, might allow us to pinpoint a specific pair of "system type" and "boundary-condition type" while excluding other combinations.

A strategy requiring so much effort and based on so many "ifs" may appear discouraging. Yet we should view it in light of the dimension of the problem, one that is perhaps best reflected in the oft-heard statement (with which, in principle, we all may be inclined to agree): "The origin of life, an event of billions of years ago, is unknowable." Yet this statement too is a preconception. Given that, in a strict sense, the origin of life may be not knowable, its type of chemistry must be reconstructable and experimentally exemplifiable.

In the introduction to this chapter, I pointed to the role of chemical synthesis as the organic chemist's major tool for understanding the molecular world. This statement specifically holds also for the organic chemist's participation in the search for the chemistry of life's origin. As organic chemistry as a whole has evolved through the study of molecules occurring in nature combined with the creation of molecules that have not existed before, the search for life's origin will be accompanied by an experimental search for self-organizational behavior in *artificial* chemical systems and, eventually, the synthesis of artificial chemical life. This research will be free of constraints such as potential naturalness and compatibility with geochemical boundary conditions, unburdened by the demands of having to refer to the past. The challenge of moving through design-induced discoveries towards a goal of such relevance will, when the time is ripe, attract mainstream chemists in larger numbers. It may indeed turn out that the most promising path toward a comprehension of the conundrum of life's origin consists in concerted efforts, launched by chemists, to create artificial chemical life forms.

References

Chaput, J. C. and Szostak, J. W. (2003). TNA synthesis by DNA polymerases. *Journal of the American Chemical Society,* **125**, 9274–5.

Crick, F. H. C. (1968). The origin of the genetic code. *Journal of Molecular Biology*, **38**, 367–9.

Eigen, M. (1971). Self-organization of matter and the evolution of biological macromolecules. *Naturwissenschaften*, **58**, 465–523.

Eschenmoser, A. (1999). Chemical etiology of nucleic acid structure. *Science*, **284**, 2118–24.

Eschenmoser, A. (2001). Design versus selection in chemistry and beyond. *Pontificiae Academiae Scientiarum Scripta Varia*, **99**, 235–51.

Ferris, J. and Hagan, W. J., Jr. (1984). HCN and chemical evolution: the possible role of cyano compounds in prebiotic synthesis. *Tetrahedron*, **40**, 1093 (review).

Gesteland, R. F., Cech, T. R. and Atkins, J. F. (1999). *The RNA World*, 2nd edn. Cold Spring Harbor, NY: Cold Spring Harbor Press.

Gilbert, W. (1986). The RNA world. *Nature*, **319**, 618.

Groebke, K. *et al.* (1998). Why pentose- and not hexose-nucleic acids? V. Purine-purine pairing in homo-DNA: guanine, isoguanine, 2,6-diaminopurine, and xanthine. *Helvetica Chimica Acta*, **81**, 375–474 (footnote 64 on p. 444).

Guerrier-Takada, C. *et al.* (1983). The RNA moiety of ribonuclease P is the catalytic subunit of the enzyme. *Cell*, **35**, 849–57.

Herdewijn, P. (1996). Targeting RNA with conformationally restricted oligonucleotides. *Liebigs Annalen der Chemie*, 1337–48.

Hyrup, B. and Nielsen, P. E. (1996). Peptide nucleic acids (PNA): synthesis, properties and potential applications. *Bioorganic and Medicinal Chemistry Letters*, **4**, 5–23.

Joyce, G. F. and Orgel, L. E. (1999). Prospects for understanding the origin of the RNA world. In *The RNA World*, 2nd edn., ed. R. F. Gesteland, T. R. Cech, T. R. and J. F. Atkins. Cold Spring Harbor, NY: Cold Spring Harbor Press, pp. 49–77.

Kempeneers, V. *et al.* (2003). Recognition of threosyl nucleotides by DNA and RNA polymerases. *Nucleic Acids Research*, **31**, 6221–6.

Kruger, K. *et al.* (1982). Self-splicing RNA: autoexcision and autocyclization of the ribosomal RNA intervening sequence of Tetrahymena. *Cell*, **31**, 147–57.

Lee, D. H. *et al.* (1996). A self-replicating peptide. *Nature*, **382**, 525–8.

Leumann, C. J. (2002). DNA analogues: from supramolecular principles to triological properties. *Bioorganic and Medicinal Chemistry*, **10**, 841–54.

Miller, S. L. (1953). A production of amino acids under possible primitive earth conditions. *Science*, **117**, 528–9.

Miller, S. L. and Orgel, L. E. (1974). *The Origins of Life on Earth*. Englewood Cliffs, NJ: Prentice-Hall.

Morowitz, H. J., Heinz, B. and Deamer, D. W. (1988). The chemical logic of a minimum protocell. *Origins of Life and Evolution of Biospheres*, **18**, 281–7.

Morowitz, H. J., Kostelnik, J. D., Yang, G. D. *et al.* (2000). The origin of intermediary metabolism. *Proceedings of the National Academy of Sciences, USA*, **97**, 7704–8.

Nielsen, P. E. (1993). Peptide nucleic acids (PNA): a model structure for the primordial genetic material. *Origins of Life and Evolution of Biospheres*, **23**, 323–7.

Orgel, L. E. (1968). Evolution of the genetic apparatus. *Journal of Molecular Biology*, **38**, 381–93.

Orgel, L. E. (1992). Molecular replication. *Nature*, **358**, 203–9.

Orgel, L. E. (2000). A simpler nucleic acid. *Science*, **290**, 1306–7.

Oro, J. and Kimball, A. P. (1960). Synthesis of adenine from ammonium cyanide. *Biochemical and Biophysical Research Communications*, **2**, 407–12.

Roberts, C. and Caserio, M. C. (1977). *Basic Principles of Organic Chemistry*, 2nd edn. Menlo Park, CA: W. A. Benjamin Inc., p. 77.

Schöning, K.-U., *et al.* (2000). Chemical etiology of nucleic acid structure: the α-threofuranosyl-($3' \rightarrow 2'$) oligonucleotide system. *Science*, **290**, 1347–51.

Sulston, J., *et al.* (1968). Nonenzymatic synthesis of oligoadenylates on a polyuridylic acid template. *Proceedings of the National Academy of Sciences, USA*, **59**, 726–33.

von Kiedrowski, G. (1986). A self-replicating hexadeoxynucleotide. *Angewandte Chemie (International Edition)*, **25**, 932. (In English.)

Watson, J. D. (1968). *The Double Helix*, 1st edn. New York: Atheneum, p. 190.

Wächtershäuser, G. (1990). Evolution of the first metabolic cycle. *Proceedings of the National Academy of Sciences, USA*, **87**, 200–4.

Woese, C. (1967). *The Genetic Code, the Molecular Basis for Genetic Expression*. New York: Harper and Row.

17

Fine-tuning and interstellar chemistry

William Klemperer

Our knowledge of the universe is steadily expanding. In large measure, this has been a result of radioastronomical observations. Among the most important is that the majority of the radiation of the universe is almost uniform and follows the spectral distribution of a thermal source at a temperature of 2.725 K.[1] This cosmic background radiation is the remnant of the initial event, the Big Bang. Although this radiation is essentially uniformly distributed in the universe,[2] the distribution of matter is highly non-uniform.

Matter, 99% hydrogen and helium, is found virtually entirely within galaxies. Galaxies occupy 10^{-7} of the volume of the universe, but contain most of the known matter. The origin of this separation of radiation and matter is a topic of much current study, as is the question of galaxy formation and the abundance and distribution of intergalactic matter. Although they are clearly fundamental to the question of the chemistry that occurs, it is not my intent or capability to discuss these most interesting questions (see Peebles, 1993). The heterogeneous distributions of matter occur universally. The *average* density within galaxies is 1 atom of hydrogen cm^3, whereas outside of galaxies estimates are of less than 1 atom of hydrogen m^3 in intergalactic regions. This sharp aggregation of matter means that the chemistry is occurring within galaxies. It is sensible to focus the discussion, therefore, primarily on the molecular abundances and the chemistry occurring in our galaxy, the Milky Way, because observations are much easier in view of the decrease of radiation intensity as the inverse square of the distance. Thus, most of the observations discussed here are of this galaxy. Although optical astronomical observations more than fifty years ago revealed the existence of molecules in the low-density matter

---

[1] For results from COBE (Cosmic Background Explorer), see http://lambda.gsfc.nasa.gov/product/cobe/.

[2] The anisotropy of the microwave cosmic background radiation is being measured by the Wilkinson Microwave Anisotropy Probe; see http://map.gsfc.nasa.gov/product/map/current/.

Fitness of the Cosmos for Life: Biochemistry and Fine-Tuning, ed. J. D. Barrow *et al.*
Published by Cambridge University Press. © Cambridge University Press 2007.

that exists between the stars (Adams, 1949), the subject has changed radically as a consequence of radioastronomical observations. A list is given in Table 17.1.

It is customary to regard chemistry and chemical processes in a utilitarian manner. Thus, the usual natural chemistry is one that occurs at the earth's surface, frequently in aqueous solution. We may further note, as seen in Table 17.2 below, that the cosmic abundance of the elements is strikingly different from the composition of the earth's crust. This difference is not simply reflected by that of hydrogen and helium. Of particular interest is the abundance ratio of carbon and oxygen, which is 1/2.3 cosmically and 1/250 terrestrially. (If carbon is the stuff of life, it would appear that the earth is not the optimal location for it.)

The techniques of analytical chemistry developed to quantitatively follow the molecular transformations that occur have evolved and have been refined over two and a quarter centuries following Lavoisier (Cobb, 2002). This natural chemistry is further aided by the vast body of empirical data built up during this period, which allows relatively facile interpolation or small extrapolation for the prediction of new molecular transformations. The jump to encompassing the chemistry of the universe is large. The techniques for the *in situ* analysis of molecular abundances are essentially those of remote sensing, in particular spectroscopy. Seeing in absorption provides the ability to determine both species and their amount. It requires a light source, generally a star, located beyond the region examined. The requirement for absorption, that the temperature of the light source be greater than that of the absorber, is generally met with ease. The requirement that light pass through the sample is frequently not met. The absorption law is $I(\nu)/I_0(\nu) \propto \exp[-\kappa(\nu)n]$, where $I(\nu)/I_0(\nu)$ is the fraction of transmitted light intensity at the frequency ν, $\kappa(\nu)$ is the absorption coefficient, and n is the column density (units cm^2). When $\kappa(\nu)n \geq 4$, the absorber is essentially opaque (at the frequency ν), making discrimination of column density, or sometimes even species, not possible.

Absorption spectroscopy in the visible and near-ultraviolet region of the spectrum dominated molecular astronomy in the three decades from 1937 to 1968. The species CH, CH^+, and CN, observed by using bright hot stars as light sources, were seen in the spectra of many stars. The spectra of these species, as shown in Figure 17.1, establish that the population in the quantum levels is strongly biased toward the lowest levels. The population distributions are fit by a temperature of 2.73 K (Bortolot *et al.*, 1969).[3]

This temperature is that of the cosmic background radiation.[1] The species have reasonably large electric dipole moments; for CH and CN, measurement gives

---

[3] The very low rotational temperature of interstellar species was pointed out by McKellar (1942). Also see Herzberg (1950), p. 496.

Table 17.1. *Observed interstellar and circumstellar molecules*

The species listed are observed by their rotational emission spectra unless otherwise noted.

Number of atoms					
2	3	4	5	6	7
H_2	C_3 v	c-C_3H	C_5 v	C_5H	C_6H
AlF	C_2H	l-C_3H	C_4H	l-H_2C_4	CH_2CHCN
AlCl	C_2O	C_3N	C_4Si	C_2H_4 v	CH_3C_2H
C_2 e	C_2S	C_3O	l-C_3H_2	CH_3CN	HC_5N
CH	CH_2	C_3S	c-C_3H_2	CH_3NC	$HCOCH_3$
CH^+ e	HCN	C_2H_2 v	CH_2CN	CH_3OH	NH_2CH_3
CN	HCO	HCCN	CH_4 v	CH_3SH	c-C_2H_4O
CO	HCO^+	$HCNH^+$	HC_3N	HC_3NH^+	H_2CCHOH
CO^+	HCS^+	HNCO	H_2CNC	HC_2CHO	
CP	HOC^+	HNCS	HCOOH	NH_2CHO	
CSi	H_2O	$HOCO^+$	H_2CHN	C_5N	
HCl	H_2S	H_2CO	H_2C_2O	l-HC_4H v	
KCl	HNC	H_2CN	H_2NCN		
NH e	HNO	H_2CS	HNC_3		
NO	MgCN	H_3O^+	SiH_4 v		
NS	MgNC	NH_3	H_2COH^+		
NaCl	N_2H^+	c-SiC_3			
OH	N_2O	CH_3 v			
PN	NaCN				
SO	OCS				
SO^+	SO_2				
SiN	c-SiC_2				
SiO	CO_2 v				
SiS	NH_2				
CS	H_3^+ v				
HF	H_2D^+				
SH v	SiCN				
HD	AlNC				
FeO					

Number of atoms					
8	9	10	11	12	13
CH_3C_3N	CH_3C_4H	$(CH_3)_2$CO	HC_9N	C_6H_6 v	HC_{11}N
$HCOOCH_3$	CH_3CH_2CN				
CH_3COOH	$(CH_3)_2$O				
C_7H	CH_3CH_2OH				
H_2C_6	HC_7N				
l-HC_6H v	C_8H				

Abbreviations: c, cyclic isomer of the C_3 hydrocarbons; l, linear isomer of the C_3 hydrocarbons; v, observed by vibration–rotation spectrum; e, observed by electronic spectrum.

Note: This table is no longer current. Important new classes of molecules have been observed. See the well-maintained Cologne Database for Molecular Spectroscopy: http://www.ph1.uni-koeln.de/vorhersagen/.

Table 17.2. *Comparison of cosmic composition and earth's crust*
The fourth column is from Cox, P. A. (1997). *The Elements*. Oxford:
Oxford University Press. The second and third columns are derived
from literature values. The ratios are original.

Element	Cosmic abundance	Number %	Earth's crust
H	330 000	—	0.22
He	46 000	—	—
C	100	24.7	0.19
N	31	7.6	0.002
O	235	58.2	46.6
F	0.01	0.002	0.06
Ne	91	22	—
Na	0.6	0.15	2.8
Mg	10.6	2.6	2.1
Al	0.8	0.19	8.1
Si	9.9	2.4	27.1
P	0.1	0.02	0.1
S	5.1	1.3	0.06
Cl	0.1	0.02	0.02
Ar	2	0.5	—
K	0.03	0.007	0.9
Ca	0.8	0.19	5.2
Ti	0.04	0.01	0.5
Cr	0.18	0.04	0.02
Fe	8	2.2	6.7
Ni	0.5	0.12	0.01

$\mu \approx 1.5$ D,[4] whereas calculation[5] estimates a somewhat larger moment for CH^+. The radiative lifetime of excited molecular levels of the two hydrides is relatively short, essentially less than one hour. At low densities, the time between effective collisions leading to excitation is much longer than the radiative lifetime. Thus, the low internal molecular temperature of the dipolar species established a low density for the interstellar translucent (diffuse) clouds. This low density couples well with a picture of atomic dominance in the interstellar medium.

The elemental composition of the universe, the cosmic abundance of the elements, is essentially 99% hydrogen and helium. The question of whether a region is atomic or molecular essentially reflects the condition of hydrogen. Among the most significant discoveries or inventions was the detection of the hyperfine transition of atomic hydrogen in 1951 at 1420 MHz (Ewen and Purcell, 1951).[6] This radio emission provided the tool for seeing cold hydrogen throughout the

[4] Phelps and Dalby, 1966: CH, $\mu = 1.46$ D; Thompson and Dalby, 1968: CN, $\mu = 1.45$ D.
[5] Green, 1974: CH^+, $\mu = 1.70$ D.   [6] These transitions were predicted by H. C. van de Hulst in 1944.

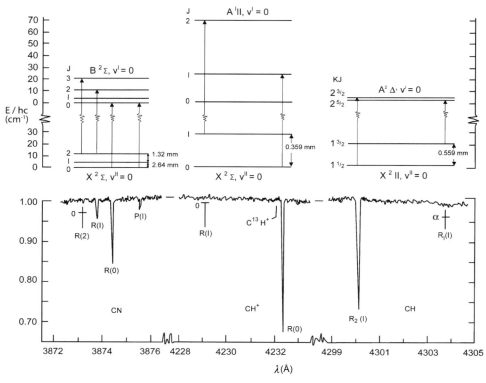

Figure 17.1. Optical spectrum of ζ Ophiuchi cloud. Reprinted with permission from Bortolot, J., Clauser, J. F. and Thaddeus, P. (1969). Upper limits to the intensity of background radiation at $\lambda = 1.32, 0.559$, and 0.359 mm. *Physical Review Letters*, **22**, 307. Copyright © 1969, American Physical Society.

universe – and, in this sense, mapping matter in the universe. The excitation of this hyperfine transition by collision at a rate faster than its radiative lifetime occurs even in regions of very low density. The picture of the universe as consisting of the highly visible stars at high densities together with a cold, low-density, essentially atomic interstellar medium fitted observations well.

An abrupt change in our picture of the universe occurred in 1968 with the detection of radio emissions from the inversion spectrum of ammonia in the direction toward the galactic center (Cheung *et al.*, 1968). Because ammonia has a large dipole moment, its rotation–inversion energy levels would be coupled to the background radiation field at low density. Thus, specific sharp line emissions could not observably occur. The discovery of high-density regions, where hydrogen is molecular and a number of molecules emit their characteristic radio frequencies, produced the present picture of the molecular universe (Rank *et al.*, 1971). It was noted that, in regions of the galaxy where considerable optical obscuration occurred, emission of the characteristic 1420 MHz of atomic hydrogen was weak or absent. These regions of the galaxy have become increasingly important.[7]

Observing the molecular universe by radioastronomy offers numerous advantages. The low energies of transitions between rotational molecular energy levels allow relatively cool regions to exhibit thermal emission. The scattering of radiation by particles varies as ν^4, so that regions opaque in the visible are quite transparent at centimeter and even submillimeter wavelengths. This provides the means of looking into the dense, visibly opaque molecular clouds. Finally, rotational transitions are theoretically the simplest molecular spectra to analyze. The spectra of semirigid species may be fitted reasonably well with three constants, the principal moments of inertia of the species (Townes and Schawlow, 1975).

Furthermore, these three constants may even be estimated, for species for which laboratory spectra do not exist, from the existing body of knowledge on molecular structure. An early example of this is shown in the story of Xogen, or HCO^+. This species was observed by a transition at 89.190 GHz (Buhl and Snyder, 1970), which did not correspond to any known laboratory spectrum; hence, it was named Xogen. The very high spectral resolution of radio-frequency spectroscopy showed no hyperfine splitting, ruling out the possibility that Xogen was either a radical or a species that contained nitrogen. The spectral transition fitted well what was expected for the lowest transition of HCO^+ (Klemperer, 1970), a species isoelectronic to the well-studied and already observed molecule HCN. The species HCO^+ is the dominant ion in dense molecular clouds. The proof that Xogen is HCO^+ was demonstrated by its laboratory spectrum (Woods *et al.*, 1975).

Radioastronomy has established the molecular character of large, important portions of the universe. Hydrogen is, of course, the most abundant molecule. Following it in abundance is carbon monoxide. The facile mapping of matter by observing molecular hydrogen is virtually impossible. The low-energy pure quadrupole rotational transitions are highly forbidden, as well as relatively high in energy, with the first transition, $J = 2 - J = 0$, occurring at 360 cm^{-1}. In addition, they occur at inconveniently high frequencies detectable in the far-infrared range and require satellite observatories. The high energy of the lowest excited emitting level, $J = 2$, then requires a heated source. For typical molecular clouds at a temperature of 20 K, the emission would be essentially negligible. It is noteworthy that the vibrational transition of H_2 is observed in strongly heated spatial regions. This transition, occurring in an atmospherically transparent region, around 4500 cm^{-1}, has been well studied in hot regions of the Orion molecular cloud (Gautier *et al.*, 1976).

---

[7] An excellent description of this discovery is available at the IEEE History Center: http://ieee.org/web/aboutus/ history_center/oral_history/oh_sh_z.html; Charles H. Townes (#143).

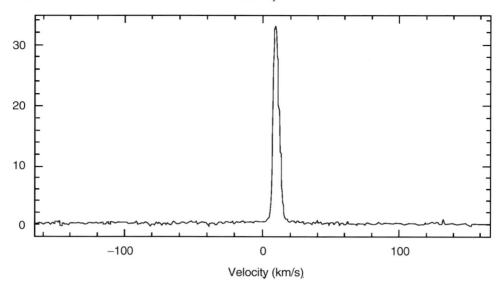

Figure 17.2. J = 1 − 0 transition of CO in Orion. The rest frequency is 115,794.24 MHz. This spectrum was obtained with a 4-foot-diameter telescope and 10 min integration. Courtesy of Patrick Thaddeus, Harvard University. Reprinted with permission.

The observation of molecular hydrogen by means of its electronic transitions in a sense follows classical optical interstellar spectroscopy. It is, however, considerably more complex, requiring essentially controlled satellite observatories. Thus, it serves to determine molecular-hydrogen column densities in translucent clouds, but cannot provide images of the dense molecular clouds. For these, carbon monoxide is the generally accepted tool. The reported results are in terms of H_2 column densities under the assumption that the H_2 : CO ratio is the accepted value of 10^4. CO is observable by means of its many isotopomers. This is extremely useful, as the common isotopomer $^{12}C^{16}O$ is frequently optically opaque, making its emission intensity independent of column density. The much less abundant isotopic forms $^{13}C^{16}O$, $^{12}C^{18}O$, and even $^{13}C^{18}O$ are then used for producing maps of molecular abundance. A final important virtue of carbon monoxide for producing radio-frequency images is its very small electric dipole moment of 0.1098 D (Muenter, 1975). This ensures the weak coupling of the CO rotational transitions to the microwave cosmic background radiation field, permitting the collisional excitation of rotational transitions in relatively low-density regions, and thus their mapping. Figure 17.2 shows the lowest-frequency rotational transition of CO observed in the Orion Molecular Cloud.

The discovery, starting in the 1970s, that the universe is richly molecular rather than sparsely atomic is a striking triumph of radioastronomy. Now more than

120 species have been positively identified by high-resolution rotational spectroscopy, where the quality factor of the line (line frequency/linewidth) frequently exceeds 10^6. This high quality factor allows very accurate species identification in comparison with laboratory spectroscopy. The polyatomic molecules attract the greatest chemical and biological interest, although in amount it is the diatomic species H_2 that is preponderant. The present largest positively identified molecular species, $HC_{10}CN$, contains 12 heavy atoms. These species are listed in Table 17.1.[8] An enjoyable reference to the discoveries of molecular astronomy is also available.[9] An abridged list of the cosmic abundance of the elements and the contrast to the earth's crust is presented in Table 17.2.

The majority of the observed polyatomic species contain carbon. Because the ratio of hydrogen to carbon in the universe is 3×10^3, the nature of the chemistry, or better chemical processes, leading to this preference for organic polyatomic species in the interstellar medium can be examined for sensitivity – or, in effect, fine-tuning. Is this a chance development or a predictable consequence of the electron structure of matter? We can further ask, in view of the abundance of hydrogen, why among the observed organic species there is such a preponderance of unsaturated compounds.

The galaxy is clearly not uniform in atomic density. The average density is 2×10^{-24} g cm^{-3} or 1 atom cm^{-3}. This rich chemistry primarily occurs in the dense molecular clouds. In developing a scheme for the formation of the species listed in Table 17.1 above, the standard method would be to first simply follow equilibrium arguments. The observed species in Table 17.1 show clearly that equilibrium thermodynamic constraints are inappropriate, because in some instances high-energy isomeric forms of species are quite abundant. Note that both HCN and its high-energy isomer HNC are observed. The observed relative abundance $\frac{[\text{HNC}]}{[\text{HCN}]}$ is near unity in cold molecular clouds. Using the canonical temperature of 20 K, the equilibrium abundance ratio $\frac{[\text{HNC}]}{[\text{HCN}]}$ predicted is 10^{-250}. Further evidence against using an equilibrium model for the prediction of the chemistry and molecular abundances is found by considering the reaction

$$CO + 3H_2 = CH_4 + H_2O. \tag{17.1}$$

At 20 K, the equilibrium constant, $K = \frac{[\text{CH}_4][\text{H}_2\text{O}]}{[\text{CO}][\text{H}_2]^3}$, is readily calculated from standard enthalpies to be 10^{490} molecules^{-2} cm^6. For $[H_2] = 10^5$ cm^{-3}, a value typical of the molecular clouds, the predicted ratio of CO/CH$_4$ is 10^{-500}. Because CO is the second most abundant molecule, this rules out arguments of chemical equilibrium, which disagree with observation by 500 orders of magnitude, for the useful prediction of the observed molecular abundances.

---

[8] The sites www.ph1.uni-koeln.de/vorhersagen/ and www.cv.nrao.edu/~awootten/allmols.html are current sources for observed interstellar molecules.
[9] See www.molres.org/astrochymist/astrochymist_ism.html.

Figure 17.3. The Horsehead nebula in Orion.

This then requires specific kinetic routes for molecule formation (Herbst and Klemperer, 1973). The specific chemistry requires an energy source since the cloud temperature is 10–20 K. The gas density is near 10^6 molecules cm^{-3} and is 99% hydrogen and helium. The beautiful Horsehead nebula in Orion, a rich molecular region, is shown in Figure 17.3; the molecular cloud Barnard 68 is shown in Figure 17.4.[10]

The clouds have high optical opacity, and thus starlight will not deeply penetrate. The initiating energy input step of the reaction scheme is volume ionization by high-energy cosmic rays ($E \geq 100$ Mev). Because H_2 and He make up 99% of the gas, only their ionization need be considered in this primary step. Quite specifically:

$$H_2 + crp = H_2^+ + e \qquad (17.2)$$
$$He + crp = He^+ + e \qquad (17.3)$$

where crp = cosmic-ray proton (or particle).

Under the assumption that the cosmic-ray flux is relatively constant throughout the galaxy and the observable universe, the ion production then is proportional to the gas density. This rate coefficient (for dense clouds, i.e. 100 Mev crp) is 10^{-17}. The secondary reactions following the initial ionization set the chemical behavior.

$$H_2^+ + H_2 = H_3^+ + H \qquad (17.4)$$

[10] These excellent pictures of the dark molecular clouds are readily available. For the Horsehead nebula in Orion, see www.noao.edu/image_gallery/html/im0661.html; for the molecular cloud Barnard 68, see http://antwrp.gsfc.nasa.gov/apod/ap990511.html.

Figure 17.4. Molecular cloud Barnard 68. Figures 17.3 and 17.4, photographs courtesy of National Optical Astronomy Observatory/Association of Universities for Research in Astronomy/National Science Foundation.

is a fast exoergic reaction occurring on every collision, and thus its rate constant is the readily calculable collision frequency $= 2 \times 10^{-9}$ s^{-1} cm^3. At the typical molecular cloud density of 10^6, H$_2^+$ is converted to H$_3^+$ in about 10 min, essentially instantaneously. H$_3^+$ reacts with CO, the second most abundant molecule, forming HCO$^+$, which, being the most abundant ion, is readily observed. H$_3^+$ is observed by its vibrational spectrum, now in many sources (see, for example, McCall *et al.*, 1999).

The highly exoergic reaction

$$\text{He}^+ + \text{H}_2 = \text{reaction products} \tag{17.5}$$

essentially does not occur at low temperatures, with limits of less than 10^{-5} per collision. Thus, the highly energetic species He$^+$ is stable in the cold hydrogen atmosphere, but reacts on every collision with the second most abundant molecule, carbon monoxide:

$$\text{He}^+ + \text{CO} = \text{C}^+ + \text{O} + \text{He} \tag{17.6}$$

The carbon ion is responsible for the rich organic chemistry observed. The production of C$^+$ by the route indicated (17.6) is at least 500 times (the ratio of He/C cosmic abundance) more efficient than direct cosmic-ray ionization of CO, the predominant interstellar carbon source – hence, the rich organic chemistry observed in the interstellar medium. The efficient production of C$^+$ has as origin the lack of reactivity of He^+ with H_2. The rich organic chemistry is a direct consequence of helium (ion) chemistry.

In now examining fine-tuning of the universe, we may examine why the rate of (17.5) is so slow. This reaction was discussed in detail by Mahan (1975), who pointed out that the equivalent reaction with argon – $Ar^+ + H_2 = ArH^+ + H$ (and other products) – is fast proceeding at the Langevin rate as expected for ion–molecule reactions. Thus, the answer to the question of whether it is possible to change the fundamental constants of the electron so that the He energy levels resemble those of argon appears to be no.

The synthesis of organic molecules by means of carbon ions can be considered. A relatively long series of essentially bimolecular reactions are suggested, as listed below. Energetics and rate constants for the species and processes listed are available (Le Teuff *et al.*, 2000). Listed below are some of the reactions in the chain leading to the formation of C_3H_n where $n = 1$ and 2. The efficiency of the synthesis depends on the ease of production of C^+ in the dense molecular clouds. The slow radiative association rate of C^+ with H_2 starts the chain of reactions:

$$C^+ + H_2 = CH_2^+ + h\nu \text{ (slow radiative association)} \qquad (17.7)$$
$$C + H_3^+ = CH^+ + H_2; CH_2^+ + H \qquad (17.8)$$
$$CH^+ + H_2 = CH_2^+ + H \qquad (17.9)$$
$$CH_2^+ + H_2 = CH_3^+ + H \qquad (17.10)$$
$$CH_3^+ + H_2 = CH_5^+ + h\nu \text{ (radiative association)} \qquad (17.11)$$
$$CH_5^+ + CO = CH_4(\checkmark) + HCO^+ \qquad (17.12)$$
$$CH_4 + C + = C_2H_2^+ + H_2 \qquad (17.16a)$$
$$= C_2H_3^+ + H \qquad (17.16b)$$
$$C_2H_2^+ + e = C_2H(\checkmark) + H \qquad (17.17)$$
$$C_2H_3^+ + e = C_2H + H_2 \qquad (17.18a)$$
$$= C_2H_2(\checkmark) + H \qquad (17.18b)$$
$$C_2H_2 + C^+ = C_3H^+ + H \qquad (17.19)$$
$$C_3H^+ + H_2 = C_3H_3^+ + h\nu \text{ (radiative association)} \qquad (17.20)$$
$$C_3H_3^+ + e = C_3H_2 \text{ cyclic}(\checkmark) + H \qquad (17.21a)$$
$$= C_3H_2 \text{ linear}(\checkmark) + H \qquad (17.21b)$$
$$= C_3H \text{ cyclic}(\checkmark) + H_2 \qquad (17.22a)$$
$$= C_3H \text{ linear}(\checkmark) + H_2 \qquad (17.22b)$$

The species labeled ($\checkmark$) have been observed in interstellar molecular clouds. The four products of dissociative recombination (17.21, 17.22a,b) are all observed. The cyclic C_3H_2 is a very abundant interstellar species.

Among the unobserved species is the highly symmetrical, very stable ion $C_3H_3^+$. The cyclopropenyl ion is regarded as the first aromatic hydrocarbon. As a consequence of its center of symmetry, it does not have a permanent electric dipole moment and therefore cannot be observed readily by its rotational emission spectrum. The deuterated species $C_3H_2D^+$ will have an observable radio spectrum, because the center of charge does not coincide with the center of mass. Although the cosmic abundance of deuterium is only 10^{-5} that of hydrogen, a very high degree of isotopic enrichment is observed in the cold molecular clouds. It is thus highly likely that the rotational spectrum of this important, but as yet not detected, reaction intermediate will soon be produced in the laboratory, allowing radioastronomical searches.

In this reaction scheme, CH_4 is produced by two steps of radiative association with slow rate constants. Because the destruction of C^+ by electron capture (radiative) is four orders of magnitude slower than the destruction of a molecular ion by dissociative recombination, there is not a rapid loss of C^+, allowing production of a saturated hydrocarbon. We recall that chemical equilibrium arguments predict preponderant conversion of carbon monoxide to methane and water. There is little evidence for this, as stated earlier. The gas-phase production of CH_4 from CO and H_2 then proceeds by a very high-energy kinetic path, namely $He^+ + CO = C^+ + O$. The production of organic hydrocarbons in this model then depends on carbon ion insertion reactions. A number of these have been measured in the laboratory and have been shown to proceed at essentially collision frequency. The production of C_3 hydrocarbons starting with C^+ is delineated in the set of reactions 17.7–17.22 listed above.

A natural question is whether this epoch is unique with respect to the rich organic chemistry observed. At first sight, based on the above discussions, this is primarily a question of the abundance of helium and carbon. Without detailed examination, it appears that the production of C^+ from CO – or, for that matter, from other carbon-containing species – will occur for a wide range of He and C interstellar abundances. This is probably an oversimplification. The ion–molecule chain of reactions suggested would be significantly interrupted by reaction with gas-phase oxygen (in the form of H_2O, CO_2, O_2, or probably even O). The present cosmic abundance ratio of O/C is 2.3. With gas-phase carbon held essentially in the carbon monoxide repository, the excess of oxygen is large and, if present in the gas phase, would disrupt the growth of carbon chains. The quickest remedy to this disruption of the synthesis scheme is to bind the oxygen into non-volatile substances, essentially refractory oxides (Scappini *et al.*, 2003).[11]

The chemical nature of the presumed condensed-phase repository is discussed later. We first examine measurements of abundances of the species H_2O, O_2, and CO_2. Recent observations of H_2O in cold molecular clouds from the Submillimeter

---

[11] See isowww.estec.esa.nl/science/ for many infrared spectra of the interstellar medium. In many of these, features of the interstellar dust are clearly shown.

378 — William Klemperer

Wave Astronomy Satellite (SWAS)[12] show the abundance to be less than 0.1% that of CO. At this level, it does not interfere with organic synthesis. With elemental evolution, an increase of the O/C ratio is expected. If this is not readily incorporated into the refractory solid phase, production of organic species in the interstellar molecular clouds could well be reduced. CO is an abundant molecule in strongly red shifted quasars (Downes and Solomon, 2003) ($z = 2.6$–6.4). Thus, it would be expected that its reaction products also are present, but harder to observe. SWAS is instrumented to measure O_2 abundances and has not observed any. (O_2, although lacking an electric dipole moment, has magnetic dipole transitions.) Note that it has probably not been observed in dense molecular clouds.[12]

The power of ion–molecule chemistry can be further illustrated by showing the detection of non-polar species by radio observations. The non-polar species N_2, isoelectronic and quite similar to CO, is readily observed by its polar protonated form HN_2^+ (Turner, 1974; Green et al., 1974). The reaction forming this species $N_2 + H_3^+ = HN_2^+ + H_2$ is identical to that forming HCO^+. HN_2^+ is an abundant molecular ion detected in many regions of the galaxy, as well as in other galaxies (Solomon et al., 1992). It provides a means of estimating the abundance of its non-polar parent N_2, which is the dominant form of nitrogen in molecular clouds. In a similar manner, CO_2 abundance also may be determined by observing its protonated form HCO_2^+ (Thaddeus et al., 1981). CO_2 is a potential repository of the excess oxygen. Thus, directly determining it in cold molecular clouds is of considerable importance. HCO_2^+ is formed by the reaction $H_3^+ + CO_2 = HCO_2^+ + H_2$. The reactions of both HCO_2^+ and HN_2^+ with CO are exothermic and fast. In dense molecular clouds, H_3^+ is primarily destroyed by CO, as HN_2^+ and HCO_2^+ will be. The very low observed abundance of HCO_2^+ in dense cold molecular clouds establishes that gas-phase CO_2 is not an important repository for excess oxygen.

Finally, the question of atomic oxygen's being the dominant oxygen species in dense cold molecular clouds can be examined. This would effectively require a ratio O/CO ≥ 1. The reaction chain

$$O + H_3^+ = OH^+ + H_2 \quad k = 8 \times 10^{-10}$$
$$OH^+ + H_2 = OH_2^+ + H \quad k = 1 \times 10^{-9}$$
$$OH_2^+ + H_2 = OH_3^+ + H \quad k = 6 \times 10^{-10}$$

leads to easy production of the very stable ion OH_3^+, an observed interstellar species. The further products of its dissociative electron attachment OH and H_2O are not observed as major species in dense molecular clouds.

The gas-phase constituents that are likely candidates to be the excess oxygen repositories have been examined and found to be of insignificant abundance in

[12] See cfa-www.harvard.edu/swas/. SWAS is designed to survey O_2 and H_2O in the interstellar gas clouds.

Table 17.3. *A suggested budget for the removal of excess*
oxygen

Element	Relative abundance	Compound	Stored oxygen
C	100	—	—
O	235	—	—
Mg	10.6	$MgCO_3 \cdot 3H_2O$	63
Si	9.9	$Si(OH)_4$	40
S	5.1	$H_2SO_4 \cdot H_2O$	25
Fe	8.9	Fe_2O_3	13

molecular clouds. If the excess oxygen is not in the gas phase, it is in the condensed phase. Thus, the budget for forming refractory oxides is considered. The major elements that will serve as suggested refractory repositories for the excess oxygen can be extracted from Table 17.3.

Note that these standard substances are adequate to effectively remove the excess oxygen. Sulfuric acid is included as a refractory substance. It has not been observed as a gas-phase species, but might be observed in infrared spectra of clouds. It is likely that most of the chemistry forming these species occurs in hot, dense stellar envelopes during their mass loss.

The chemistry in strongly heated regions changes appreciably (Walmsley, 1993). Table 17.4 shows the dramatic effect of heating in Orion. The strong reduction of HCO^+ shows the likely proton transfer to H_2O, forming H_3O^+.

Water vapor is abundant in the heated region of Orion. Of particular interest is the three orders of magnitude increase of H_2S in Orion's hot core. This is most likely the result of the reaction of H_2SO_4 with hot atomic hydrogen. Finally, the large increase in HCN relative to HNC is plausibly accounted for by the exothermic reaction $HNC + H = HCN + H$. The essential equality in abundance of HNC and HCN in cold molecular clouds is the consequence of the dissociative electron attachment process forming both, $HCNH^+ + e$. The $HCNH^+$ ion is an observed interstellar species. That in the strongly heated Orion core the high-energy isomer, HNC, decreases relative to the ordinary stable form HCN shows that the thermodynamic arguments of LeChatelier's principle are not applicable in a strongly kinetically determined chemistry. It appears likely that the converting reaction is the exothermic transformation $HNC + H = HCN + H$. It would be interesting to see laboratory evidence such as a comparison of the isotopic exchange reactions $HNC + D = DCN + H$ with $HNC + D_2 = DCN + HD$.

The cosmic ray ionization driving dark cloud chemistry could be different. The production and energy distribution of cosmic rays are thought to depend on a high-energy event, such as a supernova or accretion onto a black hole. How fortuitous

Table 17.4. *Comparison of molecular composition in cold and warm interstellar regions*

	Orion ridge	Orion hot core	TMC1	L183
T/K	40	200	10	10
H_2 cm^{-3}	10^6	10^7	10^4	10^4
N[CO] cm^{-2}	8×10^{18}	—	8×10^{17}	7×10^{17}
Log $N/[nH_2]$				
Log[CO/H_2]	-4	-4	-4	-4
Log[SiO/H_2]	-9.3	-6.7	<-11.6	<-11.4
SO	-8.7	—	-8.3	-7.7
SO$_2$	-8.4	—	<-9	-8.5
CS	-8.4	<-9.5	-8.0	-9.1
H_2S	-8.7	-5.3	<-9.3	-8.5
HCN	-8.0	-6.0	(-8.0)	-8.4
HNC	-8.6	(-7.9)	-7.3	-8.2
HCO$^+$	-8.5	<-9.5	-8.1	8.1
NH$_3$	-7.5	-6.2	-7.4	-6.9
H_2CO	-7.5	-7.6	-7.7	-7.7

Source: Walmsley, C. M. (1993). Abundance anomalies in hot and cold molecular clouds. *Journal of the Chemical Society Faraday Transactions*, **89**, 2119. Reprinted with permission.

the present production rates are is beyond the scope of this essentially chemical discourse. The chemistries are relatively stable with respect to ionization flux. Because the ions are ultimately destroyed by their very fast dissociative recombination with electrons, this leads to an ion abundance varying as the square root of the ionizing flux, which is taken to mean a highly stable system. Although the abundance of a specific ion will show variations depending on cloud conditions, the total ion abundance in dense molecular clouds should be 10^{-5} [H_2]$^{1/2}$.

The picture of a gas-phase chemistry in the dense molecular clouds, initiated by cosmic ray ionization of H_2 and He, has been greatly strengthened by the direct observation of the very early reactant H_3^+ (Oka, 1992). The identification and determination of the spectrum of H^+ have been a great triumph of laboratory studies (Oka, 1980). This ion is now observed in absorption in many sources by means of its vibrational spectrum (McCall *et al.*, 1998), as well as in planetary atmospheres in emission (Oka, 1992). Its abundance allows critical exploration of ionizing fluxes, primarily cosmic rays, and further provides an exceptional sight into dense molecular clouds.

Perhaps the most dramatic consequence of removing CO from the gas phase by freezing of the dust grains in very cold prestellar regions is the observation of extremely enhanced deuterium abundances in molecules. CO destroys H_3^+ by

Table 17.5. *Molecules observed in other galaxies*

Number of atoms				
2	3	4	5	6+
CO	HCN	NH_3	C_3H_2	CH_3OH
CH	HNC	HNCO	HCCCN	CH_3CN
CN	HCO^+	H_2CO		CH_3CCH
OH	HN_2^+			
SO	H_2O			
SiO	OCS			

the efficient reaction forming HCO^+. The reactions $HD + H_3^+ = H_2 + H_2D^+$, $HD + H_2D^+ = H + HD_2^+$, $HD + HD_2^+ = D_3^+ + H_2$ are exothermic. In these extremely cold regions, although the cosmic abundance ratio is $D/H = 2 \times 10^{-5}$, the highly deuterated forms of H_3^+ become the most abundant (Roberts *et al.*, 2003). The asymmetrically deuterated species H_2D^+ and HD_2^+ have appreciable dipole moments with respect to the center of mass and thus provide allowed electric dipole rotational transitions. H_2D^+ (Stark *et al.*, 1999) and H_2D^+ (Vastel *et al.*, 2004; Hirao and Amano, 2003) have been observed in emission in dense, cold star-forming regions of molecular clouds. Perhaps this is the most prosaic way of dramatically changing organic interstellar chemistry – by simply freezing out the carbon repository, carbon monoxide. In these cold regions (Barnard Cloud 1), the species ND_3 (Lis *et al.*, 2002) is observed at an abundance near 10^{-3} of NH_3. This is a dramatic display of the isotopic enrichment afforded by facile ion molecule reactions. The formation of N^+ is by the helium ion reaction $N_2 + He^+ = N^+ + N + He$. Both the destruction rate of the H_3^+ species and that of the He^+ are appreciably reduced with decreased CO.

I have sketched a small portion of the gas-phase chemistry that produces the observable species. The extremely high quality factor of the radioastronomical observed spectral lines coupled with the large number of observable transitions for each species has resulted in unambiguous identification of more than one hundred molecular species. In further confirmation, isotopic variants, designated "isotopomers" for lack of a better word, are found for virtually every relatively abundant species. The observations and species listed are within the galaxy and do not give the isotopomers. That this chemistry is universal is shown by the extensive list of molecules observed in other galaxies (Table 17.5) (Bertoldi *et al.*, 2003).

This list is certainly similar to the abridged list in Table 17.1. The species HCN, HNC, HCO^+, and HN_2^+ are indicative of the essential role of ion molecule chemistries. Because the above species are observed at epochs other than the present, it appears likely that no special fine-tuning is required for the molecular chemistry observed throughout the galaxy and the universe.

The anthropomorphic question of the connection and relevance of interstellar chemistry to the existence and origin of life on earth is not touched on in this discussion, although a high premium has been placed on establishing this connection and thereby creating relevance. The earth has conditions and chemistries vastly different from those dealt with in this chapter. Whether the molecular forms synthesized by clearly abiotic pathways survive the condensation of the solar nebula, or whether molecular synthesis will occur essentially from the elements, is an interesting question on which to conclude.

References

Adams, W. S. (1949). Observations of interstellar H and K, molecular lines, and radial velocities in the spectra of 300 O and B stars. *Astrophysical Journal,* **109**, 354.

Bertoldi, F., Cox, P., Neri, R. *et. al.* (2003). High-excitation CO in a quasar host galaxy at $z = 6.42$. *Astronomy and Astrophysics*, **409**, L47.

Bortolot, J., Clauser, J. F. and Thaddeus, P. (1969). Upper limits to the intensity of background radiation at $\lambda = 1.32$, 0.559, and 0.359 mm. *Physical Review Letters*, **22**, 307.

Buhl, D. and Snyder, L. E. (1970). Unidentified interstellar microwave line. *Nature*, **228**, 267.

Cheung, A. C., Rank, D. M., Townes, C. H. *et al.* (1968). Detection of NH_3 molecules in the interstellar medium by their nicrowave emission. *Physical Review Letters*, **21**, 1701.

Cobb, C. (2002). *Magick, Mayhem and Mavericks*. Amherst, MA: Prometheus Books.

Downes, D. and Solomon, P. M. (2003). Molecular gas and dust at z=2.6 in SMM J14011+0252: a strongly lensed ultraluminous galaxy, not a huge massive disk. *Astrophysical Journal,* **582**, 37.

Ewen, H. I. and Purcell, E. M. (1951). Radiation from hyperfine levels of interstellar hydrogen. *Astronomical Journal,* **56**, 125.

Gautier, T. N. III, Fink, U., Treffers, R. P. *et al.* (1976). Detection of molecular hydrogen quadrupole emission in the Orion nebula. *Astrophysical Journal,* **207**, L129.

Green, S. (1974). Quoted in S. I. Chu and A. Dalgarno. Rotational excitation of CH+ by electron impact. *Physical Review*, A**10**, 788.

Green, S., Montgomery, J. A., Jr. and Thaddeus, P. (1974). Tentative identification of U93.174 as the molecular ion N_2H^+. *Astrophysical Journal,* **193**, L89.

Herbst, E. and Klemperer, W. (1973). The formation and depletion of molecules in dense interstellar clouds. *Astrophysical Journal*, **185**, 505.

Herzberg, G. (1950). *Spectra of Diatomic Molecules*, 2nd edn. New York, NY: D. van Nostrand.

Hirao, T. and Amano, T. (2003). Laboratory submillimeter-wave detection of $D2H^+$: a new probe into multiple deuteration? *Astrophysical Journal*, **597**, L85.

Klemperer, W. (1970). Carrier of the interstellar 89.190 GHz line. *Nature*, **227**, 5264.

Le Teuff, Y., Millar, T. J. and Marwick, A. J. (2000). UMIST database for astrochemistry. *Astronomy and Astrophysical Supplement Series*, **146**, 157. http://www.rate99.co.uk.

Lis, D. C., Roueff, E., Gerin, M. *et al.* (2002). Detection of triply deuterated ammonia in the Barnard 1 cloud. *Astrophysical Journal*, **571**, L55.

Mahan, B. (1975). Electronic structure and chemical dynamics. *Accounts of Chemical Research,* **8**, 55.

McCall, B. J., Geballe, T. R., Hinkle, K. H. *et al.* (1999). Observations of H_3^+ in dense molecular clouds. *Astrophysical Journal,* **522**, 388.

McCall, B. J., Hinkle, K. H., Geballe, T. T. *et al.* (1998). H_3^+ in dense and diffuse clouds. *Faraday Discussions,* **109**, 267.

McKellar, A. (1942). Molecular lines from the lowest states of diatomic molecules composed of atoms probably present in interstellar space. *Publications of the Dominion Astrophysical Observatory, Victoria B.C.,* **7**, 251.

Muenter, J. S. (1975). Electric dipole moment of carbon monoxide. *Journal of Molecular Spectroscopy,* **55**, 490.

Oka, T. (1980). Observation of the infrared spectrum of H_3^+ in laboratory and space. *Physical Review Letters,* **45**, 531.

Oka, T. (1992). The infrared spectrum of H_3^+ in laboratory and space plasmas. *Reviews of Modern Physics,* **64**, 1141.

Peebles, P. J. E. (1993). *Principles of Physical Cosmology.* Princeton, NJ: Princeton University Press.

Phelps, D. H. and Dalby, F. W. (1966). Experimental determination of the electric dipole moment of the ground electronic state of CH. *Physical Review Letters,* **16**, 3.

Rank, D. M., Townes, C. H. and Welch, W. J. (1971). Interstellar molecules and dense clouds. *Science,* **174**, 1083.

Roberts, H., Herbst, E. and Millar, T. J. (2003). Enhanced deuterium fractionation in dense interstellar cores resulting from multiply deuterated H_3^+. *Astrophysical Journal,* **591**, L41.

Scappini, F., Cecchi-Pestellini, C., Smith, H. *et al.* (2003). Hydrated sulfuric acid in dense molecular clouds. *Monthly Notices, Royal Astronomical Society,* **341**, 657.

Solomon, P. M., Downes, D. and Radford, S. (1992). Dense molecular gas and starbursts in ultraluminous galaxies. *Astrophysical Journal,* **387**, L55.

Stark, R., van der Tak, F. F. S. and van Dishoeek, E. (1999). Detection of interstellar H_2D^+ emission. *Astrophysical Journal,* **521**, L67–70.

Thaddeus, P., Guélin, M. and Linke, R. A. (1981). Three new "nonterrestrial" molecules. *Astrophysical Journal,* **246**, L41–5.

Thompson, R. and Dalby, F. W. (1968). Experimental determination of the dipole moments of the X(2SIGMA+) and B(2SIGMA+) states of the cyanide molecule. *Canadian Journal of Physics,* **46**, 2815.

Townes, C. H. and Schawlow, A. L. (1975). *Microwave Spectroscopy.* New York, NY: Dover.

Turner, B. E. (1974). U93.174 – a new interstellar line with quadrupole hyperfine splitting. *Astrophysical Journal,* **193**, L83.

Vastel, C., Phillips, T. G. and Yoshida, H. (2004). Detection of D_2H^+ in the dense interstellar medium. *Astrophysical Journal,* **606**, L127–30.

Walmsley, C. M. (1993). Abundance anomalies in hot and cold molecular clouds. *Journal of the Chemical Society Faraday Transactions,* **89**, 2119.

Woods, R. C., Dixon, T. A., Saykally, R. J. *et al.* (1975). Laboratory microwave spectrum of HCO^+. *Physical Review Letters,* **35**, 1269.

18

Framing the question of fine-tuning for intermediary metabolism

Eric Smith and Harold J. Morowitz

Learning from our own existence

Hoyle's [1] successful prediction of the 7.6 MeV resonance of the carbon-12 nucleus, based on observation of his own carbon-based existence, established the scientific usefulness of anthropic principles. These principles have become common, if not yet standard, tools in cosmology, where theories of initial conditions may not yet exist – or, if they do exist, may admit a range of values [2, 3, 4, 5, 6]. At the same time, anthropic principles have retained a traditional role in religion and philosophy, where sensitive dependence of human existence on laws of nature that could imaginably have been otherwise is interpreted as evidence for human significance in the creation of the universe.

The tendency of life forms to make universal use of, and seemingly to depend on, specific details of natural law or historical circumstance does not end with nuclear abundances. Following decades of studying the ways in which mammalian blood achieves homeostasis by exploiting favorable regions in carbonic-acid chemistry and similar adaptations, Henderson compiled a list of such dependences in physiology [7]. Inverting Darwin's description of selective fine-tuning of organisms for "fit" to their environments, Henderson characterized his physiological sensitivities anthropically as evidence of "the fitness of the environment" for life.

The rapid growth in understanding of biology, from structures to systems, seems likely to expose many more sensitivities of life to details of chemistry, physics, and history. The question has been posed as to whether these advance Henderson's interpretation of fine-tuning of the environment or, more generally, what their anthropic significance is. In this chapter, we examine features of intermediary metabolism, whose universality and historical persistence suggest that they are not arbitrary products of chance.

Fitness of the Cosmos for Life: Biochemistry and Fine-Tuning, ed. J. D. Barrow *et al.*
Published by Cambridge University Press. © Cambridge University Press 2007.

So little is understood about the significance of specific features of metabolism, even within biology, that we find it more confusing than helpful to project the narrow and partly extrascientific interpretive framework of fine-tuning onto a first analysis. Instead, we start with three wider questions that can be answered independently and that later can be combined in a clearer framing of the question of fine-tuning for metabolism.

1. In what sense is information about the environment represented in the observation of any specific biological structure or process? How do we distinguish features that are "essential to life" from adaptations to convenient environmental circumstances? Are the meaningful sensitivities properties of abstractions, even from structures we see universally? If so, how do we abstract correctly?
2. What are the structure of anthropic arguments and their range of interpretations? What other inputs besides universality or sensitivity of life's processes contribute to assigning meaning?
3. What specific presuppositions create the meanings of "fine-tuning" or Henderson's "fitness"? How do conclusions of this type depend on whether we study sensitivity of features or relations?

We begin in the following subsections with careful definitions of anthropic arguments, fine-tuning, and fitness in terms of the general problem of inferring information about generative processes by using their output as evidence. We note that sensitivity in the output is necessary for useful inference, but that it carries different information in complex and simple systems, where it has different frequencies.

The wealth of biological evidence available today, together with growing understanding of self-organization in physics and chemistry, cautions us that we must incorporate life's variations with its universals, and its robustness with its sensitivity, at all levels in our analysis. A pragmatic anthropic principle guides us: the idea that the universals in metabolism, structure, and regulation today allow us to infer the thermodynamic and chemical forces responsible for life's emergence.

We argue that much of biological order comes from arrangement and augmentation of near order in the underlying chemical world and that the uniqueness of life is often found in this augmenting relation, rather than in particular biological structures. What seems familiar and lawful to us about life is often the lawfulness of the underlying physical and chemical world, with which we have broad experience, as that order is expressed transparently through the living process. We are seeing the environment *through* life; the meaningful category distinctions are defined not by specific molecular structures, but by specific relations to the opportunities for structure formation in chemical and energetic relaxation. In later sections, we turn to the analysis of specific features of metabolic chemistry and energetics and show how their information in the anthropic interpretation differs from that in the Neo-Darwinian explanation for emergence and universality, a sort of anthropic null model.

Anthropic principles, fine-tuning, and fitness

Living systems combine a strictness of regularity with a profusion of innovation of structures, to a degree that seems unequaled in any other single class of phenomenon we recognize. They are uniform in many ways, but strikingly diverse in others. All organisms ever alive on earth, taken together, account for a tiny fraction of the conceivable physicochemical structures of comparable complexity, and their uniformity implies that, in a statistical sense, the realized instances are drawn over and over (redundantly sampled) from that small subset of possibilities. The specification of the actual within the conceivable defines a measure of statistical information in the forms we see. As formalized by Shannon and Weaver [8], information measures surprise and is defined only when we have specified an ensemble of comparable forms from which the observed forms are drawn [9]. We may hope to use the variability within observed living systems to identify the ensemble of possible process networks from which they are selected, and thus something about the nature of life within the environment.

A few informative properties of life come from easy category distinctions, such as the fact that all known life makes essential use of carbon and carbon–oxygen–nitrogen molecules in liquid water solution. The seemingly trivial observation that such "carbaquist" chemistry is ruled out if astrophysical carbon abundance lies below a certain threshold enabled Hoyle [1] to predict the 7.6 MeV carbon-12 (^{12}C) nuclear resonance with remarkable precision because the discovery of the triple-alpha reaction synthesis of ^{12}C in stars happens to be a bottleneck for stellar nucleosynthesis of all the heavy elements.[1] The pragmatic information in this prediction is easy to measure because it guided experimental characterization of ^{12}C nuclear structure where the existing computational capabilities could not. Similar sensitive dependence of the physical state of water has been used to define a "habitable zone" in planetary physics [10], which is not predictive in the same sense as carbon abundance (we already knew where the earth's orbit lies), but which creates a useful filter in the search for extraterrestrial life.

The "Anthropic Principle" is the name given to a collection of related arguments of this type, in which observations about the existence or form of life are used as evidence to constrain theories of the natural laws from which life arises. The different forms of the Anthropic Principle are conventionally divided by philosophers

---

[1] In Appendix 18.1, we explain why a natural measure exists for the precision of this prediction and also the methodological limitations relative to which this natural measure generates a surprising prediction about nuclear structure.

into "weak" and "strong" [4], according to whether they merely make other consequences of the same laws conditional on our existence as observers or predict laws
or parameter values that are believed to be unique in nature, but for which no other
basis of selection is known.

Anthropic reasoning is a subset of the more general reductionist inference of
natural laws from observations of their effects [11], distinguished by its use of
life as evidence (in some particular cases, human life and thought, as the name
suggests). The program of model validation from posterior evidence is formalized
in a probabilistic construction known as Bayes' theorem [12]. We show in Appendix
18.1 that anthropic principles have a common logical structure of Bayesian updating
and that the distinction between weak and strong principles lies essentially in the
probability interpretation of the prior beliefs.

Bayesian formalization makes clear that refinement of prior beliefs with posterior
evidence does not by itself assign meaning to those beliefs. Thus, the same evidence
can be claimed in support of different meanings in different contexts. It can provide
a pruning rule [11, 13] to solve an intractable problem, as Hoyle's prediction did for
Fowler and Whaling [1], or a way to work around the lack of theoretical specification
of initial conditions, as in some cosmological proposals [4, 5, 6]. These are both
weak anthropic arguments whose distinct meanings are defined within the context
of scientific prediction.

In contrast, strong interpretations generally have meanings originating outside
science, precisely because initial conditions by assumption precede cause and effect
or the criteria of repeatability that validate causal theories [14]. The concept of
fine-tuning originates in an anthropomorphic interpretation of the Bayesian prior
distribution for such initial conditions. "Tuning" suggests intervention with purpose, a presupposition incompatible with the notion of causality in natural law
[15], but neither obviously inappropriate nor obviously useful for theories of initial
conditions.

For Bayesian inference to aid prediction, the posterior observable must depend
non-trivially on the prior variable being estimated. Fine-tuning presumes a natural
measure of sensitive dependence, such as a small half-width of nuclear resonance
in relation to its center frequency. We will find, however, that the weak anthropic
use of life's current forms to study its origins requires us to abstract away from the
most facile category distinctions of life from non-life. Simple categorizations based
on specific amino or nucleic acids, which may depend sensitively on history as well
as chemistry, can include adaptations that are not restrictive on either the environment or the essential nature of life and ignore its robustness under environmental
fluctuations.

Adaptation produces sensitive dependence in living forms quite generally
because it arises from growth with heritable variations. Differential growth rates

(including survival and fecundity effects) appear exponentially with time in the population frequencies of inherited features, leading to sensitive dependence of population samples of phenotypes on small differences in their relations to environments [16]. Under situations of resource constraint, this can result in competitive exclusion, through which not only the population mean, but all of its instances, may be strongly biased by small differences in viability. The relative growth rates themselves, which may not depend sensitively on environment by any natural measure, are termed "fitness" in Darwinian population dynamics [17]. Darwinian fitness is expressed through selection of individuals – each one a relatively complex package of adaptations – in response to the often complex characteristics of their environment. Therefore, it frequently leads to sensitive dependence of complex wholes on complex wholes.

Henderson appears to have appealed to this latter aspect of fitness in characterizing the laws of chemistry and physics, as well as the earth's environmental composition, as "fit for life" [7]. The observed physico-chemical environment is sensitive to his criterion that it be able to support life's complexity and also its need for homeostasis, rather than being the result of any dynamic process that produces exponential dependence on initial conditions, and in that respect is unrelated to Darwinian fitness.

Because of its variability, adaptability, and robustness, it seems likely that life will admit relatively few easy category distinctions, such as the carbaquist sensitivity to carbon abundance. Most of the "information" in the structure of life, about either its necessary circumstances or its generating processes, will likely come from more specific structures, which typically emerge at higher levels of complexity. Thus, in addition to understanding the logic of anthropic argument and the flexibility in its use of empirical sensitivities, we must understand the different kinds of surprise carried by sensitivity in complex and simple systems.

Sensitivity and antisensitivity in complex systems

Linear or near-linear response is common in simple systems with few degrees of freedom [18], averages of simple systems with few global constraints and central-limit theorem statistics [19], and closed systems near stable equilibria, not driven by persistent forces in the environment. Systems with intermediate numbers of degrees of freedom and global configurational constraints (such as nuclei or atomic electron shells) tend to display few and isolated sensitive features, such as resonances about ground states governed by Pauli exclusion. Experience with non-linear dynamics and chaos has taught us, however, that sensitive dependence on initial conditions and details of dynamics becomes a generic property of the observables of complex systems with long-range constraints [20].

Whereas we might extract information about a generative process from an obvious feature such as a parameter value (frequency of a nuclear resonance) in a system of moderate complexity, the equivalent capacity to inform us (reduce our uncertainty) about the generating process may reside only in a category of features in a complex system. Which feature we see may depend sensitively on historical accidents that are irrelevant to the generating process itself, but only when we understand how to see past this contingency – which would affect any realization of the process – do we identify the informative regularity of the whole category.

Convergence, of central empirical importance to the structuralist interpretation of Darwinian evolution [21, 22], can indicate sensitive selection of organism features for fit to the environment. Yet this sensitivity is not restrictive either of the "aliveness" of the organism or of the environment, because we recognize convergence precisely where there exist diverse solutions matched to diverse environments, as well as viable pre-convergent starting points that were not so specialized. The torpedo shape of fish is a strongly convergent trait selected for swimming at high Reynolds number. Yet phytoplankton, which must move and feed in the opposite extreme of low Reynolds number, display completely different but similarly universal traits [23, 24].

Resilience and robustness can come from the same ability to track the environment that leads to convergence or from a quite different ability to absorb its variations, leading to a kind of antisensitivity. Such antisensitivity is not exclusive to life; the atmosphere generates many negative feedbacks that confer stability against geological and biogenic shocks. Indeed, the pH stability of blood arises from its positioning at a stable region of carbon dioxide solution chemistry in water. The fitness of the environment for life may depend on its richness in such stable regions, but not necessarily on the ability of natural selection to exploit those regions.

The ability of life to accommodate environmental variation, either through complementary adaptation or through compensation, argues against naive characterization of life in terms of specific biochemical pathways or structures, even universal ones. We must first understand why they are universal and what relation to or distinction from the abiotic environment they exemplify.

Accounting for biological universality

Cellular life has existed on earth for between 3.8 and 4 billion years [25]. As there was essentially no molecular oxygen in either the atmosphere or oceans for the first two billion years [26, 27], at least the great bulk of life inhabited neutral or reducing environments and appears to have been exclusively prokaryotic. Although photosynthesis is not ruled out during this period, the large-scale photolysis of water does seem unlikely, so that the free energy to drive life had to come from geochemical reductant, abiotically generated organic compounds (of atmospheric or astrophysical origin), or low-energy photosynthesis (involving sulfur or other weak oxidants). Over the interval from 2.4 to 1.6 billion years ago, first the atmosphere and then the oceans gradually became loaded with oxygen, believed to have been generated by oxygenic photosynthesis [28] and to have accumulated once buffers against oxidation, such as reduced iron in the oceans, were exhausted. Around the same time, eukaryotes emerged through endosymbiosis. Even today, the largest number of major taxa (whether grouped by 16S rRNA phylogeny or other criteria based on way of life, which more naturally reflects the horizontal gene transfer that continued for other metabolic pathways after ribosomal RNA was effectively fixed within lineages [29, 30]), are prokaryotes [31], and all intermediary metabolic pathways are used within this group.

Prokaryotic life is found over a temperature range exceeding 0–100 °C, a pH range of 1–10 or more, a NaCl concentration ranging from 0.01 M to saturation, and an oxygen saturation ranging from 0 to 20%. In laboratory settings, cells have remained viable under deuteration of from ambient 0.001 to 0.99 of the solvent water and under replacement of water with 0–35% glycerol. Prokaryotes occupy all environments on earth that harbor life and define the limits of its robustness and metabolic adaptability as far as we know.

Throughout this remarkable range of conditions, all known life shares a common chart of intermediary metabolism. In reducing autotrophs, energy is derived from environmental redox couples by either the reductive tricarboxylic acid (rTCA) cycle, the reductive acetyl-CoA cycle, or the Calvin–Benson cycle, and all biomass is synthesized from carbon dioxide, water, ammonia, and simple inorganic salts and acids, or from more reduced inorganic chemicals produced from these, such as acetate. In photoautotrophs, reductant produced by photosynthesis either drives ATP formation and biosynthesis directly (as in some purple Archaea such as *Halobacterium halobium*) or is combined with photosynthetically produced oxygen in oxidative TCA (or Krebs)-cycle respiration. Heterotrophs implement subsets of the same metabolic networks as autotrophs, combining to produce entire networks in ecologies, which are always autotrophic.

Heterotrophy is possible because all organisms are composed of roughly the same 300 small molecules (molecular mass < 500 Dal), into which all food is broken down before being used directly or being reassembled into several thousand kinds of polymer inside the organism. All major classes of these biomolecules are synthesized from the eleven carboxylic acids of the rTCA or TCA cycle, although sugars can also be photosynthesized from 3-phosphoglycerate by an alternate pathway. A diagram of this universal synthetic core is given in Figure 18.1. Lipids come from acetate, sugars come from pyruvate or 3-phosphoglycerate, amino and nucleic acids come from oxaloacetate or α-ketoglutarate, and other molecules are

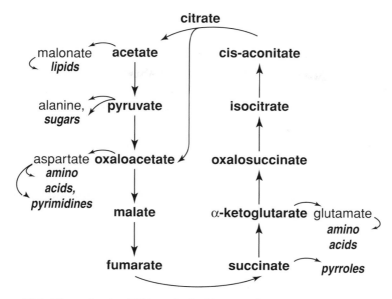

Figure 18.1. The reductive TCA cycle (bold roman font) as an engine of synthesis of the major classes of biomolecules (bold italic). Synthesis of categories usually begins with a specific molecule (lightface).

synthesized from these starting compounds. The porphyrins, an important class of ring molecules that bind transitional metals in cofactors, chromophores, and electron-transfer proteins, are generated from succinate.

All reactions among these metabolites are catalyzed in modern organisms by protein enzymes or ribozymes, all of whose sequences are encoded in DNA. Yet extensive horizontal transfer of genes in both metabolic [32, 33] and photosynthetic [34] networks has taken place, whereas the metabolites and chromophores are essentially universal. (Some variability exists in chlorophylls, rhodopsins, and some augmenting pigments [35], possibly because of their relatively high molecular mass, which is not found in the simpler metabolites.)

How are we to account for this universality within the remarkable diversity both of environments and of relative reaction kinetics within organisms? Two interpretations exist, one based on an extreme controlling role for gene-encoded catalysis and pure Darwinian selection, and the other based on a continuation from physical chemistry to core metabolism.

Gene as gatekeeper and happy accidents: the anthropic null hypothesis

One of many unique features of modern life, with its great complexity, is the ability to faithfully pass genetically encoded traits from generation to generation through reproduction. The fidelity of the inheritance mechanism is not perfect; the small incursion of mutations enables the tracking of lineages of descent [36]. The accuracy of inheritance is remarkably high, however, and traits that are not actively removed by selection can persist indefinitely in large populations, as well as proliferate across lineage boundaries by horizontal gene transfer. At the same time, mutation is blind with respect to future selective advantage, allowing even favorable traits to be lost forever [16].

Depending on how much we think of genes as the arbiters of the possible for life forms, we choose between two qualitatively different interpretations of universality and robustness. It is possible that the universal reagents and pathways of intermediary metabolism are not qualitatively distinct from other organism traits that we observe as variable or that have not been preserved, such as the number of digits on vertebrate limbs throughout history. The robustness of all these traits preserved through evolutionary time could be conferred arbitrarily by the robustness of the inheritance mechanism; indeed, this is more likely for metabolically central features than for peripheral features such as morphology simply because of the number of other pathways that evolve to depend on them exclusively. A characteristic of the neo-Darwinian commitment to the central dogma – that information flows from genes to mRNA to proteins to pathways – is that genes become the gatekeepers by which all living functions must be discovered and maintained.

Note that this is not a presumption of selective neutrality or pure genetic drift; adaptation and even convergence lie within this paradigm. Rather, it is the more subtle commitment to "accidentalism," which is forced on us when we realize that the number of combinations of pathways that are in principle consistent with enzymatic regulation by any large genome far exceeds the number of species (more generally, connected gene-swapping populations) that have ever existed. We interpose the accident of discovery between the fitness advantages potentially inherent in the environment and any opportunity for selection to find organisms that capture them, and thus we exclude "the arrival of the fittest" [37] from our realm of prediction. This second-order consequence of neo-Darwinism drastically reduces the information we are willing to admit from universal biological traits about their environment, because first and foremost we presume that they must have been discovered in a common ancestor for any but the most superficial mutations.

If our priors – as they are called in the domain of Bayesian inference, for which see Appendix 18.1 – admit a very large number of imaginable genetically regulated metabolisms (say, a number that scales exponentially with genome size, such as the number of possible combinations of enzymes), a strict neo-Darwinian view of metabolic universals becomes almost an anthropic null model, because any particular combination has near-zero probability of discovery and is preserved either because no competitive scheme was ever found by the genome or as a result of its centrality. Redundant discovery of a single solution is as improbable as multiple independent original discoveries, and we lock ourselves combinatorially into the interpretation that modern organisms are tied to primordial organisms primarily through descent. Their observed forms are then not permitted to refine the prior probabilities for the pathways selected at life's emergence. The only question in this view is how far back descent enables us to see. Gene-first models of the origin of life [38] impose Darwinian selection as a pure paradigm back to the originating events, leaving both emergence of the genome and the subsequent innovation of the pathways favored by genetic regulation a "happy accidents" [39].

Structuralism in evolution [21], the predictable selection among those competing solutions that do arise, survives on the margins of the domain of accidental discovery in neo-Darwinism. Practically, it is only a predictive theory (as opposed to a mere rationalization) for competition among nearby variants, whose probabilities are sensibly calculable. (This is also why gradualism has been so heavily emphasized in "theories" of evolution, beginning with Darwin. Even Darwin did not presume that large changes were impossible; there was simply nothing in the nature of scientific explanation that he could say about them, a point that has been assigned deep philosophical significance by Monod [40].) The tremendous complexity and diversity of the genetic regulators of metabolic pathways appear to fall well outside the domain of small variations, leaving the genomic discovery of a metabolism able to support all living structures so improbable that selective competition cannot be inferred from current forms. (Remember that the probability for the emergence of at least one meaningfully independent competitor is presumed to equal the joint probability for at least two independent originating events for life.) Although no logical objection exists to Bayesian priors that place the discovery of metabolism outside the domain of scientific prediction, such priors seem practically at odds with the assumption that the genome somehow found a metabolism so productive that cellular life could emerge within 0.2 billion years after persistent oceans formed, after which that life was so stable that it has survived to the present.

The neo-Darwinian commitment to gene as gatekeeper, and with it the unfortunate buy-in to accidentalism, appears to be a reaction against teleological interpretations of progress in evolution [41, 42, 43] and an attempt to rule out such interpretations methodologically. Yet the absence of teleology in the dynamics of mutation and recombination cannot rule out systematic progression, and that progression can be predictable if selective bias is visible above sampling bias in population dynamics. Any case of predictable, systematic progression is open to interpretation as progress by those who choose an endpoint for the sake of that definition. Thus, giving precedence to a particular macromolecule (DNA) – for the sake of its susceptibility to historical accident – doesn't overcome the true problem of prejudiced interpretation [44], and meanwhile it blinds evolutionary theory needlessly to the pressures for regularity at all the other levels of life, which may not be so subject to accident.

Seeing the environment through life

An alternative interpretation is that at least some of the empirically universal features of life *are* qualitatively different from variable features, and that their persistence is driven by forces also recognizable in abiotic ordering. This interpretation includes structuralism in Darwinian evolution, but locates it within a continuum of progressions that give predictable shape to the environment and the life within it. The genome is simply another emergent level of structure, neither sufficient to define life nor capable of precluding or overriding the many other chemical and physical forces that engender organization or stability at multiple levels. The distinction between biological and abiotic order ceases to be defined with the emergence of genes or any other single molecular class and must be sought as a system property of the relation among the various molecular functions.

This moderated view of the role of the genome allows robustness and resilience to have energetic and statistical origins, which may affect many levels of structure, from the reaction network of metabolites themselves to the centrality and order of emergence of catalysts regulating them. It is compatible with extensive lateral gene transfer [45] and with experimental investigations on the origin of the genome [46], which call into question the necessity of DNA and RNA as specific category features of life. This view is valid where systematic bias remains visible over sampling bias in population dynamics. Unlike sampling bias, systematic bias acts predictably, both at the emergence of life and during its subsequent evolution. In this interpretation, the universality of current forms can constitute evidence for the strong thermochemical biases on life and may be used anthropically to refine the possible scenarios for its emergence. The requirements on such an interpretation are that we identify the sources of systematic progression and that we demonstrate that sampling bias is not sufficient to mask these sources.

We thus characterize life as a collection of physical and chemical processes of environmental constituents, augmented by biomolecules that are rare or absent in abiotic environments. Different levels of living structures, depending on their complexity, can behave more or less like the common reaction networks in the abiotic environment. In particular, those involving many reactions of small molecules that are strongly distinguished by free energies of formation or kinetics of functional groups [47, 48] may thoroughly and redundantly sample all allowed reactions with one another (a kinetic generalization of the notion of ergodic sampling in equilibrium statistical mechanics), selecting by familiar statistical means the favored species and pathways. More complex structures such as macromolecules – with a flatter energy landscape, more kinetically equivalent combinations, and lower turnover in reactions – are both more susceptible to accident and, as a result, more eligible to record information within the organism about the environment to which it must respond [49].

Genes, compartments, and catalysts are regulatory structures that must be built from free energy and materials made available by metabolic reactions. The metabolites are smaller and simpler than the regulators, more of them are present in the ambient environment, and the possible reaction networks among them are more densely sampled than the possible networks producing complex structures. Thus, the reaction network of core metabolism is expected to be more nearly a bulk chemical process [50] than the combinatorics of either nucleic acid or amino acid polymers.

The sources of systematic progression that once drove the emergence of life and now determine its stable forms can be the same ones that generate stable order in abiotic statistical systems. Thus, we ask how life could have emerged systematically and yet without teleology and how the pathways for emergence may have exploited the tendency toward self-organization in certain chemical regimes.

Progression away from the abiotic state

A deep conceptual division separates the notions of evolution in biology and physics, which we argue is in some respects a historical artifact and should be overcome. From the origin of natural selection as a variation on human selection of traits by breeding, and the later association of one gene – one trait in the first formal population genetic models, a trait in biology has come to mean either an independent degree of variation in the genome [51, 52] or an independent axis of selection by the environment (and sometimes the conflation of the two). The fine structure of biological evolution is traced through changes in genotype and phenotype (the collection of traits, by whichever definition).

In statistical physics, nothing like the "trait" as a salient degree of freedom is recognized. Rather, ensembles of configurations, rapidly and chaotically generated by their own internal dynamics, are assumed to be maximally disordered, subject to constraints such as ensemble-averaged energy or chemical constitution. These averages are the state variables of the system. They need not correspond to degrees of freedom of the dynamics, but rather are defined as the coarse-grained boundary conditions through which the average properties of the ensemble interact exclusively with those of its environment. Maximal randomness of a statistical ensemble subject to its environmental constraints corresponds to a condition in which the microconfigurations that produce those state variables are the most numerous in the ensemble [53].

Systematic progression in statistical systems is a property of changes in the values of the state variables, which need not look anything like the diverse observables available in the rapidly fluctuating, microscopically specified configurations. The direction of progression can have a simple description toward end states that are stable equilibria or away from initial states that are unstable. The following observations indicate that the abiotic state of those environments that harbor life is unstable, or at least metastable relative to more probabilistically favored states. The idea that evolution proceeds by punctuated equilibrium, invoked to explain the dearth of intermediate forms in the fossil record compared with persistent stable forms [43], would be predicted if metastability were a common property of ecosystems, suggesting a complex landscape in which the initial departure from past states is easier to predict than the likely futures. (Thus, for example, the nucleation event that leads to the first vapor-bubble formation in a volume of superheated water is easier to understand and model than the turbulent dynamical state [boiling] to which the system subsequently settles.)

All environments that harbor life are stressed by a density of unequilibrated free energy. Trapped free energy is necessary to drive biomass away from the Gibbs equilibrium state [54] and exists only if all the processes active in the system are insufficient to create the equilibrium distribution (equivalently, to ergodically sample the state space in which the equilibrium distribution is the most disordered and thus favored by probability.)

Environments with stable energetic stresses are frequently divided into nearly decoupled spatial or compositional subsystems. This is true of quasi-stable energetic redox couples at hydrothermal vents and of the weakly coupled 6000 K spectrum of solar visible light and 300 K terrestrial thermal black body [55]. The separate components may constitute internally near-equilibrium subsystems, defined individually by simple ensemble constraints.

Internally equilibrated subsystems, which act as free energy reservoirs, are already as random as possible given their boundary conditions, even if they are not in equilibrium with one another because of some bottleneck. Thus, the only kinds of perturbation that can arise and be stabilized when they are coupled are those that make the joint system less constrained than the subsystems originally were. (This is Boltzmann's H-theorem [9]: only a less constrained joint system has a higher maximal entropy than the sum of entropies from the subsystems independently and can stably adopt a different form.) The flows that relax reservoir constraints are thermochemical relaxation processes toward the equilibrium state for the joint ensemble. The processes by which such equilibration takes place are by assumption not reachable within the equilibrium distribution of either subsystem. As the nature of the relaxation phenomenon often depends on aspects of the cross-system coupling that are much more specific than the constraints that define either reservoir, they are often correspondingly more complex than the typical processes in either of the decoupled subsystems. Thus, the reducing bacteria draw energy from a large collection of dissimilar environmental redox couples [56, 57]. As an ensemble of environments, these may be constrained only by pyrolysis at lava–water interfaces, followed by a partial quench in passing to the cooler environment of hydrothermal vents. Yet the process for redox relaxation by all organisms is transmembrane transduction to apparently uniquely suited phosphodiester bonds [58], and the mechanisms that have evolved to enable this are quite complex [35].

As free energy stress is unstable in the presence of cross-system coupling, some (generally complex) relaxation process must become the short-term steady state. The principle that stabilized relaxation processes arise where they can reduce the impact of a free energy bottleneck has been likened by Eschenmoser to a dynamic Le Chatelier principle [46].

Living processes contribute to thermochemical relaxation, but they are not the only ones to do so. Many processes in convective weather [59, 60, 61], chemistry [62], and engineered systems [63, 64] are examples of self-organized production of conduits for the transport of energy and transport or generation of entropy. Progression away from the unstable, unorganized state is frequently exponential [63, 64], giving rise to a physical form of competitive exclusion similar to that seen in Darwinian population dynamics at a higher level of complexity. (Indeed, it is natural from a physical point of view to regard competition for bulk resources to build a metabolic energy-transducing channel as the origin of directional evolution in biology. Darwinian evolution is distinguished by the non-linearities it injects into this bulk process, making the genome the unit of inheritance and the individual the unit of selection.)

As we expect many biological structures to achieve stability by exploiting statistical stability in the underlying chemical networks, so we expect biological organization to be most likely where it follows dynamically stable thermochemical relaxation pathways. Their stability can be driven by the free energy stress they relieve, by their use of near-equilibrium chemicals and reaction networks, or by the redundancy of random relaxation pathways leading to them. Thus, a continuation exists from physical self-organization to the energetically predictable biases to Darwinian fitness.

Core metabolism and relaxation

> *. . . and their work was as it were a wheel in the middle of a wheel.*
> *(Ezekiel 1.16, Authorized version)*

Reducing metabolism

We now examine specific features of core metabolism, and its reductive pathways in particular, as augmented relaxation channels for redox stress. We begin with reductive pathways for two reasons. First, in a non-oxidizing environment [26, 27], the only energy sources are abiogenic reductant, photosynthesized reductant, and abiotically produced organics. As primordial heterotrophs would have a very limited pool from which to draw reagents, and photosynthesis is a difficult mechanism to build from nothing, we suppose that the most reliable energy source for the first-half-billion or more years of life was geophysically produced reductant and that the universal features to emerge during this time hold the most immediate information about origins. Reducing environments such as volcanic trenches, and the metabolic pathways of the organisms that inhabit them, are likely to be the least changed in the modern world from their primordial forms. Second, the chemical simplicity, network topology, and energetics of the reducing metabolism identify it naturally as a relaxation mechanism in a purely chemically stressed environment. Recognizing the same relations in photosynthetic life requires going outside chemical free energy and considering much more complex metabolic pathways.

We observe that biomass lies between the energetic redox couples that constitute food in a reducing environment and the most reduced, lowest-energy waste molecules with the same element compositions. The chemical species characterizing one redox relaxation channel created by chemoautotrophic bacteria (realized entirely in the single organism *Hydrogenobacter thermophilus* [59, p. 254] are shown in Figure 18.2, in which CO_2 and H_2 are inputs and CH_4 and H_2O are outputs. The eleven acids of the rTCA cycle (oxalosuccinate not shown) lie in a narrow range of free energy per carbon atom along the pathway of progressive reduction of carbon from CO_2 to CH_4. We expect this free energy of formation per carbon to be characteristic of the more general CHO sector of biomass [65], which as noted forms from these acids by side-reactions from the cycle.

The acids of the rTCA cycle are low-energy forms at their respective hydrogen reduction stoichiometries. For instance, formaldehyde has higher free energy of formation per carbon than acetate, which has the same chemical composition as two formaldehyde molecules. Thus, in any relaxation pathway for full reduction of CO_2, likely intermediate states are the rTCA compounds.

Next, we present the reaction sequence of the rTCA cycle in Figure 18.3. We observe that topologically it is a loop with an external regenerative pathway. If we consider the primary loop the recycling of oxaloacetate, this four-carbon compound functions as a network catalyst for the synthesis of $CH_3COOH + 2H_2O$ from $2CO_2 + 4H_2$. This synthesis is net exergonic, and acetate is the most reduced rTCA compound in Figure 18.2.

When the regenerative pathway from acetate to oxaloacetate is included, the rTCA cycle develops the possibility of self-production, increasing the density of catalytic oxaloacetate molecules and hence the bulk rate of acetate synthesis from CO_2. The gain in this cycle enables an exponentially growing relaxation process to

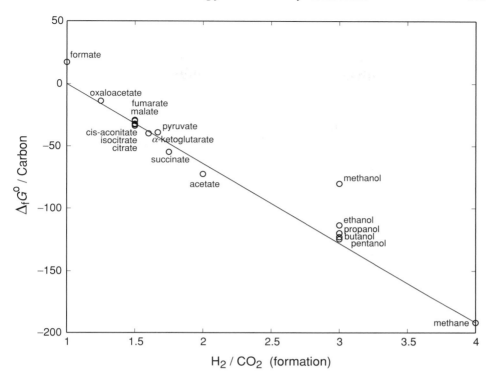

Figure 18.2. $\Delta_f G^0$ in kJ/mol per carbon atom of TCA cycle intermediates from $CO_2(aq)$ and $H_2(aq)$, together with reference molecules on the reduction pathway to $CH_4(aq)$. For the reaction $x CO_2 + y H_2 X + z H_2 O$, reducing potential per carbon taken from the environment to form species X is plotted as y/x on the abscissa. $\Delta_f G^0$ values for cycle intermediates from [66] and for other organic compounds from [67].

nucleate from few seed molecules, a common property also of dielectric breakdown, Dicke superradiance [69], or engineered self-organizing thermal engines [63, 64].

Cycles with this topology are called "network autocatalytic" because the chemical that is the network catalyst participates in the generation of more copies of itself. The notion of network autocatalysis has been explored as a mathematical abstraction of polymerization [70, 71, 72] and as an alternative to the template autocatalysis of RNA world, particularly as a mechanism to realize Fox's protein-first schema for the origin of life [73]. In contrast to the abstract simplicity of polymerization, which occurs at the price of assuming complex (activated!) inputs, the network catalysts of rTCA are structurally slightly heterogeneous, but act by enhancing condensation of simple molecules into a carrier backbone. This cycle is the first case we have seen demonstrated as a concrete example realized within a simple biological pathway.

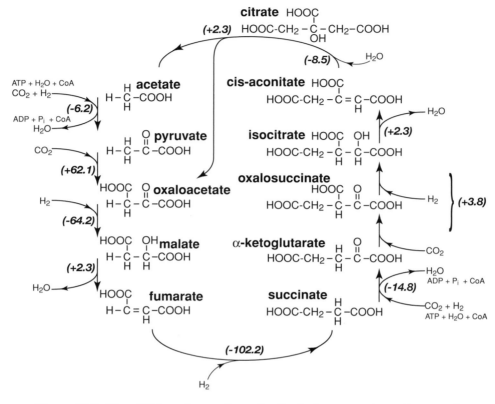

Figure 18.3. The rTCA cycle reactions. Synthesis from acetate → fumarate is repeated from succinate → cis-aconitate, with CH_2COOH replacing H as end group. $\Delta_f G^0$ values (kJ/mol, bold italic) are for reactions from $CO_2(aq)$ and $H_2(aq)$ in equilibrium with gases at 1 atm partial pressure, and $H_2O(l)$. Computed from $\Delta_f G^0$ values in [66] using $\Delta_f G^0 = -386$ kJ/mol for $CO_2(aq)$, $+17.7$ kJ/mol for $H_2(aq)$, and -237.2 kJ/mol for $H_2O(l)$. $\Delta_r G^0 = -34$ kJ/mol for ATP hydrolysis corresponds to a local stationary region around pH 6 and pMg 1.5 in [68].

Whether rTCA cycling can cross the autocatalytic threshold depends on the rate of parasitic side reactions per regeneration of a second oxaloacetate by each loop of the cycle. In an enzymatically regulated organism, autocatalysis can be enhanced by speeding the rate of reactions within the cycle. Whether the threshold for auto-catalysis can be reached in a pre-catalytic environment is the subject of ongoing laboratory experiments. Wächtershäuser has suggested, however, that the geometry of surface reactions may be required to reduce the energy barriers of key syntheses [74, 75].

A final observation about the compounds of the rTCA cycle is that they are rich in carbonyl groups from incompletely reacted CO_2. Such structures are both accessible in relatively few steps when CO_2 is an input to the reaction network and highly reactive at the carbonyl moieties [47, 48]. Enhanced reactivity is favorable both as a property of a network-catalytic relaxation pathway and for the synthesis of biomolecules that can feed back to enhance the rate of the cycle. We will return to this latter relation below.

Photochemical relaxation by photosynthesis

Unlike metabolism in a reducing world, oxidizing metabolism must build biomatter from substrates of lower free energy; anabolism and catabolism are opposed. Thus, only in the presence of a non-equilibrium photon spectrum is oxidizing metabolism a relaxation process.

Oxidative life, like reductive life, occupies a free energy bottleneck for the same reason that visible light from the sun reaches the surface of the earth, rather than thermally equilibrating with matter in the atmosphere. Visible photons couple almost exclusively to electronic transitions in small gas molecules, whereas the terrestrial microwave background couples to vibrational and rotational transitions. Quantum selection rules prevent cross-coupling, both of light and of the internal transitions, making the atmosphere an effective barrier to cross-band transduction of light energy. Photons absorbed in either band are re-emitted preferentially within the same band in steady state [55].

Inelastic visible photon absorption by larger molecules on exposed mineral surfaces occurs principally by photodissociation [76]; the resulting relaxation of excited vibrational levels of the fragments is partly responsible for the earth's elevated surface temperature. However, photodissociation removes an electronic transition with each absorption, limiting its own rate. Thus, the albedo of the earth is controlled mostly by water, through reflection from clouds (not wavelength-transducing) and absorption in the oceans, the latter of which combines the robustness of small molecules with the unusual strong vibrational coupling of molecular clusters in liquid water.

The emergence of photosynthetic life is defined chemically by biosynthesis of the chromophores, which can absorb visible photons repeatedly. Before it can be re-emitted, the light energy is converted to internal reductant, which is used by core metabolism, with oxygen (or, in fewer taxa today, other oxidants such as sulfur) released to the surroundings [28]. Biosynthesis distributes light energy among lower-energy chemical bonds, from which it is then dissipated either mechanically through motion or in vibration and rotation through degradation of biomass. The less than 30% conversion effciency of light to redox potential [77] in chlorophyll-based photosystems (for sugars, under the best of conditions), with the remainder dissipated directly as heat, both makes a pigmentation effect the major impact of biomass on changing ocean absorption and suggests that the major technical problem of rapid distribution of vibrational stresses is solved by these molecules. Additional inefficiencies in the consumption of ATP and reductant in more complex biosynthetic processes also dissipate energy directly as heat in solution.

Limitations other than conversion efficiency attest to the difficulty of transducing light energy abiotically, even in modern chromophores, according to an interpretation by George Wald [78]. Both chlorophylls and rhodopsins absorb away from the frequency of maximum light intensity at the earth's surface, the so-called water hole of atmospheric scattering [79]. An absorption band centered in the water hole would produce more free energy for self-reproduction. Yet such a solution has not been found even by modern organisms, suggesting that the universal chromophores are locally optimal solutions to a hard relaxation problem. The smallest self-sustaining metabolic networks that produce them are equivalent to the networks of the photoautotrophs, of considerably greater complexity than the autocatalytic rTCA core for redox relaxation.

The major chemical transitions and robustness

All of these observations combined lead us to believe that the development of modern life as a steady-state relaxation process in fact took place through the sequential emergence of two separate channels. The first in time, and the simple, was the emergence of reductive metabolism through autocatalytic networks either identical or similar to the rTCA cycle. All its reagents are small molecules that are selected by simple kinetic and physical properties from the complete set of CHO molecules of comparable size [65], and the reaction networks involving them are relatively densely sampled, either within the cycle or in the side-reactions that generate biomass from it.

The reductive metabolic core reactions are close enough to bulk physical chemistry to be studied with the statistical mechanics of complete reaction networks of small molecules, yet produce the biomass necessary to support the full complement of compartments, catalysts, prosthetic groups, and genes. The scenario requiring minimal happy accidents is one in which most of the complexity of cellular life developed around this metabolism over the first 0.5–2 Gy.

Reductive metabolism captures free energy ultimately produced by the fission of uranium, thorium, and potassium-40 in the earth's mantle, but makes no use of the richer free energy stress from solar fusion reactions, other than exploiting liquid water as a solvent in the habitable zone. Photosynthesis captures this independent fusion energy source, but appears to have become accessible only with the molecular complexity of modern cells. It therefore evolved to be self-supporting by artificially generating reductant to synthesize critical components such as the porphyrins from molecules provided by the rTCA cycle.

It is a remarkable example of life's robustness that the compounds in the rTCA cycle survived the poisoning of the earth's atmosphere by oxygenic photosynthesis to remain the core of biosynthesis. Photosynthesis of 3-phosphoglycerate as well as reductant enabled the direction of the cycle to be reversed, from self-generation to self-consumption, becoming the oxidative Krebs cycle. In some modern organisms, parts of this cycle are now replaced with other pathways, but the synthetic core has not been lost, and the complete cycle is still preserved in some chemoautotrophs. This robustness may be due in part to the centrality of the cycle and the genetic commitment to complex synthetic networks by the time the transition occurred. Yet the fact that a plausible primordial form has not been lost suggests to us that its chemical constitution remains a source of selective advantage.

At the same time, the oxygen loading of the atmosphere is a useful demonstration that mutation, and even short-term selective bias, are not forward-looking with respect to slower degrees of freedom, such as the ability of the atmosphere–volcanism cycle to reduce molecular oxygen. Photosynthesis is energetically selected even if it produces only reductant, in which case the by-product, oxygen, is a toxin. An additional complex pathway, apparently sharing a common ancestor with the pathway for nitrogen fixation, is required for photosynthetic production of 3-phosphoglycerate, which in turn is required to produce phosphoenolpyruvate to support reversal of the rTCA cycle to TCA. Current organisms use the atmosphere as a buffer for oxygen, which is then consumed jointly by photosynthesizers and the oxidizing heterotrophs. The high O_2 saturation of this buffer, however, requires that these organisms evolve defense mechanisms against oxidizing radicals, as in the compartment structure of the C4 photosynthesizers [28] and possibly originally endosymbiosis. Specific enzymes have also evolved for protection from peroxides, such as catalase for general scavenging of H_2O_2, and peroxidases for other alkyl peroxides [35]. Notably, both of these enzymes employ heme as the active group.

The epistasis represented by persistence of the TCA core in the face of such endogenous shocks is not easily attributed to any particular attributes of physical chemistry or geology. It is more naturally understood as the Le Chatelier-type stability of a relaxation process in the face of continuous free energy stress. Complex solutions initially present as reductive metabolism, even if they become untenable or marginalized on earth, provide exploratory seeds for generating other complex solutions to new problems that arise [70, 71, 72]. Complexity does not simply beget complexity indiscriminately, however. The solutions are strongly culled for their ability to cluster around statistically favored thermochemical channels.

Complexity hierarchy and feed-down

. . . for the spirit of the living creature was in the wheels.
(Ezekiel 1.20, Authorized Version)

We have argued that metabolic pathways are statistically favored relaxation channels in energetically stressed environments and that their universality and stability result at least partly from this function. Yet the only places we see these pathways capture a significant fraction of element abundances is within organisms. Once modern organisms exist, their greater efficiency than abiotic processes can scavenge useful reagents from the environment, lowering the residual energetic stress below the threshold to spontaneously induce life, so the dominance by organisms of these relaxation structures may not in itself be surprising [80]. However, the essentially regulatory superstructure, which emerges with the complexity of cellular life to enhance efficiency, distinguishes living from all non-living relaxation phenomena that interact with the same reservoirs by means of the same active chemical bond types.

Phosphates, geometry, and topology

Universality in these regulatory structures exists beyond that in the metabolic core, which is also likely to be informative at least about the fit of life to energetically stressed environments. As noted above, all organisms use a variety of redox couples, whether environmentally generated or photosynthesized. Some of these are consumed for reducing potential directly, such as NADH. Many reactions in cells, however, proceed by dehydrations in aqueous environments, and in all cases this difficult reaction is produced by hydrolysis of pyrophosphate [35]. Polymerization of amino acids, sugars, and nucleotides proceeds in this way, and the only reaction in the rTCA cycle requiring net energy input by an environmental molecule uses ATP for dehydration (acetate $\rightarrow$ pyruvate and succinate $\rightarrow$ α-ketoglutarate) (see Figure 18.3). Almost all modern organisms store the energetic pyrophosphate bond for these reactions in ATP, although cases are known where it is stored in crystalline polyphosphate [81].

All pyrophosphate bonds are produced ultimately from redox energy via a membrane-mediated reaction sequence. The sequence first converts redox energy to electrochemical proton potential, which then drives ATP synthesis. This three-stage process involves the interaction of chemically structured energetic transitions (redox reactions and phosphodiester bond formation) with the spatial motion of protons in pH and voltage gradients. Proton electrochemical potential is a continuous-valued energy currency made possible by the macroscopic nature of geometric separation across membranes, but also requires the topology of enclosure to support pH and voltage drop in the presence of diffusion. Membrane interaction with phosphates may have originated with the natural proton-semiconducting properties of phospholipid bilayers in water [82], and this may have been the first requirement of life for membranes, as pre-enzymatic rTCA has rate kinetics determined by bulk processes [83].

This proposal for the first association of membranes with reductive metabolism is consistent with their later use as containers that enhance reaction rates involving enzymes produced by side reactions from the core cycle. It is a natural sequence for the cellularization of primordial autotrophs, in contrast with heterotrophic origin stories where macromolecules are responsible for the first imprinting of heritable structure, and in which catalysts first become enclosed in membranes more or less by accident [84].

We note that phospholipid bilayer geometry essentially gives spherical topology as a consequence of physical chemistry [85], so once again life requiring geometry for energy transduction takes the thermodynamically favored path by being spherical. Yet in the association of variants of metabolism with enclosing vessels, we find the first forms of individuation. In growth and fissioning by metabolic production of membrane lipids, we find a crude form of inheritance of variations. Once again, it is not clear at what point these processes, often taken as defining characteristics of life, become surprising if energy transduction to phosphates has already been invoked to bind membranes to metabolism.

The requirement for cell walls or some other form of osmotic stabilization is similarly an almost immediate consequence of containment of any non-ambient metabolic process. In prokaryotes, it is provided by a universal class of structures with cross-woven amino acid and amino sugar chains, which are produced inside the cell, moved across the membrane, and assembled on the outside by a sophisticated process of insertion [86]. The more recent device of cholesterol stabilization in conjunction with a tensile cytoskeleton may have emerged to enable endocytosis, and as a result has been amenable to endosymbiosis.

Minimal genomes

Structures provide physico-chemical regulation of metabolic reactions; catalysts provide rate-kinetic regulation. Among catalysts, as among structures, an identifiable functional taxonomy, up to homology, is present.

Current prokaryotes have between 480 and 8000 genes, of which core components define their trophic relations to their environments. *Mycoplasma genitalium*, a metabolically minimal organism because it depends on the organically richest environment, has 480 genes in the wild type and an apparently necessary core of about 250 in ideal circumstances [87]. Of these, certain invariant groups, such as 95 genes associated with protein synthesis, are known to have homologs in the necessary cores of all other organisms whose minimal genomes have been isolated.

In contrast, the smallest genomes known in autotrophs have about 1500 genes. As the universal inventory of small molecules numbers about 300, it may be possible for an autotroph to function in ideal circumstances with as few as 750 genes. To identify the necessary complexity an organism must have in order to sustainably regulate metabolism in an osmotic container, it is probably necessary to experimentally isolate such minimal genomes. Because the metabolic load of genes is similar to that of catalysts (they are not consumed in providing templates for enzyme synthesis), the cost of maintaining excess genetic material is much less than that for suboptimal metabolic pathways. With the observed frequency of lateral gene transfer, it is therefore expected that most organisms have maintained so-called extraneous genes, either to enable phenotypic plasticity in fluctuating environments or as historical artifacts from ancestors living in different conditions. A functional taxonomy of genes thus provides a more reliable input to fine-tuning arguments than a purely empirical inventory. It at least enables a separation of those functions that are necessary to all organisms under all circumstances, from those that are only "necessary" for robustness or resilience (although this distinction may be only one of timescale with respect to the problem of persistence through geological time).

Regulatory feed-down onto core metabolism

All the regulatory structures we have discussed have at least a qualitative category distinction from the reagents whose pathways we have argued are most like those of abiotic chemical networks. To function, all the regulators require polymerization of small molecules in the minimal set of 300. Although cell membranes form spontaneously by surface energy minimization, they are only stable with the addition of either amino-sugar or cellulose cell walls or with the addition of membrane-dissolved cholesterol and the cytoskeleton. Amino and nucleic acids are both generated along short pathways from the TCA core, but to be useful as catalysts and templates they must be polymerized with particular sequences.

A qualitative difference arises between the somewhat sparse but orderly sampling of the reaction network among all small metabolites and the much sparser and more clearly contingent sampling of the space of synthesized polymers. The reactivity of biomass, a consequence of its reduction stoichiometry as shown in Figure 18.2, also induces small free energy differences among different sequences, making the energetic landscape of sequence space flat compared with that of the metabolites themselves.

Polymer sequences are therefore much more likely to be governed by sampling bias in evolutionary history than are metabolites [88]. One consequence of this degeneracy under permutations of sequence is that neutral models of population genetics can provide good first approximations for the evolution of traits that depend on complex syntheses, although this has also reduced most Darwinian arguments to rationalizations and led to the complaint that Darwinian evolution is not like other scientific theories. An important second point, however, is that only such neutral molecules are eligible, by possessing a degenerate configuration space with many states, to carry mutual information *within the cell* about those characteristics of the environment that it needs to anticipate [49]. Although we have argued that it is an error to identify life with polymer chemistry and sparse sampling, we agree that a qualitative distinction of the informational and regulatory character of life emerges with this class.

The regulatory structures have a universal relation to core metabolism that appears to be a distinguishing feature of life. First, they are of low metabolic load and can therefore exist in greater diversity than the metabolites they regulate. Thus, a typical cell contains as many as tens of thousands of kinds of polymers, most in small numbers, but only several dozen metabolic reagents, in amounts that scale with the mass of the organism, and perhaps 50 building blocks, cofactors, and prosthetic groups that are shared among the polymers.

Second, whereas core metabolism generates net currents and an arrow of time from non-equilibrium thermal boundary conditions, regulatory structures inherit this arrow of time from metabolism, which generates their building blocks as raw materials for combinatorics. Thus, the plausible abstractions for metabolism are based on microscopically reversible thermodynamics and chemistry, with non-equilibrium boundary conditions. The more natural abstractions for regulatory structures are cellular automata or the growth-and-culling models of Darwinian population genetics [16]. It is known that important prohibitions on the formation of order, such as the absence of phase transitions in one-dimensional systems, which apply to near-reversible processes, are not binding for cellular automata because of their time-reversal asymmetry and constructive dynamics [89, 90]. The opportunities for encoding stable information are naturally larger in these driven structures for dynamical, as well as sampling, reasons.

Finally, because regulatory structures usually do not flow between the organism and its environment, and because they act catalytically within the organism, selection of these structures takes place only through their impacts on the rate of core metabolism and their ability to efficiently draw energy and material from it for self-reproduction. (It is implicit here that the fit participation of the organism in its *living* environment is almost always a leading factor impacting the net core metabolism of the ecology as a whole.) Those structures enabling greater bulk free energy transduction, more efficient synthesis, or reduced thresholds to autocatalysis either survive in expanded environments or exclude less efficient solutions in existing niches.

We designate the reciprocal relation among components, having these three properties, as "feed-down" of regulation onto core metabolism. This is a universal relational property of all living systems that provides an energetic foundation for Darwinian fitness and governs the emergence of complexity and innovation in evolution. Feed-down determines selection bias on catalytic schemata competing intraspecifically to be the surviving metabolic strategies of autotrophs and operates through material cycling, as well as energy capture, at the level of ecologies. To us, these three relational features of regulators to substrates – sparse sampling that leads to history dependence, but also to the ability to carry information about the environment; a limited gatekeeper role over the forms of organisms; and selection through feed-down reciprocity that operates both prior and subordinate to the gatekeeper role – are more fundamental to life than any single chemical class or physical structure.

Shift in biological anthropic thinking

We have emphasized relaxation, regulation, and feed-down in this analysis to account for the remarkable universality, persistence, and specificity of biochemical processes, and at the same time for the robustness and resilience of life as a whole. We expect that many universal pathways depend sensitively on the fine structure of chemistry and that these pathways would no longer be used by life if that structure were changed, perhaps even slightly. However, that does not lead us to conclude that some other structures would not arise in their place to fulfill similar functions. Thus, we try to avoid the first fallacy in a sensitivity analysis: confusing adaptation *to* the environment with singularity *of* the environment.

The sensitivity of the categories of function we have identified is much more diffcult to estimate than the sensitivity of one or another class of chemical species, both on the structure of chemistry and on geophysical history. Clearly, relaxation requires non-equilibrium thermal boundary conditions. But what makes a relaxation pathway especially amenable to regulation? Informational structures require a large configuration space with a relatively flat energetic bias and are clearly achieved by the combinatorics of the reactive CHO and CHON molecules with characteristic hydrogen saturation and free energy of formation. But what kind of periodic table, other than a vacant one, would rule out the existence of any such combinatorial regime anywhere in the space of chemicals?

These are the questions required in a sensitivity analysis of the category distinctions we have made. Given the limitations of synthetic and computational chemistry, even sampling the space of structures permitted by the known periodic table is just becoming imaginable. Sampling the space of structures of counterfactual chemistries thoroughly is far beyond our collective current capability. Yet we expect this kind of sensitivity analysis to be appropriate for anthropic reasoning about the fitness of the environment from biology, a departure from the method used in cosmology or astrophysics. In principle, questions about the richness of the laws of nature in opportunities for biological complexity are answerable, although difficult.

This categorization, which makes strong anthropic reasoning diffcult, does permit a form of weak anthropic reasoning that we consider very useful. Primordial forms of life were probably different in many respects from all modern forms because they were not tuned by four billion years of first chemical, and then Darwinian, selection; conversely, they were not in competition with any modern organism. They were likely driven by environments with higher energetic stress because the relaxation phenomena that operate throughout the globe today did not exist, although we do not envision an environment so rich in free energy and organic raw materials [38] as to enable gene-first emergence. Despite these differences of degree, decomposing metabolism into the parts likely to have been governed by invariant laws of statistics, and those likely to have been limited by sampling and biased by selection, allows us to reason uniformly about emergence and persistence. The experience we gain through attempts to apply weak anthropic reasoning to practical and accessible problems is likely to be our best ladder to a conceptual system capable of framing strong anthropic questions.

APPENDIX 18.1 ANTHROPIC PRINCIPLES AND THEIR INTERPRETATIONS

The ambiguous role of anthropic bias

Even when they make use of quantitatively expressed sensitivities, anthropic arguments have tended to be discursive rather than formal [1, 5, 6, 7]. The demonstration that some aspect of life depends sensitively on the form of a natural law or value of an initial condition may seem surprising enough to set a course of research, and can thus be useful without formalization. Hoyle's argument for the 7.6 MeV resonance of ^{12}C set an experimental agenda for Fowler and Whaling [91] without the logical grounds' needing to be specified for surprise at the abundance of astrophysical carbon.

However, sensitivity of observables to inputs, by itself, carries neither surprise nor meaning, and the same demonstration of sensitivity can be an input to anthropic arguments with a range of meanings. We show in this appendix how the same sensitivity of carbon abundance to nuclear structure can be used in anthropic principles with three distinct interpretations.

First, it can generate a pruning rule [11, 13] to simplify a combinatorially intractable, but scientifically conventional, problem. This is the minimal, pragmatic interpretation generated by the experimental problem of nuclear characterization. The observation of astrophysical carbon abundance was a proxy for the detailed and untenable (in 1954) calculation to predict nuclear structure from the masses and scattering cross-sections of elementary particles. Since such a calculation was, in principle, possible from independently measured values, the anthropic prediction of a 7.6 MeV resonance was falsifiable theoretically as well as experimentally, or alternatively could act as a constraint on less certain inputs to the constructive calculation. Second, in cosmological scenarios that admit a range of values for either matter densities or the parameters in the standard model [4, 5, 6], anthropic reasoning can act as a filter for the set of all observations we will ever be able to make. In this capacity, predictions for the distribution of initial values are not falsifiable by assumption. The Anthropic Principle is not an input to reductionist inference, either of the existence of universes with other parameters or of the theory predicting them.

Although different, both interpretations are categorized as weak anthropic principles whose meaning is defined by the refinement they make in the predictions of observables. Informationally [8], the surprise in an anthropic prediction is measured relative to the refinement of the same prediction without the condition of our existence.

The prediction of cosmological initial conditions is a different business, however, from the validation of causal theories. Initial conditions by assumption have no cause that can be varied to check their dependence, and the cosmos is by assumption a single instance of these values, which we cannot compare experimentally to anything else to test for repeatability. Weak anthropic principles are a conservative response to this dilemma, in which the issue of boundary conditions is put off to a larger theory determining a distribution of pseudo-initial conditions for the universe we see [2, 3]. A radical response is to propose that only one universe exists and to make predictions for its initial conditions on some criterion other than causality, such as a form of mathematical consistency [92].

Strong anthropic principles provide alternative criteria to mathematical consistency for the declaration of initial conditions: whatever selects them does so for consistency with our later existence. Pragmatically, there is no difference between the predictions made by strong and weak anthropic principles, so the philosophical difference attached to their interpretation must come from outside scientific epistemology. The informativeness or surprise in the strong anthropic condition must similarly be measured with respect to some other source of uncertainty, since it is not compared with the predictive specificity of causal theory. Some strong interpretation, in which human existence has significance for the selection of cosmological initial conditions, seems to have been favored by Hoyle and Wickramasinghe [93] and obliquely by Henderson [7].

Bayesian structure of anthropic reasoning

The common element that separates anthropic argument from constructive logic is the use of posterior observations (specifically, involving the existence of human or other life) to infer boundary conditions that from the perspective of construction would be regarded as prior knowledge. Such inputs to constructive logic are conventionally referred to as "priors." Bayesian inference [12] nicely captures the logical role of anthropic bias, while remaining general enough to admit the range of interpretations we have described. Formally, Bayes's theorem relates posterior to prior probabilities for observables from a stochastic process with well-defined joint probabilities. If the conditional probabilities that appear in it are interpreted as the index functions of constructive logical arguments, however, the posterior probability can be interpreted as the likelihood of a generative model, given the data. Such an interpretation makes explicit the ontological commitments that have been made by the modeler in the form of Bayesian priors. The important conclusion of such probabilistic representations of theory building is that non-trivial inputs always remain unjustified by the data used. The differences in anthropic interpretation hinge on where these inputs originate.

For very simple problems of machine learning and pattern discovery, Bayesian inference provides practical, quantitative measures of the relative likelihood of models [94]. Obviously, any Bayesian representation of human theory building in nuclear physics, cosmology, or biochemistry (much less theology) is at best a cartoon, a sort of pseudo-code to capture the logical structure of the actual process whose implementation we do not nearly understand. None the less, it is not ridiculous to write down, if one remembers that the purpose of scientific statements is to guide behavior by offering some rationale for betting. A formal representation of anthropic argument that captures the change in rational behavior generated by the posterior observation can be valid without presuming any description of beliefs.

We therefore formalize Hoyle's prediction as an example, first in the minimal form of a pruning rule, and then extended to the conservative and radical cosmological forms. We identify the source of predictive specificity in anthropic argument and separate from that the source of interpretation of the priors in different applications.

The minimal Bayesian structure of Hoyle's proposal

Suppose nuclear physics offers a denumerable set of rules that we can denote by N, in principle sufficient to predict the states and transitions of all the chemical elements. This set may be well approximated by a few particle masses and quantum numbers, and S-matrices for the interactions of nucleons and π mesons. Then suppose we are seeking some description of the states and resonances of the ^{12}C nucleus, which we may denote $S(^{12}C)$. For convenience here, we project the description onto the positions of resonances relevant to stellar nucleosynthesis.

By assumption on N, we can write a prediction of the ^{12}C structure in the form of a conditional probability:

$$P(S(^{12}C)|N)$$

$P(S(^{12}C)|N)$ is simply the index function taking value one on the structure predicted by N, and zero elsewhere. (Given the purpose and limitations of the example, it is a digression to distinguish a discrete index function from a Dirac δ-functional if N and the resonances have real-valued parameters, or the more realistic case where inputs and outputs have probability measures generated exogenously by measurement precision.)

At any time, the completeness or accuracy of calculations of nuclear structure is limited. In the 1950s, the limitations were severe. We can measure the degree of intractability by identifying a proper subset $\hat{N} \subset N$ of rules and parameters usable in calculations and the appropriately normalized index function (no longer having one solution) of structures admitted by $\hat{N}$:

$$P(S(^{12}\text{C})|\hat{N})$$

$\hat{N}$ may be as crude as the Fermi liquid theory for nucleons, plus classical estimates of electromagnetic energy density, and $P(S(^{12}\text{C})|\hat{N})$. may have such broad support that it places essentially no restriction on stellar nucleosynthesis. For this reason, we introduce Hoyle's posterior observation of estimated carbon abundances.

Denote by $\hat{A}(M)$ the estimated cosmic metal abundance given the observation of life on earth, where "metal" conventionally denotes carbon and all heavier elements. Suppose cosmology and astrophysics offer some set $\hat{\chi}$ of inputs to tractable calculations for generating $\hat{A}(M)$, which includes primordial nuclear abundances, but also models of star and planetary-system formation. For simplicity, we first treat $\hat{\chi}$ and $\hat{N}$ as independent, as they effectively were for Hoyle, but later come back and revise this assumption to represent more recent cosmological anthropic arguments.

A range of parameters may exist for $\hat{\chi}$, because of computational intractability such as currently afflicts planetary-system formation [15] or even of the lack of a theory about the existence of a unique putative characterization χ [4, 5, 6]. Unlike the case for the usable subset $\hat{N}$, identified uniquely by the Standard Model and methodological constraints, it makes sense to sum over the parameters of $\hat{\chi}$ with some distribution $P(\hat{\chi})$. The nuclear and astrophysical inputs provide a joint distribution:

$$P(\hat{A}(M)S(^{12}\text{C})|\hat{N}) = \sum_{\hat{\chi}} P(\hat{A}(M)S(^{12}\text{C})|\hat{\chi}\hat{N})P(\hat{\chi}) \tag{18.A1}$$

$P(\hat{A}(M)S(^{12}\text{C})|\hat{\chi}\hat{N})$ is the index function for admitted values of ^{12}C structure and metal abundance given $\hat{\chi}$ and $\hat{N}$.

Constructively, the ^{12}C structure is prior in principle to the metal abundance, but the procedure to extract it is intractable, whereas the empirical metal abundance is readily accessible. Thus, we factor Equations (18.A1) in two different ways, using the chain rule for probabilities:

$$P(\hat{A}(M)S(^{12}\text{C})|\hat{N}) = \sum_{\hat{\chi}} P(\hat{A}(M)|S(^{12}\text{C})\hat{N}\hat{\chi})P(S(^{12}\text{C})|\hat{N})P(\hat{\chi})$$

$$= P(S(^{12}\text{C})|\hat{A}(M)\hat{N}) \sum_{\hat{\chi}} P(\hat{A}(M)|\hat{N}\hat{\chi})P(\hat{\chi}) \tag{18.A2}$$

where the conditional probability $P(S(^{12}\text{C})|\hat{\chi}\hat{N})$ formally required by the chain rule has been shortened to $P(S(^{12}\text{C})|\hat{N})$ in keeping with our assumption that nuclear structure is conditionally independent of cosmological parameters given N and that the maximal constructively usable subset of N is $\hat{N}$.

Expanding the probability for abundances in Equation 18.A2 as

$$\sum_{\hat{\chi}} P(\hat{A}(M)|\hat{N}\hat{\chi})P(\hat{\chi}) = \sum_{\hat{\chi},S'} P(\hat{A}(M)|S'(^{12}\text{C})\hat{N}\hat{\chi})P(S'(^{12}\text{C})|\hat{N})P(\hat{\chi}) \tag{18.A3}$$

Bayes's theorem is the rearrangement

$$P(S(^{12}\text{C})|\hat{A}(M)\hat{N}) = \frac{\sum_{\hat{\chi}} P(\hat{A}(M)|S(^{12}\text{C})\hat{N}\hat{\chi})P(S(^{12}\text{C})|\hat{N})P(\hat{\chi})}{\sum_{\hat{\chi},S'} P(\hat{A}(M)|S'(^{12}\text{C})\hat{N}\hat{\chi})P(S'(^{12}\text{C})|\hat{N})P(\hat{\chi})} \tag{18.A4}$$

Hoyle observed that non-trace metal abundances result only from those approximate calculations $P(\hat{A}(M)|S(^{12}\text{C})\hat{N}\hat{\chi})$ where $S(^{12}\text{C})$ has a resonance within a relatively narrow range around 7.6 MeV. The actual structure of this prediction has since been refined [95]. If $P(S(^{12}C)|\hat{N})$ and $P(\hat{\chi})$ are comparatively featureless and bounded in the range where these forward predictions are made, then the support of the posterior distribution for $S(^{12}C)$ is a narrow region around 7.6 MeV.

Minimal interpretation: a pragmatic pruning rule

Hoyle's argument narrows $P(S(^{12}C)|\hat{A}(M)\hat{N})$ relative to $P(S(^{12}C)\hat{N})$ to the extent that $P(\hat{A}(M)|S'(^{12}C)\hat{N}\hat{\chi})$ is a sharply peaked prediction of metal abundance. $(S'(^{12}C)|\hat{N})$ is an augmented nuclear model based on a constructively prior but unknown input. As values of $S'(^{12}C)$ are not tightly constrained by $P(S'(^{12}C)|\hat{N})$, the unaugmented rules $\hat{N}$ simply coarsely bound the search space for this input. Similarly, if $P(\hat{\chi})$ is generated from observations of the local galactic neighborhood and the Copernican assumption of mediocrity, it is primarily limited by confidence in planetary-system models that do not alter terrestrial relative to cosmic abundances by many orders of magnitude. Thus, in this application it could be replaced by some "given" value of $(\hat{\chi})$ (as were the "given" rules $\hat{N}$) without changing the results.

Hoyle's posterior observation of $\hat{A}(M)$ thus generated a pruning rule [11, 13] by which the experimentally likely configurations for $S(^{12}C)$ were narrowed for Fowler and Whaling. Its predictive specificity comes from the quality of the index $P(\hat{A}(M)|S'(^{12}C)\hat{N}\hat{\chi})$, but the commitment to priors $\hat{N}$ and $P(\hat{\chi})$ comes from outside this problem. Such exogenous Bayesian priors can always (in principle) undermine the specificity of any model $P(\hat{A}(M)|S'(^{12}C)\hat{N}\hat{\chi})$ that depends on them non-trivially, a formal recognition of the fact that scientific conclusions inevitably depend on philosophical (or, more neutrally, methodological) commitments external to scientific reasoning [15].

In this problem, $\hat{N}$ and $\hat{\chi}$ may themselves be regarded as posterior to the collective validation sets of nuclear physics and astrophysics and cosmology, with distributions vastly narrowed relative to the true ontological priors. Furthermore, the preferred augmented model $(S(^{12}C)\hat{N})$ in this pragmatic interpretation of Hoyle's anthropic argument was provisional, required either to be derivable from N with improved computations or to act as an experimental constraint on improved estimates of N in keeping with the reductionist program [11]. Weakening of both features in multiverse scenarios [5, 6] and Henderson's conception of environmental fitness [7] significantly changes the interpretation of both the posteriors and anthropic reasoning.

The conservative cosmological extension

A representative anthropic argument from carbon abundance relevant to multiverse cosmology uses the same variables with a slightly different interpretation. Suppose now that nuclear physics depends on the cosmological model through parameters (e.g. arbitrary features of consistent solutions to some more fundamental grand unified theory) and that $\hat{N}$ is the subset of parameter-independent features (rather than those that are simple to use). Then expand $\hat{\chi}$ to include the cosmological parameters that are theoretically variable, and suppose that structures such as $S(^{12}C)$ are predictable from $\hat{N}$ and $\hat{\chi}$ with some index function

$$P(S(^{12}C)|\hat{N}\hat{\chi})$$

For the sake of this model, suppose that the interesting variations in $\hat{\chi}$ do not depend constructively on $\hat{N}$, so that

$$P(\hat{\chi}|\hat{N}) = P(\hat{\chi}) \tag{18.A5}$$

The theory may imply that the distribution of nuclear, cosmological, and astrophysical properties

$$P(S(^{12}C)\hat{\chi}|\hat{N}) = P(S(^{12}C)|\hat{N}\hat{\chi})P(\hat{\chi}) \tag{18.A6}$$

is no longer sharp because $P(\hat{\chi})$ is broad, even if $P(S(^{12}C)|\hat{N}\hat{\chi})$ is sharp. The best we can do then is factor

$$P(\hat{A}(M)S(^{12}C)\hat{\chi}|\hat{N}) = P(\hat{A}(M)S(^{12}C)|\hat{N}\hat{\chi})P(\hat{\chi}) \tag{18.A7}$$

possibly extending $\hat{A}(M)$ to several features on which life appears to depend sensitively.

The anthropic Bayesian posterior becomes

$$P(S(^{12}C)\hat{\chi}|\hat{A}(M)\hat{N}) = \frac{P(\hat{A}(M)|S(^{12}C)\hat{\chi}\,\hat{N})P(S(^{12}C)\hat{\chi}|\hat{N})}{\sum_{\hat{\chi}',S'} P(\hat{A}(M)|S'(^{12}C)\hat{\chi}'\hat{N})P(S'(^{12}C)\hat{\chi}'|\hat{N})} \tag{18.A8}$$

The constructive model $P(\hat{A}(M)|S(^{12}C)\hat{\chi}\,\hat{N})$ acts as before, as a filter on the prior $P(S(^{12}C)\hat{\chi}|\hat{N})$ output from the cosmological theory. In this respect, the anthropic bias is indistinguishable from the pruning rule generated by Hoyle; but as it is driven by incomplete theoretical specification rather than mere intractability, the posterior distribution is not provisional and cannot act as a constraint on reductionist refinements of $\hat{N}$ or $\hat{\chi}$. It is not a substitute for a theory of particular initial conditions, because it adds nothing axiomatic to the structure of the causal theories represented by $P(\hat{A}(M)|S(^{12}C)\hat{\chi}\,\hat{N})$ and $P(S(^{12}C)\hat{\chi}|\hat{N})$, and it places no constraints on the latter if the former is sufficiently sharp. At most, it elaborates the structure of conditional posterior distributions that we can practically sample.

The "fine-tuning" interpretation

The difference between the conservative and radical interpretations of anthropic arguments for cosmological initial conditions corresponds roughly to the difference between frequentist and Bayesian interpretations of the distribution $(S(^{12}C)\hat{\chi}|\hat{N})$ we have used as a prior in Equation 18.A8. (This terminology should not mask the fact that Bayes's theorem relates prior to posterior distributions under successive observations in either interpretation.)

Cosmological theories that generate a distribution of universes correspond to a frequentist interpretation in which all possible universes $(S(^{12}C)\hat{\chi})$ are "really" instantiated. Even if we are unable to sample them in direct experience, denying us the strict frequentist interpretation, our commitment to a theory (made for whatever exogenous reasons) generates a frequentist definition for $P(S(^{12}C)\hat{\chi}|\hat{N})$. Under this interpretation, anthropic bias limits the range of posterior observables sampled by our experience relative to those generated by the process that defines the distribution.

The alternative "Bayesian" interpretation of $P(S(^{12}C)\hat{\chi}|\hat{N})$ retains the idea of a single, entire universe in which values of initial conditions $(S(^{12}C)\hat{\chi})$ should be regarded as fixed aspects of a unique N, and no maximal invariant proper subset $\hat{N} \subset N$ exists. Although theories of initial conditions are by assumption not causal, they are expected to commit to specific forms the same way they currently specify the algorithmic structure of natural laws. A non-sharp $P(S(^{12}C)\hat{\chi}|\hat{N})$ is then associated by Bayesians with a range of "beliefs," of which all but the correct $(S(^{12}C)\hat{\chi})$ are interpreted as so-called counterfactuals.

If the index function $P(\hat{A}(M)|S(^{12}C)\hat{\chi}\hat{N})$ of the generative model is sharp, as it is for carbon nucleosynthesis, then a well-defined posterior $P(S(^{12}C)\hat{\chi}|\hat{A}(M)\hat{N})$ is consistent with initial conditions only in the support of that index function. If $\hat{A}(M)$ is constructively posterior, and the constructive priors admitted counterfactuals $(S(^{12}C)\hat{\chi})$ relative to the posterior $P(S(^{12}C)\hat{\chi}|\hat{A}(M)\hat{N})$, then a Bayesian interpretation of the theories of initial conditions claims that a constructive prior is missing, and the theory is incomplete. The requisite constructive prior is called "finely tuned" with respect to the posterior anthropic observation $\hat{A}(M)$ if the support of $P(\hat{A}(M)|S(^{12}C)\hat{\chi}\hat{N})$ has some small measure relative to the support of the current prior $P(S(^{12}C)\hat{\chi}|\hat{N})$. An intrinsic measure of sensitivity in Hoyle's prediction is the half-width of the resonance, a characteristic scale for nuclear features, divided by the center frequency of the resonance.

In the Bayesian interpretation, the "belief system" that generates priors $P(S(^{12}C)\hat{\chi}|\hat{N})$ originates outside scientific methodology for discovering causal laws. Thus, the interpretation of additional finely tuned principles that omit counterfactuals from the prior, while leaving $\hat{A}(M)$ posterior to the causal dynamics, is extrascientific.

One cannot, however, sensibly define a constructive prior to choose initial conditions "only" to produce the observed posterior observations $\hat{A}(M)$. If the condition of equality between $P(S(^{12}C)\hat{\chi}|\hat{A}(M)\hat{N})$ and $P(S(^{12}C)\hat{\chi}|\hat{N})$ does not affect the causal dynamics $P(\hat{A}(M)|S(^{12}C)\hat{\chi}\hat{N})$ (miraculous intervention excluded from conventional natural law [15]), then it is a tautology within the anthropic argument, imposing no selection on prior beliefs other than what is computed in the posteriors, and thus having no constructive role.

A non-trivial constructive prior would be some theory of initial conditions that produces equivalent $P(S(^{12}C)\hat{\chi}|\hat{A}(M)\hat{N})$ and $P(S(^{12}C)\hat{\chi}|\hat{N})$ and also places constraints on other posterior observables. The theological overtones often given to fine-tuning arguments, even when not directly stated [93], presumably intend to establish a link between initial conditions permitting the emergence of carbon-based life and those implying a favored moral system (thus also in principle observable) through some extra-human construct such as Kant's categorical imperative [96].

The choice of anthropic evidence

The posterior evidence variable $\hat{A}(M)$ has three distinguishable roles in the three anthropic interpretations we have presented. It is chosen for predictive specificity in the generation of a pruning rule; for appropriateness as a constraint on the observable ensemble in multiverse cosmology; and for selectivity of those aspects of initial conditions we regard as relevant to modification of beliefs in the Bayesian cosmological interpretation.

In all three applications, the quantity we really want to compute is (some generalization of)

$$P(S(^{12}C)\hat{\chi}|\hat{N}) = P(S(^{12}C)\hat{\chi}|\hat{A}(M)\hat{N})P(\hat{A}(M)) \tag{18.A9}$$

In the original work of Hoyle and Fowler, $P(\hat{A}(M))$ represents estimation uncertainty, from measurement precision and model diversity, of a single character selected for its sensitivity and relevance to a specific experimental problem. In cosmological models of either type, $P(\hat{A}(M))$ can also represent a distribution over alternative characters regarded as substitutes for identifying the posterior distribution we wish to specify.

To extend the former role from Hoyle's example to biochemistry, we must specify what intractable problem we wish to prune in order to choose the relevant and readily

416 Eric Smith and Harold J. Morowitz

measurable anthropic characters. To extend the latter role, we must specify what ensemble or belief structure is to be filtered, what alternative forms of anthropic evidence $\hat{A}(M)$ we consider equivalent to specify "life," and possibly the relative weights in a distribution $P(\hat{A}(M))$ we wish to give to alternative characters.

References

[1] Hoyle, F. On nuclear reactions occurring in very hot stars. I. The synthesis of elements from carbon to nickel. *Astrophysical Journal Supplement*, **1** (1954), 121–46.
[2] Coleman, S. (1988). Why there is nothing rather than something: a theory of the cosmological constant. *Nuclear Physics*, B**310** (1988), 643–68.
[3] Hawking, S. W. The cosmological constant is probably zero. *Physics Letters*, **134B** (1984), 403–4.
[4] Barrow, J. D. and Tipler, F. J. *The Anthropic Cosmological Principle*. New York, NY: Oxford University Press (1986).
[5] Davies, P. *God and the New Physics*. New York, NY: Simon and Schuster (1984).
[6] Davies, P. *Are We Alone? Philosophical Implications of the Discovery of Extraterrestrial Life*. New York, NY: Basic Books (1995).
[7] Henderson, L. J. (1913). *The Fitness of the Environment: An Inquiry into the Biological Significance of the Properties of Matter*. New York, NY: Macmillan. Repr. (1958) Boston, MA: Beacon Press; (1970) Gloucester, MA: Peter Smith.
[8] Shannon, C. E. and Weaver, W. *The Mathematical Theory of Communication*. Urbana, IL: University of Illinois Press (1971).
[9] Gell-Mann, M. and Lloyd, S. Information measures, effective complexity, and total information. *Complexity*, **2** (1996), 44–52.
[10] Kasting, J. F. Habitable zones around low-mass stars and the search for extraterrestrial life. *Origins of Life and Evolution of Biospheres*, **27** (1997), 291–307.
[11] Morowitz, H. J. *The Emergence of Everything: How the World Became Complex*. New York, NY: Oxford University Press (2002).
[12] Savage, L. J. *The Foundations of Statistics*, 2nd rev. edn. New York, NY: Dover Publications (1972).
[13] Holland, J. H. *Emergence: From Chaos to Order*. Reading, MA: Addison-Wesley (1998).
[14] Popper, K. Reduction and the incompleteness of science. In *Studies in the Philosophy of Biology*, ed. F. Ayala and T. Dobzhansky. Berkeley, CA: University of California Press (1974).
[15] Fry, I. *The Emergence of Life on Earth: A Historical and Scientific Overview*. New Brunswick, NJ: Rutgers University Press (2000).
[16] Hartl, D. L. and Clark, A. G. *Principles of Population Genetics*, 2nd edn. Sunderland, MA: Sinauer Associates (1988).
[17] Grant, V. *The Evolutionary Process*. New York, NY: Columbia University Press (1985).
[18] Wilson, K. G. and Kogut, J. The renormalization group and the ε expansion. Physics Reports. *Physics Letters*, **12C** (1974), 75–200.
[19] Huang, K. *Statistical Mechanics*. New York, NY: Wiley (1987).
[20] Alligood, K. T., Sauer, T. and Yorke, J. A. *Chaos: An Introduction to Dynamical Systems*. New York, NY: Springer (1997).
[21] Gould, S. J. *The Structure of Evolutionary Theory*. Cambridge, MA: Belknap Press, (2002).

[22] Conway-Morris, S. *Life's Solution: Inevitable Humans in a Lonely Universe.* Cambridge, UK: Cambridge University Press (2003).

[23] Karp-Boss, L. and Jumars, P. A. Motion of diatom chains in steady shear flow. *Limnology and Oceanography*, **43** (1988), 1767–73.

[24] Karp-Boss, L., Boss, E. and Jumars, P. A. Motion of dinoflagellates in a simple shear flow. *Limnology and Oceanography*, **45** (2000), 1594–602.

[25] Morowitz, H. J. *Beginnings of Cellular Life*. New Haven, CT: Yale University Press (1992).

[26] Fenchel, T. and Finlay, B. J. *Ecology and Evolution in Anoxic Worlds*. New York, NY: Oxford University Press (1995).

[27] Fenchel, T. *Origin and Early Evolution of Life*. New York, NY: Oxford University Press (2002).

[28] Blankenship, R. E. *Molecular Mechanisms of Photosynthesis*. Malden, MA: Blackwell Science (2002).

[29] Woese, C. R. Interpreting the universal phylogenetic tree. *Proceedings of the National Academy of Sciences, USA*, **97** (2000), 8392–6.

[30] Woese, C. R. On the evolution of cells. *Proceedings of the National Academy of Sciences, USA*, **99** (2002), 8742–7.

[31] Margulis, L. and Schwartz, K. V. *Five Kingdoms: An Illustrated Guide to the Phyla of Life on Earth*. New York, NY: W. H. Freeman (1998).

[32] Doolittle, W. F. Evolution, uprooting the tree of life. *Scientific American*, **282** (2000), 90–5.

[33] Jain, R., Rivera, M. C., Moore, J. E. *et al.* Horizontal gene transfer in microbial genome evolution. *Theoretical Population Biology*, **61** (2002), 489–95.

[34] Raymond, J., Zhaxybayeva, O., Gogarten, J. P. *et al.* Whole-genome analysis of photosynthetic prokaryotes. *Science*, **298** (2002), 1616–20.

[35] Stryer, L. *Biochemistry*, 4th edn. New York, NY: W. H. Freeman (1995).

[36] Hillis, D. M., Moritz, C. and Mable, B. K., eds. *Molecular Systematics*, 2nd edn. Sunderland, MA: Sinauer Associates (1996).

[37] Fontana, W. and Buss, L. W. The arrival of the fittest: toward a theory of biological organization. *Bulletin of Mathematical Biology*, **56** (1994), 1–64.

[38] Miller, S. L. and Orgel, L. E. *The Origins of Life on the Earth*. Englewood Cliffs, NJ: Prentice-Hall (1974).

[39] Crick, F. *Life Itself*. New York, NY: Simon and Schuster (1984).

[40] Monod, J. *Chance and Necessity*. Glasgow: Collins (1974).

[41] Lewontin, R. *The Triple Helix: Gene, Organism, and Environment*. Cambridge, MA: Harvard University Press (2000).

[42] Lewontin, R. C., Rose, S. and Kamin, L. J. *Not in Our Genes: Biology, Ideology and Human Nature*. New York, NY: Pantheon Books (1984).

[43] Gould, S. J. *Wonderful Life: The Burgess Shale and the Nature of History*. New York, NY: W. W. Norton (2000).

[44] Morowitz, H. J., Srinivasan, V. S., Copley, S. *et al.* The simplest enzyme revisited. *Complexity*, **10** (2005), 12–13.

[45] Morowitz, H. J. Phenetics, a born-again science. *Complexity*, **8** (2003), 12–13.

[46] Eschenmoser, A. Chemistry of potentially prebiological natural products. *Origins of Life and Evolution of Biospheres*, **24** (1994), 389–423.

[47] Weber, A. L. Chemical constraints governing the origin of metabolism: the thermodynamic landscape of carbon group transformations under mild aqueous conditions. *Origins of Life and Evolution of Biospheres*, **32** (2002), 333–57.

[48] Weber, A. L. Kinetics of organic transformations under mild aqueous conditions: implications for the origin of life and its metabolism. *Origins of Life and Evolution of Biospheres*, **34** (2004), 473–95.

[49] Adami, C., Ofria, C. and Collier, T. C. Evolution of biological complexity. *Proceedings of the National Academy of Sciences, USA*, **97** (2000), 4463–68.

[50] Williams, R. J. P. and Fraústo da Silva, J. J. R. Evolution was chemically constrained. *Journal of Theoretical Biology*, **220** (2003), 323–43.

[51] Shpak, M., Stadler, P. F., Wagner, G. P. *et al.* Aggregation of variables and system decomposition: applications to fitness landscape analysis. *Theory in Bioscience*, **123** (2004), 33–68.

[52] Shpak, M., Stadler, P. F., Wagner, G. P. *et al.* Simon Ando decomposability and fitness landscapes. *Theory in Bioscience*, **123** (2004), 139–80.

[53] Jaynes, E. T. *Papers on Probability, Statistics, and Statistical Physics*, ed. R. D. Rosenkrantz, Boston, MA: Kluwer, (1983).

[54] Schrödinger, E. F. *What Is Life?: The Physical Aspect of the Living Cell*. New York, NY: Cambridge University Press (1992).

[55] Hoelzer, G., Pepper, J. and Smith, E. On the logical relationship between natural selection and self-organization. *Journal of Evolutionary Biology* **19** (2006), 1785–94.

[56] Lengeler, J. W., Drews, G. and Schlegel, H. G. *Biology of the Prokaryotes*. Stuttgart: Thieme (1999).

[57] Schlegel, H. G. and Bowien, B. *Autotrophic Bacteria*. New York, NY: Springer-Verlag (1989).

[58] Westheimer, F. H. Why nature chose phosphates. *Science*, **235** (1987), 1173–8.

[59] Peters, O., Hertlein, C. and Christensen, K. A complexity view of rainfall. *Physical Review Letters*, **88** (2002), 1–4.

[60] Peters, O., Hertlein, C. and Christensen, K. Rain: relaxations in the sky. *Physical Review*, E**66** (2002), 36120–9.

[61] Peixoto, J. P. and Oort, A. H. *Physics of Climate*. New York, NY: Springer-Verlag, (1992).

[62] Glansdorff, P. and Prigogine, I. *Thermodynamic Theory of Structure*. New York, NY: Wiley (1971).

[63] Smith, E. Carnot's theorem as Noether's theorem for thermoacoustic engines. *Physical Review*, E**58** (1998), 2818–32.

[64] Smith, E. Statistical mechanics of self-driven Carnot cycles. *Physical Review*, E**60** (1999), 3633–45.

[65] Morowitz, H. J., Kostelnik, J. D., Yang, H. *et al.* The origin of intermediary metabolism. *Proceedings of the National Academy of Sciences, USA*, **97** (2000), 7704–8.

[66] Miller, S. L. and Smith-Magowan, D. The thermodynamics of the Krebs cycle and related compounds. *Journal of Physical and Chemical Reference Data*, **19** (1990), 1049–73.

[67] Plyasunov, A. V. and Shock, E. L. Thermodynamic functions of hydration of hydrocarbons at 298.15 K and 0.1 MPa. *Geochimica et Cosmochimica Acta*, **64** (2000), 439–68 (data available at http://webdocs.asu.edu).

[68] Alberty, R. A. Standard Gibbs free energy, enthalpy, and entropy changes as a function of pH and pMg for several reactions involving adenosine phosphates. *Journal of Biological Chemistry*, **244** (1969), 3290–302.

[69] Andreev, A. V., Emelyanov, V. I. and Ilinskii, Y. A. *Cooperative Effects in Optics: Superradiance and Phase Transitions*. Philadelphia, PA: Institute of Physics Publications (1993).

[70] Kauffman, S. A. Autocatalytic sets of proteins. *Journal of Theoretical Biology*, **119** (1986), 1–24.

[71] Kauffman, S. A. *The Origins of Order: Self-Organization and Selection in Evolution.* New York, NY: Oxford University Press (1993).

[72] Fontana, W. Algorithmic chemistry. In *Artificial Life II*, ed. by C. G. Langton *et al.* New York, NY: Addison-Wesley (1991), pp. 159–209.

[73] Fox, S. W. Life from an orderly cosmos. *Naturwissenschaften*, **67** (1980), 576–81.

[74] Wächtershäuser, G. Evolution of the first metabolic cycles. *Proceedings of the National Academy of Sciences, USA*, **87** (1990), 200–4.

[75] Wächtershäuser, G. Groundworks for an evolutionary biochemistry: the iron-sulphur world. *Progress in Biophysics and Molecular Biology*, **58** (1992), 85–201.

[76] Woodruff, W. H. (personal communication). For a coarse-grained comparison of the resulting albedo of land and water, see [61], Sections 6.3 and 6.7.

[77] See [35], p. 677, where the author estimates *c.* 30% conversion effciency from 8 absorbed photons per CO_2 reduced to hexose. In [28], p. 35, the author refines this slightly, correcting for quantum efficiencies. In all cases, this is an upper bound under idealized conditions. See Spanner, D. C. *Introduction to Thermodynamics.* New York, NY: Academic Press (1964) for some discussion of the complexity of this calculation, as well as Clayton, R. K. *Light and Living Matter, vol. 1: The Physical Part.* New York, NY: McGraw-Hill (1970), pp. 91–6, 98–103.

[78] Wald, G. Fitness in the universe: choices and necessities. *Origins of Life and Evolution of Biospheres*, **5** (1974), 7–27.

[79] Jackson, J. D. *Classical Electrodynamics*, 2nd edn. New York, NY: Wiley (1975).

[80] Darwin, F. *The Life and Letters of Charles Darwin*, vol. 3. New York, NY: Johnson Reprint Corp. (1969), p. 18.

[81] Baltascheffsky M. and Nyren, P. The synthesis and utilization of inorganic pyrophosphate. In *Bioenergetics*, ed. L. Ernster. Amsterdam: Elsevier (1984), pp. 187–206.

[82] Morowitz, H. J. Proton semiconductors and energy transduction in biological systems. *American Journal of Physiology*, **235** (1978), R99–114.

[83] Smith, E. and Morowitz, H. J. Universality in intermediary metabolism. *Proceedings of the National Academy of Sciences, USA*, **101** (2004), 13168–73.

[84] Hanczyk, M. M., Fujikawa, S. M. and Szostak, J. W. Experimental models of primitive cellular compartments: encapsulation, growth, and division. *Science*, **302** (2003), 618–22.

[85] Božič, B. and Svetina, S. A relationship between membrane properties forms the basis of a selectivity mechanism for vesicle self-reproduction. *European Biophysics Journal*, **33** (2004), 565–71.

[86] Koch, A. L. How did bacteria come to be? *Advances in Microbial Physiology*, **40** (1998), 353–99.

[87] Fraser, C. M., Gocayne, J. D., White, O. *et al.* The minimal gene complement of *Mycoplasma genitalium. Science*, **270** (1995), 397–403.

[88] King, J. L. and Jukes, T. H. Non-Darwinian evolution. *Science*, **164** (1969), 788–98.

[89] Gacs, P. Reliable cellular automata with self-organization. *Journal of Statistical Physics*, **103** (2001), 45–267.

[90] Gray, L. Introduction to Gacs's positive-rates paper. *Journal of Statistical Physics*, **103** (2001), 1–44.

[91] Burbidge, E. M., Burbidge, G. R., Fowler, W. A. *et al.* Synthesis of the elements in stars. *Reviews of Modern Physics*, **29** (1957), 547–650.

[92] Hartle, J. B. and Hawking, S. W. Wave function of the universe. *Physical Review*, **D28** (1983), 2960–75.

[93] Hoyle, F. and Wickramasinghe, C. *Evolution from Space*. London: J. M. Dent and Sons (1981).

[94] Jordan, M. I. *Learning in Graphical Models*. Cambridge, MA: Massachusetts Institute of Technology Press (1999).

[95] Livio, M., Hollowell, D., Weiss, A. *et al.* The anthropic significance of the existence of an excited state of 12C. *Nature*, **340** (1989), 281–4.

[96] Kant, I. *Groundwork for the Metaphysics of Morals*, ed. A. W. Wood. New Haven, CT: Yale University Press (2002).

19

Coarse-tuning in the origin of life?

Guy Ourisson*

Like others (see for instance Oparin, 1968; Monod, 1970), I too am convinced that the origin of life will be eventually understood in scientific, rational terms. However, I recognize, of course, that so far it remains an unsolved problem, and even an unfathomable one. We scientists are playing with partial explanatory hypotheses and have few hard facts at our disposal. Therefore, we must remain humble in attempting to describe each scenario and be careful to discuss each one's constraints, recognizing where the scenarios may be flexible to synthesis with other hypotheses. Thus, my present discussion explores some areas in which there may indeed be some flexibility in terms of what solutions could potentially lead to living systems – in other words, where some *"coarse-tuning"* might be tolerable.

Specifically, I address the following series of questions:

1. Is life necessarily based on carbon?
2. Must the "bricks of life" have originated by some process closely related to Miller's "spark tube" experiment, or do other possibilities exist?
3. Is it necessary that any kind of life be associated with water from the start?
4. Is it necessary for any kind of life to be cellular and for the cells to be bounded in water by membranes?
5. Are the *n*-acyl phospholipids of "classical" membranes, or the more recently discovered archaeal lipids, plausible constituents of primitive membranes?
6. Could early membrane-forming amphiphiles have simply been polyprenyl phosphates, following the "Strasbourg scenario" (Birault *et al.*, 1996)?
7. Are there automatic consequences of the self-organization of amphiphiles (such as polyprenyl phosphates) in water into membranes, and does this lead to novel properties? Specifically, I consider the available evidence for the following features of membranes:

* Sadly, Professor Ourisson passed away while this book was in production. While the editors have striven to ensure the accuracy and completeness of this chapter's content, some of the author's comments may be somewhat out of date.

Fitness of the Cosmos for Life: Biochemistry and Fine-Tuning, ed. J. D. Barrow *et al.*
Published by Cambridge University Press. © Cambridge University Press 2007.

- the automatic concentration of lipophiles within the membranes;
- the concentration-aided reactions of some lipophiles in the membranes;
- the orientation, within the anisotropic membranes, of anisotropic amphiphiles;
- the automatic self-complexification of membrane-bound vesicles by induction of vectorial properties, simply because the inside leaflet of the vesicles becomes intrinsically different from their outside one.

8. Is it possible to obtain closed vesicles that contain at least some of the components necessary for a living cell (such as nucleic acids)? Can these remain functional?

Against this background, I will examine how "coarsely tuned" these simultaneous effects can be – that is, what variations can occur and what flexibility does this enable? I shall finish close to where I started, with a confession: that major problems remain and gaps in our current understanding exist. Put frankly and simply, nobody knows yet how life originated. As I concentrate on the fine- and coarse-tuning aspects of life, I will refer to some of the many detailed reviews of the topic of prebiotic chemistry for more specific discussions (see, for example, Miller and Orgel, 1974; Sutherland and Whitfield, 1997).

Is life necessarily based on carbon?

All known living organisms use a carbon-based chemistry, and the huge variety of constituents of living organisms stems from the incredible flexibility of organic chemistry. Yet, against the possibility of building an infinite variety of carbon-based building blocks, the universal biochemistry of life as we know it involves a relatively limited set of "bricks of life," universal constituents of all living organisms – amino acids, nucleotides, sugars, lipids, and so forth – to which I add terpenes (see more below). Put another way, "secondary metabolites" are . . . well, secondary.

Based on what is known of silicon chemistry to date, a silicon-based life looks rather improbable. Some science fiction has "described" living organisms based on silicon,[1] but no really complex silicon derivative has been prepared so far. However, this restriction may well be temporary, as the possibility of the existence of complex silicon derivatives has been established by experiments in which high-energy molecular beams of dihydrogen and dinitrogen were focused onto a silicon (or silica) target in an atmosphere of dioxygen at very low pressure (Devienne *et al.*, 1998, 2002) (see Figure 19.1).

The resulting products were extracted directly into a mass spectrometer to study their fragmentation in an argon collision chamber, and the atomic composition of the peaks obtained in this manner was checked by their satellites because of the simultaneous existence of three isotopes of silicon: ^{28}Si (92%), ^{29}Si (5%), and ^{30}Si (3%). An unexpectedly large variety of molecular peaks have been observed, the

---

[1] For a "silicon zoo" containing artifacts "lurking" in electronic constructions, see for instance the *Molecular Expressions* website: http://micro.magnet.fsu.edu/creatures/.

$$\underset{\text{Sila-glycine}}{H_2Si\overset{\overset{\displaystyle NH_2}{|}}{\rule{0pt}{0pt}}\!\!-\!\!\!-\!\!\!-SiO_2H} \qquad\qquad \underset{\text{Sila-alanine}}{H_3C\overset{\overset{\displaystyle NH_2}{|}}{\underset{\underset{\displaystyle H}{}}{Si}}\!\!-\!\!\!-SiO_2H}$$

Figure 19.1. The possibility of the existence of complex silicon derivatives has been established by experiments in which high-energy molecular beams of dihydrogen and dinitrogen were focused onto a silicon (or silica) target in an atmosphere of dioxygen at very low pressure (Devienne *et al.*, 2002).

fragmentations of which suggest that they include sila-glycine, sila-alanine (see Figure 19.1), sila-threonine, sila-uracile, sila-cytosine, sila-valine, sila-glutamine, and other sila-analogs of some of the most classical carbon-based biological molecules. These identifications remain tentative, as they are based on an interpretation of the fragmentation patterns for which no comparison standards are available. None of these sila-analogs has been isolated; so far, this is the only, limited indication of the potential existence of a silicon-based chemistry of a complexity similar to that of carbon, and we must provisionally accept that only carbon can support life. However, further study of complex silicon analogs of bio-organic molecules would be interesting and helpful to justify such views one way or the other.

Must the "bricks of life" have originated by some process closely related to Miller's "spark tube" experiment, or do other possibilities exist?

Assuming carbon-based life, we may turn now to the origin of the "bricks of life" themselves, the universal constituents of all living organisms.

Just over fifty years ago, a major breakthrough occurred: the celebrated Miller "spark tube" experiment (Miller, 1953), in which sparks (simulating electric storms) in a putative prebiotic atmosphere of methane, ammonia, dihydrogen, and water produced, under recycling conditions, a complex mixture of products containing in particular a variety of α-amino acids (identical to those used by all living organisms) (Miller, 1953).

Some critics assert that Miller's atmosphere was "too reducing" to be credible as a prebiotic model (Bada and Lazcano, 2003). Dihydrogen, with its very low molecular mass and resultant high average speed, higher than the "escape velocity" on a planet the size of earth, could not have been a permanent constituent of our atmosphere because it would escape into space. However, other recent experiments have shown that by having the right elements (H, C, N, O), providing them with enough energy, and quickly cooling the system, a wide variety of bricks of life can be obtained, not only by "Miller sparks," but also by very different processes, such as:

- UV irradiation of ice particles loaded with impurities providing the necessary atoms. This procedure had been designed explicitly to simulate the possibility that the bricks of life had been produced by solar irradiation on "dirty" ice crystals in the primitive earth stratosphere (Bernstein *et al.*, 2002; Muñoz Caro *et al.*, 2002).
- High-energy molecular beams of the appropriate atoms hitting a graphite target simulating interstellar grains (Devienne *et al.*, 1998). Note that in this case, unlike in the silicon case mentioned earlier, a *positive* proof of the identity of the molecules formed was provided by the very convincing direct comparison of their mass spectrometric fragmentations with those of authentic materials. Several of the proteinous amino acids, as well as several nucleotides, have thus been identified.

These experiments could all be interpreted in the same way. In all cases, the required atoms have to be provided in some molecular form. They must be raised to a high energy level by some physical process (sparks, photo-excitation, high-energy molecular bombardment), and the resulting excited mixture ("soup," or plasma) must be quenched to produce the products observed, including the required bricks of life. This interpretation also allows us to predict that any process in which one could inject enough energy into simple molecules containing the right atoms, followed by quick cooling, might work just as well. This suggests that a water solution of, for example, formamide, sodium carbamate, or urea, sonicated under conditions energetic enough to lead to cavitation (and therefore to very high local temperatures followed by very fast cooling), might also lead to a spark-tube-like process. This prediction has not yet been tested.

All these abiotic syntheses yield, of course, only racemic mixtures of all the chiral products obtained. A word is in order at this stage about the "chirality problem." All proteinous α-amino acids are of the same L-chiral series and are "homochiral." Much work has been devoted to the development of processes leading to homochirality, and these are well-established and of the utmost importance in synthetic organic chemistry. However, one may anticipate that a natural homochiral system containing the D-antipodes, and only them, would be equally viable – but would be just a "through the looking glass" world. By contrast, of course, in a world where the L- *and* D-antipodes were simultaneously involved, intractable complexities would result. For instance, L-Ala–L-Ala is a substance different from D-Ala–L-Ala. A very early chance step leading to an L,L dimer might have been all that was needed to orient our world toward the homochiral set of molecules we know today. One must also take into account the fact that once some molecules are enriched in one of their homochiral varieties, they can display catalytic properties enabling them to "seed" chirality into the products obtained. The enantiomeric excesses observed are, for the time being, very low; but the number of cases studied is also quite limited (Pizzarello and Weber, 2004).

Figure 19.2. Water and formamide.

Furthermore, does an intrinsic reason exist in our world to exclusively use the
L-amino acids instead of the D-series? It is remarkable that the amino acids isolated from extraterrestrial organic matter from meteorites that are at least somewhat carbonaceous are slightly enriched in the enantiomers of the same L-series as
those present in the amino acids of living organisms (Pizzarello, 2004). This would
become a serious question only once we have discovered *a few other life-forms in
the universe* and found them *all* to be based on L-amino acids. I know that others
have a different point of view and consider the *origin* of chirality and of homochirality to be an important problem. I think it is not, in contrast with some related but
simpler questions, such as that of the *propagation* of homochirality.

Is it a necessity that any kind of life be associated with water from the start?

Life as we know it is so intimately associated with water that the detection of water
on a planet would be considered a prerequisite, and even a possible indicator, of
the presence of some life. This view is linked in particular with the hypothesis
that living organisms must be cellular (see below), and the formation of cells is, as
far as current knowledge tells us, linked with the special properties of water: very
high dielectric constant, considerable cohesive forces, and formation of "hydrogen
bonds." These factors add up to produce what is known as the "hydrophobic effect"
(Tanford, 1978; Blokzijl and Engberts, 1993; Lemieux, 1996). Indeed, we think that
we understand why, in a liquid of high dielectric constant (such as water), properly
built amphiphilic molecules can become self-organized into surface monolayers,
micelles, and eventually vesicles (liposomes) in which "inside water" is separated
from "outside water" by a membrane (Israelachvili *et al.*, 1980). However, is water
itself specifically and exclusively required?

It has been suggested (although as far as I know only in science fiction!) that
liquid ammonia might be an equivalent of water. I know of no experimental results
suggesting that ammonia might display "organizing" properties similar to those of
water, but related substances do. In formamide, for example, the elements of water,
H and O, are present and separated by the insert N–CH (see Figure 19.2).

Formamide exhibits a high dielectric constant, is strongly H-bonding, and is
a highly organizing solvent. Micelles, and even vesicles, have been obtained in

formamide with the same phospholipids as in water (Rico and Lattes, 1986). The properties of formamide and of other water mimics certainly deserve more study, even though the possibility that they might support other forms of life looks rather thin. Formamide would have to be primarily formed, and we do not see how at present.

Is it necessary for any kind of life to be cellular, and for the cells to be bounded in water by membranes?

Leaving aside the possible limitations mentioned below, living organisms are cellular. Cells are limited by membranes, and the singular importance of this cellular structure has been emphasized repeatedly, most forcefully and perceptively by Harold Morowitz, who is represented in Chapter 18 of this volume (Morowitz *et al.*, 1988; Morowitz, 1992). The widely accepted model of membranes is that of Singer and Nicolson (1972), in which the cells are considered to be limited by a "fluid mosaic" membrane built of phospholipids arranged as a double layer. Intrinsic membrane proteins float within this lipid bilayer, to which are hooked, by lipophilic tethers, membrane-bound appendages such as sugars, proteins, and various receptors. Inside the cells, the cytoskeleton ensures stability and displays a variety of important functions. Cells are complex microscopic units defined by such membranes, and the only way to define separate compartments in an aqueous environment is to separate "inside water" from "outside water" by means of such a water-insoluble semi-permeable wall.

However, the following points are worthy of consideration. First, several groups of organisms are not "cellular" in the usual sense of the word. For example, the well-known hardy sea algae *Caulerpa* spp., and probably other Siphonales species, are characterized by a plurinucleate macroscopic structure that resembles a "normal" green alga, but contains no individual cells.

Second, and perhaps more fundamental, is that at least one conceivable alternative system with "water inside" separated from "water outside" could theoretically exist, namely a very large dendrimer. Dendrimers are molecular structures built of branches filling up more and more of the available space progressively as they diverge. Known dendrimers (all fully synthetic) are much too small to define a biologically useful interior space. However, this would not necessarily apply to extremely large dendrimers, built like a boxwood topiary, in which the successive branchings would leave enough space inside to accommodate water and analogs of cytoplasmic structures, terminated at the periphery in a dense outer surface of "twigs" that effectively would create a semipermeable barrier functionally equivalent to a membrane. Even the most complex dendrimers so far synthesized fall very short of the required complexity! I mention this model as a theoretical alternative

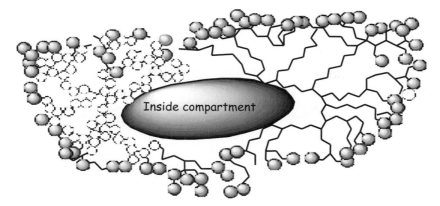

Figure 19.3. A dendrimeric "cell" without membrane.

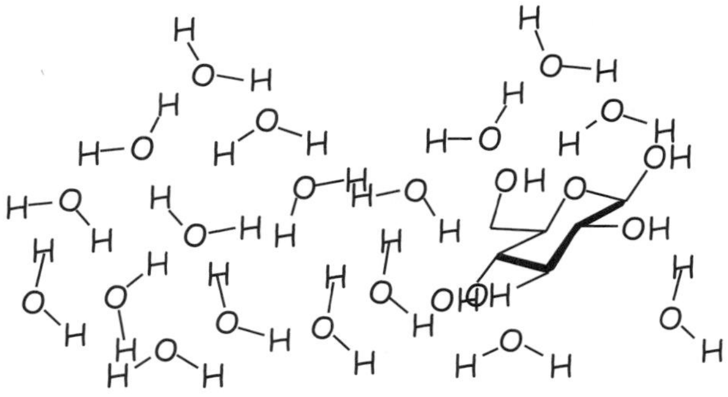

Figure 19.4. Schematic structure of water containing a molecule of glucose. The hydroxyl groups of the substrate can engage in hydrogen bonding and cause little distortion of the network of H bonds.

to the Singer model, but its molecular complexity renders it utterly improbable that it could be an archaeobiotic system. (See Figure 19.3.)

Are the *n*-acyl phospholipids of "classical" membranes, or the more recently discovered archaeal lipids, plausible constituents of primitive membranes?

Let me begin by reminding the reader of the physico-chemical basis for the formation of membranes in water from amphiphilic molecules. The physical characteristics of water lead to the solubility of salts and of molecules able to participate in extensive hydrogen bonds, such as sugars (see Figure 19.4). Hydrocarbons and other non-polar molecules are by contrast insoluble. To be dissolved, they would require

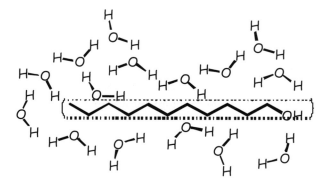

Figure 19.5. Dissolving a long-chain alcohol in water would require "boring a hole" in the system of H bonds.

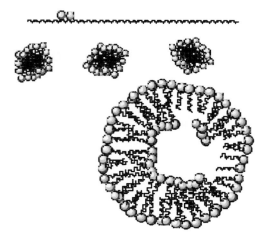

Figure 19.6. Amphiphilic molecules at the surface forming a monolayer (or in bulk liquid forming micelles or vesicles).

the uncompensated destruction of a large number of hydrogen bonds between individual water molecules, forming a large "hole" in the hydrogen bond network (see Figure 19.5). Thus, hybrid molecules (such as fatty acids, alcohols, and more specifically phospholipids), which possess a polar head-group and a non-polar side-chain, interact with water according to conflicting criteria.

Instead of dissolving, hybrid molecules self-organize. Small amounts of these amphiphilic molecules produce surface films; larger amounts lead to bulk assemblies, in which clusters of molecules minimize their destructive effect on the network of hydrogen bonds by forming micelles or vesicles (liposomes) with a specific topology that depends on the concentration and specific shape of the amphiphilic molecules (see Figure 19.6).

This self-complexification leads to a major consequence: vesicles are the simplest objects in which an inside compartment is formed spontaneously through the interplay of physical factors that we understand at least qualitatively. Such a spontaneous compartmentalization is, of course, an attractive candidate as a prototype of universal cellular organization in living organisms. It defines an inside compartment, separated from the outside medium by a semipermeable fluid membrane. Also, the dimensions of spontaneously formed vesicles are of the same range as those of living cells. But what can we say about the nature of the *most primitive* amphiphiles?

In fact, this point has received little attention, despite the fact that classical membrane lipids display all the hallmarks of modernity. Their amphiphilic character is due to the combination of a deceptively "simple" *n*-acyl straight chain and of a large variety of polar head-groups. The structure of the C_{14-20} *n*-acyl chains may be easy to memorize and to draw, but it is very difficult to synthesize: no *in vitro* process has been found to give such part-structures from very small synthons. The Fischer–Tropsch reaction is often invoked, but it gives mixtures of *n*- and branched hydrocarbons or of their functionalized derivatives, but not C_{14-20} *n*-acyl derivatives. It is used to make diesel oils (the South African *Sasol* process), but not (to my knowledge) edible fats with *n*-acyl chains.[2] The biosynthesis of C_{14-20} *n*-acyl chains also implies remarkably complex processes despite their simple structures, calling as it does for seven to ten successive additions and reductions of C_2 units (acetate in the activated form of acetyl-coenzyme-A) and a selective control of the chain length. Furthermore, all "modern" phospholipids require *separate* biosyntheses for their glycerol and phosphocholine head groups (or variants). I think it is obvious that *n*-acylphospholipids are "modern."

Interestingly, bacteria very often contain not only these *n*-acyl amphiphiles, but largely, or even mostly, mixtures of analogs with branched chains (e.g. iso- or anteiso-, or cyclopropyl- or ω-cycloalkyl-substituted chains). However, the biosynthetic processes required are again very complex.

Some carboxylic acids have been isolated from meteorites, the sodium salts of which have produced vesicle-like structures in water (Deamer, 1986). Their aromatic structures are, however, very different from those of any known constituent of biomembranes, and an import from the cosmos by way of meteorites does not look like a reasonable hypothesis of the origin of the phospholipids of biomembranes. Zhang's contribution to the present volume (Chapter 20) may well offer a major step forward in positing plausible prebiotic membranes.

---

[2] Shortly before World War II, Reichsmarschall Hermann Göring had proclaimed that Germany would produce "Kanonen statt Butter." If it had been possible to use the Fischer–Tropsch process to produce *n*-acyl chains, the Third Reich could have produced both cannons *and* butter.

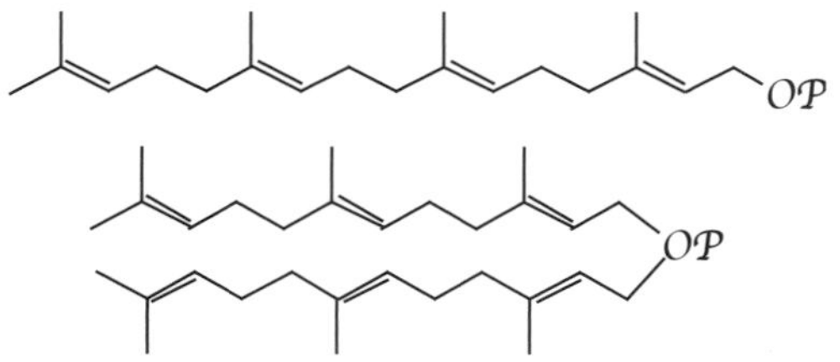

Figure 19.7. Geranylgeranyl phosphate and difarnesyl phosphate.

Could early membrane-forming amphiphiles have simply been polyprenyl phosphates, following the "Strasbourg scenario"[3]?

The long chains required to impart the required lipophilic character to membrane-forming lipids might rather have been, in an early world, polyprenyl, identical with those found in Archaea. This would require much simpler biosyntheses: just three or four identical condensations of C_5 units, by reactions that can be simulated on clays, can produce such molecules (Désaubry *et al.*, 2003). One could even conceive of such a cascade of C_5 condensations terminated selectively at the third (C_{15}) or fourth (C_{20}) step by physical segregation because of spontaneous vesicle formation. Thus, polyprenyl chains are convincingly plausible as potential "primitive" lipids in that they can be synthesized *in vitro* without enzymes. However, the actual archaeal lipids are themselves still much too complex to be prototypical; for example, they are actually more diverse than their eukaryotic counterparts in their polar head-groups. This is why we have been led to formulate the hypothesis that simple phosphates of polyprenyl alcohols, with only a simple phosphate group as a polar head and without any further complicating feature, might produce vesicles in water. This hypothesis has been fully validated, both for molecules containing a single polyprenyl chain and for the diesters (Pozzi *et al.*, 1996; Birault *et al.*, 1996).

Specifically, polyprenyl phosphate vesicles form easily, provided that their "lipophilic balance" (the ratio of the lipophilic part to the hydrophilic part) is adequate. Single-chain polyprenyl phosphates require at least 20 carbons (geranylgeranyl, phytyl, or phytanyl phosphates). Double-chain polyprenyl phosphates require at least two C_{15} (farnesyl) chains to generate vesicles, although this sometimes requires the help of a small amount of the corresponding free polyprenol (Pozzi *et al.*, 1996) (see Figure 19.7.)

At present, polyprenyl phosphates, with C_{15-25} chains, are the simplest vesicle-forming phospholipids known, and they may have been the most primitive membrane constituents. We have been pleased to discover that the properties of vesicles

---

[3] See Birault *et al.* (1996).

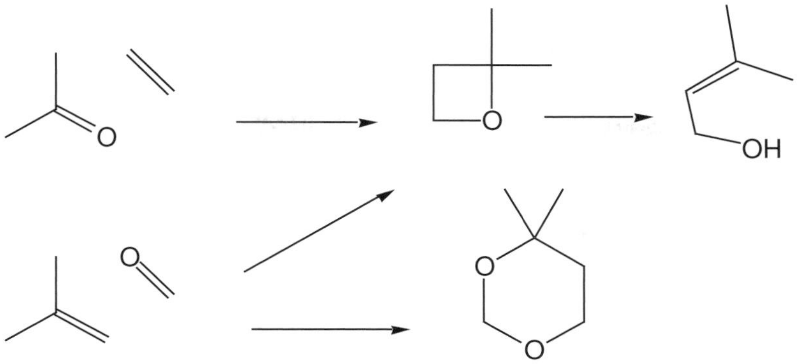

Figure 19.8. "Simple" reactions leading to the C$_5$ unit of isopentenol.

made of these polyprenyl phosphates are remarkably similar to those of vesicles made of the usual phospholipids.

However, the origin of these polyprenyl derivatives is still not obvious. The abiotic origin of the C$_5$ isoprenic units is very problematic, and the C$_{5-25}$ chains are not as easy to obtain from suitable C$_5$ precursors, as one could deduce from first-year organic chemistry. Indeed, although one can conceive of several reactions leading *on paper* to isopentenol from simpler precursors, none withstands close scrutiny as a plausible prebiotic process. For example, the acid-catalyzed Prins reaction of isobutene and formaldehyde (Arundale and Mikeska, 1952; Blomquist and Verdol, 1955; Brace, 1955) has been described and is appealing because formaldehyde is a precursor favored by prebiotic chemists for its (relative) abundance in interstellar space, and reactions on an acidic rock as a catalyst are conceivable. However, isobutene has never been identified in space or on earth, except as a minor constituent of natural gas. The same is true for acetone, which might have been the substrate of a Paterno–Büchi photochemical reaction (Paterno and Chieffi, 1909; Büchi *et al.*, 1954), and for ethylene, both of which are also unlikely precursors (see Figure 19.8).

Furthermore, the head-groups of archaeal phospholipids are much too complex to be archaic, and I have postulated that *polyprenyl phosphates* might have been the most primitive membrane constituents. These phosphates can be biosynthesized by quite simple reactions, which, as we have shown, can be simulated *in vitro* from their C$_5$ precursors (Désaubry *et al.*, 2003).

The initial proposal of this scheme for the evolution of terpenes (Ourisson and Nakatani, 1994) was graced with a full page "Comment" from the editor of *Nature* (Maddox, 1994). It also led to the recognition that *all living organisms require and use terpenes*: farnesyl and geranylgeranyl tethers link proteins to membranes; ubiquinones, with their polyterpene chains, serve as transmembrane electron transporters; cholesterol or α, ω-dipolar carotenoids serve as membrane reinforcers; steroids are involved in mechanisms as diverse as calcium metabolism, potassium/sodium balance, sex hormones, attractants, defense substances, and so forth. Terpenes are not "secondary" metabolites!

Are there automatic consequences of the self-organization of amphiphiles (such as polyprenyl phosphates) in water into membranes, and does this lead to novel properties?

The self-organization of amphiphiles (e.g. polyprenyl phosphates) in water leads to important consequences and to novel properties, such as the following.

- It makes possible a *selective concentration* of lipophiles in the membranes. Vesicles made of *n*-acyl lipids or of polyprenyl phosphates extract selectively into the membrane any lipophilic substance: lipophilic pigments (Nile Red), cholesterol, polyprenols, carotenoids, etc. This is particularly important for those lipophilic substances that play a role in stabilizing the membrane: cholesterol, hopanoids, polyprenols, and carotenoids (Bisseret *et al.*, 1983; Milon *et al.*, 1986; Lazrak *et al.*,1988; Krajewski-Bertrand *et al.*, 1990).
- The increased concentration leads in turn to *novel reactions* of some lipophiles in the membrane. Ringsdorf and colleagues (Folda *et al.*, 1982) and Luisi and colleagues (Blocher *et al.*, 1999) have shown that some lipophilic derivatives of amino acids are selectively extracted into the membranes and can spontaneously condense into lipophilic derivatives of oligopeptides without requiring any biocatalyst such as a peptidase. However, this potentially exciting finding must be tempered with caution: this condensation does not lead to the spontaneous formation of polypeptides larger than di- or tripeptides, and it occurs only when specific, modified, and activated amino-acid derivatives are presented. In short, it seems unlikely that this concentration effect may have led to a primitive synthesis of proteins. Perhaps more interesting is that analogous condensations have been observed with long-chain esters of α-amino acids in mono- and multilayers adsorbed on a solid surface, where again the concentration is locally increased (Fukuda *et al.*, 1981).
- On a different note, anisotropic molecules, once extracted fully or partly into the membrane, become *selectively oriented*. By itself, the inside of the membrane is a highly anisotropic, hairbrush-like, lipophilic solvent. Any anisotropic lipophilic molecule selectively extracted into the membrane must therefore automatically become oriented to avoid the unfavorable energetics of disrupting this ordered state. Indeed, it has been shown experimentally (by specific labeling with a transmembrane photosensitive probe) that cholesterol is oriented *perpendicularly* to the membrane in closed vesicles (Nakatani *et al.*, 1996). By contrast, the linear hydrocarbon β-carotene has been shown (by linear dichroism spectroscopy on flat bilayer stacks) to lie *parallel* to the surface (Nordén

et al., 1977). In other words, membranes are a class of structure that can encourage the formation of order from disorder, a theme that is central to understanding the origins of life.

- The membrane becomes spontaneously more complex by the automatic induction of *vectorial properties*. The membrane is made of all identical molecules of amphiphiles. However, those forming the exterior, convex leaflet are less compressed than those forming the inside, concave one. The two groups, although formed of identical molecules, become different by virtue of their different surroundings (Chrzeszczyk *et al.*, 1977). This difference has been explored with further experimentation, for instance by the separation of the ^{31}P NMR signals of the polar head-groups of the inside and of the outside layers (identified by their different intensities because of the smaller number of molecules contained in the inner half of the membrane). Moreover, these vectorial properties engendered by self-organization extend to produce segregation of different phospholipids when mixtures of amphiphiles are used (Swairjo *et al.*, 1994; Lee *et al.*, 2002).

- Further complexification can be obtained beyond the lipid bilayer itself by the formation of a pseudo-cell wall. Specifically, if pullulan (a fungal metabolite with a long polysaccharide chain) is anchored onto the membrane by a cholesterol or a polyprenyl substituent as a tether and labeled with a fluorescent tag, this spontaneously leads to coating of the outside surface of the vesicles by the polysaccharide, as can be observed by optical microscopy (Ueda *et al.*, 1998; Ghosh *et al.*, 2000). In other words, a pseudo-cell wall of polysaccharide spontaneously forms, although this process is efficient only with a good match between the chains of the phospholipid and the lipidic tethers (a cholesteryl tether for *n*-acyl-phospholipid vesicles, a polyprenyl one for polyprenyl phosphate vesicles).

- Most significantly, self-organization can include a form of self-replication. It is possible to obtain self-multiplying closed vesicles or vesicles as long as they contain at least some of the essential elements of a living cell. Specifically, a micelle that contains enzymes hydrolyzing additional precursors of the membrane amphiphiles is automatically increasing the total volume of the membrane. At first, this leads to expansion of the single micelle, but as further lipids are added the micelle self-replicates by a process that resembles bacterial multiplication by budding (Bachmann *et al.*, 1990, 1991).

- This brings us to the topic of whether and which biological macromolecules can be incorporated into micelles. Much work has focused on the incorporation of DNA into lipid vesicles since Jay and Gilbert (1987) first showed that basic protein can enhance this process (see, for example, Monnard *et al.*, 1997; Szostak *et al.*, 2001). Indeed, a polymerase chain reaction (the standard method for duplicating nucleic acid sequences *in vitro*) can be run in closed vesicles (Oberholzer *et al.*, 1995a), and enzymatic RNA replication has been achieved in self-replicating vesicles (Oberholzer *et al.*, 1995b). It is possible to obtain closed vesicles containing at least some of the essential elements of a living cell, for instance nucleic acids.

Giant DNAs or RNAs observable with an optical microscope after being labeled with a fluorescent tag can be injected into these vesicles with laser tweezers, a process akin to artificial fecundation or simply trapped during vesicle formation by hydration, a surprisingly efficient process (Nomura *et al.*, 2001). They remain functional once inside the vesicles, as shown by the synthesis of the green fluorescent protein, and are protected from outside DNases/RNases (Nomura *et al.*, 2003).

A provisional summary

Our knowledge of which type of biochemistry is and is not possible in this universe is hampered by an obvious limitation: at present, we know of only one version of life. As a result, when considering the possible "biocentric" nature of our universe, we must use clues from synthetic chemistry and physics to infer what aspects of this terrestrial life are necessary and whether alternative scenarios could be built, ensuring in different but convergent ways the major features of life: self-organization and evolution despite homeostasis. In this chapter, I have collected and discussed some of the most prevalent ideas and insights from the "origin of life" field of research to sketch a framework for introducing this topic. I have shown that although silicon-based life might be possible, current knowledge indicates that life is carbon-based. I have also shown that although water-alternatives might support life, we can suspect that a strongly dipolar, hydrogen-bonded liquid is probably necessary for life, and it seems obvious that water is by far the most likely candidate. In addition, I have shown that at least a subset of the fundamental chemical building blocks from which life is constructed are likely to be widely distributed throughout the universe and are intuitive "choices" for life elsewhere. Finally and most importantly, I have tackled the all-important topic of how independent, self-replicating entities might form, given these building blocks.

My view is that amphiphilic molecules are the key to understanding how life could – and would – emerge from such a biochemically interesting universe. Their self-organization into membranes leads to closed vesicles, which define outer and inner compartments, and spontaneously to self-complexification by the accretion of novel properties. This self-complexification, in turn, renders membranes highly germane to exploring the notion of a "biocentric" universe.

The above seems to constitute the beginnings of a potentially fertile scenario. However, we must acknowledge the existence of some major problems.

Some problems

Some problems tend to be hidden, but deserve to be highlighted. A general problem with all scenarios for the origin of life appears to be that of local concentrations of the initial molecular precursors. The image of the "warm little pond" by the side of the sea, concentrated by evaporation and replenished by wave splashes, had been proposed by Darwin (1859) and has been espoused by Miller and Orgel (1974) and many others. This image is appealing, but difficult or impossible to submit to experiment; therefore, it remains just "comfortable" and impossible to prove (Shapiro, 1986). Bernal suggested that concentration might have been achieved by adsorption onto clay particles, a chemically very reasonable hypothesis, that has been developed by Cairns-Smith (Cairns-Smith and Hartmann, 1986) into his theory of a primitive "surface life" on appropriate minerals. This is again appealing, but not compelling; personally, I have never understood how a two-dimensional life form might have originated on a clay particle – and even less how it could later have exploded into the third dimension.

A particularly poignant problem is that life as we know it involves phosphates as polar groups, and there is no obvious major source of phosphates. Although Westheimer (1987) has given very powerful arguments to show why nature has "chosen" phosphates instead of sulfates or some other polar group, this is in contrast to the fact that phosphorus is a relatively rare element in the universe:[4]

Universe	7 ppb
Crustal rocks	1,000 ppb
Seawater	0.070 ppb
Humans	11,000 ppb

Apart from phosphate rock originated in the deposition and fossilization of marine organisms, no large reservoir of phosphate is known, and natural abiotic sources are few, dilute, and very localized (e.g. volcanic springs).[5]

Furthermore, phosphates and polyphosphates display a very low reactivity in phosphorylation; this leads to the difficulty of producing organic phosphates from the corresponding alcohols, without enzymes, and therefore of involving them in the origin of life. (It also leads to difficulties in hydrolyzing them, which is a good thing for life, but a bad one for water-treatment plants, where one must resort to a bacterial treatment to hydrolyze the polyphosphates.) A plausible way out of this difficulty might be to rely on the activation of phosphates by carboxylates (Biron and Pascal, 2004). One should refer in this connection to the much-too-neglected work of Baltcheffsky and Baltcheffsky (1992), showing that sodium polyphosphates may have been precursors of activated phosphates such as adenosine triphosphate.

Conclusion

A valid conclusion of this review is, of course, "We do not know!" Furthermore, currently we have no efficient search plan. However, I feel that a large number of

---

[4] See www.webelements.com/webelements/elements/text/P/.
[5] See www.universetoday.com/am/publish/printer_meteorites_ provided_phosphorus.html.

possible tracks have been left unexplored, some elaborated here, but many more left to the imagination of others – and these may well lead to unexpected and fruitful results. I hope that by presenting explicitly some hitherto little considered possibilities, I may have helped others to unravel answers to that weighty question: How did life begin?

Bibliography

Arundale, E. and Mikeska, L. A. (1952). The Prins reaction. *Chemical Reviews*, **51**, 505–55.

Bachmann, P. A., Walde, P., Luisi, P. L. and Lang, J. (1990). Self-replicating reverse micelles and chemical autopoiesis. *Journal of the American Chemical Society*, **112**, 8200–1.

Bachmann, P. A., Walde, P., Luisi, P. L. and Long, J. (1991). Self-replicating micelles: aqueous micelles and enzymatically driven reactions in reverse micelles. *Journal of the American Chemical Society*, **113**, 8204–9.

Bada, J. L. and Lazcano, A. (2003). Origin of life – some like it hot, but not the first biomolecules. *Science*, **300**, 745–6.

Baltcheffsky, M. and Baltcheffsky, H. (1992). Inorganic pyrophosphate and inorganic pyrophosphatases. In *Molecular Mechanisms in Bioenergetics: New Comprehensive Biochemistry*, ed. L. Ernster. Amsterdam: Elsevier, pp. 331–48.

Bernstein, M. P., Dworkin, J. P., Sandford, S. A. *et al.* (2002). Racemic amino acids from the ultra-violet photolysis of interstellar ice analogues. *Nature*, **416**, 401–3.

Birault, V., Pozzi, G., Plobeck, N. *et al.* (1996). Di(polyprenyl) phosphates as models for primitive membrane constituents: synthesis and phase properties. *Chemistry – a European Journal*, **2**, 789–99.

Biron, J.-P. and Pascal, R. (2004). Amino acid N-carboxyanhydrides: activated peptide monomers behaving as phosphate-activating agents in aqueous solution. *Journal of the American Chemical Society*, **126**, 9198–9.

Bisseret, P., Wolff, G., Albrecht, A.-M. *et al.* (1983). A direct study of the cohesion of lecithin bilayers: the effect of hopanoids and dihydroxycarotenoids. *Biochemical and Biophysical Research Communications*, **110**, 320–4.

Blocher, M., Liu, D., Walde, P. *et al.* (1999). Liposome-assisted selective polycondensation of α-amino acids and peptides. *Macromolecules*, **32**, 7332–4.

Blokzijl, W. and Engberts, J. B. F. N. (1993). Hydrophobic effects. Opinions and facts. *Angewandte Chemie International Edition*, **32**, 545–79. (In English.)

Blomquist, A. T. and Verdol, J. A. (1955). The thermal isobutylene-formaldehyde condensation. *Journal of the American Chemical Society*, **77**, 78–80.

Brace, N. O. (1955). The uncatalyzed thermal addition of formaldehyde to olefins. *Journal of the American Chemical Society*, **77**, 4566–8.

Büchi, G., Inman, C. G. and Lipinsky, E. S. (1954). Light-catalyzed reactions. I. The reaction of carbonyl compounds with 2-methyl-2-butene in the presence of ultraviolet light. *Journal of the American Chemical Society*, **76**, 4327–31.

Cairns-Smith, A. G. and Hartmann, H., eds. (1986). *Clay Minerals and the Origin of Life*. Cambridge, UK: Cambridge University Press.

Chrzeszczyk, A., Wishnia, A. and Springer, C. S. Jr. (1977). The intrinsic structural asymmetry of highly curved phospholipid bilayer membranes. *Biochimica et Biophysica Acta*, **470**, 161–9.

Darwin, C. (1859). *On the Origin of Species by Means of Natural Selection*. London: J. Murray.

Deamer, D. W. (1986). Role of amphiphilic compounds in the evolution of membrane structure on the early earth. *Origins of Life and Evolution of Biospheres*, **17**, 3–25.

Deamer, D. W. and Barchfeld, G. L. (1982). Encapsulation of macromolecules by lipid vesicles under simulated prebiotic conditions. *Journal of Molecular Evolution*, **18**, 203–6.

Deamer, D. W., Harang Mahon, E. and Bosco, G. (1994). Self-assembly and function of primitive membrane structures. In *Early Life on Earth*, Nobel Symposium no. 84, ed. S. Bengtson. New York, NY: Columbia University Press, pp. 107–23.

Désaubry, L., Nakatani, Y. and Ourisson, G. (2003). Toward higher polyprenols under "prebiotic" conditions. *Tetrahedron Letters*, **44**, 6959–61.

Devienne, F. M. and Barnabé, C. Synthesis of biological compounds in quasi-interstellar conditions. *Comptes Rendus de l'Académie des Sciences, Paris*, Series IIc, 435–9.

Devienne, F. M, Barnabé, C., Couderc, M. *et al.* (1998). Synthesis of silicon–oxygen derivatives in quasi-interstellar conditions. *Comptes Rendus de l'Académie des Sciences, Paris*, Series IIc, 435–9.

Devienne, F. M., Barnabé, C. and Ourisson, G. (2002). Synthesis of further biological compounds in interstellar-like conditions. *Comptes Rendus de l'Académie des Sciences, Chimie*, **5**, 651–3.

Fischer, A., Franco, A. and Oberholzer, T. (2002). Giant vesicles as microreactors for enzymatic mRNA synthesis. *ChemBioChem*, **3**, 409–17.

Folda, T., Gros, L. and Ringsdorf, H. (1982). Formation of oriented polypeptides and polyamides in monolayers and liposomes. *Macromolecular Rapid Communications*, **3**, 167–74.

Fraley, R., Subramani, S., Berg, P. *et al.* (1980). Introduction of liposome-encapsulated SV40 DNA into cells. *Journal of Biological Chemistry*, **255**, 10431–5.

Fukuda, K., Shibasaki, Y. and Nakahara, H. (1981). Polycondensation of long-chain esters of α-amino acids in monolayers at air/water interface and in multilayers on solid surface. *Journal of Macromolecular Science, A: Pure and Applied Chemistry*, **15**, 999–1014.

Ghosh, S., Lee, S. J., Ito, K. *et al.* (2000). Molecular recognition on giant vesicles: coating of phytyl phosphate vesicles with a polysaccharide bearing phytyl chains. *Chemical Communications*, pp. 267–8.

Israelachvili, J. N., Marcelja, S. and Horn, R. G. (1980). Physical principles of membrane organization. *Quarterly Review of Biophysics*, **13**, 121–200.

Jay, D. G. and Gilbert, W. (1987). Basic protein enhances the incorporation of DNA into lipid vesicles: model for the formation of primordial cells. *Proceedings of the National Academy of Sciences, USA*, **84**, 1978–80.

Krajewski-Bertrand, M. A., Hayer, M. and Wolff, G. (1990). Tricyclohexaprenol and an octaprenediol, two of the "primitive" amphiphilic lipids, do improve phospholipidic membranes. *Tetrahedron*, **46**, 3143–54.

Lazrak, T., Wolff, G., Albrecht, A. M. *et al.* (1988). Bacterioruberins reinforce reconstituted Halobacterium lipid membranes. *Biochimica et Biophysica Acta*, **939**, 160–2.

Lee, S., Désaubry, L., Nakatani, Y. *et al.* (2002). Vectorial properties of small phytanyl phosphate vesicles. *Comptes Rendus de l'Académie des Sciences, Chimie*, **5**, 331–5.

Lemieux, R. U. (1996). How water provides the impetus for molecular recognition in aqueous solution. *Accounts of Chemical Research*, **29**, 375–80.

Maddox, J. (1994). Origin of the first cell membranes? *Nature*, **371**, 101.

Miller, S. J. (1953). A production of amino-acids under possible primitive earth
    conditions. *Science*, **117**, 528–9.
Miller, S. L. and Orgel, L. E. (1974). *The Origins of Life on Earth*. Englewood Cliffs, NJ:
    Prentice-Hall.
Milon, A., Wolff, G., Ourisson, G. *et al.* (1986). Organisation of carotenoid–phospholipid
    bilayer systems: incorporation of zeaxanthin, astaxanthin, and their C_{50} homologues
    into dimyristoyl-phosphatidylcholine vesicles. *Helvetica Chimica Acta*, **69**,
    12–24.
Monnard, P. A., Oberholzer, T. and Luisi, P. L. (1997). Entrapment of nucleic acids in
    liposomes. *Biochimica et Biophysica Acta*, **1329**, 39–50.
Monod, J. (1970). *Le Hasard et la nécessité*. Paris: Le Seuil.
Morowitz, H. J. (1992). *The Beginnings of Cellular Life*. New Haven, CT: Yale University
    Press.
Morowitz, H. J., Heinz, D. and Deamer, D. W. (1988). The chemical logic of a minimum
    protocell. *Origins of Life and Evolution of Biospheres*, **18**, 281–7.
Muñoz Caro, G. M., Meierhenrich, U. J., Schutte, W. A. *et al.* (2002). Amino-acids from
    ultra-violet irradiation of interstellar ice analogues. *Nature*, **416**, 403–6.
Nakatani, Y., Yamamoto, M., Diyizou, Y. *et al.* (1996). Studies on the topography of
    biomembranes: regioselective photolabelling in vesicles with the tandem use of
    cholesterol and a photoactivable transmembrane phospholipidic probe. *Chemistry – a
    European Journal*, **2**, 129–38.
Nomura, S. M., Yoshikawa, Y., Yoshikawa, K. *et al.* (2001). Towards proto-cells:
    "primitive" lipid vesicles encapsulating giant DNA and its histone complex.
    ChemBioChem, **2**, 457–9.
Nomura, S. M., Tsumoto, K., Hamada, T. *et al.* (2003). Gene expression within cell-sized
    lipid vesicles. *ChemBioChem*, **4**, 1172–5.
Nordén, B., Lindblom, G. and Jonás, I. (1977). Linear dichroism spectroscopy as a tool
    for studying molecular orientation in model membrane systems. *Journal of Physical
    Chemistry*, **81**, 2086–93.
Oberholzer, T., Albrizio, M. and Luisi, P. L. (1995a). Polymerase chain reaction in
    liposomes. *Current Biology*, **2**, 677–82.
Oberholzer, T., Wick, R., Luisi, P. L. *et al.* (1995b). Enzymatic RNA replication in
    self-reproducing vesicles: an approach to a minimal cell. *Biochemical and
    Biophysical Research Communications*, **207**, 250–7.
Oberholzer, T., Nierhaus, K. H. and Luisi, P. L. (1999). Protein expression in liposomes.
    Biochemical and Biophysical Research Communications, **261**, 238–41.
Ohnishi, T., Hatakeyama, M., Yamamoto, N. *et al.* (1978). Electrical and spectroscopic
    investigations of molecular layers of fatty acids including carotene. *Bulletin of the
    Chemical Society of Japan*, **51**, 1714–16.
Oparin, A. I. (1968). *Genesis and Evolutionary Development of Life*. New York, NY:
    Academic Press.
Ourisson, G. (1986). Vom Erdöl zur Evolution der Biomembranen (Heinrich-Wieland
    Lecture). *Nachrichten aus Chemie, Technik und Laboratorium*, **34**, 8–14.
Ourisson, G. and Nakatani, Y. (1994). The terpenoid theory of the origin of cellular life:
    the evolution of terpenoids to cholesterol. *Chemistry and Biology*, **1**, 11–23.
Ourisson, G. and Nakatani, Y. (1999). Origins of cellular life: molecular foundations and
    new approaches. *Tetrahedron*, **55**, 3183–90.
Paterno, E. and Chieffi, G. (1909). Light-induced chemical synthesis. Part 2. Reaction of
    unsaturated hydrocarbons with aldehydes and ketones. *Gazzetta Chimica Italiana*,
    39, 341–50.

Pizzarello, S. (2004). Chemical evolution and meteorites: an update. *Origins of Life and Evolution of Biospheres*, **34**, 25–34.

Pizzarello, S. and Weber, A. L. (2004). Prebiotic amino acids as asymmetric catalysts. *Science*, **303**, 1151.

Pozzi, G., Birault, V., Werner, B. *et al.* (1996) Single-chain polyprenyl phosphates form primitive membranes. *Angewandte Chemie, International Edition*, **35**, 177–79. (In English.)

Rico, I. and Lattes, A. (1986). Krafft temperatures and micelle formation of ionic surfactants in formamide. *Journal of Physical Chemistry*, **90**, 5870–2.

Rohmer, M., Knani, M., Simonin, P. *et al.* (1993). A novel pathway for the early steps leading to isopentenyl diphosphate. *Biochemical Journal*, **295**, 517–24.

Rohmer, M., Seemann, M., Horbach, S. *et al.* (1996). Glyceraldehyde 3-phosphate and pyruvate as precursors of isoprenic units as an alternative non-mevalonic pathway for terpenoid biosynthesis. *Journal of the American Chemical Society*, **118**, 2564–6.

Schwartz, A. W. (1996). Did minerals perform prebiotic combinatorial chemistry? *Chemistry and Biology*, **3**, 515–18.

Shapiro, R. (1986). *Origins: A Skeptic's Guide to the Creation of Life on Earth*. New York, NY: Simon and Schuster.

Singer, S. J. and Nicolson, G. L. (1972). A mosaic model of cell membranes. *Science*, **175**, 720–2.

Sutherland, J. D. and Whitfield, J. N. (1997). Prebiotic chemistry: a bioorganic perspective. *Tetrahedron*, **53**, 11493–527.

Swairjo, M. A., Seaton, B. A. and Roberts, M. F. (1994). Effect of vesicle composition and curvature on the dissociation of phosphatidic acid in small unilamellar vesicles – a ^{31}P-NMR study. *Biochimica et Biophysica Acta*, **191**, 354–61.

Szostak, J. W., Bartel, D. P. and Luisi, P. L. (2001). Synthesizing life. *Nature*. **409**, 387–90.

Tanford, C. (1978). The hydrophobic effect and the organization of living matter. *Science*, **200**, 1012–18.

Tsumoto, K., Nomura, S. M., Nakatani, Y. *et al.* (2000). Giant liposome as a biochemical reactor: transcription of DNA and transportation by laser tweezers. *Langmuir*, **17**, 7225–28.

Ueda, T., Lee, S. L., Nakatani, Y. *et al.* (1998). Coating of POPC giant liposomes with hydroxylated polysaccharide. *Chemical Letters*, 417–18.

van de Ven, M., Kattenberg, M., van Ginkel, G. *et al.* (1984). Study of the orientational ordering of carotenoids in lipid bilayers by resonance-Raman spectroscopy. *Biophysical Journal*, **45**, 1203–10.

Walde, P., Wick, R., Fresta, M. *et al.* (1994). Autopoietic self-reproduction of fatty acid vesicles. *Journal of the American Chemical Society*, **116**, 11649–54.

Weissbuch, I., Bolbach, G., Leiserowitz, L. *et al.* (2004). Chiral amplification of oligopeptides via polymerization in two-dimensional crystallites on water. *Origins of Life and Evolution of Biospheres*, **34**, 79–92.

Westheimer, F. H. (1987). Why nature chose phosphates. *Science*, **235**, 1173–8.

Yamagata, Y., Watanabe, H., Saitoh, M. *et al.* (1991). Volcanic production of polyphosphate under primitive Earth conditions. *Nature*, **35**, 516–19.

Yu, W., Sato, K., Wakabayashi, M. *et al.* (2001). Synthesis of functional protein in liposomes. *Journal of Bioscience and Bioengineering*, **92**, 590–3.

Zhou Z., Okumura, Y. and Sunamoto J. (1996). NMR study of choline methyl group of phospholipids. *Proceedings of the Japan Academy*, **72B**, 23–7.

20

Plausible lipid-like peptides: prebiotic molecular self-assembly in water

Shuguang Zhang

Introduction

Life as we know it today completely depends on water. Without water, life would be either impossible or totally different. Thus, a deep understanding of the relationship between water and other simple building blocks of life is crucial to gain insight into how prebiotic life forms could have originated and evolved and whether the physical laws of this universe are in any way predisposed to the emergence of life (Henderson, 1913, 1917; Eisenberg and Kauzmann, 1985; Ball, 2001).

It is unlikely that under prebiotic conditions the complex and sophisticated biomacromolecules commonplace in modern biochemistry would have existed. Thus, research into the origin of life is intimately associated with the search for plausible systems that are much simpler than those we see today. However, it is also plausible that these simple building blocks of life might have been amphiphilic molecules in which water could have had an enormous influence on their prebiotic molecular selection and evolution, because water can either form clathrate structures or drive these simplest molecules together (Ball, 2001).

Structure of water

Water is both simple and complex. It is simple because it consists of only one oxygen atom and two hydrogen atoms (see Figure 20.1). But at the same time it exhibits highly complex molecular behavior (far exceeding the multibody problem in mathematics, planetary science, and astrophysics) wherever numerous water molecules interact dynamically (Eisenberg and Kauzmann, 1985; Ball, 2001). Indeed, this behavior becomes even more complex when water molecules interact with other atoms and molecules (Eisenberg and Kauzmann, 1985; Ball, 2001).

Fitness of the Cosmos for Life: Biochemistry and Fine-Tuning, ed. J. D. Barrow *et al.*
Published by Cambridge University Press. © Cambridge University Press 2007.

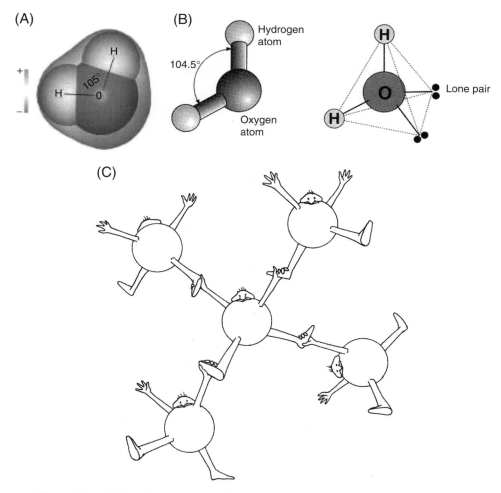

Figure 20.1. Molecular structure of water. (A) Each water molecule has a typical dipole with one oxygen atom covalently sharing bonds with two hydrogens at 104.5–105° in an asymmetric manner. (B) The remaining two lone pairs of the oxygen atom can form additional weak hydrogen bonds with other water molecules or other substances. (C) Thus, each water molecule can form four hydrogen bonds: two covalent bonds as hydrogen donors, like hands; and two non-covalent bonds as hydrogen acceptors, like feet. (D) Structure of ice formation. When temperature decreases, water molecules repack themselves to form a tighter structure. (The images and drawings are courtesy of Philip Ball and reprinted with permission.)

The root of this complexity is that the oxygen can form up to four chemical bonds with other atoms or molecules. In water, oxygen forms two covalent bonds with hydrogen, leaving two electron lone pairs that can form hydrogen bonds with the hydrogen atoms of neighboring water molecules or with other atoms. In other words, water is a strongly dipolar molecule, wherein negatively charged oxygen

(D)

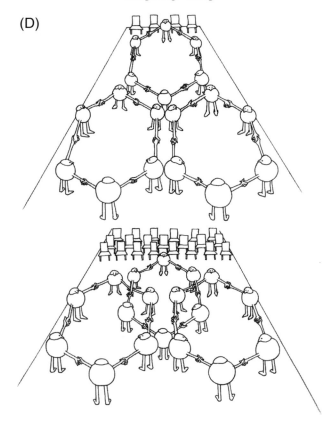

Figure 20.1. (*cont.*)

attracts positively charged hydrogens with an angle of 104.5–105° (Ball, 2001). It is this dipole property that makes water very interactive with other neighboring molecules.

The human form serves as an analogy for representing the structure of a single water molecule: the arms represent hydrogen donors that can hold things, and the legs are hydrogen acceptors that can be held by other "arms" (Ball, 2001). Wherever numerous water molecules co-exist, they form a dynamic network, constantly connecting and disconnecting with extreme rapidity, much like constantly and rapidly exchanging dance partners at a densely populated dance party (see Figure 20.1).

When temperature decreases, the length of bonds between different atoms diminishes. The water molecules pack together more and more tightly and eventually form an ice structure (see Figure 20.1). On the other hand, when temperature increases, bond lengths increase as atoms are stretched farther and farther apart, like receding stars and galaxies in the universe. Beyond a certain threshold, the hydrogen bonds

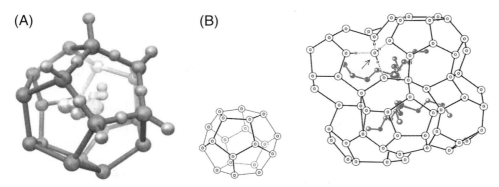

Figure 20.2. Structure of water clathrates. When water molecules interact with non-water substances, water molecules form a cage-like structure to surround them. Depending on the size of the substance, water molecules pack in different ways to solubilize or to form an interface with the materials.

between water molecules eventually break, and water forms vapor. Without a doubt, this "physics of water" operated billions of years ago, just as it does today. Thus, cyclical temperature fluctuations not only could have driven numerous prebiotic molecular interactions in an unpredictable way, but could also have generated predictable synthesis reactions (Henderson, 1913, 1917).

As water molecules exhibit a dipolar moment that allows them to interact with one another and with other atoms and molecules, they can encase other molecules within cage-like ("clathrate") structures. These multifaceted balls resemble viral protein-coats by encapsulating molecules within them (see Figure 20.2). In fact, we now know of diverse clathrate structures that vary considerably, depending on the non-water molecules involved. When non-water molecules become too large, water molecules cease to encase them, such that they aggregate and either precipitate out of water or form macromolecular interfaces with water. Precipitated aggregates exhibit little structure, but the latter systems (multimolecular interfaces between water and biopolymers) are often found in biological membranes. Many researchers have pointed out that some simple kind of membrane would have been a requirement for the earliest metabolism to form partitions and enclosures that optimized specific reactions or sequestered important metabolites (see, for example, Morowitz, 1992; Chakrabarti and Deamer, 1994).

Simplest enclosure system in water

In the summer of 1992 at the "Origin of Life" Gordon Conference in New Hampshire, I asked two simple questions. What might be the simplest amphiphilic biopolymers that could self-organize to enclose other biological molecules in a primarily aqueous prebiotic environment? Could such structures be constructed from the simplest of molecular building blocks? I reasoned that the biopolymers could be neither phospholipids nor nucleic acids because they are relatively complex multicomponent molecules containing several distinctive parts (Hargreaves *et al.*, 1977). Likewise, proteins seem unlikely because the polymerization of any specific (or even semi-specific) sequence from a "soup" of possible monomers brings in unavoidable problems of combinatorial mathematics (moreover, most extant proteins also require chaperone molecules to facilitate folding into the correct three-dimensional shape). In short, abiotic syntheses of all three types of complex molecule in the prebiotic environment seem rather unlikely. I therefore asked whether it were plausible that the simplest peptides, comprising just two or three of the simplest amino acids, could function in this membrane/enclosure-forming capacity. Although several groups had studied the chemistry of various amino acids and amino acid biopolymers (Fox and Harada, 1958; Brack and Orgel, 1975; Brack and Spach, 1981; Yanagawa *et al.*, 1988), distinctive structures and enclosures were not reported or observed at that time.

Simplest amino acids

Glycine with a side chain (R=H), alanine (R=CH$_3$), and aspartic acid (R=CH$_2$COOH) are among the chemically and structurally simplest amino acids. They are of particular interest to prebiotic molecular evolution because of their presence not only in the products of biochemical simulations of earth's presumed prebiotic environment (Miller, 1953; Miller and Urey, 1959; Oro and Kamat, 1961; Bada *et al.*, 1994), but also in the CI-type carbonaceous chondrites, including the Orgueil, Ivuna, and Murchison meteorites (Kvenvolden *et al.*, 1970; Wong and Bronskill, 1979; Anders, 1989; Chyba and Sagan, 1992; Ehrenfreund *et al.*, 2001). Specifically, glycine is the simplest possible amino acid (it is an achiral molecule without any true side chains and is universally indicated to have been the most abundant amino acid in abiotic environments).

Beyond synthesis of the amino acids themselves, it has been experimentally demonstrated that these amino acids (and their derivatives) can form peptides when subjected to repeated hydration–dehydration cycles under microwave heating, in aqueous ammonia, or on heated clays that mimic various hypothesized conditions of early life on the planet (Oro and Guidry, 1961; White *et al.*, 1984; Yanagawa *et al.*, 1988). Indeed, oligo-glycine appears even more easy to produce abiotically, as it has been synthesized by subjecting glycine monomers to extended exposure (more than forty hours) of supercritical water conditions, that is under high temperature (several hundred degrees celsius) and high pressure (tens of atmospheres) (Goto *et al.*, 2004).

These high-temperature and high-pressure conditions are similar to those found in deep-sea volcanoes and hydrothermal vents, a favorite hypothesized venue for the origin of life. It is plausible that some of the simplest biochemical building blocks could have produced complex life forms over eons of natural selection and evolution. The challenge, however, is to explain how sufficiently complex proteins, or ribozymes, could have been produced in the lipid membranes necessary for the metabolism of their own catalysis.

If, instead of lipid membranes, simple peptides with hydrophobic tails and hydrophilic heads (made up of merely a combination of these robust, abiotically synthesized amino acids) could self-assemble into nanotubes or vesicles, they would have the potential to provide a primitive enclosure for the earliest RNA-based (Beaudry and Joyce, 1992; Wilson and Szostak, 1995) or peptide enzymes and other primitive molecular structures with a variety of functions.

In other words, if such structures can be demonstrated to exist, then this makes plausible the idea that in the prebiotic world lipid-like peptides of various lengths could form and self-organize into distinct vesicles and tubes that could act as naturally formed enclosures, isolated from the broader environment, for prebiotic rudimentary enzymes and ribozymes to accumulate. From this starting point, it is far easier to envisage how a diverse population of peptides and RNA not only could condense into complex structures, but also could evolve increasing sophistication, stimulating their own synthesis and replication and evolving ultimately into the wondrously efficient chemical and biological catalysts we encounter today (Zhang and Egli, 1994, 1995).

Lipid-like peptides that form nanotubes and nanovesicles

To this end, I focused on designing a class of simple amphiphilic lipid-like peptides that consist exclusively of plausible prebiotic amino acids. One class of these molecules comprises peptides that exhibit lipid-like or surfactant properties (Vauthey *et al.*, 2002; Santoso *et al.*, 2002a,b; von Maltzahn *et al.*, 2003; Yang and Zhang, 2006; Nagai *et al.*, 2007). I designed such peptides with computer modeling, linking amino acids together one at a time to achieve a length and shape similar to there of lipids. Although individual chemical species within this population of peptides have completely different composition and sequence, they share a crucial common feature: a hydrophilic head comprising one or two charged amino acids and a hydrophobic tail comprising four or more consecutive hydrophobic amino acids (see Figure 20.3). In other words, although it would be possible to design peptides with non-charged hydrophilic heads using serine and threonine, those that

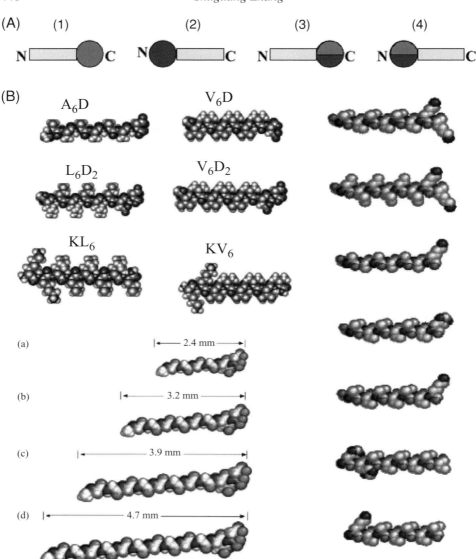

Figure 20.3. Models of the simplest biphase peptides. (A) Schematic illustrations of four types of biphase (hydrophobic tail and hydrophilic head) peptide. Hydrophilic heads can exihibit either negative charges (aspartic acid and glutamic acid) or positive charges (lysine, arginine, and histidine). (B) Selected molecular models of diphase peptides: G_4D_2, G_6D_2, G_8D_2, $G_{10}D_2$, A_6D, V_6D, V_6D_2, V_6K_2, L_6D_2, KL_6, KV_6, A_6H, HA_6, and H_2A_6. Note: The glycine tail has no side chain, as glycine's R-group is a single hydrogen atom. Alanine's tail has a methyl side chain, and valine's tail has an *n*-isopropyl side chain. The hydrophobicity increases as the hydrocarbon side chains become large, as in this case. G_6D_2, A_6D, and V_6D have a negatively charged head, but V_6K_2 has a positively charged head.

Table 20.1. *Aspartic acid head at N- or C-termini with variation in hydrophobic tails*

When an aspartic acid is placed at the C-terminus of a peptide, the resulting molecule possesses two negative charges, one from the side chain and the other from the C-terminus. On the other hand, when an aspartic acid is placed at the N-terminus of the peptide, it bears one negative charge from the side chain and one positive charge from the N-terminus.

Name	Sequence	Name	Sequence
G_6D	GGGGGGD	DG_6	DGGGGGG
A_6D	AAAAAAD	DA_6	DAAAAAA
V_6D	VVVVVVD	DV_6	DVVVVVV
I_6	IIIIIID	DI_6	DIIIIII
L_6D	LLLLLLD	DL_6	DLLLLLL
F_6D	FFFFFFD	DF_6	DFFFFFF

Table 20.2. *Lysine head at N- or C-termini with variation in hydrophobic tails*

When a lysine is placed at the C-terminus of a peptide, it possesses one positive charge from its side chain and one negative charge from its C-terminus. On the other hand, when lysine is placed at the N-terminus of a peptide, it bears two positive charges from the side chain and a positive charge from the N-terminus.

Name	Sequence	Name	Sequence
G_6K	GGGGGGK	KG_6	KGGGGGG
A_6K	AAAAAAK	KA_6	KAAAAAA
V_6K	VVVVVVK	KV_6	KVVVVVV
I_6K	IIIIIIK	KI_6	KIIIIII
L_6K	LLLLLLK	KL_6	KLLLLLL
F_6K	FFFFFFK	KF_6	KFFFFFF

I designed resemble lipids (or other organic surfactants) in their possession of a hydrophobic tail and a charged hydrophilic head (see Figure 20.3).

Not only do the shape and physical structure of these lipid-like peptides resemble lipids and other organic surfactants, but their chemical properties do as well. For example, peptides have six hydrophobic either alanine or valine residues from the N-terminus, followed by a negatively charged aspartic acid residue (A_6D = Ac-AAAAAAD; V_6D Ac-VVVVVVD); thus, they possess two negative charges, one from the charged terminal side chain and the other from the C-terminus (Vauthey *et al.*, 2002). In contrast, several simple peptides, G_4DD (Ac-GGGGDD), G_6DD (Ac-GGGGGGDD), G_8DD (Ac-GGGGGGGGDD), have four, six, and eight

Table 20.3. *Aspartate ($^-$) head with various hydrophobic tail lengths at N- or C-termini*

When an aspartate is placed at the C-terminus of a peptide, it has two negative charges, one from the side chain and the other from the C-terminus. On the other hand, when an aspartate is placed at the N-terminus of a peptide, it bears one negative charge from its side chain and one positive charge from the N-terminus.

Name	Sequence	Name	Sequence
G_4D	GGGGD	A_3D	AAAD
G_5D	GGGGGD	A_4D	AAAAD
G_7D	GGGGGGGD	A_5D	AAAAAD
V_2D	VVD	I_2D	IID
V_3D	VVVD	I_3D	IIID
V_4D	VVVVD	I_4D	IIIID
V_5D	VVVVVD	I_5D	IIIIID
L_2D	LLD	F_2D	FFD
L_3D	LLLD	F_3D	FFFD
L_4D	LLLLD	F_4D	FFFFD
DG_4	DGGGG	DA_3	DAAA
DG_5	DGGGGG	DA_4	DAAAA
DG_7	DGGGGGGG	DA_5	DAAAAA
DV_2	DVV	DI_2	DII
DV_3	DVVV	DI_3	DIII
DV_4	DVVVV	DI_4	DIIII
DV_5	DVVVVV	DI_5	DIIIII
DL_2	DLL	DF_2	DFF
DL_3	DLLL	DF_3	DFFF
DL_4	DLLLL	DF_4	DFFFF

glycines, followed by two aspartic acids with three negative charges (Santoso *et al.*, 2002a,b). Similarly, A_6K (Ac-AAAAAAK) or KA_6 (KAAAAAA) has six alanines as the hydrophobic tail and a positively charged lysine as the hydrophilic head (von Maltzahn *et al.*, 2003). These lipid-like peptides can self-organize to form well-ordered nanostructures, including micelles, nanotubes, and nanovesicles in water. Furthermore, the structure formation is concentration-dependent: namely, at low concentration, there are no defined structures. These structures spontaneously form at a critical aggregation concentration (CAC) (Yang and Zhang, 2006; Nagai *et al.*, 2007), in a way similar to that of lipids and other surfactants.

Six amino acids of varying hydrophobicity (Gly, Ala, Val, Ile, Leu, and Phe) can be used to generate the non-polar tails. Such hydrophobic tails never exceed six residues, so that the total length of the peptide detergents will be seven, about 2.4 nm in length; interestingly, this is a size similar to that of the phospholipids

Table 20.4. *Lysine head with various hydrophobic tail lengths at N- or C-termini*

When a lysine is placed at the C-terminus of the peptide, it has one positive charge from its side chain and one negative charge from its C-terminus. On the other hand, when a lysine is placed at the N-terminus of a peptide; it bears two positive charges from its side chain and another from the N-terminus.

Name	Sequence	Name	Sequence
G_4K	GGGGK	A_3K	AAAK
G_5K	GGGGGK	A_4K	AAAAK
G_7K	GGGGGGGK	A_5K	AAAAAK
V_2K	VVK	I_2K	IIK
V_2K	VVVK	I_3K	IIIK
V_4K	VVVVK	I_4K	IIIIK
V_5K	VVVVVK	I_5K	IIIIIK
L_2K	LLK	F_2K	FFK
L_3K	LLLK	F_3K	FFFK
L_4K	LLLLK	F_4K	FFFFK
KG_4	KGGGG	KA_3	KAAA
KG_5	KGGGGG	KA_4	KAAAA
KG_7	KGGGGGGG	KA_5	KAAAAA
KV_2	KVV	KI_2	KII
KV_3	KVVV	KI_3	KIII
KV_4	KVVVV	KI_4	KIIII
KV_5	KVVVVV	KI_5	KIIIII
KL_2	KLL	KF_2	KFF
KL_3	KLLL	KF_3	KFFF
KL_4	KLLLL	KF_4	KFFFF

abundant in membranes, although this in part reflects the fact that the first lipid-like peptide was designed by modeling the peptide using the phosphatidylcholine as a size guide. However, when more than six hydrophobic residues (except glycine) are used, the lipid-like peptides themselves become less soluble in water. Although Tables 20.1–20.4 list only Asp ($^-$) and Lys ($+$) as the hydrophilic head groups, it must be emphasized that Glu ($^-$), Arg ($+$), and His ($+$) can also be used the same combinatorial ways. Therefore, they can broaden the spectra of variations and increase the possible number of lipid-like peptides.

Moreoever, similar to the dynamic behavior of phospholipid vesicles and other microstructures (Wick *et al.*, 1996), these simplest of peptide nanostructures appear to behave as dynamic entities in water: they can fuse, divide, and change shape as a function of time and environmental influence (see Figures 20.4–20.7) (Vauthey *et al.*, 2002; Santoso *et al.*, 2002a,b; von Maltzahn *et al.*, 2003).

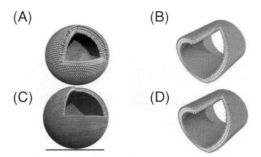

Figure 20.4. Molecular models of cutaway structures formed from the lipid-like peptides with negatively charged heads and glycine tails. Each peptide is *c.* 2 nm in length. (A, C) Peptide vesicle with an area sliced away. (B, D) Peptide tubes. The glycines are packed inside the bilayer away from water, and the aspartic acids are exposed to water, much like other lipids and surfactants. The modeled dimension is 50–100 nm in diameter. Preliminary experiments suggest that the wall thickness may be *c.* 4–5 nm, implying that the wall may form a double layer, similar to phospholipids in cell membranes.

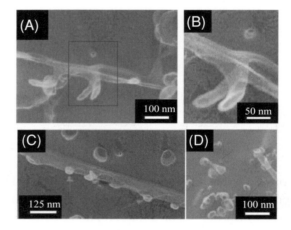

Figure 20.5. Transmission electron microscopic (TEM) images of lipid-like glycine peptide enclosures. Glycine tail and aspartic acid head peptides formed tube and vesicle structures. Note the growth of the tube opening (A, B, C) and the presumed vesicle division (D). If these dynamic enclosures can encapsulate other biomolecules, this may be one step closer for prebiotic molecular evolution.

It is thus plausible that in the prebiotic world, under the influence of water, lipid-like peptides of various lengths might self-organize into distinct vesicles and tubes (regardless of sequence) that could enclose prebiotic rudimentary enzymes to isolate them from the environment. Thus, a diverse population of peptides and RNA might condense into complex structures that evolve to perform different functions.

(A)

(B)

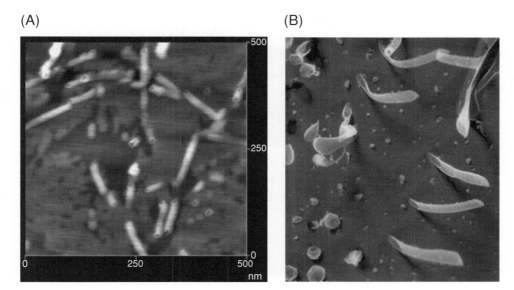

Figure 20.6. Images of lipid-like alanine and valine peptide enclosures. (A) Atomic-force microscopic (AFM) image of alanine tail and lysine head lipid-like peptide; note the tube structures. (B) Transmission electron microscopic (TEM) image of valine tail and aspartic acid head lipid-like peptide; note the tube structure with open and closed ends as well as vesicles.

(A)

(B)

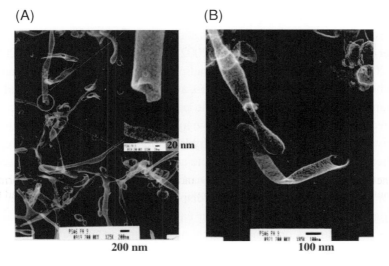

Figure 20.7. pSA6, a lipid-like peptide. Images of lipid-like phosphoserine head and alanine tail peptide enclosures. (A) Low magnification of structures; insert shows the high magnification of the single tube with opening. (B) High magnification of structures; note the life-like complexity and the vesicles.

(A)

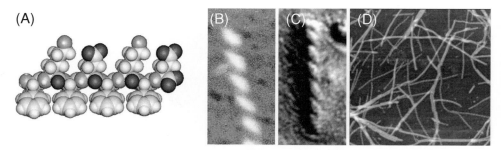

Figure 20.8. An amphiphilic peptide FKE8 (FKFEFKFE). A single FKE8 peptide is shown with a hydrophobic side, phenylalanine and hydrophilic side, lysine (positive charge), and glutamic acid (negative charge) (A). Atomic-force microscopic (AFM) (B) and transmission electron microscopic (TEM) (C) images of FKE8; note the distinctive left-handed helices in high magnification and overall fibrous structure with defined diameter (D).

Concluding remarks

Other simple amphiphilic peptides

A number of simple alternating peptides have a few amino acids. These peptides, like Lego® bricks, have two distinctive side-pegs and holes. The hydrophilic part bears charged residues, either positive charges – lysine, arginine, histidine – or negative charges – aspartic acids and glutamic acids. On the other side are hydrophobic residues – alanine, valine, isoleucine, leucine, phenylalanine, tyrosine, and tryptophan. Some of them can form well-ordered helical and other fibrous structures (Zhang *et al.*, 1993, 2002; Marini *et al.*, 2002; Hwang *et al.*, 2003; Zhang, 2003) (see Figures 20.8–20.10). Others form non-helical fibrous structures (Zhao and Zhang, 2004). They form many stable structures, as either double-layered helical or non-helical tapes that sequester hydrophobic parts because the hydrophobic residues must move away from water and leave the hydrophilic side exposed to water. Water is the driving force that sequesters the hydrophobic part of a molecule, regardless of whether it is a protein, a lipid, a nucleic acid, or some other small molecule.

To summarize, this provides one vision of a crucial bridge between the physicists' claim of a "fine-tuned universe" (one predisposed to the production of water, carbon, and nitrogen) and the reality of life on earth. The existence of stable nanotube structures demonstrates how biochemical molecular fine-tuning could give rise to complex entities, presumably through a process of prebiotic molecular selection applied to primitive, quasi-living autocatalytic networks.

Put more simply, when considering prebiotic selection and evolution in the context of the origin of life, the enormously powerful force of water must never be underestimated. As all life is based on water, all molecules in living systems interact with it, and water likely has driven molecular evolution from the very beginning, here on earth or plausibly elsewhere in the universe – or multiverse.

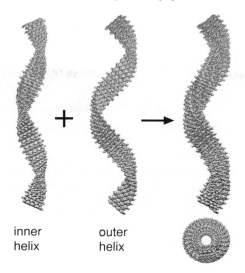

inner
helix

outer
helix

Figure 20.9. Molecular simulations of numerous FKE8 peptides, with hydrophobic phenylalanine on one side and hydrophilic lysines and glutamic acids on the other, undergo self-assembly to form left-handed helical fibers that contain 97 peptides per helical turn, 7 nm in diameter with a 19 nm pitch. The molecular simulated structures are consistent with the experimental observations. Water drives the hydrophobic phenylalanine pack inside the left-handed peptide double helix (Marini *et al.*, 2002; Hwang *et al.*, 2003).

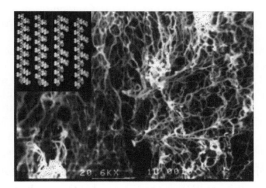

Figure 20.10. Amphiphilic ionic self-complementary peptides. This class of peptides has 16 amino acids, *c.* 5 nm in size, with an alternating polar and non-polar pattern. They form stable β-strand and β-sheet structures; thus, the side chains partition into two sides, one polar and the other non-polar. They undergo self-assembly to form nanofibers with the non-polar residues inside; positively and negatively charged residues form complementary ionic interactions, like a checkerboard. These nanofibers form interwoven matrices that further form a scaffold hydrogel with a very high water content (>99.5%). The simplest peptide scaffold may form compartments to separate molecules into localized places where they can not only have high concentration, but also form a molecular gradient, one of the key prerequisites for prebiotic molecular evolution.

References

Anders, E. (1989). Pre-biotic organic matter from comets and asteroids. *Nature*, **342**, 255–7.

Bada, J. L., Bigham, C. and Miller, S. L. (1994). Impact melting of frozen oceans on the early Earth: implications for the origin of life. *Proceedings of the National Academy of Sciences, USA*, **91**, 1248–50.

Ball, P. (2001). *Life's Matrix*. Berkeley, CA: University of California Press.

Beaudry, A. A. and Joyce, G. F. (1992). Directed evolution of an RNA enzyme. *Science*, **257**, 635–41.

Brack, A. and Orgel, L. E. (1975). Beta structures of alternating polypeptides and their possible prebiotic significance. *Nature*, **256**, 383–7.

Brack, A. and Spach, G. (1981). Enantiomer enrichment in early peptides. *Origins of Life and the Evolution of Biospheres*, **11**, 135–42.

Chakrabarti, A. C. and Deamer, D. W. (1994). Permeation of membranes by the neutral form of amino acids and peptides: relevance to the origin of peptide translocation. *Journal of Molecular Evolution*, **39**, 1–5.

Chyba, C. and Sagan, C. (1992). Endogenous production, exogenous delivery and impact-shock synthesis of organic molecules: an inventory for the origins of life. *Nature*, **355**, 125–32.

Ehrenfreund, P., Glavin, D. P., Botta, O. *et al.* (2001). Extraterrestrial amino acids in Orgueil and Ivuna: tracing the parent body of CI type carbonaceous chondrites. *Proceedings of the National Academy of Sciences, USA*, **98**, 2138–42.

Eisenberg, D. S. and Kauzmann, W. (1985). *The Structure and Properties of Water*, 2nd edn. Oxford: Oxford University Press.

Fox, S. W. and Harada, K. (1958). Thermal copolymerization of amino acids to a product resembling protein. *Science*, **128**, 1214.

Goto, T., Futamura, Y., Yamaguchi, Y. *et al.* (2005). Condensation reactions of amino acids under hydrothermal conditions with adiabatic expansion cooling. *Journal of Chemical Engineering of Japan*, **38**, 4: 295–9.

Hargreaves, W. R., Mulvihill, S. J. and Deamer, D. W. (1977). Synthesis of phospholipids and membranes in prebiotic conditions. *Nature*, **266**, 78–80.

Henderson, L. J. (1913). *The Fitness of the Environment: An Inquiry into the Biological Significance of the Properties of Matter*. New York, NY: Macmillan. Repr. (1958) Boston, MA: Beacon Press; (1970) Gloucester, MA: Peter Smith.

Henderson, L. J. (1917) *The Order of Nature: An Essay*. Cambridge, MA: Harvard University Press.

Hwang, W., Marini, D, Kamm, R. *et al.* (2003). Supramolecular structure of helical ribbons self-assembled from a beta-sheet peptide. *Journal of Chemical Physics*, **118**, 389–97.

Kvenvolden, K., Lawless, J., Pering, K. *et al.* (1970). Evidence for extraterrestrial amino-acids and hydrocarbons in the Murchison meteorite. *Nature*, **228**, 923–6.

Marini, D., Hwang, W., Lauffenburger, D. A. *et al.* (2002). Left-handed helical ribbon intermediates in the self-assembly of a beta-sheet peptide. *NanoLetters*, **2**, 295–9.

Miller, S. L. (1953). A production of amino acids under possible primitive earth conditions. *Science*, **117**, 528–9.

Miller, S. and Urey, H. C. (1959). Organic compound synthesis on the primitive earth. *Science*, **130**, 245–51.

Morowitz, H. J. (1992) *Beginning of Cellular Life*. New Haven, CT: Yale University Press.

Nagai, A., Nagai, Y., Qu, H. *et al.* (2007). Dynamic behaviors of lipid-like self-assembling peptide A₆D and A₆K nanotubes. *Journal of Nanoscience and Nanotechnology*, **7**, 2246–52.

Oro, J. and Guidry, C. L. (1961). Direct synthesis of polypeptides. I. Polycondensation of glycine in aqueous ammonia. *Archives of Biochemistry and Biophysics*, **93**, 166–71.

Oro, J., and Kamat, S. S. (1961). Amino-acid synthesis from hydrogen cyanide under possible primitive earth conditions. *Nature*, **190**, 442–3.

Santoso, S., Hwang, W., Hartman, H. *et al.* (2002a). Self-assembly of surfactant-like peptides with variable glycine tails to form nanotubes and nanovesicles. *NanoLetters*, **2**, 687–1.

Santoso, S., Vauthey, S. and Zhang, S. (2002b). Structures, functions, and applications of amphiphilic peptides. *Current Opinion in Colloid and Interface Science*, **7**, 262–6.

Vauthey, S. Santoso, S., Gong, H. *et al.* (2002). Molecular self-assembly of surfactant-like peptides to form nanotubes and nanovesicles. *Proceedings of the National Academy of Sciences, USA*, **99**, 5355–60.

von Maltzahn, G., Vauthey, S., Santoso, S. *et al.* (2003). Positively charged surfactant-like peptides self-assemble into nanostructures. *Langmuir*, **19**, 4332–7.

White, D. H., Kennedy, R. M. and Macklin, J. (1984). Acyl silicates and acyl aluminates as activated intermediates in peptide formation on clays. *Origins of Life and Evolution of Biosphers*, 14, 273–8.

Wick, R, Angelova, M. I., Walde, P. *et al.* (1996). Microinjection into giant vesicles and light microscopy investigation of enzyme-mediated vesicle transformations. *Chemical Biology*, **3**, 105.

Wilson, C. and Szostak, J. W. (1995). In vitro evolution of a self-alkylating ribozyme. *Nature*, **374**, 777–82.

Wong, J. T.-F. and Bronskill, P. M. (1979). Inadequacy of prebiotic synthesis as origin of proteinous amino acids. *Journal of Molecular Evolution*, **13**, 115–25.

Yanagawa, H., Ogawa, Y., Kojima, K. *et al.* (1988). Construction of protocellular structures under simulated primitive earth conditions. *Origin of Life and Evolution of Biospheres*, **18**, 179–207.

Yang, S. and Zhang, S. (2006) Self-assembling behavior of designer lipid-like peptides. *Supramolecular Chemistry*, **18**, 389–96.

Zhang, S. (2003). Fabrication of novel materials through molecular self-assembly. *Nature Biotechnology*, **21**, 1171–8.

Zhang, S. and Egli, M. (1994). A hypothesis: reciprocal information transfer between oligoribonucleotides and oligopeptides in prebiotic molecular evolution. *Origins of Life and Evolution of Biospheres*, **24**, 495–505.

Zhang, S. and Egli, M. (1995). A proposed complementary pairing mode between single-stranded nucleic acids and b-stranded peptides: a possible pathway for generating complex biological molecules. *Complexity*, **1**, 49–56.

Zhang, S., Holmes, T., Lockshin, C. *et al.* (1993). Spontaneous assembly of a self-complementary oligopeptide to form a stable macroscopic membrane. *Proceedings of the National Academy of Sciences, USA*, **90**, 3334–8.

Zhang, S., Marini, D., Hwang, W. *et al.* (2002). Design nanobiological materials through self-assembly of peptide and proteins. *Current Opinion in Chemical Biology*, **6**, 865–71.

Zhao, X. and Zhang, S. (2004). Design molecular biological materials using peptide motifs. *Trends in Biotechnology*, **22**, 470–6.

21

Evolution revisited by inorganic chemists

R. J. P. Williams and J. J. R. Fraústo da Silva

Introduction

If we inspect our surroundings on earth, we will see a myriad of materials, objects, natural and artificial constructions (many resulting from the activities of the human species), and, of course, an enormous variety of living organisms, from the simplest bacteria to the most complex animals. We can also inspect the sky and observe other planets, stars, galaxies, and clusters of galaxies far away in our expanding universe, leaving us wondering whether life can also be found elsewhere. Is it all an inevitable product of such a finely tuned construct as the universe seems to be, given appropriate local conditions? At a very basic level, it must be. We are an evolved species co-existing with many other simpler species on which we depend. All are made of the same chemical elements resulting from that fine-tuning that allowed their kinetically controlled formation in big stars and created our planet and the fields to which all life is exposed. Life must be a possibility included in a finely tuned cosmos – we sense it in our minds. We also know that it has evolved and diversified here on earth, although the reasons for this – that is, the factors that determined evolution – are not so obvious. This question has intrigued philosophers and scientists through the ages, and it reached public awareness toward the end of the eighteenth century. The names Lamarck and, especially, Darwin are perhaps most widely known, but many others prepared the ground for earlier developments and for those that have followed. The progress in knowledge and understanding has been enormous, and inquiry has extended even to the origin of life itself. Nevertheless, no accepted explanation of the origin of life exists, nor does a rational chemical description of evolution. In their book *The Major Transitions of Evolution*, Maynard Smith and Szathmáry [1] go so far as to state that evolution has resulted in an increase in complexity, but that this increase "is neither universal nor inevitable." The book

Fitness of the Cosmos for Life: Biochemistry and Fine-Tuning, ed. J. D. Barrow *et al.*
Published by Cambridge University Press. © Cambridge University Press 2007.

arrives at this conclusion with reference to the development of species, but it does not take an ecological/chemical viewpoint, as no mention of the environment is to be found. We consider that without reference to the environment and its changes, evolution of life cannot be understood. The environment is essentially composed of inorganic chemicals.

Our purpose then in this chapter is to tackle evolution, not the origin of life, seen, however, from the point of view of the inorganic and physical chemistry of the combination of the environment and organisms. The elucidation of the three-dimensional structure of DNA by Watson and Crick and publication of the seminal paper of Stanley Miller on prebiotic synthesis, both in 1953, as well as the beginning of the systematic study of the biological role of the chemical elements [2], have made such an approach possible. A major part of our argument is, then, that the chemistry of life is not just "organic" – far from it – and that life cannot be studied in isolation from the original and changing inorganic environmental surroundings in which it emerged and evolved. We must bring together the nature of DNA, the prebiotic and subsequent environmental chemistry, as well as life's chemistries. We will show why we believe that the resultant overall ecological development was inevitable, but we cannot give a reason for its timing. It could have been stuck for a very long time in a given condition.

To structure the presentation, we show in Figure 21.1 a rather simplistic scheme of the requirements necessary for the formation and functioning of a living cellular organism: (1) *raw materials*, necessarily obtained from the environment; (2) *available (usable) energy*, obtained from the earth's core and mantle or from the sun's radiation; (3) *confinement* (or closure), to localize the reaction space and to avoid dispersion of raw materials and products; (4) *operational instructions* ("information"), to allow internal self-regulation and reproduction with a minimum of error (which must not be identified with or just reduced to a genetic *code*, which, by definition, can code only something that already existed or developed simultaneously), to which we add (5) the ability to distinguish favorable from unfavorable conditions in a changing environment. The environment will be seen to be a second primary source of information (see "Steady states and final conditions" below). Because it is extremely complicated, the idea of information in a cell will be treated later. In essence, the first four are the requirements for a reproductive organization, while the fifth allows an initial system to evolve. Some parallels with the evolution of industrial plants can be seen.

In this chapter we will concentrate first on the raw materials of life, starting from their constitutive chemical elements (see Figure 21.2), trying to show that changes in their environmental bioavailability influenced decisively the possible increase in organizational complexity of living organisms, causing what de Duve [3] calls "vertical" evolution – that is, the major overall steps of evolutionary change [4].

(a)

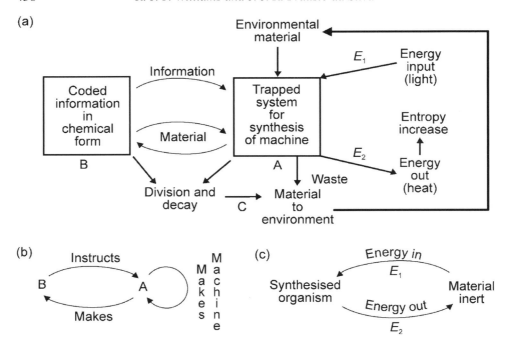

(b)

(c)

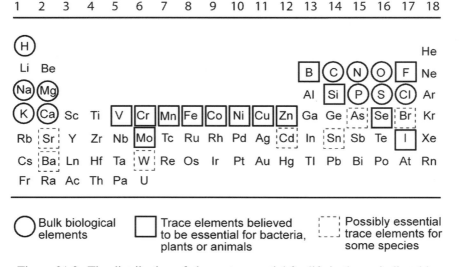

Figure 21.1. To generate a living system, (a) it is necessary to have sources of energy and material, a coded program (B), and a machine (A). The essence of the machine is that it provides routes to synthesis (as irreversible reactions) because synthesis has to absorb energy, E_1, and decay generates it, E_2; see (c). (b) The code and the machine evolve together with the environment.

Figure 21.2. The distribution of elements essential for life in the periodic table.

Table 21.1. *The stages of evolution*

A. Prebiotic protocells, including a variety of energized flow systems of chemicals leading to precursors of the primary chemicals of later stages.

B. Prokaryote cells came first and are of many more than one variety today. The one major compartment, the cytoplasm, is contained by one major membrane. The cell activities are already coded and concerted. The extreme variety of these cells – anaerobes and aerobes are different chemotypes – and their metabolism have increased with time. Their sensing of the environment is not advanced.

C. Single eukaryote cells have many internal compartments, vesicles, and organelles, and many types of such cells exist. The basic metabolism in the cytoplasm is very like that of prokaryotes. They are all aerobic large cells with a much increased organization internally and an increased ability to recognize environmental factors.

D. Multicellular eukaryotes, often classified as fungi, plants, and animals, are all aerobes. Organizational complexity and signaling between differentiated cells and organs have greatly increased. They have an increased ability to sense the environment.

D. Animals with brains – the development of the nervous system with a brain allowed the animal to be informed about the environment and to remember experiences. The fast responses were increasingly independent of reference to DNA. Organization in groups seen in patterns of behavior developed.

F. Humankind – the further development of the brain led to the understanding of the environment and many features of organisms. Hence, the environment became usable in constructs independent of inheritance in the DNA. External equipment could perform many desired functions. Information was passed down the generations through external and internal recording. Organization expanded enormously externally.

Source: Maynard Smith & Szathmáry [1].

Table 21.1 indicates the stages in evolution we will be discussing. Although this approach seems to us to make intuitive good sense, it is curiously absent from most theories and discussions of biological evolution, perhaps because evolution has been, through the years, a field to which many biologists, biochemists, paleontologists, cosmologists, physicists, organic chemists, philosophers, and others have given their attention, largely to species, but to which inorganic chemists have seldom contributed [1]. Again, by limiting our analysis to what in effect are the chemical classes of life – we shall call them "chemotypes" as opposed to genotypes – we avoid many of the difficulties associated with species. We consider that the concentration on species hides the general progression, denied in reference [1], of the ecosystem.

Of course, life is not just a question of chemicals. Therefore, as stated, we will also discuss topics that are equally relevant, but on which misunderstandings and differences of opinion persist, such as the question of the "quality" of energy and its usefulness in achieving a possible optimal cyclic steady state; the confusion around the concept of entropy and the applicability of the second law of thermodynamics to structured organizations such as living organisms; the concept of information; and the nature and function of the whole genetic apparatus [5]. In the later part of this chapter we will attempt to impart a better understanding of these ideas, together with the propositions that synergism (including symbiosis) is a major factor of biological evolution and that the evolution of cooperative activity in one ecosystem superimposes on and dominates the evolution of organisms and species. It is the ecosystem that has a logical development, a view that we share with others [4, 5] and that may well force a change of paradigm to allow a clearer vision of a fundamental problem for humankind: the nature of life and its sustainability in an ecosystem, of which we are the dominant and probably the most vulnerable part. We turn first to the element content of organisms so as to stress the role of inorganic – apart from the well-described organic – chemicals in organisms [6, 7].

The biological elements

Life seems to have emerged on earth about 3.5 to 3.8 billion years ago, when the environment was essentially reducing [8] and free oxygen was not present in an appreciable amount. The high reactivity of oxygen had forced it to combine with hydrogen, carbon, and many less abundant elements, such as phosphorus, silicon, aluminum, calcium, iron, magnesium, sodium, and potassium, but generally not the later transition metals (Co to Zn), which, in the absence of oxygen, combined preferably with sulfur to give sparingly soluble sulfides in the primordial, some-what acidic (HCl, H_2S) conditions. No one knows the way in which the first living organisms arose in this environment, and it is doubtful that we ever will. Many theories exist, but no fully satisfying evidence has been found concerning the origi-nal synthesis of the primordial building blocks – organic compounds together with required inorganic elements – on the earth or in the interstellar space to give emer-gence (in a "warm little pond" or in the deep-sea volcanic trenches) to the first forms of cells.

The same can be said of the many theories concerning the origins of organization: self-assembly (autopoiesis), self-catalyzed cycles, dissipative systems, and order-out-of-chaos – all just theories without experimental verification. Because the initial conditions for the emergence of life are not known with any certainty, it is unlikely that we will ever manage to go beyond this nebulous thinking unless we are able to generate life itself. We will therefore begin from a chemical analysis of organ-isms, starting from the nature of life in the most primitive cells, which existed, as recognized by what is found today, in what we take to be a primitive environment.

The facts are that living organisms exist; their current diversity is enormous; they are all cellular; their organizational complexity has increased in evolution [1]; the remnants of the presumed most primitive organisms co-exist with the modern

Table 21.2. *Maintained pathways throughout evolution*

Pathway	Example
Syntheses and degradation of saccharides	glycolysis (Mg)
Dicarboxylic-acid reaction sequence	CO_2 incorporation in incomplete cycle; completed later to give energy capture (i.e. Krebs cycle) (Fe, Mg)
Amino-acid synthesis	products of glycolysis and Krebs cycle + NH_3 (Fe)
Protein synthesis	formyl initiation (Fe) and methionine initiation (Fe, Co)
DNA, RNA synthesis	nucleic-acid pathways (Mg, Zn, Fe/S, B_{12}, or Fe_2O)
Synthesis of fats	β-carbon oxidation/reduction (flavin, Fe)
Nitrogen incorporation	formation of NH_3 (Mg, Fe, V, Mo) in symbiotic bacteria
Hydrogen reactions	H_2 as a reductant (Fe, Ni) in anaerobic Archaea
Energy: electron/proton flow in membranes	energy capture related to ATP (Fe, NADH, flavin, quinones)
Exchange of ions in membranes	osmotic and electrolyte control (Na, K, Mg, Ca, H)
Light capture	chlorophyll and heme (Mg, Fe)
Pumps (ATPases)	most elements as cations or anions
Redox balance	glutathione, Fe

ones, cooperating actively (see below); and they all have in common a particular chemical characteristic in their cellular cytoplasm. All cells contain a basic set of organic chemicals, proteins, lipids, nucleic acids, and polysaccharides made from the non-metal elements H, C, N, O, P, and S. To make them environmental, carbon compounds, such as CO and CO_2, had to be reduced. Their basic cytoplasmic metabolic reactions have remained, reducing even until now – that is, throughout 3.5 billion years (see Table 21.2). The reason for these facts is more or less obvious: at the fixed, almost neutral pH at which seas settled, these are the set of reactions giving rise to biopolymers that provided the kinetic stability and activities of the organization we observe today. We know of no others, and it is likely that none exists. However, we need to add to the list of essential elements in these molecules an equally large group of metal ions (see Figure 21.2 and Table 21.3) because polymers have to be synthesized from different non-metal raw materials, mainly CH_4, H_2O, CO, CO_2, H_2S, NH_3, (HCN), or derived compounds resulting from prebiotic chemistry (see Table 21.2). It was the reduction of some and the reactions of all of these small non-metal molecules that required catalysts, and these could not be made from organic elements alone. Thus, certain metal elements had to

Table 21.3. *The absolute minimal element content of primitive life*

Element[a]	Source	Use
H	H_2S (air), HS^- (sea)	organic molecules, energy capture
C	CH_4, CO, or CO_2	organic molecules
N	NH_3, HCN (sea)	organic molecules
O	H_2O, CO, CO_2	organic molecules
Na^+, K^+, Cl^-	sea salts	electrolyte balance, osmotic control
Ca^{2+}, Mg^{2+}	sea salts	structure stabilization, weak acid catalyst (Mg^{2+})
$P(HPO_4^{2-})$	sea salts	organic molecules, energy transfer
S	H_2S (air), HS^- (sea) $/S^{2--}$ (sea)	element transfer, energy metabolism
Fe	Fe^{2+}/Fe^{3+}	catalysis

[a]These 12 elements were inevitably incorporated into any vesicle formed in the sea in the period around 3–4 $\times$ 10^9 years ago. Others that were present in reasonable amounts but perhaps not incorporated of necessity initially were Al, Si, V, Mn, Co, Ni, W, and perhaps Se and Br. The primitive organisms we know, such as archaebacteria, have approximately 20 elements.
Source: See [6, 7].

be obtained from the environment. Other essential inorganic chemicals besides catalysts are also required for the stability of cells, and we describe these below.

All of the elements had to be selected initially from among those most available in the prevailing primitive environmental conditions (probably somewhat acidic with a high content of H_2S). From analysis, the total basic list of the necessary elements for all forms of life, and therefore for the emergence of life, we deduce to be C, H, N, O, S, P, Se, (Cl), (Na), K, Mg, (Ca), (Mn), Fe, {Zn}, and probably a little Ni, Co, V, and W (see Table 21.3) [6, 7]. We know of life only in which the vast majority, if not all, of these elements are present. The elements in parentheses (M) are at lower free concentrations in cells than in the environment because they were largely rejected. The requirement for Zn, shown in braces, is uncertain. The reasons for this "natural" selection of the elements are known [6, 7]; here we will limit ourselves to passing comments. The selection is dependent on binding constants of the available metal elements to organic molecules that conform in fact to well-defined sequences, e.g. the Irving–Williams series, as most of these interactions come to equilibrium [7, 9] and on the functional capacity of the resultant complexes.

The organic molecules in cells are mainly anionic. Sodium, as Na^+, cannot be allowed to neutralize these negative charges of the non-metal compounds inside cells because its external concentration (in the sea) is too high (10^{-1} M) and its entry would result in too great a cellular osmotic pressure. For this reason, Cl^- also had to be largely rejected. Instead, potassium, K^+, is selected for the uptake into the cytoplasm. Neither of these cations binds to organic molecules. Calcium, as Ca^{2+}, is also deleterious inside cells because at its environmental concentration (10^{-3} M) it binds to many negative centers and forms gels or precipitates with them. Manganese, as Mn^{2+}, competes easily with Mg^{2+}, which is necessary inside cells to neutralize negative charge and to act as a structuring element or as a catalyst. Mg is lowered in concentration in the cytoplasm. (Note that Mg is also indispensable for the cell energetic processes; ATP should be written Mg–ATP.) The need to neutralize charge is, as stated, due to the fact that, on average, the biological polymers and substrates are negatively charged because they have groups that at pH 7 are deprotonated, e.g. carboxylate and phosphate. The last element, in braces above, is Zn. Its very low presence or absence in the primitive cells was due to its very low availability from its sulfide (see below). Cells cannot tolerate high concentrations of this or any free-transition-metal ions M^{2+} as they would compete for useful Mg^{2+} and Fe^{2+} sites, but such concentrations were unlikely initially because of the insolubility of their sulfides [7].

Iron and magnesium are the preferred catalysts in primordial organisms; both were abundant and available (although Fe forms a slightly insoluble sulfide, it also gives soluble Fe/S clusters). Mg^{2+} became the major ion for condensation and Fe^{2+} the major ion for redox processes. Note that iron binds preferentially to N and S donors, such as amines and thiolates, whereas Mg binds to O donors, such as carboxylates and phosphates. Thus, given their concentrations, they can be complexed selectively. Iron can then be used for redox catalyses and magnesium for acid–base catalyses. Apparently, vanadium (V) and tungsten (W) also played an important role as catalysts in primitive or slightly later biological chemistry (V in the nitrogenase of bacteria and W in redox processes occurring especially in extremophile Archaea). They were both displaced when Mo was liberated from its sulfide in aerobic conditions. Although Ni and Co are of low abundance and form rather insoluble sulfides, they occur in today's bacteria and Archaea in catalysts of reactions involving H_2, CH_4, CO, and CH_3. Their presence in primitive organisms may have been due to more acidic environmental conditions where their sulfides are more soluble. Table 21.4 summarizes the roles of the elements, which relate to the minimal set required for primordial life (Table 21.2). Note the absence of copper and the very small amounts of zinc in primitive anaerobes.

Of course, the relative concentrations of all these elements in the cell cytoplasm had to be controlled. Therefore, some had to be captured and some rejected, by using a variety of processes, including pumping, which requires energy, Mg–ATP, or proton gradients. In part, this control is managed through the genetic machinery, but it also involves simple questions of chemical equilibrium, e.g., stability constants of complexes, solubility products of salts, etc., which are not related to genes [6, 7, 9]. The intimate link to the environment, source of materials, and energy is obvious in all these cases. We consider that the free concentrations of Mg^{2+}, Fe^{2+}, phosphates, and other substrates also act in feedback loops involving the control of their cellular

Table 21.4. *Major functions of elements in primitive cells*

Elements[a]	Major functions
H, C, N, O, P, S	formation of polymers from H_2O, CO_2, NH_3, HPO_4^{2-} and S^{2-}
Na^+, K^+, Cl^-	electrolytic and osmotic balance
Mg^{2+}	mild catalysis (phosphate compounds); structural in RNA, DNA, etc.
Mn^{2+}, (Zn^{2+})	some stronger acid–base catalysis
V, Fe^{2+}, Co^{2+}, Ni^{2+}, Se, Mo, (W)	redox catalysts, often devoted to reduction
Ca^{2+}, Si	strengthening of outer structures

[a]Note the absence of Cu, Br, I, the low content of Zn, and the reduced state of iron, sulfur, and selenium.

concentration and their own enzymes by connection to the DNA, the pumps for their uptake, and the environment. These substances acted not just as active agents, but as messengers to maintain an "informed system" (see "Information" below). Most important of all controls linked to energy and material uses is Mg–NTP, where N is a nucleotide, e.g. Mg–ATP.

For many purposes, the simple metal ions Mg^{2+} and Fe^{2+} are not ideal because they exchange too easily from complexes. Again, nickel and cobalt are not easily incorporated into cells, given the insolubility of their sulfides, which increased as the seas approached neutral pH. An advantageous development was the synthesis of organic molecules capable of incorporating a few metal ions in irreversible processes. Once so incorporated, the metal ions do not exchange, so that the species formed behave effectively as new "metals," different from those that were sequestered. The main molecules of this kind were the porphyrins, all derived from uroporphyrin. The metals so incorporated were Mg (in chlorophyll), Fe (in heme), and small amounts of Co (in cobalamin) and Ni (in the F-430 factor). All have considerable thermodynamic and kinetic stability and do not equilibrate with the free-metal ions. One of these complexes, chlorophyll, enabled the capture of solar radiation, a new source of energy, and then the development of photosynthesis [10]. It is this photosynthesis, using H_2O to give hydrogen and rejecting dioxygen and aided by Fe and Mn enzymes, that forced evolution. Such photosynthesis arose, we consider, in an inevitable drive toward optimal energy capture in steady states (see "Steady states and final conditions" below), supposing life started using chemical energy trapped in inorganic chemicals during earth's formation. (Note that some Fe/S centers are also formed irreversibly, but others can exchange Fe.)

Externally, calcium and manganese are the major active elements; the first, initially a poison, became one of the most important biological elements, as we will stress later, and the second, bound on the outer surface of the membrane, gave access to the production of free molecular oxygen, O_2, from water in the photosynthesis process, which changed the nature of life and of the environment, i.e. of the whole earth's ecosystem.

A feature of these first, and all subsequent, organisms is that they live in and are "informed" by external gradients of different kinds – gravitational, magnetic, and chemical [8] – as well as by internal feedback loops. The chemical gradients are partly self-generated because as the cells take in chemicals they create a lowered concentration around themselves. Again, the very act of creating a cell membrane with the discharge of other ions produces an electrical gradient, particularly of Na^+, K^+, Cl^-, and Ca^{2+}. Disturbances of any of these gradients are a potential source of information about the outside environment, and if they can be recognized the cell can search the environment by swimming to advantage. It is then a matter of how the sensing of the environment is connected to the cell's physico-chemical apparatus. The connection in prokaryotes is very largely to the synthesis of coded sensor units, proteins. Once the proteins are in place, the cell is informed by the gradients and need not refer to the DNA for immediate response.

We shall see that a major step in evolution concerns DNA indirectly, but is directly linked to this metabolic awareness of the environment and the ability to respond. Bacteria respond to chemicals, food gradients, through a special messenger in the cytoplasm – production of c-AMP – but not to these ionic gradients. They show chemotaxis, but have little other sense of the environment; for example, they cannot alter their shape. Their life cycle is short, and energy is overwhelmingly directed to reproduction.

Note that in this primordial biological chemistry some fifteen elements are essential, but the list does not include metals that became progressively important, especially molybdenum and copper, which form insoluble sulfides and so were not available initially. Availability changed as sulfide was oxidiyed and water supplied useful hydrogen but waste oxygen.

The road to unicellular eukaryotes

When oxidizing conditions developed, access to many essential non-metals and metals alike diminished, while others became available (see Figure 21.3). Carbon was now only present as CO_2, hydrogen as H_2O, nitrogen as N_2, sulfur as sulfate, and so on among non-metals. Now, apart from these changes, many sulfides were oxidized to soluble sulfates to give increasingly available, but poisonous, Zn^{2+}, Cu^{2+}, Co^{2+}, Ni^{2+}, and molybdate, which organisms had not met in quantity before. Among metals, iron was oxidized to Fe^{3+} and precipitated as $Fe(OH)_3$. In the cytoplasm, new metabolic paths were therefore essential (see [6, 7]). For example, to obtain essential iron, the cell's machinery had to develop ways to synthesize special

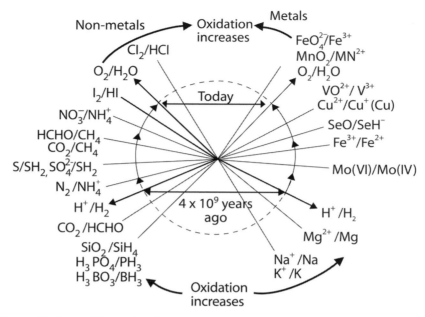

Figure 21.3. An illustrative diagram to show the way in which the environment has changed as it moved inevitably from reducing to oxidizing conditions because of the rise in oxygen in the atmosphere (as shown in Figure 21.4). The standard oxidation/reduction potentials at pH = 7.0 are used on a scale of H_2/H^+ at -0.4 V and H_2O/O_2 at $+0.8$ V. The environment is assumed to change close to the equilibrium redox potential set by the partial pressure of oxygen. Further data are provided in references [5, 6].

external complexing agents, siderophores, capable of binding Fe^{3+} and bringing it inside the cell. Slowly, the nature of the environment changed, from reducing to oxidizing, and the cell's internal cytoplasmic chemistry became therefore even more reducing relative to the environment because, as stated, the fixed reduction is imperative to enable the synthesis of the biological polymers.

How could the extant, simple, anaerobic, unicellular organisms survive and reproduce in such conditions? From what we observe, the answer is straightforward: many of them did not and died; some found shelter in anoxic niches where they could be (and still are) viable; others *adapted* to the prevailing oxidizing conditions, which required considerable changes; and, finally, other cells underwent dramatic change. Our thesis is that *organisms had to evolve in a particular chemical way, forced by the new environmental conditions.* They could evolve in two ways: by adaptation to give aerobic prokaryotes or by radical change to aerobic eukaryotes. We have described the change to aerobic bacteria elsewhere [6, 7]; here we wish to stress why and how the eukaryotes arose and led to huge ramifications, as seen

Table 21.5. *Levels of self-organization*

A. Prokaryotes (3.5×10^9 years ago to the present)
1. Single-membrane containment
2. Differential, inside–outside distribution of elements
3. Polymer–polymer synthesis and interactions
4. Relatively simple messenger-based coordination

B. Eukaryotes, single cells (2 to 2.5×10^9 years ago to the present)
5. Multiple membrane-limited in the compartments and later organelles
6. Differential distribution of elements and large molecules in 5
7. More complex polymer–polymer interactions including filaments, flexible membranes, and external mineralization, which generate very selective shapes
8. Complicated messenger systems linking compartments and the environment, especially using calcium

NB. Eukaryotes evolved from the archaebacteria plus organelles from eubacteria.

C. Multicellular organisms (about 1.0×10^9 years ago)
The extra organization is now of eukaryotic cells, so we add to (5) to (8) of (B) above:
9. Multiple cell-limited activities, organs
10. Differential distribution of all chemicals in different cells
11. Increasing control over extracellular body fluids, filaments, and minerals
12. Messenger systems linking cells (nerves) via transmitters, hormones, etc., organic in nature, followed by nerve connections in late animals
13. Central co-ordination (the brain) in more advanced animal organisms ($<0.5 \times 10^9$ years ago)

The reader may wish to add other features to this table.

---

in the present. Note that the anaerobic and aerobic prokaryotes did not disappear, but gradually became deeply involved in a total ecosystem (see below). In fact, as stressed in reference [1], all species have developed during evolution to some degree, but our interest is not with them in particular. Rather, it is in the overall development of an ecosystem with new types of cellular organization arising to meet new environment conditions more effectively within them, although in other ways, e.g., rate of reproduction, they were at a disadvantage. Much of this disadvantage is overcome by symbiosis.

The *major necessary strategy* of eukaryotes to reconcile a reductive cell's cytoplasm with an oxidizing environment was the development of new membranes and compartments separated from the cytoplasm inside cells and where oxidizing chemistry could take place (Table 21.5) [6, 7]. (Such a compartment outside the cytoplasm is also the periplasmic space between the internal and the external membranes, seen in aerobic bacteria.) Vesicles inside cells probably originated from invagination of the new external membranes, incidentally inverting all their pumps so that their contents became similar to the external environment relative to the cytoplasm. For example, they have high concentrations of Ca^{2+}, Na^+, Mn^{2+}, and

Cl⁻, as well as other oxidized chemical species that were hardly available before the emergence of oxygen, such as bound Zn^{2+} and then Cu^+/Cu^{2+}, but low concentrations of K^+, Fe^{2+}, and Mg^{2+}. In addition, the rejection of Ca^{2+} to the environment and then to internal compartments allowed the formation of, for example, organized calcium carbonate to be used in external skeletons (as in coccoliths). The export of Ca^{2+} and Zn^{2+} enzymes enabled acid–base hydrolytic digestion not only outside cells, but more generally in vesicles.

It is important to note that one apparently essential change that occurred to the primitive unicellular, eukaryote anaerobic organisms was the incorporation of *cholesterol*, especially in their outer cell membranes [7]. Cholesterol is a product of the oxidation of squalene derivatives, giving the membranes greater strength so that they could do without a wall, but yet through greater flexibility could allow vesicle formation within them and, generally, enlargement of a cell. The evolution of persistent internal filaments, the changed membranes, and the loss of cell walls allowed the increase in size of such organisms and also the first possibility of synergism through the engulfing and incorporation in metabolism of smaller organisms – bacteria, leading to the formation of cellular organelles – mitochondria (from a methanogenic bacterium) and chloroplasts (from a cyanobacterium). These developments utilized the flexibility of the cell membrane. Each of these organelles has a DNA different from that found today in its host (the nuclear DNA), evidence of their separate origin and elaboration of control [11]. In effect, the organelles are just symbiotic compartments in an enlarged organization [4] (see also [19]).

Of course, the creation of more compartments in one organization implies the simultaneous development of a new network of messengers, such that the entire cellular chemistry functions harmoniously (see Table 21.5 and reference [12]). One major new messenger is Ca^{2+} from the environment, which can be used in the cytoplasm in only brief impulses, given its tendency to bind rather strongly at high concentration to other essential chemicals, blocking their action. Calcium-containing vesicles became a kind of accessory "condenser" in these electrolytic circuits, controlling a series of processes linked especially to the use of chemical energy in the form of Mg–ATP, namely mechanical processes such as the tension of internal filaments that maintain and adjust the cell's shape and the compartments in place, and the motion of cilia and fiber contraction. Another important messenger is internal, cyclic AMP, synthesized in bacteria and linked to such processes as sensing food, which now became a calcium-controlled process. Calcium and various other organic phosphates, often in coupled systems, came to be used generally to inform the cell of the environment, thereby gaining a great advantage for cell survival. Note that the rejection of calcium by bacteria generated the message possibility, but without a receiver for it (see below). Their short life made such information transfer of little use. In evolution, DNA increased in size as one source of information, but sensors increased information from the environment. The calcium cell sensors act on metabolism in fast response, independently of DNA response. This leads eventually to the development of the phenotype (see below). We make the general statement that *activity of organisms is increasingly linked to information from the environment whereas information in the gene dominates reproduction*, but we need to discover how the two became interactive (see "Information" below).

Simultaneously with these developments, the cellular DNA became positioned behind an internal separate membrane (the nuclear membrane), which ensured higher protection and stability. In effect, a new compartment was created in which the DNA became progressively associated with a series of proteins, the histones, in a more compact structure. The "reading" of this code obviously became a safer but far more complicated process because the parts to be "read" demand that they are exposed to the reading machinery. As the DNA is exposed and unfolds, it has a higher local vulnerability and, consequently, mutation rate. The DNA was now split into reading (exons) and non-reading (introns) frames. At the same time, new transcription factors and messengers in the cytoplasm arose. We note particularly the increasing use of zinc in zinc fingers. We will come to these aspects of evolution later, but note that this gross evolutionary process to more complex organisms was indeed enforced by the presence of an oxidized environment and directed by the environmental changes, for example of available zinc. It was not just the accidental consequence of random mutations, as much as these are required to create the synthesis to match this environment. But were even the mutations, in fact, to some degree directed? (See the discussions of epigenetic possibilities in [13, 14].)

We may therefore say that the evolution from prokaryotes to eukaryotes was achieved through: (1) the creation of new flexible membranes; (2) the development of internal compartments, including the separate nucleus; (3) the development of internal persistent filament structures; (4) the use of oxidizing chemical processes and new elements in cells and particularly in vesicles (note especially copper and zinc released from their sulfides); (5) the use of new messengers to ensure communication between all component compartments, including the environment, the essence of organization; and (6) the beginnings of synergism (symbiosis) between different organisms. These changes propelled vertical evolution (Table 21.1), but they did not stop at unicellular organisms because considerable advantages in material and energy capture result in multicellular eukaryotes. However, they have the disadvantage of great complexity, and their existence came to depend on unicellular organisms in direct or indirect symbiosis.

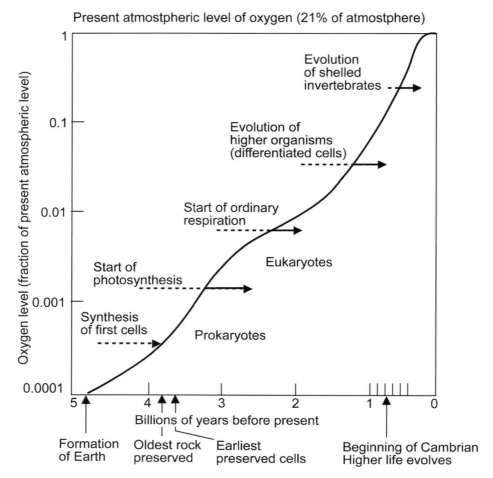

Figure 21.4. The rise in oxygen partial pressure with time and an indication of the correlation with the rise of organization which we take it was the way to manage the changed chemical environment of Figure 21.3. We use the word "higher" here to indicate an increase in organizational complexity, but we have to include symbiosis.

Multicellular eukaryotes

As the production of O_2 increased, the possibility of special evolution of novel organisms to ensure improved survival grew (see Figure 21.4). This novelty gave rise to increased light capture by plants and increased foraging for food by fungi, plants, and animals different from unicellular organisms [1]. These two developments of multicellular organisms also of necessity involved much greater organization. We saw that complexity of a cell had generally increased in unicellular eukaryotes, but it was alleviated by effectively combining cells with internal bacteria, the organelles. In this way, organisms were effectively multicellular, using the organelles, mitochondria, and chloroplasts to supply energy, while the cytoplasm managed most synthesis and uptake of chemicals. Multicellular eukaryotes evolved further to generate individual cell (and organ) specialization by differentiation and growth, alleviating complexity to a further degree in any particular cell. This evolution on a large scale used many of the novel features of unicellular eukaryotes, but with the introduction of new constructs including cell–cell binding, extra sensing of the environment, and new external and internal (cell–cell) messengers – all, of course, dependent on parallel development at the DNA level for controlled reproduction and production of proteins. The organisms required relatively rigid external filament structures to hold cells together, in positions; this can be achieved only through *oxidized cross-linking of polymers*. The earliest cross-linking chemistry in plants used free-radical reactions with the participation of O_2 or H_2O_2, mainly in enzymes binding Fe-protoporphyrins, and rejected manganese (in peroxidases and ligninases). Remember that the organisms now depended on lower organisms for many chemical elements.

Very similar peroxidases are used by plants to break connective tissue and allow growth. This particular procedure is not safe in animal growth because of the risk of mutation and cell migration, and we find that cross-linking is mostly by *copper enzymes* (oxidases) in extracellular fluids, not by using radicals, to produce, in particular, collagen in animals. On the other hand, zinc acquired new essential functions, such as external hydrolysis of collagenases, elastases, etc. to disrupt external filamentous structures so as to allow growth, migration, and reassembling of animal cells. Such changes were possible only through the extended use of these newly available elements: zinc first, and later copper, coinciding – one billion years ago – with the oxidation of their sulfides (Table 21.6). Simultaneously, new external chemical messengers arose to allow cell–cell communication and maintain the coordinated control and organization of the system.

These messengers were now organic compounds, hormones, and other substances, including transmitters, again largely produced by oxidative processes. Note that the simple ion messengers – phosphate, calcium, magnesium, and iron – were already used, so that to avoid confusion such organic messengers had to be called into use. Those such as adrenaline and amidated peptides were synthesized in vesicles by using *copper enzymes*. They acted on the outside of cells and connected across the cell membrane through calcium pulses, whereas those such as sterols connected to zinc transcription factors directly in the cell. In the oxidative processes, we stress the relevance of iron (heme) and copper in the production, reception, and destruction of such organic messengers in both plants and animals. Meanwhile, zinc also found greatly increased use internally in transcription factors (five percent of

Table 21.6. *New element biochemistry after the advent of dioxygen*

Element	Biochemistry
Copper	most oxidases outside higher cells; connective tissue finalization; production of some transmitters and hormones, dioxygen carrier, N/O metabolism
Molybdenum	two-electron reactions outside cells; NO_3^-, SO_4^{2-}, aldehyde metabolism
Manganese	higher oxidation state reactions in vesicles, organelles, and outside cells; lignin oxidation (note especially plants); O_2 production; activity of Mn in the Golgi
Nickel	virtually disappears from higher organisms
Vanadium	new haloperoxidases outside cells
Calcium	calmodulin systems; γ-carboxyglutamate links; general value outside cells and in cell–cell links
Zinc	zinc fingers connect to hormones produced by oxidative metabolism
Selenium	detoxification from peroxides, de-iodination?
Halogens	new carbon–halogen chemistry, poisons, hormones
Iron[a]	vast range of especially membrane-bound and/or vesicular oxidases; peroxidases for the production of hydroxylated and halogenated secondary metabolites; dioxygen carrier and store

[a]Note that iron can act as an agent responding to redox potential changes in the cytoplasm (e.g. in P-450).

the total in some higher animals), and exchangeable zinc may act as an inorganic hormone, connecting together cell–cell growth via zinc fingers and zinc carriers (see [6, 7]).

It is essential to note the way environment changes predicated the development of the chemistry of organisms outside the cytoplasm (copper and iron with oxygen and oxidized chemicals and zinc in hydrolysis) while internally the development was not so much of metabolism, but of organization, especially of zinc transcription factors and calcium-dependent messengers. Now increased protection was needed against O_2 and its reduction products, O_2^-, H_2O_2, and OH·, which are even more active. For this purpose, superoxide dismutase (first in prokaryotes Fe and Mn, or Ni, but note later in eukaryotes the safer Cu/Zn enzyme), catalase, and free-radical (OH·) scavengers emerged. These very reagents also became used in protection against invasive organisms and chemicals, especially heme and peroxides in plant and animal immune vesicle systems. Again, the new possibility of introducing oxidized states of selenium alone or selenium and iodine generated further protection, and the synthesis, with heme iron, of the thyroid hormones (see their oxidation/reduction potentials in Figure 21.3).

In the opposite sense, the involvement of H_2, CH_4, CO, etc., present in primitive atmospheres, became of little value as these gases were oxidized away. As time

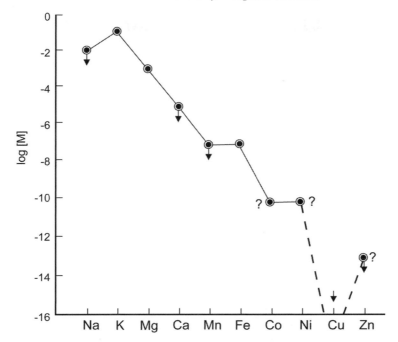

Figure 21.5. The proposed free-metal ion concentrations in the cytoplasm of all cells. The values are required for the well-being of the reductive organic chemistry in cell cytoplasm. The order of divalent ions of Mg > Mn ≥ Fe > Co Ni > Cu < Zn is dictated by complex ion equilibria managed by organic ligand synthesis, which protects the function of each metal ion. This order is the reverse of the order of binding, the Irving–Williams order, in most complex ion equilibria [5, 6].

passed, nickel – which at first had a considerable role in this chemistry – became almost absent in animals and is found only in urease in plants. Similarly, cobalt, valued early in carbon transfer, is not used by higher plants with the complete replacement of the functions of coenzyme B_{12} in plants, often by zinc enzymes, and the requirement for it as a little-used vitamin in animals. Note that vitamins are synthesised by symbionts or are obtained from food (indirect symbiosis). Table 21.6 summarizes the changes in the roles of the elements after the advent of free dioxygen. However, we observe (see Figure 21.5) that, throughout evolution, the free-metal ion concentrations in the cytoplasm of cells hardly changed. This restriction is due to the poisonous nature of many free ions controlled in cells by complex ion equilibria in order that the essential cytoplasmic functions, constant in all life, can be maintained.

Many of these considerations developed as organisms invaded the land when the level of free molecular oxygen in the atmosphere fortunately allowed the formation of an ozone layer that filtered the noxious UV radiation from the sun. This occurred some 500 million years ago (see Figure 21.4). Before this, life may have been possible only in the sea or perhaps in niches well protected from UV radiation.

In the development that ensued – which required greater and greater organization in multicellular organisms (see Table 21.4) – we wish to draw attention to the increasing interdependence and the development of necessary cooperation between chemotypes (positive synergism) to overcome the problems of organization. Animals became dependent on plants (and bacteria) for diverse essential products, and plants became dependent on bacteria, for example for nitrogen fixation. The general rule is that as complexity increases it is better to delegate different functions to separate compartments to increase efficiency and to avoid confusion. Division of labor is the rule – the biosphere is then essentially cooperative, even though similar species may be in competition. Modern industrial plants are based on a similar principle. Adam Smith in his *The Wealth of Nations* (1776) was adamant in this respect; we could say that his view had a clear biological basis. The naive approach to the "survival of the fittest" among species cannot be looked on in this way because it refers to species of similar kinds. The survival of the most complex in an ecosystem is perhaps curiously dependent on the presence of the least complex, and individual fitness of species is not relevant. We may well believe that globalization of human organizations in a cooperative way is the solution to many ills by removing competition.

Development of effective use of energy and resources is one aspect of division of activity into separate units in space, but they must be coordinated in each animal and as far as possible in groups of animals. Coordination requires greater and greater reliance on information transfer. Clearly, animals have increasingly linked activity to information from the environment, more so than plants, while the DNA information dominated reproduction only. In very large animals, this information transfer had to be fast as they had to scavenge quickly and had to protect themselves, and a new way of conveying information appeared in the form of a nervous system. This system linked distant cells in multicellular animals by cooperative electrolytic messages. Note that the conducting ions are Na^+, K^+, Cl^-, and sometimes Ca^{2+}. Their gradients were created in the original prokaryote form of life, but fast transmission of messages became of value only when large mobile organisms developed. Coordination of muscles and senses by nerves was then critical. Na^+ and K^+ are ineffective as binding agents, but produce a cooperative voltage switch, not an individual relay of ions, to affect organic message systems by way of Ca^{2+} input, so they had no value until they could be coupled to Ca^{2+} and organic messengers. By themselves, they do not carry information (see below).

The senses evolved with the development of touching, seeing, hearing, and smelling. The nerves, in turn, had to be co-ordinated, and we see the rapid development of the brain, totally dependent on the above electrolytic currents. An end-product of this evolution was humankind. In passing, note how step after step in evolution is linked to either the introduction of novel elements, elements in novel oxidation states, or just to new uses of elements – i.e. basic inorganic chemistry. It is useful to see evolution as the inevitable introduction of new, but dependent, chemotypes in an unavoidable chemical sequence.

Humankind

The final step of this vertical evolution – until now – was the emergence twenty million years ago of hominids, which developed into the modern human beings. Obviously, human beings are "higher" animals, but they have a particularly strongly evolved nervous system with a self-conscious brain, capable of storing vastly more "information" than all other organisms and of using it as a kind of second code, quite different from DNA. It differs in that it uses information stores of extensive, concentration-dependent properties of something approaching twenty molecules and ions instead of the intensive, concentration-independent four symbols in linear DNA.

Information is handed down by copying. The man is father to the child pheno-typically, and this succession is open to rapid innovation, unlike the functioning of DNA, as it is increased generation on generation and not by species change. Thus, although not reproduced, the knowledge in the brain can be inherited and gives increased knowledge in humankind's evolution by a different method: recording and transmission. The response of the brain is fast and relates to the environmental information, not to the reproductive information in DNA. The functioning of the brain can be understood only in systems language and not in that of molecular structures; compare, for example, the shapes of clouds [22].

We shall not elaborate on this topic here because the chemistry of the brain is scarcely known [15]. The brain processes the information and reacts accordingly, sending instructions through nerves and then transmitters produced by glands that link to the calcium network. It is also curious to note that the brain is immersed in a special protective fluid – the cerebrospinal fluid – the composition of which differs somewhat from the other external body biological fluids [7, 15]. The form in which the human brain works requires a quite new approach to the use of energy and the environment in the chemistry associated with organisms.

The big change humankind has made is that through an understanding of physics and chemistry, stored in the brain, we are now able to manipulate extensively the use of the environmental resources. We not only tap the full range of use of the chemical elements available on earth – all 92 (see Figure 21.2) – but can carry out syntheses in compartments outside cells and can employ energy from new sources. We are a quite new chemotype. The development of massive external activities in industry is resulting in the redistribution of elements (and chemicals) in quite a novel way. Humankind has also initiated all kinds of transport and message systems that use unique equipment. An addition to this activity is the computer, an extra information processor; but note that information in a computer is quantized: intensive, as in DNA, and not extensive, as in the brain. We detail these changes elsewhere [6, 7].

Finally, humans can manipulate DNA so that the information network internal to all cells is also open to exploitation. Although we may say that this is not biological evolution, in effect it is exactly so if, instead of basing our view of evolution on DNA, or genetic changes in species of organisms, we show how systems of materials energized from outside advance in chemotype organization. We analyze this possibility below in "Steady states and final conditions."

Looking to the future

So far in this chapter, we have described the evolution of chemotypes from protocells to humankind. Before we look more directly at the present situation and the possible future, as far as we can see it as a progression of evolution, we summarize our thesis as follows.

Any system will advance toward optimal use of all the components, energy, and material that can be incorporated into the system toward a final state that can no longer evolve unless the environment changes. The implication is that no matter what the system is, in the end it exploits its accessible resources to the full, including its early waste. The timing of evolution of organisms is unknowable, but its general nature is determinate and inevitable in chemotypes because of the basic reductive chemistry essential for the origin of life and its waste (oxygen) – and despite the problems associated with the random nature of species [1].

In evolution, seen in this ecological geochemical–biochemical fashion, the limitation can be that not all materials that enter the system can become fully incorporated. We saw this in the case of the effect of oxygen that generated metal ions in the environment, such as cadmium, lead, and mercury, none of which has a biological use and all of which are toxic. The activities of humans could also lead to an environment with chemical elements or chemicals not compatible with the nature of life, and we must take care to prevent such damaging effects. In the next sections, we look briefly at this concern and those of the two other main strands in our description of evolution: possible energy and organizational (information) changes. We shall then make a final summary.

The influence of the environment

The influence of the (inorganic) environment in all life processes is so extensive and so clear that it is indeed surprising to see that many texts concerning biological evolution omit it. Even more surprising is to see that this still happens in times when ecological concerns have become so widespread and "preservation" of the environment is a major political issue. Everyone agrees that the "deterioration" of the environment affects living organisms, among which are vulnerable, dependent species.

What can be asked is how new environmental changes caused by humankind, such as the introduction into the environment of newly available elements, can affect future evolutionary steps. Recent examples of disturbances – acid rain, the greenhouse effect, the ozone hole, photochemical fog, radioactive wastes, etc. – show that they can affect the environment in many ways. Acid rain, for example, can erode clays and release aluminum, which is transported to rivers and lakes, becoming a permanent component of the aquatic media at lower pH. How will aquatic organisms react to the presence of a new toxic element in their habitat? The answer is always the same – history here repeats itself. Many organisms will not survive and species will become extinct; some will hide in protected niches (difficult to imagine in this case); and others will adapt to the new conditions and in time will make use of the new element. As history shows, the trend is: toxic element rejection → messenger of risk → incorporation → utilization (metabolism, new structures, new messengers, new forms to obtain energy). As we see it historically, this has driven evolution in the exploitation of elements such as calcium, sodium, copper, zinc, and even chlorine. To take advantage, the evolution of DNA must *follow* in a step-wise succession. DNA is a code and not a direct determinant.

The important and timely question is whether a selfish species such as ours is willing to take steps to avoid the potential risks of the extremely rapid change in environmental composition, a duel between "matter" and "mind" – that is, between the struggle to satisfy the basic animal needs of food, shelter, defense, and repro-duction, to which humankind adds comfort and pleasure, and reason and good common sense, which impose caution and restraint. Once there is no doubt about the balance that must be achieved, we will be better prepared; but this requires much public scientific education and awareness. All life is not at risk – it has never been in almost four billion years – but the human species may disappear unless it adapts to the new prevailing conditions. In time, it could be that a different chemotype survives toward a final cyclic steady state (see below).

Evolution of energy sources

At this stage of our account of evolution, we have described the way in which past and future changes in availability of the elements in the environment, which could be put to use, have changed the nature of living systems internally and externally (outside cells), with an ever-increasing efficiency of use of energy and elements in ever-more complex organization. The sequence arose through the effect of "waste" (oxygen) from organisms. In essence, in this biological evolution, organisms used only those available energy sources that could be applied directly to internal chemical changes, namely to create non-equilibrated chemicals using internally the energy of the crust of the earth and light from the sun. We need to stress that with the coming of intelligent human beings, additional potential extra sources of energy have been recognized, as have materials for storing it, now that we understand that energy is a resource. Untouched until humankind arrived was, for example, the energy of the non-equilibrated nuclear abundances on earth. Now, two nuclear projects are under development: fission and fusion. Humankind has also realized that biological evolution lost a great amount of carbon and hydrogen in its own debris – coal, oil, and much of natural gas – which was then not easily incorporated into organisms. These chemicals are the bases of modern fuels. Again, energy can be captured by humankind's devices from gravitational fields (river waters and tidal power) and temperature gradients (air and sea currents). Some of these uses are perhaps risky, but the drive of evolution will always be toward the optimal use of all available sources of energy as well as of all available elements. We must realize that there is an end-point to this evolution quite separate from the risks of material waste, which could send it backward. The increase of energy use is linked to heat production, and in turn this is linked to temperature change. A very high temperature may not allow eukaryote life.

We shall now describe the limit to which energy and materials can be exploited to optimize a flow system.

Steady states and final conditions

Let us now return to the driving force of evolution as we see it. A basic diagram (see Figure 21.1) exists for all systems that absorb energy. They generate heat while inducing patterns of flow into the objects absorbing the radiation. Many physical examples may be given, including the formation of winds, sea currents, clouds, and rivers. The chemicals irradiated will also be forced into energized flow conditions as they release energy as heat. The flows are driven by the overall increase of thermal entropy as small numbers of large energy quanta (light) become many small energy quanta (heat), but a certain amount of order and/or of organization is created. Such a system could become completely cyclic, and in such a condition it will have absorbed energy optimally and created the greatest degree of organization, not order [6, 16, 22]. Such a cyclic system, like an equilibrium state, cannot evolve, although it will also create a re-occurring or reproductive form. The flow arises from the environmental chemicals, (A). A second possibility, however, is that the flow only enters a steady state and produces material waste (B) not initially involved in the cycle Energy+Material (A) → Return Flow + Waste (B) + Heat.

This general state of the interaction of energy and material in a system was inevitable on earth 3.5×10^9 years ago; the system still exists today, but in a new organization because of biological evolution. As waste builds, the environment is changed. Therefore, the system can evolve with the new environment as it adapts or changes to use it. In fact, it is the only way it can evolve unless new energy or new materials are introduced in other fashions. Let us assume that no new energy source is possible. Thus, if the waste (B) is by its very nature interactive either with (A) or with the process producing flow, then the system evolves as the environment, (A) + (B) + subsequent products of reactions of (A) + (B), evolves. It becomes an evolving ecosystem that inevitably struggles to become fully cyclic, the end-point of any evolution other than the collapse of energy supply when the system reverts toward equilibrium. (Evolution is dependent on the energy of the sun, and the sun's collapse will destroy life on earth.)

Next we must recognize that several different products of such a system as (A) + (B) + energy may exist. They are not strictly reproducible unless the flows are organized by a central control system, say DNA. We will then see the changing environment – new inorganic chemistry, as described above – as forcing organized biological evolution toward optimal energy capture and in altered organized flow. The DNA is forced to change. Now the nature of (B) and its products can be best developed by increasing use of compartments and informed activity, including message systems. The compartments keep incompatible, but useful, chemical reactions separate as they become newly available. However, do not forget that this enforcement is due to the very nature of the early steps of energy-capture by chemicals, whether abiotic or biotic, controlled largely by genes until eukaryotes developed, when the environment became more and more a source of information that could be stored in the brain and used externally. Eventually, this led to quite new chemical and energy sources being used by humankind and quite a new possibility for a control command over all life, but the end-point remains clear.

We turn next to look at information transfer, which is necessary if we are to understand organization of energy and material input in systems and the way they have developed and can develop further. This is a further limitation on evolution.

Information

In biological chemistry, "information" [5] is a term usually restricted to the potential instructions stored in the coded sequence of the four nucleotides of DNA. In our opinion, this is too limited a view of what information is and of the kinds of information effectively used in biological chemistry (see [5, 12, 14, 17, 18]). To open the discussion, we require a comprehensive definition of "information," not an easy task because it is an elusive concept. Looking into dictionaries or encyclopedias for a proper definition is of little use because these definitions are restricted to particular cases, such as "that conveyed in the spoken communication between two persons" or, in computer programs, "represented by a particular arrangement of numbers or symbols." In a more general operational definition, other factors or agents must be considered. These assume both the existence of an energized source of information (the sender) and the recognition of the message (information) by a receiver, but we can only tell that the receiver has been informed if it acts as a consequence, again using energy. Information is only in the message in an "informed system" and does not arise directly just from the DNA sequence or the flows from or to the donor or receiver – they do not alone create change in the system. (Note that the dictionary definition involves only people. People can interpret DNA because they know how to read its sequence, but the cell cannot. To read its sequence, the cell needs metabolites and proteins to interact with the DNA. The cell also receives proteins from the DNA, generating reactions – that is, metabolism changes. We see that DNA actually is just a part of an informed cell system.) Let us start again from a definition.

We can see that an "informed system" is more readily defined as a system with an energized receiver that has the capability to react to a stimulus from a message sent by an energized source. The message as information is described only in terms of properties of the whole sequence – donor, message, receiver – which uses energy.

For simplicity, let us start considering systems not involving living organisms, but that are abiotic and are not informed. Consider the sun, the earth, and the present oxygenated atmosphere. The sun as a source of radiation sends out energy in part captured by the receiver, atmospheric molecular oxygen, O_2, and so creates the cyclic steady balance $3O_2 \rightleftharpoons 2O_3$ (not an equilibrium); but note that oxygen responds only to that fraction of radiation for which it has electronic transitions. On reacting, O_2 forms ozone, O_3, which is heavier than oxygen and more attracted by the earth's gravity field, so that it falls toward the earth. Ozone, through the effect of a different radiation, decomposes back to oxygen, a lighter molecule. The process is repeated again and again until a cyclic steady state is eventually reached. The gravity field generated by earth maintains the relative positions of O_2 and O_3 so that a persistent confined ozone layer is created by it and by a UV gradient at a sufficiently low temperature and height. We can therefore say that the O_2/O_3 flow system *is formed* by direct interaction both with the energy supplied by the sun in the form of radiation and with the gravity field of the earth, which together have a structuring function on O_2/O_3 flow. We see immediately that all energy or material gradients can interact to cause flow and then generate a formed system. Other formed systems are the clouds in the sky, the planetary system, and the meanderings of rivers. Life must have started from such very low-level abiotic formed chemical flows, but we do not call such systems *informed* because all the components are separate parts of the systems' activity.

Information is not required in forming a system, but acts between two or more formed systems, donors and receivers, to create a higher level of activity in the receiver with the "demand" transferred from the donor. Consider a cell. All of us agree that the DNA is for us a coded form of information when acting as a donor, but by itself it is meaningless in a cell. It therefore requires information and interpretation (reading) and becomes informative when it gives an energized message to the directed flow of energy or material, for example into the metabolic synthesis of new DNA and/or RNA. The sender of this information is the polymerase system + DNA + energy, and the message is in the sequence of monomer bases after being read. The polymerase reads out the DNA code (also a message) and the new message formed is messenger RNA (mRNA). In the next step, mRNA is translated as proteins, which the ribosomal machinery and energy plus this RNA create as a further message to be passed to substrates.

The proteins, with energy, act as catalysts in the cytoplasm in metabolic events. Thus, the proteins are carriers involved in information to the primary small molecules leading with energy to products of a guided kind. Continuing down the chain, the primary products have arisen in part by diffusion, together with pumping of material from outside the cell. The internal metabolism is thus constrained in a selected way by the environment, the membrane, and, distantly, the DNA. The environment is now an equal participant in information, but from the outside of the cell, causing it to start intake or not, also according to its content, while the cytoplasm allows intake, or restricts it, according to its content, both by the membrane. The whole, environment and metabolizing chemicals/DNA(RNA), is an interacting network.

Now let us work up the same chain of events. The environment supplies basic materials and energy. The resultant internal concentrations of substrates and later proteins can act directly on uptake mechanisms by generating or stopping activity, initiated at DNA, of scavenger or pump molecules. They can also stop initiation by binding to the DNA synthesis machinery through transcription factors. At the same time, other transcription factors influence the RNA machinery through the concentrations of synthesized substrates of many kinds, switching protein synthesis on or off. Finally, active proteins, in one metabolic path of a substrate – or even in a path in which this substrate is not involved – are informed by feedback flow from the substrate concentrations so as to stop or start catalysis. In a sense, the donors of information have become the receivers of it in a loop. We see that the whole cellular organization is criss-crossed by signals linked to the environment as well as to DNA. In fact, cells and the environment have been increasingly knitted into one organization [19, 22], which includes all the elements and structures in the environment, as the amount of interaction with the outside has increased with time in the more organized organisms.

Table 21.7. *Some messengers of primitive and advanced cells*

(A) Primitive messengers

Messenger	Functional control over
Mobile coenzymes	distribution of metabolic fragments H^-, CH_3^-, $-COCH_3$, etc.
Nucleotide triphosphates, Mg^{2+}	distribution of energy
Fe^{2+}, $2RS-/(RS)_2$	distribution of electrons; redox state balance
Some simple substrates (feedback)	metabolic products (e.g. glutamine, nucleotide bases, amino acids) and expression
Phosphorylation of proteins, Fe^{2+} (Mn^{2+})	gene expression

(B) Oxidized organic messenger chemicals[a]

Messenger	Source
NO	arginine
Sterols	squalene (cholesterol)
Adrenaline	tyrosine
8-OH tryptophan	trytophan
Amidated peptides	terminal glycine of peptides
C_2H_4	(?)
Plant hormones	indoles
Retinoic acid	retinol (vitamin A)
EGF-peptides	hydroxyaspartate

[a]Note the receptors dependent on Cu (C_2H_4) or Zn (sterols, etc.) and the metabolism of peptides dependent on Zn.

If we examine the new living systems appearing along the time axis, we see that the needs and kinds of "information" required or received by such systems have increased dramatically and systematically. Examples start from the chemotactic gradients in prokaryotes and increase with the calcium gradients in single-celled eukaryotes, then the organic messengers and hormones in multicellular eukaryotes (see Table 21.7), then the Na^+/K^+ gradients in the nervous system of animals, etc. These gradients across membranes act on receiver proteins built by cells coded by DNA. The genome plus the cell activity guided by external inputs poses a very difficult problem for the description of cooperative activity of organisms. The messenger activities are a systematic, not a random, advance following the environmental sequence of changes and the potential consequences [22].

Two further complications arise. Sending messages from the DNA through RNA through proteins to metabolism is slow. Sending of messages from the environment to the metabolism is fast. The more the cells sense the environment quickly, the more their behavior depends on the external environment, and the more they react to it rapidly. A major progression in evolution is speed of response. More and more, this helps survival despite the slow reproduction rate. As information concerning the environment increases, so the environment itself can be used to influence the organism, and the organism can react in turn in an effort to control its environment. (Note that survival, reproduction, diversification, etc. are all forms of ensuring the persistence of life, a particular mechanism to degrade energy.)

To this view of informed systems we must add the influence of the brain, which is clearly a store of information based on genes only for its initial very incomplete formation. We may look on this curious malleable organ in the following way. The genetic apparatus generates a vast bundle of neurons, perhaps 100 billion of them, in no particular overall connected array, although some localized regions are different from others. It has very little information – that is, contact with itself or the environment. This "jumble" is connected to the sense organs, from which it receives information very quickly about physical or chemical conditions in the environment that help it to build images. The neurons then grow to make contacts with other neurons connected to other body parts and so organize possible responses. A large increase of brain capacity arises by making a complex network of contacts that reflect, in a coded form, the distribution in four dimensions of outside events. As a result, images are created that persist. The organism associates these images with other sensory impacts that relate to food, danger, etc. and acts accordingly. The coded information in this case is not like that in systems based on computers or in the DNA code. Unlike DNA, the brain is connected to the entire historical features of the organism's life through the further functional connections of the nerves to glands, organs, and so on down to all cells through chemical bindings and growth, much independent of DNA. Connections are not all or none (binary 1 or 0), but are based on quantitative extensive properties: concentration, binding strength, and local dynamic structure. The whole malleable brain is not fixed at birth and develops through all of life. This is found increasingly in animals up to humankind. Moreover, from the knowledge humankind has gathered, ways have been found of expressing information as output in the form of language – spoken, written, computerized, etc. Information (now in the dictionary sense), which is not in DNA, is then passed down through generations, making today's society possible.

The resultant activity is in part *scientific information* applied to all kinds of material and energy resources, but it is only a vast development of the increasing involvement of the environment with organisms throughout evolution. However, our society is limited in that it cannot flout the ultimate universal rules of the second law of thermodynamics or its corollary, the rule of the optimization of energy resources in chemical constructs. There is also a further twist: the individual observes itself and others, and hence individuality evolved strongly. The arrival of self-consciousness should allow people to manage earth, but the downside can only be a reduction in individual use of energy and material. Already, too many people are acquiring goods for a comfortable steady state at the highest level of individual demand.

Despite the wondrous nature of the brain, *Homo sapiens* is a vulnerable animal, being dependent on hundreds, if not thousands, of species – even chemotypes – for all kinds of chemical supplies. In fact, humans are an internal ecosystem and the product of a much larger external ecosystem. As mentioned earlier, biological evolution is not concerned with this or any other species; it is about survival of the fittest *ecosystem* as described above. As the complexity increased with "information" from the environment in organisms, they decreased in self-sufficiency for good reasons: Complexity is difficult to manage in one chemotype. However, it is complexity that human beings must manage in the future.

Evolution and DNA

Now a major problem arises. If the DNA represents a well-adapted species in a given environment, how does the DNA change when the environmental conditions become hostile for the organisms so that the slowly acquired adaptive modifications are transmitted to offspring – by random mutations of DNA and natural selection of the fittest, the official story – or by mutations influenced by external environmental conditions? And to what extent are epigenetic changes [13, 14] affected or directed by the external environment? Is it really necessary to involve all of the DNA or the structure of chromatin in a random way in response to change, or can there be change in just selected parts to be transcribed [12, 16, 17]? Is selective mutation the answer to the seemingly contradictory views of Darwin and Lamarck [20]? Can we learn from the study of the way in which the immune system responds [21]?

These are momentous problems, and currently many research efforts to elucidate them are being made, which may also change our views on evolution. In this context, we have to offer some explanation of the origin of species (see below) that differs from the idea of a systematic development of the major features of evolution of chemotypes.

Species

If we ask in the context of vertical evolution, "What is the origin of species?" the only answer we can give, seeing the large variety of organisms of equal survival strength around us, is that they probably arise from *chance bifurcation* multiplied time and again. Systematic advance can then be based, not on species individually, only on relationships between species [1]. Species, we therefore agree, may arise through chance mutation throughout the DNA and Darwinian natural selection. Also, distinct series of major mutations exist that are beneficial in that the organization within certain organisms gains some special advantage from novel interaction with the environment. We then must view the species in themselves as belonging to what de Duve [3] called "horizontal evolution."

Thus, the drive to gain from a change in the environment, inhibited by the conservative nature of the required cytoplasmic chemistry and DNA, awaits imposed mutational change. The solution to this problem (which resembles a complex equation with many variables) is a switch in kind, say from non-photosynthesizing to photosynthesizing – that is, eventually a switch in a large number of related, equally adapted species within several different groups, or chemotypes. The change in chemotypes is forced by the change in the environment, and to bring about this development of the whole ecosystem the DNA must adapt its sequences. This is the process of evolution, but it is deeply dependent on an ecosystem. We suggest that there must be a more direct effect of the changing environment on the genes than random variation of the whole genome. The most likely possibility is localized genetic variation dependent on novel chemicals forcing exposure of localized gene regions that then mutate most rapidly.

Summary

If we are correct, the inevitability of evolution exists on the large scale; this is a matter of organization of chemicals in a physico-chemical succession. The process is in accord with the second law of thermodynamics in that it leads to an overall increase of thermal entropy production. The energy degradation that causes this increase of entropy will of necessity increase in rate until no more energy can be absorbed. Thus, the rate of entropy production can be said to strain to reach an optimum. This driving force is of necessity linked to the increase of absorption of useful energy in synthesized or concentrated chemicals. The system incorporates more and more materials to optimize energy absorbed, retained, and eventually converted to heat (see Table 21.8) until there can be no further change. We observe that it is guided as a flow system within boundaries. The final condition, as stated, is the fully cyclic steady state when evolution ceases of necessity.

The increasing involvement of energy and chemicals can only arise if the initial energized chemical process created waste, for it is waste (e.g. oxygen) that in turn creates new opportunities for energy absorption until they too are incorporated in the cycle. The environment changes – mainly inorganic – are faster than changes in organisms and lead the way. Organisms die, hide in niches, or must adapt. We cannot examine only genomes to understand evolution, but must look at the presence of organic and inorganic elements, e.g. in the metabolome and the metallome, as descriptive of evolution.

Table 21.8. *Involvement of elements in homeostasis during evolution*

Primitive prokaryotes (anaerobic)	Early single-cell eukaryotes (aerobic)	Later single-cell and multicellular eukaryotes (aerobic)
H, C, N, O, P, S, Se substrates and polymers	⟶	
H^+, Na^+, Mg^{2+}, Cl^-, K^+, Ca^{2+} exchangers	⟶	
Ca^{2+} structural	⟶	
H^+, P, S, Fe signals	⟶	
W enzymes ⟶		
Mn, Fe, Mo, low-potential enzymes ⟶		high-potential enzymes
Ni enzymes (H_2, CO) ⟶		
Ni (urease) ⟶	---------- ⟶	plants only
Co (B_{12}) ⟶	---------- ⟶	animals only
Ca^{2+} ATP-ases	⟶	
(Zn enzymes)? ⟶	Zn enzymes ⟶	in vesicles and extra-cellular
		⟶ Zn signaling (DNA)
Ca^{2+} rejected ⟶	Ca^{2+} in vesicles and filaments and inner signaling (calmodulin) ⟶	S-100, annexin ⟶
Na^+, K^+ osmotic and charge balance ⟶		outer filaments and signaling Na^+/K^+ between cells, Na^+/K^+ ATPase
		organic hormones iodine hormones
	⟶	Cu enzymes

Clearly, we must consider inorganic as well as organic biological chemistry to understand evolution of organisms together with environmental change. Starting from the beginning – the process of initial absorption of energy – any (inorganic) system, conditioned by pre-existing gradients, or fields, becomes formed, organized in a confined space (see our early examples). Efficiency of energy capture requires that different formed systems cooperate. Any small, confined system open to fluxes of energy and material, such as a protocell, can, in principle, come to reproduce itself, but this requires a blueprint that takes the form of a code. The code is a memory of a previous existence, predating the DNA code, which can be connected to an energized source to maintain and reproduce internal gradients in flows of chemicals. These flows can feed back to the energized activity associated with the code, and they interact with the external gradients in the environment. The system then becomes informed as an ecosystem and can sustain itself. Such a reproductive organism/environment system greatly strengthens its ability to capture energy (and degrade it), and it will of necessity (see above) evolve using any changes in the environment, including those self-generated by waste (e.g. O_2). In some manner or other, the DNA must follow any major environment change in chemotypes so as to keep reproducibility.

At the same time, a huge variety of species, genotypes, arise without rational explanation, and similar ones compete. Just as we may say "there will be wars and rumors of wars, but the end is not yet," there may be no end to evolution of species, as such evolution lies in full cooperation between the environment and the chemotypes (no matter what the species) in an optimal flowing, interactive ecosystem, aiming at an ideal cyclic steady state. However, such a state may be unattainable. Humankind must become aware of the nature of such energized systems that dictate the way life evolved and can evolve on earth. It is the whole ecosystem that evolves – that is, environment and organisms. It requires deep thought by humankind to preserve the system of which humans have become the last source of innovation and the guardians of it all in a very short period of five hundred years. It is just as possible to go backward or forward as to maintain a cycle.

We can now answer the two parts of the question posed by the initiators of this book. (1) Was life destined to happen in this universe? (2) Was life destined to lead inexorably to greater and greater complexity?

Probably, life was destined to happen because it is an effective (and efficient) way to degrade available energy. Once it started, we believe evolution was also inevitable, as an ecosystem would of necessity develop with greater and increasing complexity; but this development is of the ecosystem of organisms *plus* the environment, not just a development of organisms. The ecosystem evolved in the way it did because of the required chemistry of effective energy capture. Many forms of life evolved, but only the multifarious mixture of them is optimal. Within the development of life, no logic is apparent in the minor species variations that have appeared; they probably arrived by chance.

But one species is now different, for its development is due to rational thought, not gene changes, to the degree that it is a separate chemotype. Human activity represents a logical end-point of exploitation of the material elements, of energy, and of life, while remaining dependent on a multitude of other species. However, humankind must be careful. The human species is interventionist in that, through understanding, it can to a large extent dictate ecological evolution with little left to chance, at least in principle. However, restraint in uses of resources must be accepted to sustain a favorable environment for survival of the present ecosystem.

Note and acknowledgment

This article is expanded in a full book by us on our general chemical approach to the evolution of ecosystems [22]. Details of the metal proteins in the proteome are given in [23] and [24].

We thank Professor Stephen Freeland, co-editor of this volume, for helpful criticism and a useful exchange of views.

References

[1]   J. Maynard Smith and E. Szathmàry. *The Major Transitions of Evolution* (New York: W. H. Freeman, 1995).

[2]   R. J. P. Williams. Metal ions in biological systems. *Biological Reviews*, **28** (1953), 381–406.

[3]   C. de Duve. *Life Evolving* (New York: Oxford University Press., 2002).

[4]   H. J. Morowitz. *The Emergence of Everything* (New York: Oxford University Press., 2002).

[5]   P. A. Corning. Synergy and self-organization in the evolution of complex systems. *Systems Research*, **12** (2) (1995), 89–121. Control information. *Kybernetes* **30** (2001), 1272–88.

[6]   R. J. P. Williams and J. J. R. Fraústo da Silva. *The Natural Selection of the Chemical Elements – The Environment and Life's Chemistry* (Oxford: Clarendon Press, 1996).

[7]   J. J. R. Fraústo da Silva and R. J. P. Williams. *The Biological Chemistry of the Elements – The Inorganic Chemistry of Life*, 2nd edn (Oxford: Oxford University Press, 2001).

[8]   J. F. Kasting and J. L. Siefert. Life and the evolution of Earth's atmosphere. *Science*, **296** (2002), 1066–8.

[9]   K. A. McCall and C. A. Fierke. Probing determinants of the metal ion selectivity in carbonic anhydrase using mutagenesis. *Biochemistry*, **43** (2004), 3979–86.

[10] R. E. Blankenship. *Molecular Mechanisms in Photosynthesis* (Malden, MA: Blackwell, 2002).

[11] S. G. Andersson, O. Karlberg B. Kanback. On the origin of mitochondria: a genomics persepctive. *Philosophical Transactions of the Royal Society of London*, B**358** (2003), 165–77.

[12] L. Wolpert. *The Triumphs of the Embryo* (Oxford: Oxford University Press, 1991).

[13] B. M. Turner. *Chromatin and Gene Regulation – Molecular Mechanisms in Epigenetics* (London: Blackwell Science, 2001).

[14] E. Jablonka and M. Lamb. *Epigenetic Inheritance and Evolution* (New York, NY: Oxford University Press 1995).

[15] R. J. P. Williams. The biochemical chemistry of the brain and its possible evolution. *Inorganica Chimica Acta*, **356** (2003), 27–40.

[16] R. J. P. Williams and J. J. R. Fraústo da Silva. Evolution was chemically constrained. *Journal of Theoretical Biology* **220** (2003), 323–43 and references therein.

[17] K. Kull. Organisms can be proud to have been their own designers. *Cybernetics and Human Knowing*, **7** (1) (2000), 45–55.

[18] H. J. Morowitz. *Beginnings of Cellular Life* (New Haven, CT: Yale University Press, 1992).

[19] L. Margulis. *Symbiotic Planet* (New York, NY: Basic Books, 1998).

[20] L. H. Caporale. *Darwin in the Genome* (New York, NY: McGraw-Hill, 2004).

[21] M. S. Neuberger, R. S. Harris and J. M. Di Noia. Immune system changes by deamination. *Trends in Biochemical Sciences*, **28** (2003), 305–12.

[22] R. J. P. Williams and J. J. R. Fraústo da Silva. *The Chemistry of Evolution: The Development of Our Ecosystem*. (Amsterdam: Elsevier, 2006).

[23] C. L. Dupont, S. Yang and B. Palenik. Modern proteomes contain imprints of ancient shifts in ocean chemistry. *Proceedings of the National Academy of Sciences, USA*, **103** (2006), 17822–7.

[24] R. J. P. Williams. The evolution of calcium biochemistry. *Biochimica et Biophysica Acta* **1763** (2006), 1139–46 and references therein.

Index